THE TRANSPORTATION EXPERIENCE

The Transportation Experience

SECOND EDITION

William L. Garrison and David M. Levinson

OXFORD
UNIVERSITY PRESS

Oxford University Press is a department of the University of
Oxford. It furthers the University's objective of excellence in research,
scholarship, and education by publishing worldwide.

Oxford New York
Auckland Cape Town Dar es Salaam Hong Kong Karachi
Kuala Lumpur Madrid Melbourne Mexico City Nairobi
New Delhi Shanghai Taipei Toronto

With offices in
Argentina Austria Brazil Chile Czech Republic France Greece
Guatemala Hungary Italy Japan Poland Portugal Singapore
South Korea Switzerland Thailand Turkey Ukraine Vietnam

Oxford is a registered trademark of Oxford University Press
in the UK and certain other countries.

Published in the United States of America by
Oxford University Press
198 Madison Avenue, New York, NY 10016

Library of Congress Cataloging-in-Publication Data
Garrison, William. L., 1924- author.
The transportation experience / William L. Garrison and David M. Levinson. – Second edition.
pages cm
"A history of the development of transportation systems, with suggestions for further efficiency"
–Provided by publisher.
ISBN 978–0–19–986271–9 (alk. paper)
1. Transportation and state—United States—History. 2. Transportation and state—Great Britain—History.
I. Levinson, David M., 1967- II. Title.
HE203.G37 2014
388.0973—dc23
2013023316

Contents

PART FIVE | WAVE THREE: 1890–1950

List of Figures

List of Tables

THE TRANSPORTATION EXPERIENCE

1 Wave One: 1790–1851

Let no one who wishes to receive agreeable impressions of American manners, commence their travels in a Mississippi steamboat.

—FRANCES TROLLOPE, *Domestic Manners of the Americans*

1 Rivers of Steam

1.1 Steam Boats and Stream Boats

ON AUGUST 26, 1791, Thomas Jefferson, head of the new US Patent Board, awarded John Fitch, James Rumsey, Nathan Read,[1] and John Stevens patents associated with inventing various steamboat technologies. All had desired a patent on the entire steamboat. Solomonically dividing the patent spoils kept any from dominating the market, or developing a large enough market to succeed.[2] The idea of automating water travel was not original. A century in advance of the market, in 1696, Thomas Savery applied for a patent for rowing ships using paddle wheels and a capstan, but this was dismissed following a negative report from the UK admiralty.

Before steamboats, floating from Pittsburgh to New Orleans took 4–6 weeks and and 4 months to walk back. The idea of applying steam to solve transportation problems was "in the air." Benjamin Franklin was interested in the subject, but discouraged paddle wheels in favor of jet propulsion in 1785 (a bit ahead of the technology), sending Fitch and Rumsey down an unsuccessful technology path. John Fitch began out of Philadelphia on the Delaware, James Rumsey out of western Virginia (now West Virginia) on the Potomac. Rumsey obtained support from Franklin and met with Matthew Boulton and James Watt.

Other inventors (and their rivers) included Isaac Briggs and William Longstreet (Savannah) who were issued a 1788 patent by the state of Georgia, as well as Oliver Evans (Delaware), Samuel Morey (Connecticut), Nicholas Roosevelt (Ohio and Mississippi), and John Stevens (Hoboken).

John Fitch (1743–1798) with partner (and rival for the affections of his wife) Henry Voigt developed a steamboat, *Perseverance*, that plied the Delaware River beginning in 1787 using steam-powered oars. After a few years in Philadelphia, he decided to move to France where he also obtained a monopoly. Finding the Reign of Terror in full force, he moved to England, and then Kentucky (aiming to serve the Ohio River market), seeking fortune, before his death.

George Washington's Patowmack Company hired James Rumsey (1743–1792) as a manager, and he set to develop his steamboat to solve river navigation problems. Rumsey at the time was in the unique position of being able both to modify the boat to run on the river and to modify the river to better accommodate the boats. He fell out with Washington over money and in December 1787 demonstrated his steamboat on the Potomac in Shepherdstown. His innovation was a form of jet propulsion. Rumsey acquired some patents in England after moving there in 1788. His admirers formed the Rumseian Society to help fund his work.

The United Kingdom had a parallel, though related set of developments. Funded by Lord Dundas, who was a major backer of the Forth and Clyde Canal in Scotland, William Symington (1763–1831) constructed the *Charlotte Dundas* (see Figure 1.1) in 1803 for use as a canal tugboat. It could pull two 70-ton barges 32 km (20 miles) at 5.3 km/h. We know little about Symington. He had built some stationary engines and at least one unsuccessful tugboat prior to the *Charlotte Dundas*, and by 1801 had taken out patents on the use of a crosshead and slide bars to control piston movement in a horizontal cylinder producing rotary motion.

Symington used a double-acting engine, and a single 10-horsepower paddle wheel was set within the stern of the boat in order to minimize generation of waves that would damage the canal bank. The knowledge that had been developed about water wheels was directly applicable to boats.

The *Charlotte Dundas* was used on the Forth and Clyde Canal for a year or so and then was converted to a dredge. It seems to have been successful, but not successful enough to

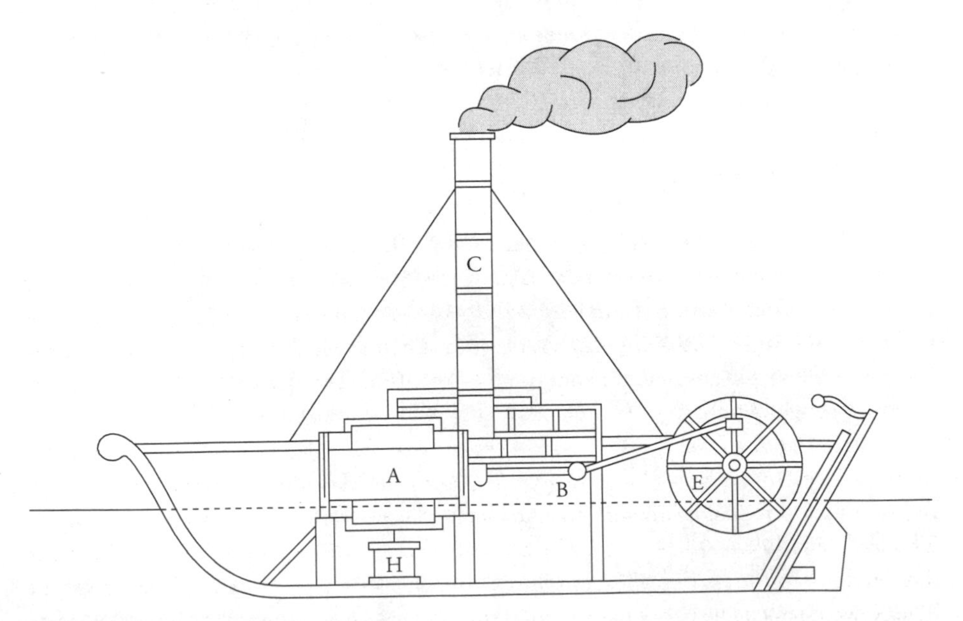

FIGURE 1.1 The *Charlotte Dundas*, 1802
Source: Thurston, 1878

drive widespread adoption. It came close. The Duke of Bridgewater ordered a number of boats from Symington for use on his canals, but Bridgewater died in 1803 and the order was withdrawn. No other orders came along, and Symington drops out of our story.

Robert Fulton (1765–1815) is recognized in the United States as the innovator of the steamboat. The son of a farmer in the Philadelphia area, Fulton was largely self-educated and early in life showed talent as a painter. This took him to Europe in the 1790s, first to London where he studied with an American painter, Benjamin West, and met James Rumsey. He had some funds from his painting, and moved in society with those who were stylish and good at conversation and who visited on country home social circuits. Innovation, industry, and commerce were much discussed, and Fulton's peer group extended beyond the landed gentry. He also met the Duke of Bridgewater, and while in France lived with poet/diplomat Joel Barlow and his wife Ruth.

He introduced steamboat service on the Hudson River in 1807, and he and that date are associated with the beginning of the steamboat era. Fulton's application was not so advanced as Symington's. Fulton used a Watt-Bolton engine, side paddle wheels, and a rather complex arrangement to obtain rotary motion. Fulton's steamboat also used sails. The engine was neither fuel efficient nor powerful. Even so, he found the market for passengers, and finding the market is part of the innovation puzzle.

Robert Fulton did not invent the steamboat; he did not even get funded until the original steamboat patents of 1791 expired. Fulton made it a success with financial partner Robert Livingston; together they obtained a 20-year navigation monopoly on the Hudson River, a better market than the Delaware. "Fulton knew that navigation monopolies were better than patents,"[3] although his particular navigation monopolies were later thrown out by the US Supreme Court as unconstitutional because only the federal government could regulate interstate commerce (which travel on the Hudson River was) and by the 1830s there were more than 100 steamboats on the Hudson, including ones owned by Cornelius Vanderbilt.

The steamboat could not have occurred without the steam engine, and of course would have been irrelevant without inland waterways and canals.

1.2 The Steam Engine

Christiaan Huygens (1629–1695), a Dutch mathematician, experimented with piston engines, using gunpowder to create a vacuum (1673) and drive water pumps at a garden in Versailles. Later, steam was condensed to move the piston. In 1679, his Huguenot exile student Denis Papin (1647–c. 1712) built a steam digester, which extracted fats from bones. The bones were then ground into bone meal. He continued work on what is credited as the first steam engine, but it was never practical.

In 1696, Thomas Savery (c. 1650–1715) patented a steam engine, which he described in his book *The Miner's Friend; or An Engine to Raise Water by Fire*, that was aimed at draining water from mines. While not technologically successful, the patent proved valuable. Thomas Newcomen (1663–1729), a blacksmith, and his partner, John Cawley, a plumber, did develop a practical engine (about 1704), which was marketed under Savery's patent. The Newcomen engine began to be used in tin, copper, and lead mines in 1712. Coal mine use came later. The exact linkage between Newcomen and Papin is not clear, but Reid[4]

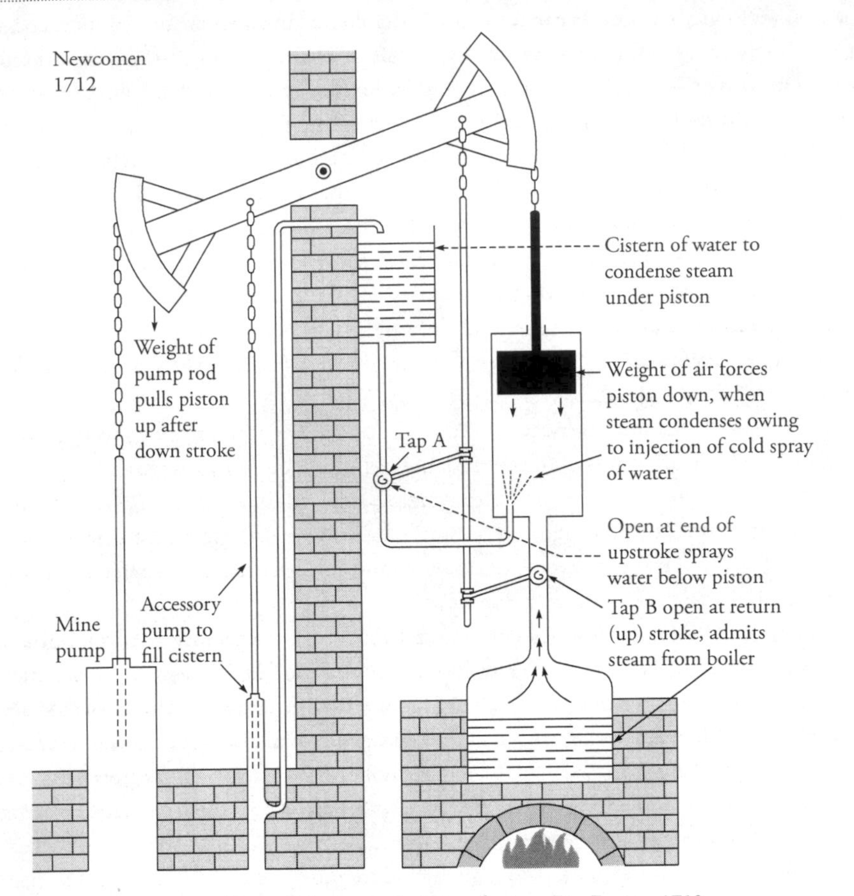

FIGURE 1.2 Diagrammatic View of Newcomen's Atmospheric or Fire Engine, 1712
Source: Csele, 2003

argues that Newcomen discussed the matter with Dr. Robert Hooke, who was well aware of Papin's work. You can see elements of Papin's engine inside Newcomen's larger device. The path between science and technology was two-way, each learning from the other.

Newcomen's engine (see Figure 1.2) worked this way: Rigged to a weight, the piston rested at the top of the cylinder. Steam from a boiler could then be introduced into the cylinder, and then cooled by a water spray. The resulting vacuum would pull the piston down. The reintroduction of steam then breaks the vacuum, and the piston returns to its rest position until reintroduced steam is cooled by the reintroduction of water. This was a low steam pressure, atmospheric engine. Steam pressure was not sufficient to push the piston upward; atmospheric pressure pushed the piston down.

Automatic control was one key to Newcomen's success: control of water and steam was automatic, and a centrifugal governor controlled velocity. Such governors had been long used on windmills.

John Smeaton (1724–1792) seems to have been the first engineer to consider improvements in Newcomen's steam engine. About 1770 he examined a number of engines and their efficiency. The engines used more steam than was required to fill the cylinder, and

Smeaton reasoned that this was because the water cooled the wall of the cylinder, and that three-fourths of the steam was wasted in reheating the wall. Using analysis, he developed some differing geometric shapes for the boiler and the cylinder, and these had the potential of being more than twice as efficient as the Newcomen design.

Smeaton was England's first engineer-analyst. Trained as an instrument maker, he became interested in larger scale engineering and studied engineering works on the continent. His first major commission was the lighthouse on Eddystone rock near Plymouth. To solve the problem of a cement that would harden despite being flooded by tides, he engaged in careful experimentation and measurement and developed a successful hydraulic cement, the first since Roman times.

This was an example of rational analysis, and Smeaton applied similar analysis to water wheels and other topics of current interest. Smeaton had studied Newton's *Principia* (1687) and engaged in careful measurement and comparison of theory and reality. He wrote extensively and was in contact with like workers on the continent. Through the model provided by his work and his writing, he stimulated rational analysis. One result was the founding of the Society of Civil Engineers in 1771, and early members were referred to as Smeatonians. Smeaton and his associates provided the know-how for the extensive civil engineering works of the times.

James Watt (1736–1819) also began his career as an instrument maker, and, like Smeaton, he became interested in the steam engine and measured the excessive steam consumption because of the wall heating requirement. In 1765, Watt had the idea of a separate condenser. That is, a condenser was attached to the cylinder and kept cool. Access to the condenser was blocked off as steam filled the cylinder, and opened when the cylinder was full. The steam then condensed in the condenser, providing the vacuum for the downstroke of the piston in the cylinder.

The condenser concept was patented in 1769, and in 1774 Watt entered into an agreement with Matthew Bolton, a Birmingham businessman, for the construction of Watt-Bolton engines. The factory constructed some engines. It provided designs and erection assistance to others, and payment was received via a royalty for the use of the condenser.

The mine pumping market for the engine was growing, and extensive developments in other sectors were important. It is of interest that pumps began to be used in urban water supply projects.

The factory market used rotational power from water wheels, and to extend his engine to that market, Watt began to work on what would now be called the kinematics of machinery. He developed flywheels and sun-and-planet gears to transform the reciprocating motion of the piston to rotary motion. He also developed other kinematic devices, and he protected all of his inventions by patents. Watt's developments mooted Smeaton's work with the steam engine.

Watt, Smeaton, and other engineers were by this time in touch with developments on the continent, and there seems to have been a good bit of two-way information exchange, and exchange to and from engine and construction engineering and within factory engineering.

Society changed greatly within the sweep of time we are viewing. Factory development, Corn Laws, enclosure, and the American and French revolutions all were causes and reflections of those changes. Smeaton's ideas of civil engineering drew attention to the social impacts of and the social services engineering might provide, and the developing cadre of engineers responded. Watt was regarded as somewhat of a political radical, and some engineers worked on designs for workers housing and fireproof factories. A few were engaged in the debate over child labor laws. (Watt's radical political leanings died from disappointment with the French Revolution.)

Watt and Smeaton had similar backgrounds; both were instrument makers and capable of careful measurements. Later, both paid very careful attention to details. Smeaton wished to spin off his style of work, ideas, and knowledge. Watt wanted to protect his. He protected his intellectual property with patents, even if he had no plans to use the ideas. (For instance, he patented the double acting engine in 1782, but never used the idea nor did he plan to.)

Watt had protected himself with many patents, as mentioned, and to the extent he could, he extended that protection by obtaining new patents or minor changes in old ones. Even so, patents began to run out at about 1800, opening opportunities for other innovators. By that time, the Watt-Bolton engine was widely used on the continent. Indeed, as early as 1735, there were precursor Newcomen engines in Sweden, the mining districts of the Austrian Empire, and at Liege near Brussels. Hands-on knowledge of the engine was widely dispersed, so many had the opportunity to try new ideas.

Watt seems not to have been much interested in transportation, and the engine he marketed was so heavy relative to its output that applications were limited. Early trials in road vehicles were attempted, but the first lasting applications were in the water mode and at docks where steam engines pumped water for the operation of pneumatic systems. Stationary steam engines were also used to pump water to the heads of lock systems during periods of low water.

1.3 Bridgewater

The beginning of England's canal era was marked by the 12-km (7.5 mile) canal of Francis Egerton, third Duke of Bridgewater (1736–1803). Built by James Brindley, it connected coal mines on the Bridgewater estate at Worsley to Manchester. Construction began in 1759, and the canal opened six years later. This was not a construction first; there had long been experience on the continent, and some canals had been built previously in Ireland and England.

Bridgewater's canal had some interesting technological features. It tied into his mine drainage scheme, and the boats ran into the mine for loading. Although the canal could accommodate larger boats, the within-mine operations kept the beam of the boats to about 2.1 m (7 feet); they were 15 m (50 feet) long. In order to hold water, the canal was lined with puddled clay, and to avoid extensive lock construction, an aqueduct was constructed over the River Irwell.

Bridgewater's canal caught the imagination of the public; it was a financial success, so it also caught the imagination of developers and investors.

The result was a flurry of inland navigation acts between 1759 and 1794. Most of these were narrow canals, using boats the same width as on Bridgewater's canal. Adoption of the standardized (2.1 m) boat width kept construction costs down and saved water, a problem for many canals. (Bridgewater quickly built a canal connecting Manchester to the coast; there had been starts before, and much later a ship canal was built.) Boats ran 21 m (70 feet) in length, could carry about 30 tons, and were hauled by a horse walking along the side. Canal building yielded valuable experience with earth moving, lock structures, bridges, aqueducts, and tunnels, as well as other civil engineering tasks. Construction activities were institutionalized, and navigators became "navvies."

Although navigation acts were private acts, the policy and institutional aspects of canal building began to fall into a pattern. Canal companies were organized and issued stock. Rights of use carried over from roads, and acts began to require that anyone could operate a boat if tolls were paid. This was not the case for Bridgewater's canal. Companies, such as Pickfords, emerged to offer canal plus pickup and delivery service.[5]

Canals themselves had predecessor models. Well before canals for moving water for transportation became common, water itself was transported through artificial pipes and open channels. London's New River was established by a company aiming to supply fresh water to London, and was financed in part by adventurers (those who we would today call venture capitalists), and the various kings.[6] Open in 1613, it was ultimately municipalized in 1904 when it was acquired by the Metropolitan Water Board. It took some time for the project to find its customers, so it needed ongoing private and pubic subsidy for many years. Of course, it also provided public benefits. Aside from clean drinking water, compared with the cholera-inducing wells that were susceptible to contamination from feces,[7] water could be used to fight fires, and so on.

Canals also learned from experience in improving rivers.

While steam engines were pumping water from coal mines, new means were needed to move coal, and other commodities, from mines to markets. On account of tides, wet dock construction was needed. A good bit of organization, construction, and financing experience accumulated in response. The low value of many of the commodities moved and the high costs of land transportation urged movement of barges as far inland on rivers as practical.

River navigation posed some physical difficulties: large tides, low water levels during some seasons, dredging needs. Use conflicts developed. Mill operators had dams and resisted releasing water for navigation. Owners with riparian rights claimed tolls for improvements or for the use of the embankments for pulling boats or trans-shipment and/or could resist river improvements.

A pattern seems to have evolved. Prior to 1500, city corporations were given river development authority (for instance, the City of London started developing the River Thames in 1179). Later, the Crown gave development authority to local landowners who put forward specific development and toll schemes. The latter is of interest because it is part of the model carried over to the canal era.

The river experience yielded the first institutional form for canals, while technological experience carried over. River development utilized dredging, flashboards (structures placed alongside the water to increase the amount of water that can be contained), and locks. Lock technology, in particular, could carry over directly, as could some of the dredging technology. John Smeaton, later a famous canal engineer, obtained his first experience on river projects. The flashboard system required a good water supply, and canals could not be so wasteful of water.

Arthur (2009) describes the domain "canal world" as a "watery world of barge-horses and boatmen and locks and towpaths" which was fluid but slow.[8] It had one functionality, the movement of bulk cargoes, but had huge cost savings compared to alternative means to accomplish this task.

Canal developments were considerable on any scale, but England did not lead Europe. Building on Leonardo's invention of the miter gate with sluices, France, the Low Countries, and North Italy were well served by networks of canals by the late eighteenth century. The Canal du Midi dates from 1666. The decision to adopt standards suited for Flemish boats was made in 1810, and a general plan for the canalization of France was adopted in 1820. (Another plan was developed and implemented in 1880, at a time when inland canals were largely obsolete.)

Building on the fringe of the feasible pressed for suitable technologies, and some small tub canals were constructed. Commodities were moved in trained, 6-ton tubs. A horse could move each tub on a near-level tramway, and they could be handled easily on inclines.

Figure 1.3 provides an interesting classification of the sources of funds for canals and river improvements. Adding the tub canals, England had by about 1820 a four-level system of inland waterway improvements:

- The rivers and their improvements
- Broad canals, extending river navigation[9]
- Narrow, 2.1-m canals
- Tub canals.

James Brindley (1716–1772), fresh off the success of the Duke of Bridgewater's canal, conceived of canals to link the four great rivers of England (the Mersey, Trent, Severn, and Thames) in a Grand Cross. He engineered the 150-km Trent and Mersey Canal as

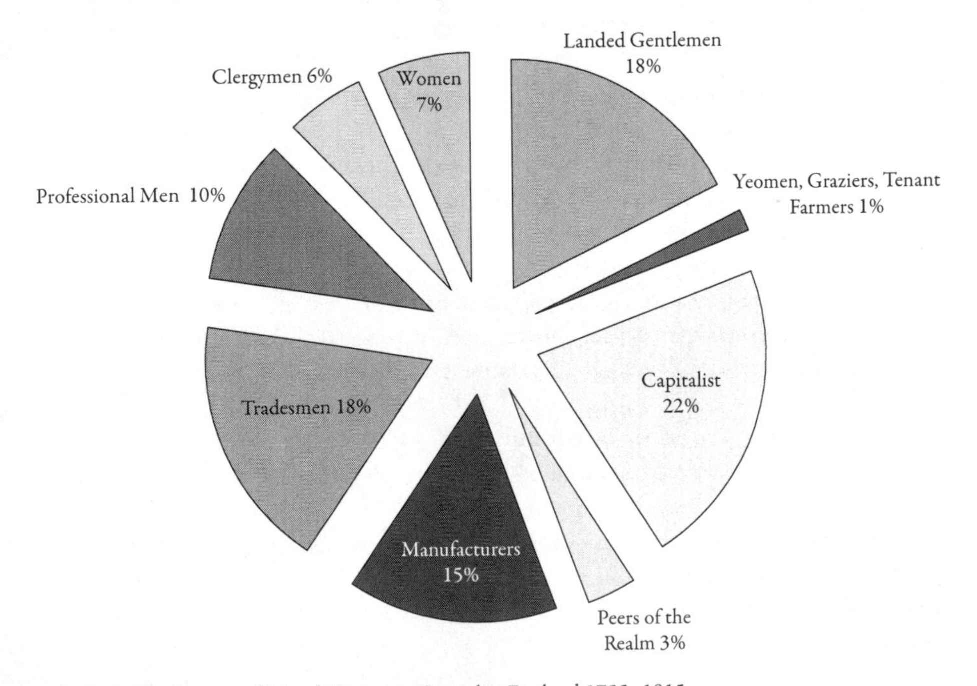

FIGURE 1.3 Sources of Inland Waterway Capital in England 1755–1815
Source: Ward, 1974

the Grand Trunk Canal, and other canals quickly followed at the hands of other engineers like Thomas Telford.

By the 1820s, the era of canal building in England was over (See Figure 1.4). In part, the system was "built out" in that the feasible canals had been built and rail competition

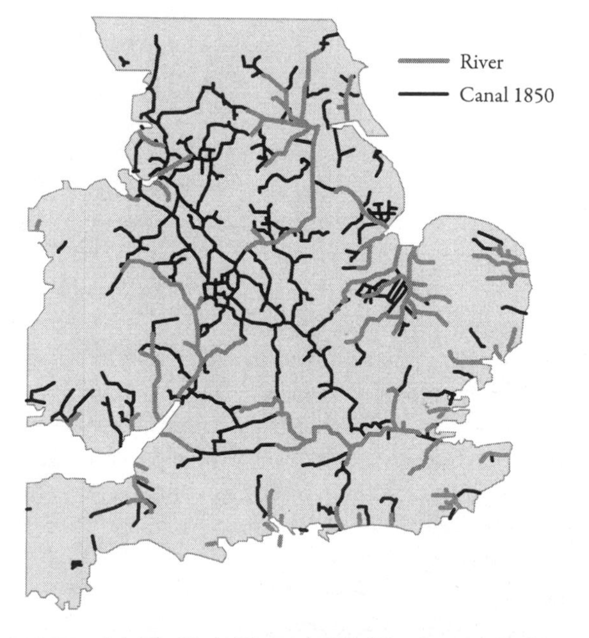

FIGURE 1.4 The Birth, Rise, and Fall of Canals in England (1750, 1850, 1950)

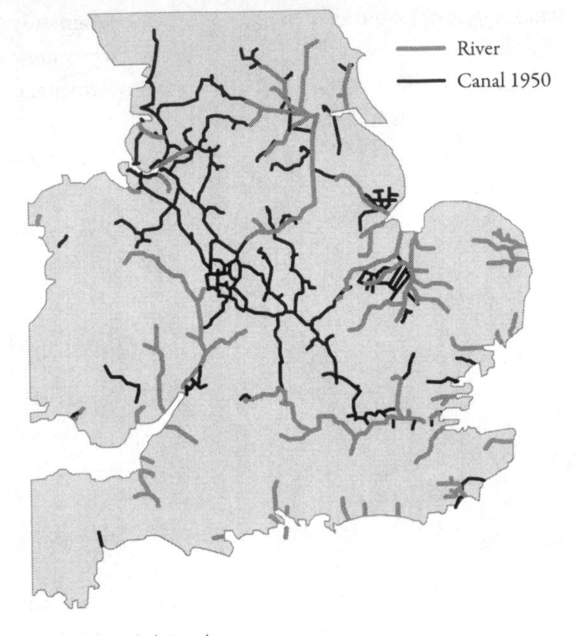

FIGURE 1.4 (*Cont.*)

came along. Prior to the canal era, England's coastal and river trade took place using flat bottom sailing barges that could move 40–80 tons. The trade mainly centered on London. Coal, cattle, grain, building materials, and other commodities were moved to serve the needs of London. Rivers served collector-distributor functions for trade routes across open seas. Though technological change was slow in both dock and barge technology, more docks and barges were constructed as the coastal trade grew.

1.4 Erie and Emulation: Canals in the United States

Throughout the United States, settlers wanted access to transportation by means of rivers and canals and roads so they could bring their produce to market.[10] In a nutshell, by 1800, the coast from Philadelphia and southward was well served by small ships in local trade, and numerous ports offered trans-Atlantic services. The Connecticut and Hudson Rivers served limited roles in New York and New England, and needs there pressed for early canal and road developments. Many early small canals were built, especially in New England. These were in the English organizational, financing, and technology style, and they used some English capital. Also, the opportunities to the west (of Appalachia) and in the interior South pressed for major long-distance, open-up-territory access improvements. King Cotton in the South benefited from downstream flows of the rivers, as bulky cotton floated downstream, and finished goods could more easily be moved upstream.

Although 1824 marked the beginning of continuing federal involvement in river improvements, early appropriations were modest and were mainly restricted to rivers regarded as of national importance. Debates about the federal role continued. The states and private interests were busy with canal and river works and there was a rising tide of demands for federal appropriations. Avoiding conflict over specific projects, the federal government made land grants to the states for internal improvements, and the states and sometimes the federal government became stakeholders in projects by purchasing stock.

Where the terrain was difficult and/or water in short supply, inclined planes were constructed. The plane shown in Figure 1.5 was constructed on the Pennsylvania Canal as part of the portage tramway across central Pennsylvania. The canal boats could be handled in sections, an unusual feature of this canal (Figure 1.6).

A benchmark, open-territory event was the opening of the 360-mile (580 km) long Erie Canal (built 1817–1825) from Albany on the Hudson to Lake Erie. Dubbed "Clinton's Folly" by skeptics in honor of project advocate New York City Mayor (and later New York State Governor) DeWitt Clinton, the Erie Canal was built with New York State support. The canal lowered costs and rates on shipments an order of magnitude, and it set off a clamor for similar investments. Due to success, in 1835, ten years after the Erie Canal opened, the New York legislature authorized an expansion that would increase the surface from 12 to 21 m (40 to 70 feet) and the depth from 1.2 to 2.1 m (4 to 7 feet). This was completed in 1862.[11]

During the canal boom, three types of canals were constructed: connecting tidewater and upcountry, crossing the Appalachians, and connecting the Ohio River system with the Great Lakes.[12] Many canals failed to recover enough funds to pay off capital. Some could not even pay operating costs.

The decision of Pennsylvania to build a mainline canal from Philadelphia to Pittsburgh (with some inclined railroads for the really steep bits), along with a set of feeder canals, was

FIGURE 1.5 Inclined Plane No. 8, by George Storm, date unknown.

FIGURE 1.6 A Sectional Canal Boat in the Hitching Shed, by George Storm, date unknown.

misguided. Topographically, Pennsylvania was far less suited than New York for canals, and as a consequence, the canal rose much higher and had many more locks than the competing Erie Canal. It also started later, and due to the feeder routes, was slower to build. While it carried traffic, it was not successful in the way the Erie Canal or later railroads were (despite being south of, and thus less frozen than, the Erie Canal). Pennsylvania sold much of the Main Line Canal system off to railroads, which were a much more suitable technology for crossing the Appalachians.

Indiana's canals were among the worst systems; some sections were open for only 7 years before shutting, the better sections for only 20 years, and the revenue raised did not come close to repaying costs.[13]

States owned an even larger share of canals than turnpikes, and raised much of the funds for those share-holdings through sales of bonds. The majority of government debt by the 1860s in many states was due to financing internal improvements like canals.

The need for the additional state support for canals is in part due to their cost. Canals (at $12,000–$18,000 per km [$20,000–$30,000 per mile]) were much more expensive enterprises than turnpikes (at $3,000–$6,000 per km [$5,000–$10,000 per mile]).[14] Railroads were typically more expensive than canals; the New York Central came in at around $18,000 per km ($30,000 per mile).

About 8.4 million people lived in the United States in 1815, comparable to the population of just the City of New York today. The economy was driven by agriculture and resource extraction, and was mostly local and rural (only 7.2 percent urban). Land transportation was difficult, much harder than by water, and the population hugged the coasts and navigable waterways. Physical mobility was limited, but social mobility was to be strived for. In contrast with England, "printer" (i.e., publisher) John Niles noted that the United

States possessed an "almost universal ambition to get forward," while in England "once a journeyman weaver, always a journeyman weaver."[15] New York attained long-term preeminence among US East Coast ports in this era (displacing Boston and Philadelphia). Albion[16] identifies four factors:

- Establishment of an attractive auction system for disposing of imports
- Organization of regular transatlantic packet service
- Development of the coastwise trade
- Building the Erie Canal.

Baltimore was also very fast growing in this period. Its merchants were able to consolidate local trade in the Chesapeake Bay region and open up to the west faster than rivals from Philadelphia, even before the landmark opening of the Baltimore and Ohio Railroad. It developed the Clipper ship to enable fast sailing. A well-developed turnpike system also emerged around the city.[17]

The dominance of a few major cities (New York, Boston, Philadelphia, Baltimore) kept secondary cities from growth. Salem, Albany, Providence, Richmond, and Norfolk were all subdued by larger neighbors that were able to achieve faster growth due to what are now called "economies of agglomeration," first with economies of scale, then with specialization.

Cities were dominated by trade and commerce, and the merchants who conducted it. "Wharves, warehouses, and stores characterized their physical appearance, and everything was tied together and directed from their nerve centers, the counting rooms [banks]. The streets of every large city led down past the warehouses to the piers. From the farms by river or road came products for export, but this was the *back country*; in 1815 every city seemed to face the sea."[18]

Business was beginning to specialize, moving from a merchant controlling finance, retailing, and transport into separate specialist firms doing these things only. Instead of owning a ship, newer merchants would contract with common carriers. The inland waterway system meets the maritime transportation system, discussed in Chapter 5.

With a lag of several years while large projects got organized, canal spending by the states expanded rapidly in the late 1830s, as did railroad construction. Reading the competitive situation and being displaced by government capital, spending by private canal companies decreased sharply. Committed to programs and projects, the public sector did not read the competitive situation. The state governments were slow to recognize the strong competition to be provided by railroads. Figure 1.7 shows the development pace in the United States.

One reason for the quick demise of canals was the unusually cold winters of the middle 1840s—there were short shipping seasons just at the time new canals were opening and facing competition from railroads.

A so-called "debt-repudiation" depression ran from 1839 to about 1847. The conventional reason given for that depression is the write-down of investments made in canals and some changes in currency and banking policy. While not dismissing those factors, the depression also arose from the clash between the new and old technology and, in particular, to the displacements the uses of the technology occasioned.[19]

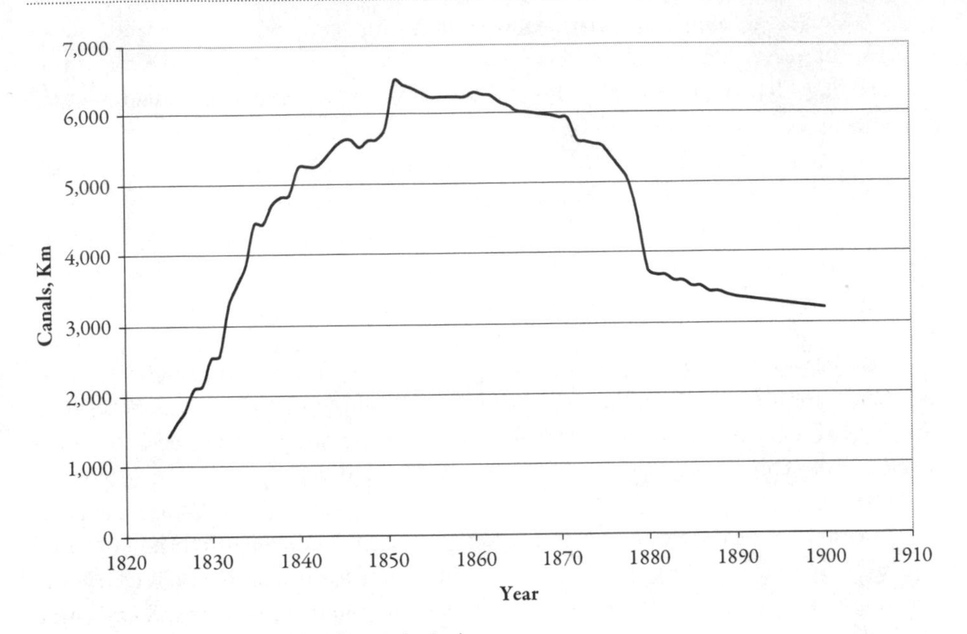

FIGURE 1.7 Canals in the United States (in km)

Early railroads were at first adjuncts to canals and then competition for them. Tramways had long been used to move products of forests and mines short distances, and they fed traffic to canals. Some years after development of the steam engine, steam replaced animal power on tramways. The locomotive was evolved, and railroad deployment began competing—first for passengers previously carried in carriages or on canals. Railroads next competed for commodities and began to expand beyond territories already served by canals.[20]

The above is a simple and well-known summary of canal development, but we should remember that it holds for only a small part of the world, particularly the eastern seaboard of the United States and portions of Western Europe. Most of the world did not share the history of road and canal building; rather, the railroad was the mode that spread the commercial revolution, consolidated the control of central governments, and so on. This *different beginnings* is an important point. It is one of the reasons that the land transportation situation in Japan differs from that in the United States, for example.

Canals withered given competition with rail, some rapidly, some not. The Delaware and Lehigh canals survived for some time.[21] There was a mixed story for coastal transportation and inland waterway transportation, depending upon topography and peculiarities of markets. The lack of bridges in the San Francisco Bay Area, for example, kept the scow (hauling hay, building materials, etc.) and ferry boat in business well into the twentieth century. Coastal transportation of bulk materials continued, and Great Lakes freight movements increased. River-based waterway transportation withered quite a bit (to be reinvented later through barge technology).

1.5 France in America: The US Army Corps of Engineers

The US Army Corps of Engineers began in 1802, the same year that Congress founded the US Military Academy at West Point. During the early years, instruction at West Point was in French, and the Corps followed the French tradition of science-based engineering and a prominent role for engineers in public works agencies. France was a leader in science and engineering, and French recruits to Washington's Army and the new nation's appetite and need for projects pressed for action. There was plenty to be done—surveying for canals and roads, navigation improvements, and construction in the new capital city.

The Corps' start explains easily, as does the Corps' continuing role in military engineering. Its early experience in surveying proposed projects and its scientific-engineering start help explain its early development of what is now called benefit-cost analysis. But what explains the continuing roles of the Corps in nonmilitary waterway, flood control, and other internal improvements?

An internal improvement role was debated at the Constitutional Convention and the provision of internal improvements was mainly left to the states. Improvement advocates Benjamin Franklin and Elbridge Gerry managed to include in the Constitution reference to "post roads and post offices" in addition to mention of military structures and navigation aids. But the states lacked fiscal means, and with private investors they pressed Congress for aid, and appropriations were first made in 1802. Small appropriations were made for a number of years and defense was sometimes cited for justification. Surveys were common and led to the General Survey Bill of 1824 authorizing surveys for improvements of commercial or military importance or for the transport of mail. Even though the political and Constitutional grounds for federal action were weak, two years later an Act was passed and funds were appropriated for improvements to the Ohio and Mississippi Rivers. The US Army Corps of Engineers provided a variety of services on inland waterways and at ports.[22]

After beginning work on important inland waterways, such as the Ohio River, the Army Corps of Engineers institutionalized its responsibilities and strengthened district offices. Actually, the district office system had been developed by the 1850s, but those offices had only sporadic work.

Following common law and tradition, federal involvement focused on ocean ports and the tidal parts of rivers in the early nineteenth century. But actions by Congress increasingly applied to inland waterways and ports. These actions were legitimized by the Supreme Court in 1870 (*Pennsylvania v. Wheeling Bridge Company*), which said, in essence, that navigable rivers (those possible to navigate) are highways of commerce over which unobstructed trade may be carried. It specified the right of regulation, and the right to improve facilities followed.

The Army Corps of Engineers backed into flood control work in part because of the need for water to augment low river flows during the summer. Dams built to augment low flows could also be used for flood control, although there is a conflict between those two tasks. The Corps began work at the headwaters of the Mississippi in 1880. However, as large dams and locks were built on the Upper Mississippi and as other developments occurred, the Corps found that it was in the recreational resources and flood control business. Not needing much water storage space to augment low flow once high dams with locks and slack pools were built, these functions could be accomplished without too much difficulty.

Elsewhere, and at other times and in different circumstances, the Corps took on flood control responsibilities; it began to do levee work to improve channels and control floods. The levee work was an extension of its work with wing dams and dredging and started early on in the New Orleans area. The high dams were built in 1950s, and later electricity began to be generated with the Corps using a private company franchise system.

Already involved in port and nearby waterway navigation improvements, the Corps became the agent for river development, and as the role of the federal government expanded, its role increased accordingly.

The Corps' also backed into water quality work. The first problem was the blocking of channels by debris, mainly from sawmills in the early days, and related problems of sewage in harbors and in slack water ponds. In response, 1899 federal legislation gave the Corps authority to control dumping of sewage and debris. But that "clean water act" was restricted in meaning by the courts. In essence, the Corps could restrict dumping that interfered with navigation, and that applied to sewage in only limited situations, example, where solids were filling a harbor. Industrialization and urban development led to a situation where one gallon of every four passing Louisville during low water had been through an untreated sewage system. A similar situation must have held on the Upper Mississippi.

The Corps got into general environmental quality work by the turn of the century. The Upper Mississippi, in particular, was regarded as a scenic and bird flyway route and the Corps made much of its role in providing supportive river improvements.

Corps programs were influenced sharply by the upturn in environmental awareness during the 1970s, just as highway and airport programs were. They also shared with those programs increased problems with land taking—they took land at fair market value but paid no relocation costs. The conflicts that resulted were a very real shock to the Corps, for it had regarded itself as a resource development, environment enhancement agency.

The Corps did not charge tolls for the use of facilities. That's a tradition that goes back to river development in Europe and the notion that waterways ought to be free to all. (Though that is certainly not true everywhere in Europe; tolls were common on the Rhine, for instance). But that's murky. Canal operators charged tolls, and a river improved with dams and locks is much like a canal.

The operative argument against tolls went something like this. We start improvements with a single dam. It's not reasonable to charge tolls or expect tolls to cover costs because waterway-related developments and traffic won't emerge until all the dams are built. The short-hand expression of that idea is "infant industry," and that expression has been much used in political debates about waterway tolls. But not all dams get built. Those opposed to tolls say the "system" isn't complete because many dams on tributaries still need to be built.

Today, the Corps expends about $5 billion per year on civil programs ranging from river and port projects and radioactive site cleanup, to dam safety and recreation site improvements (see Figure 1.8). Funds are authorized in a budget separate from the military budget. Expenditures include some matching funds, and the Corps is essentially a civil works agency managed by the Army. Armies divide geographic responsibility, so the Corps has district and local offices. It's in the news when there is an argument about wetlands, the site for

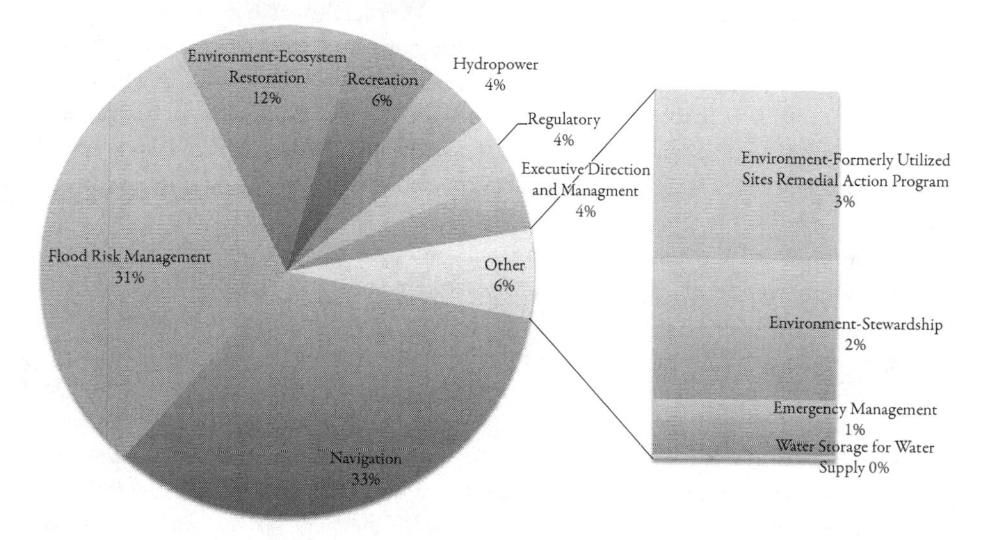

FIGURE 1.8 USACE 2011 Budget

TABLE 1.1

U.S. Army Corps of Engineers Costs to Operate and Maintain Selected Inland Waterways

River	Cost per ton-mile	Cost per ton-km
Mississippi River	$0.03	$0.02
Ohio River	$0.05	$0.03
Appalachicola, Chattahoochee, and Flint Rivers	$5.26	$3.27
Allegheny River	$9.04	$5.62
Kentucky River	$11.23	$6.97

Source: Budget of the United States Government, 2003. *Note*: See webpage of Corps of Engineers Civil Works, http://www.whitehouse.gov/omb/budget/fy2003/bud24.html. See also: http://www.amrivers.org/corpsreformtoolkit/riverbarges.htm for some similar, but different, estimates.

a new bridge, flood control–drainage projects, or when a member of Congress announces funded projects. It was found liable after Hurricane Katrina (2005) breached the levees of New Orleans.[23]

The discussion above began to indicate how the structure and performance of inland waterway activities evolved. As would be expected, flow economies of scale on important routes on the network plus traffic patterns and variations in the natural conditions of routes result in vast differences here and there in the costs of operations and the provision of facilities (Table 1.1). The Lower Mississippi and coastal facilities from the mouth of the Mississippi to the west are relatively inexpensive ($0.03/ton-mile) and those routes have lots of traffic. The tributaries of the Lower Mississippi and the Upper Mississippi and Ohio have more costly facilities and operating costs are also higher. Corps maintenance costs per ton-km for the entire system run about $0.10. That is vastly greater than the ton-km cost of rail services.

In light of these differences, one might suppose that there is much discussion of differential toll charges and network rationalization, especially abandonment of facilities. That's not the case. As a surrogate for tolls, operators pay a modest fuel tax regardless of where they operate. Congestion occurs at some locks and dams, and congestion tolls have been argued but not implemented. Taxes are collected to fund dredging, and the ports are now expected to share dredging costs. Everywhere, locals are expected to share the costs of improvements, for example, for flood control facilities.

The Corp's now has many tasks extending beyond inland waterway and coastal navigational facilities. Flood control work and related stream improvements give it a presence in every state, and the same district offices that do civil work also do military construction. Dredging work at ports is another important Corps responsibility. Recently, the Corps has taken on the job of dam inspection, and has been a player in the development of Maglev technologies. It is involved in wetland development. In part, the expanding scope of Corps work represents an institutional response to its having accomplished its major work. Expanding scope also reflects the Corps creditability. It has access to engineering expertise and is a long-term user of benefit-cost techniques (since the 1930s) in the context of multipurpose development. Its project and system plans are state of the art. Whether the Corps deserves to be so well regarded is another matter, in part because its benefit-cost studies are often questioned.

By about 1970 the growth of traffic was yielding significant delay at locks on the Ohio and Upper Mississippi. Operators seeking scale economies were creating a demand for increasing the controlling depth to 3.7 m (12 ft). The Corps responded, and plans began to emerge for dredging and the modification of locks and dams. In addition to increasing lock depth, where not already doubled, locks were to be doubled. Larger locks were proposed to avoid the breaking of tows and reduce delays. Lock and Dam 26 on the Upper Mississippi moved a bit and needed urgent repair. The Corps proposed rebuilding it to the proposed higher standards, and the project became a cause celebré in the environmental movement. The dam was rebuilt, but as a consequence of the debate, the plans for enlarged, higher capacity facilities are on the shelf. But work for the Corps is available because there is a considerable backlog of repair and rebuilding required.

The Corps' activities have some transit-policy features and some highway-policy features. The district offices do plans and programs and manage construction; Washington rations money among districts. That's highway-like. At the same time, local groups work directly with Congress to push projects in a pork-barrel style.[24] The Corps has managed to build some dubious projects. The Tenn-Tom, a project to link Mobile, Alabama, with the Tennessee River via the Tombigbee River is discussed below in Section 1.5.3.

1.5.1 OHIO-MISSISSIPPI RIVER SYSTEM

The United States acquired much of the Mississippi River system with the Louisiana Purchase of 1803. Control of the Port of New Orleans, which grew to become "the outlet of produce for more than half of America's inhabitants" was the critical driver in the Louisiana Purchase.[25]

Before the Ohio-Mississippi River system could become an effective nineteenth century transportation system, it first had to be domesticated. This required cleaning out the snags and timber blockages (rafts) that had accumulated over the centuries. Next steps included channeling, using jetties to narrow the rivers to focus flows to cut deeper channels. Steamboat traffic increased, and St. Louis, Cincinnati, and Pittsburgh were among the many inland cities served by the steamboats of the times.

It was during this period that the lasting expression "pork barrel" came into use. Farmers in the Upper South and Lake States shipped pickled pork downriver in barrels. As states and private interests sought federal funding for river and harbor improvements, they were wishing for pork barrel funds.

The Civil War devastated much of the river fleet and, much more importantly, rapid railroad expansion captured passenger and much freight commerce. That could have been the end of the story—after about 1900 modest river commerce continued in niche markets where river and competitive conditions permitted.

But that didn't happen because many recalled the valuable services performed by the river fleets before and during the Civil War, and there was concern about the growing monopoly power of the railroads absent competition. A Congressional Report in 1873 called waterways natural competitors and regulators of railway transportation. In 1888 Congress asked for and funded the construction of locks and dams along the Ohio that would provide a 1.9-m (6 ft) channel and twenty-two years later a 2.7-m (9 ft) channel was authorized and began to be funded.

Looking around today, 2.7-m (9 ft) channels are found on the Mississippi and its main tributaries, channel widths run 45 m (150 ft) or wider, and locks are 33 m (110 ft) in width and 180 to 360 m (600 to 1200 ft) in length. Locks tend to be smaller on the lesser used tributaries. There is seasonal congestion at some of the locks and need for continuing channel and lock maintenance. There are advocates for deeper channels and larger locks and dams. However, environmental groups and competitive transportation firms oppose such improvements.

1.5.2 KENTUCKY RIVER

The Kentucky River joins the Ohio River about midway between Louisville and Cincinnati. It drains the Blue Grass country and extends to the coalfields of eastern Kentucky. River development was essential to early European settlement, and Kentucky made surveys and plans for improvement as early as 1798. The appearance of steamboats on the Ohio river in the 1810s pressed for funds for channel clearance. To deal with the problem of low water several months of the year, Kentucky asked the federal government for a survey. The survey was made, but at that time the federal government invested where rivers formed state boundaries and individual state interest was blurred. Augmenting the federal survey with a state survey, the state developed a plan for fifteen dams and locks, maximum lift 4.6 m (15 ft), to manage the 69.5-m (228 ft) change in elevation of the river from the Ohio to the mountains. Considering the flow of the river and potential traffic, it was decided to construct 11.6 by 51.8 m (38 by 170 foot) locks and seek a channel depth of 1.8 m (6 feet). Work began in the 1830s, and by the 1850s the lower 193 km (120 miles) of the river had been improved. The locks were stone-masonry, but the dams were inexpensive stone-filled timber cribs.

The state issued bonds to fund the improvements; tolls were collected. However, the tolls just covered operating costs. No money was available to repair existing dams, much less extend service upriver. To cope, the state turned operations over to a private company, and it had no better success than the state. What's worse, the dams fell into disrepair and became mainly an obstruction to downstream navigation of floating logs. The state went to the courts and managed to regain control; it debated a new financing plan. At the same time, it asked Congress for help, and the Rivers and Harbors Act of 1878 was the first of a series of acts for improvements on the river.

Work began and continued for decades. To start, existing dams were rebuilt and new dams upstream were added. Interestingly, the first upstream dam was at the head of navigation, Beattyville in the three forks area. Users there wanted a slack water pool for the storage of rafts and barges. Sudden release of water (through bear-trap dams) in the spring was used to float tows to the improved part of the river.

But questions began to be asked about the improvements before the march upstream produced many new dams. The railroad was in the region by about 1850, and it had extended service to the Beattyville area by the turn of the century. The railroad captured packet boat and coal traffic; coal barges were no longer being used. The only traffic was log rafts, and dams and locks were in the way of that traffic.

Suppose the improvements were completed, would the river recapture coal traffic? Recapture was in doubt. Coal from Pittsburgh could move to Carroltown at the mouth of the Kentucky River in about the same time as the Beattyville-Carroltown move, and the latter was more expensive because of the smaller scale facilities.

The US Army Corps of Engineers pointed out the problems. People in Kentucky wanted the improvements and the march upstream continued into the 1920s, working directly with Congress. Eventually, fourteen locks and dams were constructed and about 400 km (250 miles) of the river were improved. Some off and on traffic occurred, but in small amounts. In 1970, the Corps abandoned its responsibility for locks 5 through 14, and the state took them over. Now only recreational traffic uses the locks. The Kentucky River story is one from a larger set of stories that may be explored.

The Kentucky River gave an early lesson about attempts to improve lightly used routes. But that lesson was pushed aside by political demands for improvements, and over its history the Corps improved many *Kentucky Rivers*, such as the Cumberland and Missouri Rivers. There is not much debate about these facilities.

1.5.3 TENN-TOM

The French explorer the Marquis de Montcalm first recommended that the Tennessee and Tombigbee Rivers be connected in 1760, linking the Ohio River with the Gulf of Mexico and bypassing the Mississippi River. At the time, that territory was in French hands, and such a canal would aid in French administration of that area. The canal idea reemerged several times. In 1810, a petition from Knox County, Tennessee, argued for the connection. In 1819, the new state of Alabama studied it. In 1874, another study was undertaken by the administration of Ulysses S. Grant, which concluded the Corps of Engineers could build it technically, but that it would be economically impractical. Again in 1913, it was found to be cost prohibitive. The Corps again studied the waterway in 1923, 1935, 1938, and 1945.

Congress approved the waterway in 1946. Proponents asserted that this would be a natural extension of the works constructed by the Tennessee Valley Authority. However, most people outside the region saw the Tenn-Tom project as pork barrel, and railroads opposed it as government subsidy for competing modes. President Lyndon Johnson budgeted funds for engineering and design in 1968 despite the opposition, but in the 1970s, environmental groups became opponents as well. Though construction started in 1972, it was in the courts for over seven years during the construction period.

The Tenn-Tom finally opened in 1985 at a cost of at least $2 billion. The US Army Corps of Engineers predicted 27 million tons would be shipped the first year. By 2000, however only 6 million tons per year were actually shipped. As with many large-scale engineering projects, they are ultimately technically successful but economic failures.

1.6 Discussion

We see an overriding question. The United States has to make up its mind about what it wants to do with its inland water resources. We told the Kentucky River and other stories to lay background for this point. River transportation was the name of the game in the early days. But today inland water transport is economically viable in only a few places. Large subsidies are required elsewhere.

Yet, the imprint of river transportation remains in the location of cities; old, unused, or little used dams; and in other physical remnants. The important thing is that river transport remains in the minds of those in the regions that once had those services; it's part of the culture and history. This sense of history, roots, role, or nostalgia has value, and perhaps it should be the main consideration in policy development. If we recognize that the subsidy for water transportation is not really for transportation, but rather to preserve the history of water transportation, then we can decide if the history is worthwhile.[26]

The modern modes owe much to canals. They demonstrated the payoffs from capital-intensive transportation improvements. On the hard technology side, they provided for the development of construction know-how. They also provided experience to manage, finance, and operate related institutions. This was learning by those who provide transportation. Railroad construction organizations derive from the canal experience.[27] The public learned about investment opportunities and about the off-system developments induced by transportation improvements.

Canal learning provided experience toward generic policy for transportation and public works. Of course, canals have been around for a long time. We have not addressed learning by the English from other Europeans, or learning by Europeans from the Chinese or the Egyptians (e.g., Pharaoh's Canal, 2000–610 BCE).

Canals developed following the experience with toll roads. Unlike toll roads, large upfront capital for investment was required for canals. Also, canals usually required more engineering work. A style of engineering–fiscal planning emerged, and was engaged in by early civil engineers. Although a more complex matter, canals were organized and deployed much in the way that precursor toll roads had been developed. Subject to constraints on the availability of water and suitable grades, canals tended to parallel thriving routes of commerce (as did the continuing development of coastal shipping). Although engineers of the times also worked on road plans, bridges, and harbors, the planning and construction of canals was their major activity.

The discussion treated changes that took place during a period of population growth, economic development, and renewal and change in political institutions. On reflection, we are struck by how well the transportation development story unfolds with only minimal attention to these matters. Even more, we are struck by how the story unfolds with little reference to war and revolution and the great international political and economic changes taking place.[28]

The main conclusion from the discussion above is that experience was accumulated and embedded within the modes. Know-how (rules or policy) evolved bearing on organizations, pricing, and management and also on the use of hard technologies. With these exceptions, government actions were generally reactive. Even so, the notion emerged that government ought be more active.

The water story continues in Chapter 5.

The ultimate aim of a means of communication must be to reduce not the costs of transport, but the costs of production.

—JULES DUPUIT (1844)

2 Design by Design: The Birth of the Railway

2.1 Plateways to Railways

IN THE EARLY 1800s, England's demand for coal was beginning its long rise. The Pease family owned many of the Auckland coalfields near the town of Darlington. To develop and market Auckland coal, a considerable and risky investment in transportation was required. Edward Pease (1767–1858) decided to take the risk, and secured from Parliament an enabling act for construction in 1821.[29]

The proposal was for a conventional plateway (sometimes called a tramway or wagonway). The differences between a plateway and wagonway deal with the type of track. The wagonways included wooden track, while plateways had that track plated with iron. Those plates would later be L-shaped to keep the wheel on the track. Carts drawn by horses or mules were to be pulled along rails fastened to stone or wood chairs (supports). Rails and chairs were spaced so that draft animals could walk between them. Rails at the time were either made of wood faced with an iron strip or iron plates with no wood backing. Where grades were steep, a fixed steam engine hauled cars using chains or ropes (cable cars); when practical, self-acting planes were used in which the gravity acting on full cars rolling downhill pulled empty cars uphill.

The technology for wagonways is old, having evolved first in the metal mines of central Europe and in the copper, lead, and zinc mines of England. There was a natural extension from these beginnings to hauling coal from mines to water. Propulsion was provided in near-level situations by horses. Sharp lifts were accomplished by horses turning capstans (rotating machines with a vertical axle), and by steam power when rotary power could be obtained from steam engines (about 1780). New animal-powered plateways were being built in the United States at least as late as 1826, when one opened in Quincy, Massachusetts, to haul rocks from quarry to dock.[30]

The rail evolved from iron strips fastened to wood, through L-shaped plates, to forms that begin to look like today's T-shaped rail. Because the early plateways were associated with mines, they were financed, owned, and operated by mine operators.

As canals expanded, plateways expanded. At first this resulted from the new opportunities for mining. Later, plateways were incorporated in canal designs—some served as feeders where canals could not be practically extended, and some served as sections of canals.

Plateways were poised for another round of development just before the Stockton & Darlington was developed. Thomas Telford (whom we meet in discussion of turnpikes, see Section 3.5) had proposed upgrading the road system to plateways using L-shaped edge plates. He had incorporated parallel granite stones in the Holyhead Road for use by coaches.

The investment in the Stockton & Darlington plateway was risky because the route was to be long relative to plateways of the times. Also, the high elevation of the coalfields and the topography of the route were unfavorable—lots of up and down grades. In light of these difficulties, Pease's problem was to find an engineer who could keep construction and operating costs to a minimum.

He found that engineer in George Stephenson, a self-taught "mechanic" with an excellent reputation (see Section 2.3). Stephenson did a superior job of engineering. Cuts and fills were balanced to reduce material hauling costs. Improving on "best practice" of the times, there was a combination of near level grades and short, sharper grades to be worked with self-acting planes or steam engine-rope haulage.

Stephenson had experience with steam engines at coal mines (lifts, pumps). He had also rigged some locomotives. For reasons that likely had to do with Stephenson's interests, two locomotives (Figure 2.1) were ordered. These were to be used as a horse. Just as a horse could, locomotives were to move 3- to 6-ton carts on level or near level ground.

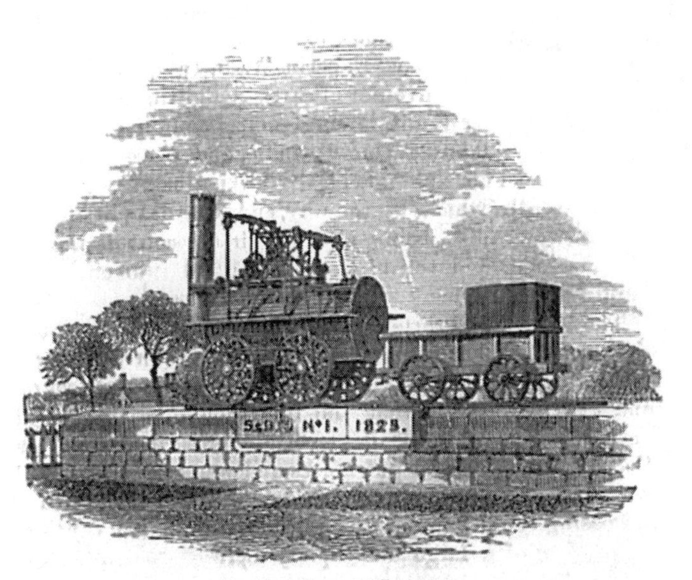

FIGURE 2.1 *Stephenson's Locomotion No. 1* (1825) Engine at Darlington. Image from *Lives of the Engineers*, 1862, digitized by Project Gutenberg 27710

The Stockton & Darlington (Figure 2.2) was a success. It demonstrated that a well thought out design could make money. Success in the London coal market showed that clearly. It also showed that locomotives were a very effective substitute for horses. Indeed, they had the power to move multiple carts. Stephenson's next engine, *Rocket*, is shown in Figure 2.3. Emulation followed.

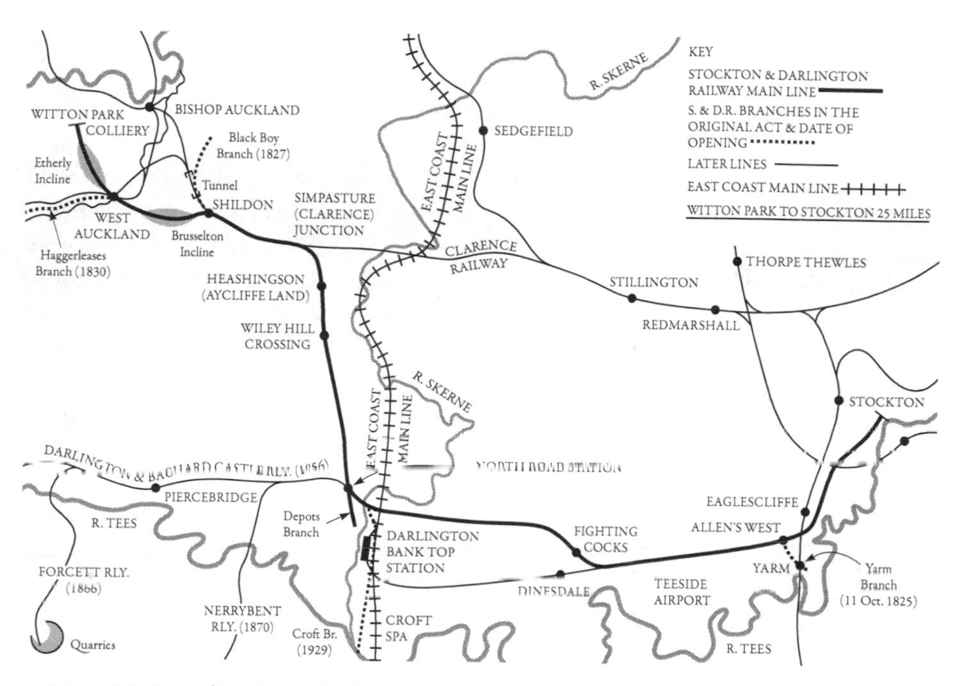

FIGURE 2.2 Map of Stockton & Darlington

FIGURE 2.3 Stephenson's *Rocket*. Image from *Mechanics Magazine* (1829)

2.2 Profile: Richard Trevithick

Richard Trevithick (1771–1833) developed the first practical locomotive. He began experiments in 1803, and his locomotives were operating on colliery plateways at Leeds and on Tyneside in 1812 well before the 1824 opening date of the Stockton & Darlington. Trevithick grew up in the Cornwall mining region where he obtained experience with Watt engines. A self-taught engineer, his first innovative effort conflicted with Watt's patents. To evade those, he developed the idea of strong (high pressure) steam. The steam expansion drove the piston, and once expanded, the steam was exhausted into the atmosphere. Watt's condenser was no longer needed.

Other ideas then came quickly. Strong steam required a strong boiler, but the boiler could also be made smaller. To increase efficiency, Trevithick made boiler improvements, and the steam was exhausted via the chimney to improve draft. Safety valves and fusible plugs were developed. The smaller, more powerful engine was much better than Watt's for transportation uses. Trevithick developed a steam road carriage in 1801 (not the first; Watt engines had been tried on vehicles in France and elsewhere), a gear driven plateway locomotive in 1804, and a dredge in 1806. Later locomotives used connecting rods to crank pins.

In addition to the carriage that was operated in London, Trevithick demonstrated a railway-like system. The single cylinder was set horizontally within the boiler. These and additional transportation efforts did not seed successful business ventures, and Trevithick lost interest in transportation and began to concentrate on mine machinery. That work yielded still another important improvement—the early cutoff of steam flow to the cylinder. Some others experimented with locomotives at the time, but Trevithick was clearly the leader.

2.3 Profile: George Stephenson

George Stephenson (1781–1848) was born in Wylam, England.[31] His family homestead was adjacent to the Wylam wagon-way, a facility with wooden tracks for horse-drawn carts, which had been built in 1748 to take the coal from Wylam to the River Tyne. In 1802 he took a job as an engine-operator at a coal mine in Killingworth, and was later promoted to engine-wright. There he developed a safety lamp that would not explode when near the volatile gases in the mines. In 1814 he stopped operating engines, and started designing them. His first traveling engine, named *Blücher*, shown in Figure 2.4, was released in 1814. The first locomotive to employ flanged wheels on a track, could haul 30 tons of coal at a time. By 1819, Stephenson constructed another 16 engines.

His first railway ran for 13 km (8 miles) between Hetton and Sunderland. This railway used gravity to move the cargo downhill and locomotives to go flat and uphill. It is considered the first railway to avoid animal power. This led to his job with the Stockton & Darlington, discussed above. The Stockton & Darlington broke some other records; the *Locomotion*, built with his son Robert, hauled an 80-ton load a distance of 14 km (9 miles), at a peak speed of 38 km/h (24 mph). It also included the first intentional passenger car. His innovations were not simply with the engine, but also with the trackage and right-of-way. Stephenson insisted upon keeping railway inclines to a minimum, using cut

FIGURE 2.4 Stephenson's *Blücher* (1814)

and fill extensively. The *Locomotion* developed for the Stockton & Darlington was a crude locomotive compared to Trevithick's. A craftsman, Stephenson's contribution was the way he placed the building blocks into a successful format. Later, and working with his son, Robert (1803–1859), Stephenson developed a successful line of locomotives and he was an important promoter and developer.

Stephenson continued as chief railway engineer with the Bolton & Leigh, the Liverpool & Manchester, Manchester & Leeds, Birmingham & Derby, Normanton & York, and Sheffied & Rotherham.

2.4 Stretching the State of the Art

The above description of the Stockton & Darlington is just that, description. Causality is explored only to the level of proximate cause (e.g., Pease had energy and took risks; Stephenson was a skilled engineer who knew cost-effective designs). Once the Stockton & Darlington was in place, emulation played a major role, and diffusion was rapid. Let's dig deeper.

One remarkable thing about the Stockton & Darlington was the way it stretched the state-of-the-transportation-art. A critic of the times might have said that it had no prospects. Existing transportation systems were built-out and mature. Canals, plateways, and roads had been built where the topography was reasonable and the economics was right. The task was that of managing what had been constructed and making marginal improvements. Indeed, John Loudon McAdam's famous book on roads and other publications were addressed to just that. They were mainly addressed to road pavements (macadam surfaces) and toll road management. Although today McAdam (who we meet in Section 3.4) is known for the macadam road, at the time his fame was as a manager.

With respect to the Auckland coalfields, a canal was out of the question. The coalfields were at a high elevation, and difficult construction and many locks would have been required. Road wagon costs were too high. A facility with some plateway features was the only option.

The text *Wood's Railway Treatise*, reprinted at the end of this chapter, emphasizes costs, features of the route, and details of engineering. Similar topics are found in the literature of the times. This was done here, that was done there, where "this" and "that" refer to engineering details and costs. Today's reader might be surprised by how much was known about the strength of metals (but not fatigue), the resolution of forces, and other enduring topics.

The author of *Wood's Railway Treatise* makes the point that the Stockton & Darlington was not the first railway. He cites some plateways in England, and we know of some existing at the time in the United States, Germany, and France. The author failed to realize that on dimensions such as scale, capability to manage throughput, institutional arrangements, and profitability the Stockton & Darlington was quite different from previous plateways. It was a new combination of old things that opened a new way to provide transportation services. The lesson is simply that the new has a lot of the old in it and that the essence of "new" is in combinations.

2.5 Design by Design

It was already mentioned that Stephenson used "best practice" from plateway and other construction. He borrowed from previous construction learning. Note that we said best practice and not standard practice, and note also that this best practice was carefully applied to the physical and market situations. Similar remarks may be made about the design and use of locomotives. The technology was tailored to the physical situation and to the market niche.

There was also borrowing and tailoring in operations, policies, and finance. In particular, the common carrier concept was borrowed from the canals where schedules of tolls were published. Anyone could bring his boat and use a canal as long as he paid the toll. That generally was not the situation on plateways that were associated with a particular property and operated by the owner. Most were short, downhill all the way in the traffic direction, and fed traffic from mines to canals.

With an expensive fixed plant, Pease needed traffic to cover costs, and the desire for traffic was the motive for the common carriage policy. Some feeder routes were built to tap mines owned by others. Stephenson and Pease developed a design in a market niche. It was the design that was new; the innovation itself was a design. With a play on words, it was a design by design (on purpose). It was a new design built from old building blocks.

Except for the design, there was nothing new about the Stockton & Darlington. There was already knowledge and policy relative to transport enterprises—their financing, construction, tariffs, and so on. The technologies were not new. (Stephenson had previously constructed a locomotive for the Killingworth Colliery, building on Watt's improvements of the steam engine. Others had built locomotives, and many were better than Stephenson's.) Yet based on the above arguments, we claim the Stockton & Darlington was not a mere plateway, it was the world's first railway.

2.6 Defining the Railway

Pease had a clear, but difficult, cost-oriented objective: keep costs down in order to make money. The problem of costs was deepened when the decision was made to use locomotives; the original plateway charter had to be revised. The political deal-making to accomplish the charter revision worsened Pease's situation. A London coal dealer and Member of Parliament, John Lambton, insisted on a constraint on coal tariffs to be trans-shipped to London. Per unit distance rates on London coal were set at one-eighth those for local destinations. Pease had to have costs lower than that seemingly impractical rate. That's partly why Pease used the common carrier format.

Pease had to learn how to make the common carrier concept work. For instance, he began by purchasing cars and wagons for lease to independent operators. He was learning about non-coal traffic: how much, how to price it, and services to be provided. In particular, and much to the surprise of managers, passenger traffic swelled and became an unexpected source of revenue. Other questions included:

- How reliable were locomotives, how many carts could they train, and what were the costs?
- Would the wheels slip under load? Did rolling resistance increase with velocity?
- What type of rail worked best?
- How to control traffic?
- What were the best mechanical trans-shipment devices?
- Should tracks be set in stone or wooden chairs?

Not everything went smoothly. At first, passengers were hauled on coal cars. The first passenger car constructed was so heavy that it could not be hauled upgrade when loaded.

Learning continued on railways that emulated the Stockton & Darlington. Learning, and change based on learning, were so rapid that *Wood's Treatise* confused the question of the first railway. In many ways, the Stockton & Darlington was a plateway with locomotives operating on level or near level grades, and some things like that had been developed earlier. But the Stockton & Darlington was a proof of concept quite different from earlier developments.

2.7 Discussion

The Stockton & Darlington experience says a good bit about how revolutionary change occurs in transportation. Much of what it says is contrary to common wisdom, both today and at the times. At the time, it was well-known that transportation was pretty much limited to tram, canal, maritime, and road systems, nothing else was needed or practical, and management of existing systems was the priority. A similar view is wisdom today. A fixed production set exists, let's manage it.

Yet experience says that revolutionary change occurs in market niches and by design. Old building blocks are arranged in new designs, and the steam engine and locomotive were among those old building blocks, for they had been available for about thirty years. Some building blocks were "hard" technology ones; others were "soft," such as the common carriage format and construction know-how. Building blocks are borrowed.

The experience says that markets and production formats are found by inquiring and learning. Change is very rapid when designs are found that bring new resources into the economy (Auckland coal) and support new activities (passenger transportation and its purposes).

Although not discussed, experience says that financing will be found if a design is successful. Pease got his first funding from Quaker-controlled banks operated by his relatives. But once the viability of railways was understood, there was an abundance of financing available. Acquiring land was not trivial, and the side payments to the powerful had to be arranged. Finally, experience says that radical change can occur in short time frames.

We view policies as rules for the control of the flow of information and materials (more about this in Chapter 31). That is a broad abstraction. To use it, let's divide policies into (1) those based on social and political consensus and established and implemented by law or social custom, and (2) those created by the modes for their own purposes and embedded in modal practice. Except for the remarks about obtaining charters from the Parliament, nothing was said about the first type of policy. Parliament was exercising policy on how to birth organizations, policy based on previous experiences.

Embedded policy was discussed—we talked about policy borrowed from previous experiences (rules for construction, common carriage, etc.) and policy created by the mode to meet its needs. The latter were design rules in the main.

One purpose for reviewing the Stockton & Darlington in the first part of *The Transportation Experience* was to illustrate how we can interpret experience. Our discussion began with description. Widening it, we saw how the Stockton & Darlington related to its context and triggered consequential change in transportation services. By "consequential" we mean that productivity at least doubled; by "change" we mean that it offered new options for production and consumption. Public policy didn't play much of a role in occasioning consequential development. Development occurred because innovators did things. Can we learn enough about policy to be proactive on consequential transportation development matters?

Our second purpose was more general. If we avoid concentrating on petty details and strive to generalize, we will find that there has been a transportation experience. The realizations of the experience have involved different actors, technologies, and geographical and temporal stages, but similarities overwhelm differences. Although we have only begun to examine the experience, it's not too early to ask where the Stockton & Darlington fits in the experience, or how it is similar to other things in the experience.

This chapter referred to the design of a new system using old building blocks. Although not yet discussed, we will see close analogies among Juan Trippe's Pan American Airlines service developments of the 1920s and 1930s (Chapter 12), Malcom McLean's Sea Land container services of the 1950s (Chapter 19), and the Stockton & Darlington. The early development of auto-highway transportation and some other developments do not compare so easily, but they do compare.

Our discussion treated the Stockton & Darlington. We saw more than the birth of the railway, for the discussion treated the search for workable institutional and technological designs. We saw the birth, system shake-out, and design revisions.

The easy lesson from the Stockton & Darlington, which began operating in 1825, was that plateways could be built at larger scale and scope than had been imagined before. Using Stephenson's steam engine and edge rail, large volumes of bulk traffic could be moved at low cost. As a consequence, fifty or so plateway proposals emerged in the decade after the Stockton & Darlington opened. Most were cable cars, powered by stationary steam engines and cables, both for incline and flat running; some were to be mainly locomotive powered.

Stockton and Darlington Railroad

The road extends from Stockton, on the river Tees, to the coal mines, are 12 miles distant from Darlington. The length of the main line is about 25 miles; and there are several branches which extend in the aggregate 15 miles. The line traverses an undulating and hilly country, and the amount of excavation and embankment was enormous. Some of the cuttings and embankments are 30 and even 40 feet from the surface. The curves on this road are abrupt and causing much friction, the repairs of the rails and wagons require unusual expense. The profile of that part of the road, where stationary power is not employed is undulating — varying from level to an inclination of 1 in 104 nearly, or 51 feet in a mile; the average is 1 in 246. There are two summits, the Etherley and Brusselton, where is passed by means of stationary engine on each, which works the two inclined planes on each side of the summit. The Etherley north plane is one half of a mile in length, and the amount is 180 feet. The engine is estimated a 30 horse power. The Etherley south plane is rather more than 1760 yards in length, and the descent is 312 feet. From the foot of the latter plane to the foot of the west Brusselton plane is four-fifths of a mile. This plane is one mile long, and the ascent is 150 feet; the steam engine on this summit is estimated at 60 horse power, the east Brusselton plane is one half of a mile in length and descends 90 feet. Thirty-two tons (including the weight of the wagons) are drawn up these planes, in one train, at the rate of 8 or 10 miles per hour. This Railroad is a single line, with four turn outs, each of 100 yards in length, in a mile; width between the tracks: 4 feet 6 inches. The rails are chiefly of malleable iron, 15 feet long, & 1.4 inches wide on the top, and weigh 28 lbs. per yard (the form of Birkenshaw's patent.) The cost of the iron was more than twice the present price of that article. The expense attending their charter was £12,000, and the land cost more than this sum. Without including these items, the cost of the main line was nearly £5,100 per mile, being a much larger sum than any single Railroad in Europe had cost. The locomotive engines on this road weighed twelve tons, and this enormous and improper load materially injured the rails, which were of the lightest pattern. The cost of traction, with these imperfect and antiquated engines, was one-fourth of a penny per ton for goods per mile; the cost and repairs of the engines were one-eighth of a penny additional. Horses were formerly employed on this road; their load was 16 tons gross each at the rate of 3 miles per hour. On the Tees there is a suspension bridge, supported by iron chairs.

Many writers have stated that this was the first Railroad intended for the purpose of general trade: this is a mistake. Several Railroads of considerable extent were made in Great Britain many years previously — The Surrey, the Sirhoway, the Cardiff and Morthyr Tydvill, and several other Railroads were intended to accommodate a general trade, and the tolls on them specified accordingly in the several acts of Parliament; although the articles conveyed on them, as well as on the Stockton and Darlington, are chiefly minerals and other heavy goods.

** This includes the profit of the contractors, toll, &c.; they also draw back the empty wagons without any charge. The prime cost of the wagons was defrayed by the Company.*

FIGURE 2.5 Stockton and Darling Railway described in *A practical treatise on rail-roads, and interior communication in general* as reprinted in an 1832 issue of *The American Railroad Journal.*

The idea extended to continental Europe. The Saint-Etienne railway, completed in 1828, is said to be the first French railway. It was a plateway with some locomotives built on the English model and linked the Loire valley with Paris to move coal. It was soon extended to Lyon and carried passengers. In the 1980s, this line was upgraded to a TGV, high-speed passenger rail.

The rail story continues in Chapter 6.

The third and last duty of the sovereign or commonwealth is that of erecting and maintaining those public institutions and those public works, which, though they may be in the highest degree advantageous to a great society, are, however, of such a nature that the profit could never repay the expense to any individual or small number of individuals, and which it therefore cannot be expected that any individual or small number of individuals should erect or maintain. The performance of this duty requires, too, very different degrees of expense in the different periods of society.

—ADAM SMITH, *An Inquiry into the Nature and Causes of the Wealth of Nations*

3 Incentivizing Investment: Roads through the Turnpike Era

3.1 Steam Cars

NATHAN READ, an early steamboat developer in Connecticut, proposed a steam-powered automobile in the 1790s. When his patent application was read aloud in the House of Representatives, members struggled to suppress laughter.[32] Nevertheless, he received the patent. Read's innovation was the multi-tubular boiler, which reduced the size of boilers by creating a large boiling surface in a small volume. He passed water through the boiler in tubes. All subsequent vehicular boilers put water in tubes through fire, or fire in tubes through water. He later served in Congress himself.

Read wasn't the only one who conceived of steam-powered cars. Charles Mahon, the third earl of Stanhope, developed a steam-powered land carriage "that had the amazing tendency to speed up hills, stall on level ground, and stop completely on downhill slopes."[33] In 1805 Oliver Evans adapted a steam engine to an amphibious land-water vehicle, the *Orukter Amphibolos* (Greek for amphibious digger), driving from Philadelphia's streets (starting at his works where later the John Wanamaker Department Store would locate) into the Schuylkill River down to the Delaware. This was both the first "car" and first working steamboat on the Delaware since Fitch's *Perseverance* 15 years earlier. Fitch also developed a prototype of the first self-propelled steam locomotive before he died, though the technology ultimately was brought to market by others. While the steam-powered automobile would wait nearly a century, the steam engine's application to rail was soon to come.

Richard Trevithick (see 2.2) developed and demonstrated a steam road carriage. Goldsworth Gurney (also the developer of limelight) developed a successful steam coach and a tow for coaches (a drag) in 1825; in 1831, and using a Gurney Drag, Charles Dance offered coach services with a 75 percent reduction in coach fares, and Thomas Telford (see 3.5) proposed widespread services using plateways (tracks consisting of flanged strips that were used in some early railroads). Walter Hancock's steam van, the unfortunately named *Autopsy*, was offering service in the London area during the 1830s.

Steam-powered road vehicles never received much of a trial; they were resisted by road providers who wanted to protect their facilities from heavy vehicles. The debate over who would get the transportation rents—the coach or wagon operators or the road providers—must have played a part. Road policy (then as now) emphasized the protection of road surfaces using limits on weights, widths of wheels, and so on. No doubt, too, providers of wagon and carriage services strived to protect their businesses from interlopers using new technology. It has even been suggested there was a conspiracy to block steam vehicles on roads on the part of the railroads and others.[34]

In spite of this resistance by road providers, someone might have found a niche and gotten the design right. But the clock ran out. The window for development was closed when markets were co-opted by rail service. That's one explanation for what happened. Less general explanations point to the low horsepower per unit of weight of then steam engines and the high rolling resistance of then road surfaces.

Thus, unlike rails and rivers in the preceding two chapters, roads were not to be plied by steam-powered vehicles en masse, they were to remain the domain of human and animal-powered motion until the end of the nineteenth century when the electric streetcar emerged.

3.2 From Trails to Roads

The first networks that humans exploited were given by nature. For transportation, these include rivers and waterways, ocean and air currents, and even deer trails (mainly connecting deer bedding and feeding sites, but also trails to mating sites, and trails for escape) on land.[35]

Despite evidence of roads elsewhere, we begin by touching on the development of Roman roads, a system built over a 500-year period, with mileage about the same as that of the US Interstate Highway System. Except for bridges, Roman engineering seems not to have been emulated very much. Reverse engineering on the Roman roads tell us how they were built—they were stone structures made of polygon-shaped rock bound by mortar (see Figure 3.1). The roads comprised long straight segments with stiffer grades than those allowed in later road building; curvature was avoided except where necessary (mountain passes, natural places to ford or bridge rivers). Comparisons of Roman and modern road mileage between places find that the Roman roads were shorter than modern roads.

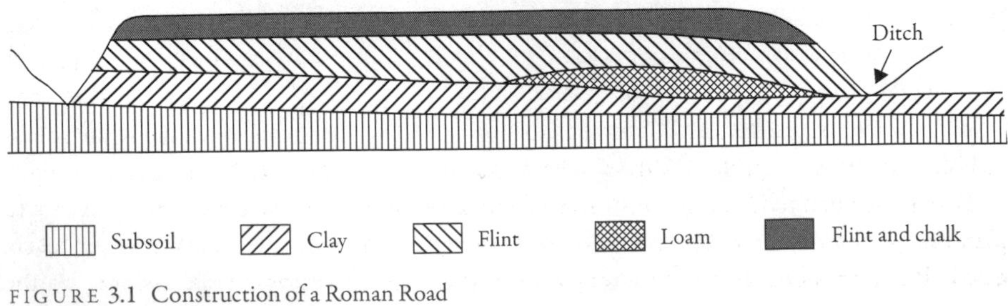

FIGURE 3.1 Construction of a Roman Road

Source: (Margary, 1993.)

It may have been that straight and sometimes steep roads allowed for rapid movement of marching troops. Wagons, which would have liked low grades, were not so common. Without front axles that turned, Roman wagons worked best on straight roads. The typical load seems not to have been more than 225 kg (500 lb). Some evidence exists that the right of anyone to use a road dates from Roman times, as does the idea of the common carrier. A good part of Roman design was symbolic—a strong, powerful artifact ignoring terrain and pushing straight ahead: Roman roads equal Roman power. The decline of Roman power led to the slow decline of Roman roads, as nature took its course. As new areas developed that had been unserved by Roman roads, new links were needed, and those were not built to the same standards (if the term "built" can be used at all).

Roman roads accommodated carts, but by about 1600 the population, and military and church institutions, were mainly served by ways, trails, traces, or paths along which people could walk, ride on horses, or drive animals. (Geoffrey Chaucer provided descriptions circa 1380.) Like the Roman roads, these routes had an "as the crow flies" character; they were paths and often had fairly steep grades, with route orientation tempered by the location of mountain passes and fords. While roads may seem an inconvenient form of transport, when compared with the rivers and coasts, which allowed fluid movement, they had their advantages. Unlike rivers and canals, roads would not freeze for the winter season, and unlike the coasts, were not severely affected by winds. This kept roads significant for inland markets that had alternative networks available.

3.3 The Corvée

3.3.1 THE CORVÉE IN ENGLAND

England's long-standing parish road system used (one hesitates to say employed) statute labor. The corvée system geared itself to local needs and relied on manorial organization. The system of statute labor seems foreign to the twenty-first century, especially in a country without a military draft. The idea is similar to taxation, except residents were taxed their time instead of their money. The difficulty is ensuring quality control; a dollar is a dollar (a pound is a pound), but an hour from a productive individual is very different from an hour from an unproductive one. The system began to be strained in the sixteenth century because of the growth of through traffic (roads were no longer a local matter) and the breakdown of the manorial organization of life. This also introduces the free-rider problem. Local residents have little incentive to maintain roads for non-local (through) travelers. Moreover, the substitution of the Church of England for the Roman Catholic Church eliminated a major source of road maintenance: monks aiding those on pilgrimages.

The decline in quality of the roads led to a revision of the early sixteenth-century system. Bridges were a technical and fiscal problem, addressed by the 1537 Statute of Bridges. This was followed by the Highways Act of 1555, resulting in the election of surveyors to plan and supervise the four days of statute labor per year (which was soon raised to six days). With revisions, this system served until the 1830s. Surveyors were assigned to the Quarter Sessions (the county's quarterly meeting of justices of the peace, the predecessor to the modern County Council), they were given powers to requisition materials, and parishes were empowered to tax for road improvements. Moreover, labor could be hired. This mattered because hired labor has more incentive to be productive than statute labor.

However, the mismatch between regional travel and local responsibility remained, and policy had not responded to technical questions, example, how to build roads. Radical changes began to emerge as the local Quarter Sessions attempted to deal with those problems. Some started charging tolls on heavy vehicles to raise money for repair; others set maximum weight limits for vehicles. Several Quarter Sessions worked together to develop toll and road programs. Others wanted to increase the widths of wheels on large vehicles. Narrow wheels place more of the wagon's load on a smaller space, creating rutting, while distributing the load across wider or more wheels increases the life of the road. In the meantime, traffic of all types continued to grow; stagecoaches and mail wagons emerged in the early seventeenth century. Packhorses and mules were less and less used; wagons increased in size and weight.

The bridge problem became acute. The Act of 1537 simply put the responsibility for bridge maintenance on the parishes, and it had few teeth in it. Technical problems had to do with the narrow width of bridges, unsuitable timber construction, and steep access and egress.

The stage was set for considerable revision of highway policy. One revision allowed and encouraged turnpike developments, beginning about 1650, but taking off at about 1700. Another was a restatement of the responsibilities of the parishes.

The parishes' first policy emerged in the 1760s and 1770s with a rationalization tone. They attempted to define the technologies needed; town access roads were to be at least 6 m (20 feet) wide. The effort to control weights and vehicle configurations continued: six horse teams hauled wagons with wheels 15 cm (6 inches) across; more horses, wider wheels. Little-used routes could be abandoned. Fees were graded so that those who used the roads the most paid the most. Some ills were cured, but poor construction, drainage, and bridge problems remained.

Another round of road development occurred prior to the development of the railroads. The evolution of toll roads built in part from the canal experience, which is discussed in Chapter 1; the stories are intertwined.

3.3.2 THE HEART OF MIDLOTHIAN

The center of activity in any significant Scottish burgh was the "tolbooth" [*sic*]. As might be expected, this is where tolls (really more akin to tariffs for entering town) were collected, but as a small well-organized financial center, these tolbooths came to be administrative centers, with courthouses, and then prisons, and police and fire stations. (In England, similar facilities might have been called the Custom House.) In Edinburgh, for instance, the Canongate Tolbooth dates from 1477, the present structure from 1591 (Clark et al., 1991). Canongate was a burgh, just down the Royal Mile from the Edinburgh Castle, eventually incorporated into the larger city of Edinburgh. The 1737 Tolbooth Riot (in which the mob aimed to free a prisoner jailed at the tolbooth) was fictionalized in Sir Walter Scott's *The Heart of Midlothian*, a novel in which the tolbooth was the *Heart*.[36]

3.3.3 THE CORVÉE IN FRANCE

The French government had been exercising considerable taxing powers, in part because of the military needs of the central government. The vision and power were there, too. In the seventeenth century Jean-Baptiste Colbert, Louis XIV's minister of finance, developed a plan for the orchestration of national life through the use of canals and roads for commercial and military purposes. A vision sometimes called Colbertism emerged. Colbertism can be summarized as the belief that the state has the wisdom, resources, and power to do the best and just things. The vision is sometimes referred to as a tradition of the Ancien Régime.

Beginning in 1716, state highway engineers, the Corps des Ponts et Chaussées,[37] prepared standardized maps for provinces, with centralized planning of roads beginning in 1747 (the year the school for roads and bridges was established). Between 1720 and 1770 the road construction budget increased eight-fold. Plans were implemented using public capital. In addition to capacity expansion plans, standards were developed for bridges and other aspects of the physical plant. For instance, all roads were to be 18 meters (60 feet) wide. By 1789, the year the French Revolution began, the country had a 27,000-km (17,000 mile) road network. The state engineers had maps for the entire country, and they saw the need for a plan with standardized facilities, regardless of local physical conditions and markets.

A high quality, state planned, financed, and operated, free road system centered on Paris had been developed prior to the emergence of the railroad. That was at a time when English major roads were toll roads improved by local capital and managed locally, and the United States had not much more than strips of mud. Despite its praise-worthiness as an engineering feat, the road network was burdensome to the poorer farmers. Roads require resources. The king appointed tax collectors to raise money, and those who couldn't pay often saw their land confiscated. Further, from 1737 the roads were maintained by a special road tax, but those who couldn't pay were required to contribute 30 days of labor per year on road crews. Farmers were inspired to overthrow their oppression of forced labor of the corvée and join the Revolution. Ultimately, the Revolution nationalized what private toll roads there were as well as the king's royal roads. However, by eliminating the dedicated sources of financing and maintenance, roads became a general public service.[38]

3.3.4 THE BEGAR IN SOUTH ASIA

Tolls were known in India, and part of everyday life.[39] Around the eleventh century, one ruler, Kulottunga I, bore the surname Sungandavirtton, that is, "the abolisher of tolls."[40] Elsewhere in South Asia, in Thailand, an early thirteenth-century obelisk (whose authenticity is admittedly doubted)[41] recalls the life during another ruler. It says in part:

> In the time of King Ram Khamhang this land of Sukhothai is thriving. There is fish in the water and rice in the fields. The lord of the realm does not levy toll on his subjects for traveling the roads; . . . .

What was notable in both examples was not the presence of tolls, but their absence.

In India, a monthly one- or two-day corvée (during Mogul times this uncompensated contribution of human and animal labor was called Begar) was often exacted from artisans, mechanics, and sudra laborers according to surviving legal texts from very early times. The

Begar was more severe in India than the corvée in Europe, as some persons were required to bear loads, not merely maintain roads. How early this began is unclear, but the corvée was mentioned in the Code of Hammurabi (c. 1700 BCE), though that document mostly deals with exemptions. The Laws of Manu (c. 1500 BCE) permitted the sovereign to demand temporary services.[42] It was noted in the seventh century by Hsüan-tsang in his *Travels*. It was established at different times in different regions, example in the thirteenth century in the Hoysala kingdom, and later by Muslim princes. It came and went; example in the Mogul Dynasty (starting in the sixteenth century) Akbar eliminated Begar, but Jehangir reinstated it and Shah Jahan abolished it again. It was widespread in places, especially near the Himalayan border. In some areas, for some persons, this lasted through the first part of the twentieth century,[43] and was not legally abolished until Indian independence in 1948. Despite the relatively poor quality of humans as beasts of burden, this practice was an important element of transportation as well as construction and maintenance in India.

3.3.5 THE CORVÉE IN JAPAN

Gishi-wajin-den, a Chinese history from the third century, provides the earliest record of Japanese roads. From this period through the Meiji Restoration in the late nineteenth century, roads served pedestrians and animal-powered transportation. As a result, maintenance was simplified and reduced, since humans and animals don't damage roads nearly as much as wheeled vehicles. The mountainous terrain was not conducive to wheeled traffic, which did not develop as it had in China and Europe. As with the islands of Greece, sea routes were more prominent than land transportation, especially for bulk cargo.

Major overland routes were reported in the *Chronicles of Japan Niho Shoki*.[44] After the Taika Era Reformation (645), a central government modeled on the Chinese Tang Dynasty reorganized the governance of Japan. A system of seven *kaido* (roads) was established connecting Honshu, Shikoku, and Kyushu islands. These roads, centered on Kyoto, were the Tokaido (eastern route), Tosando, Hokurikudo, San-indo, San-yodo (western route), Nankaido, and Saikaido. The backbone provided by those roads remains, and modern superhighways in Japan use roughly the same routes that were established when they were originally constructed.

The seven roads were used for domestic communication; a postal courier system, the "Ekiba, Tenma," modeled on a system in China and similar to the later Pony Express in the United States, was established. Every 16 km (10 miles or 30 ri), stations were placed to provide roadside services and to allow riders to change horses. The corvée was imposed to construct postal stations, and a rebellion against the corvée broke out in 1764 (East Asia in the World, 2004).

Roadside beautification, later to be popularized by Lady Bird Johnson in the Unied States, was established in the mid-eighth century with the planting of fruit trees along the seven roads. Road signs, or *ichirizuka*, were mileposts, similar to those constructed on Roman roads, provided navigational information for travelers and were popularized in the mid-seventeenth century. Road standards were put in place. At the beginning of the seventeenth century, the Edo Shogunate specified that major highways would be 11 m wide; secondary routes would be 5.5 m wide. The roads were to be gravel, and covered with sand. Residents who lived alongside the road through the Shogunate Era performed road maintenance.

Walking with its overnight stays had a social side. A 1952 translation of Ikku Jippensha's ca. 1800 *Hizakurige* provides a humorous description of travel events. During Japan's great modernization, the Meiji Era, wheeled vehicles such as carts and carriages began to be used on roads. This caused roads designed for people and animals to deteriorate. Crow (1881) wrote "The Tokaido is in a dreadfully bad state, with ruts and holes large enough almost to swallow a cart, and yet traffic is very large, both in horse and man-power vehicles." The condition of roads was exacerbated as the Meiji government privileged rail and sea travel to road travel.

The Fukushima incident of 1882, in which advocates of the Japanese Freedom Party were suppressed by the prefectural government, resulted from a local rebellion against the corvée labor tax imposed by the government to build roads (Mainichi Daily News Co., 2002). In response, road construction subsidies (from monetary taxes) were imposed. Still, it was only after World War II that Japanese roads were truly modernized.

3.4 Profile: John Loudon McAdam

John Loudon McAdam (1756–1836), dubbed the "Colossus of Roads," was born in Ayrshire, Scotland. As with many innovators, his interest began young; while in school he built a model road. At the age of fourteen he went to America to work for his uncle in New York City as an "agent of prizes"—a legal form of dealing in prize ships and cargos captured by the British prior to the end of the American Revolutionary War. By the time he returned to Scotland in 1783, he was wealthy, and became a road trustee for his district. Noticing the poor quality of the roads, and this being the era of the amateur scientist, McAdam pursued a number of techniques to improve the quality of the road. He later moved to Cornwall, and received a government grant to continue the work. His conclusions, documented in two books *Remarks on the Present System of Road-Making* (1816) and *Practical Essay on the Scientific Repair and Preservation of Roads* (1819), called for roads that were elevated compared with the adjacent ground, and covered first with large rocks, then smaller stones, and finally bound with gravel—all to improve drainage, which he identified as the major destroyer of roads.

Following a parliamentary hearing on road building, "macadamization" was adopted as the standard for construction, and McAdam was appointed Surveyor-General of Metropolitan Roads in Great Britain and was compensated for his expenses in researching the issue. The technology was adopted elsewhere.

McAdam's system, while more expensive than the dirt roads it largely replaced, was much cheaper than the similar quality of road inspired by Roman road-building, the use of stone slabs carefully laid by masons.

Richard Edgeworth improved the "macadam" method by using stone dust and water (a form of cement) as a binding agent (though this was opposed by McAdam himself). In 1854, the technique was further improved in France by using bitumen (asphalt) binding, to ensure a smoother surface than simply gravel and slag could provide. This "Tar-Macadam," abbreviated "tarmac," was very much the forerunner of the blacktop common today. For his efforts, McAdam was offered a knighthood, which he declined before his death.

3.5 Profile: Thomas Telford

McAdam's great rival, Thomas Telford (1757–1834), was born near Westerkirk, Scotland. After spending his youth aiding his uncle as a shepherd, at the age of fourteen, he was apprenticed to a stonemason. In 1780 he left for Edinburgh where he gained work as a mason, and by 1792 he was among the elite of his profession, working in London on Somerset House. As a supervisor of construction, Telford's first bridge was in Montford across the River Severn. His other works include designing and building churches.

An interesting project was a canal that was on an aqueduct across the River Dee. Rather than using clay to form the edges of the canal, which would have been too heavy, he innovated design by fixing troughs made of cast-iron plates in masonry. In this era of the engineer, it was not uncommon for an engineer to work on structures, waterworks, and highways. He aided in the professionalization of civil engineering by founding the first Institute of Civil Engineers.

He is perhaps most famous for his suspension bridge, such as the Menai Bridge linking the isle of Anglesey with Wales. He also built cast iron arch bridges and turned them into works of art.

In 1803 Telford was hired to help develop the Scottish highlands. He was in charge of rebuilding the roads from London to Holyhead and from Shrewsbury to Holyhead, as well as others, totaling 1,450 km (900 miles). He set standards for maximum grades among others.

Telford's method differed from McAdam's. First, McAdam wanted 25 cm (10 inches) of graded broken stone above the ground, while Telford used an 18-cm (7 inch) foundation base of larger stones, laid in a trench (rather than on top of the ground). McAdam argued that Telford's method was costlier and required more maintenance, and that well-drained ground would be an adequate foundation. In a sense their dispute was the first salvo in the battle between asphalt concrete, usually called asphalt (which most resembled McAdam's method) and Portland cement concrete (PCC) (which is somewhat closer to Telford) that still rages today (see Chapter 15).[45] PCC still is more expensive and takes longer to repair, but is thought to be more durable under heavy loads than asphalt concrete.

Telford is buried in Westminster Abbey.

3.6 Stagecoach

The evolution of transport vehicle technology is a long one, and subtle changes (to us in the modern world) were significant changes at the time. The transformation of freight travel from packhorses, to horses pulling carts and wains (two-wheeled vehicles), to horses pulling wagons (four-wheeled vehicles) was not sudden, but was quite significant in improving the productivity of surface transportation.[46] Four-wheeled vehicles were rare in England before 1500, and monopolies for use of wagons for transport between Ipswich and London were granted as early as 1582. The problem was that wagons would damage roads, creating ruts, and shake the foundations of bridges, and regulations were soon imposed (e.g., Kent in 1604). At least 91 wagoners were prosecuted for overweight vehicles in Star Chamber between 1614 and 1625. Other changes, such as the width of the wagon wheel, which increased in the 1700s, were in part regulated, and depended as well on the quality and size of the road surface. Larger wheels could distribute loads more evenly, doing less damage to the road, but required a wider path on which to operate.

Two-wheeled vehicles seem to have disappeared around 1623. The causes are several, and include lower costs from competing technologies (wagon and packhorse), as well disadvantages of carts themselves (difficulty of balance, difficulty in traveling downhill, difficulty in loading, danger when fording water, and smaller capacity). The carts themselves were lighter, so the amount of deadweight moved was lower. Packhorses remained, and in some markets replaced wagon services. Costs of freight travel were dominated by the costs of provisioning horses, so improving horse productivity was critical.

There were limits to wagon efficiencies. Though each wagon could carry more than a cart, and adding horses increased the weight that could be pulled, horses suffered from diminishing marginal returns; each additional horse had lower productivity. Still, a packhorse cost about one-third more than a wagon for a typical Oxford to London run;[47] but on poor roads, or for smaller loads, a packhorse might be more cost effective.

The innovation of the "flying wagon" greatly increased freight throughput.[48] The wagon was mostly kept in motion, while wagoners and teams of horses were periodically swapped out. This reduced freight travel times, enabled an early version of what we might call "just-in-time" production today, and was particularly important for perishable items (like some food items), though the dominant freight was in the cloth industry. Inns were places to feed horses and men and allow sleep for both. But they were also loading and unloading points, and innkeepers acted as agents for the carriers. The carriers had to have warehouses at either end to store goods, and factors (agents) who would collect and distribute money and goods. Over time, the freight carriers substituted their own "rest stops" for inns, so that they could provision their own fleet of horses and better control costs.

Services were scheduled and regular, and known to have occurred as early as 1737 (Frome to London), at a speed of up to 74 km (46 miles) per day (almost three km (two miles) per hour, which seems slow, but includes stopping time). The carrying trade showed great continuity, some firms lasting nearly two centuries (Thomas Russell & Co. from the 1670s to 1841), providing service on the same route. The market lasted longer, with records of regular service dating to the middle of the 1400s in the West Country to London market.

Improved roads made the stagecoach feasible in the United States for both passengers and freight traffic. Horse-powered coaches were slightly faster than walking, averaging 3 to 5 km/h (2–3 mph), meaning the trip from New York to Philadelphia could take 2 to 3 days in the early nineteenth century.[49]

By 1838, over one thousand wagons entered and left London weekly, each carrying up to six tons and pulled by six to eight horses. Ultimately the flying wagons ended as railways took over the market.

Through the 1700s and early 1800s, carriers were generally organized as partnerships, as the corporate form was not well established. This kept liability with the proprietors, and made for conflicts between partners. The historical evidence remains due to the lawsuits between contending partners of Russell's.[50]

Russell's had grown to become the second largest carrier serving London around 1800. The largest was Pickfords. There is still a moving company with the name Pickfords, which traces descent from the original company founded in 1695 near Manchester. Pickfords was founded using packhorses to carry stones to help construct turnpikes, and used what would otherwise be unloaded trips of horses back to the quarry to carry goods for others. They later moved into flying wagons as well as fly boats (water and animal-powered canal boats, which would swap out teams to keep the goods moving). They also arranged for freight cars to be pulled by engines of the commercial railways, though that business was eventually taken over by the railways themselves.

The vehicle of transportation was considered a social and economic signifier. During Anti-Federalist Riots in Baltimore in the War of 1812, the Federalists who were being arrested ("voluntarily," to avoid mob justice) wanted to travel to jail in carriages befitting their station, while the mob insisted on carts. The authorities made them walk.[51]

3.7 Turnpike Trusts

The parish roads were handled as if they were local roads serving local landowners. But the growth of mail, passenger, and freight traffic, many with non-local origins and destinations, was imposing costs on the locals. Non-local and heavy local users were demanding improved facilities. The parish system had problems enough dealing with strictly local demands.

The toll road or turnpike was the answer. Such roads predate the canal era, and early ones (e.g., in the seventeenth century) were ad hoc extensions of the parish system wherein several parishes would work out a joint project.

Of course tolls were not invented in the seventeenth century. Precursors to tolls date to the mythological world, with Charon charging a toll to ferry people across the river Styx. It was easy enough to transfer the idea of tolls for service from ferries to bridges, and then bridges to roads. Other early examples include mentions of tolls in Arabia and other parts of Asia by Aristotle and Pliny. In India, the Arthasastra identifies tolls prior to the fourth century BCE. Tolls were also known among the Germanic tribes in Europe during the late Roman Empire, who offered to guide travelers through mountain passes (let's call it a protection payment) for a fee. Later tolls were widely used in the Holy Roman Empire in the fourteenth and fifteenth centuries, and were thought to be interference in commerce.

Returning to the early eighteenth century, the idea of local administrators seeking permission for tolls from Parliament met much opposition. In the large, it was regarded as attempting to foist local responsibilities onto others. The locals objected to paying tolls on roads on which they labored and which they had to use.

This combination of opponents prevented many schemes from going forward. While some schemes were accepted, Parliament turned many down. When approved, there were often concessions for local traffic. Even so, there is a record of local uprisings against toll roads, riots, destruction of gates, and so forth. This reflected a larger conflict; toll roads are in a sense the network analogue of the enclosure movement facing England at the same time. Fenced fields and tollgates were redefining property relationships and class rights.

Although development of turnpikes was slow in the first half of the 1700s, a pattern of development had emerged by about 1750. "Town-centered" sums it up. The effects on roads of traffic to London and provincial centers required action, and successful acts and schemes prior to about 1750 yielded a map of rather disconnected, urban centered routes (Figure 3.2). By then, there were 143 trusts managing about 4,800 km (3,000 miles) of road. Later acts then filled in the map. By 1830, there were more than 1,000 trusts and 32,000 km (20,000 miles) of road.

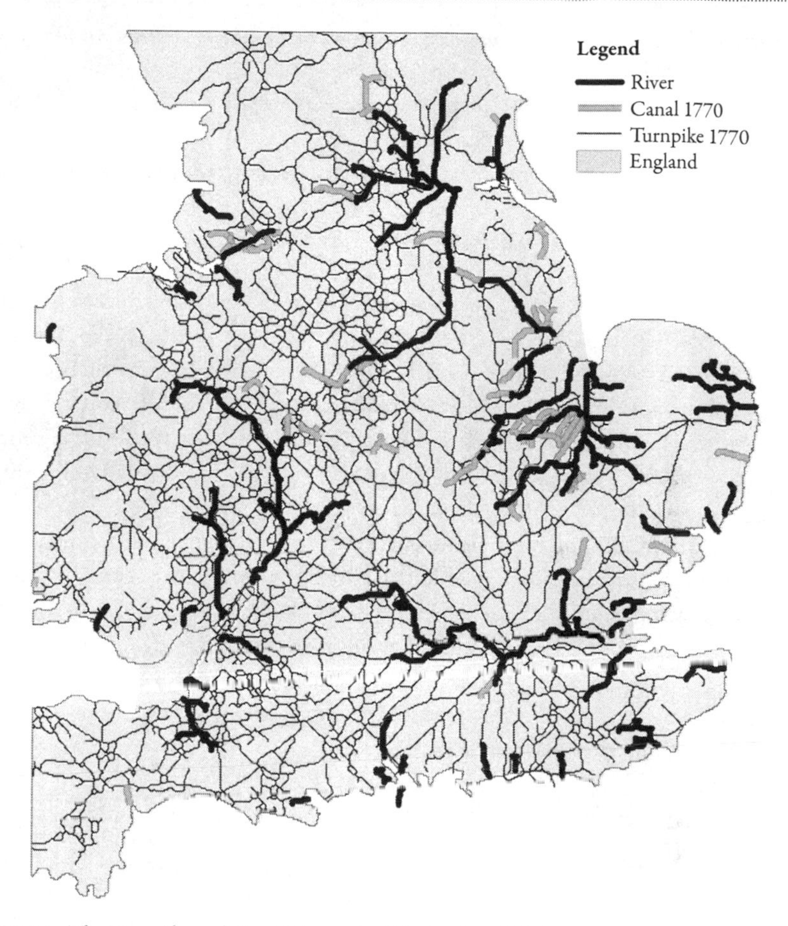

FIGURE 3.2 The Turnpike and Canal Network of England and Wales in 1770

Road trustees were local men (and they were almost always men) of substance. Enabling acts specified powers, accounting and meeting requirements, and provided for posts of treasurers and surveyors. The early pattern was for tolls to be collected at gates (by toll farmers), and after expenses were met, reinvested in repair. After experience, mortgage funding, guaranteed by gate income, began to be common.

Toll schedules were very complex. They reflected geographical (or spatial) equity, the principle that those who impose costs ought to pay. The tolls thus aimed to protect roads from damage, by placing high charges on excessive loads and on vehicles with narrow wheels.

Levy of tolls using the Dartford to Northfleet turnpike road, 1760:

> *For every Coach, Berlin, Landau, Chariot, Chaise, Calash, Caravan, Hearse, Waggon, Wain, Cart, Dray, or other carriage, drawn by six horses, or other cattle, or more, the sum of one shilling.*
>
> *If drawn by two horses, or other cattle, the sum of six pence.*
>
> *If drawn by one horse, or other beast, the sum of four pence.*

For every horse, mare, gelding, mule or ass, laden or unladen, and not drawing, one penny.

For every drove of oxen, cows, or neat cattle, eight pence by the score, and so in proportion, for any greater or lesser number.

For every drove of calves, sheep or lambs, five pence by the score, and so in proportion, for any greater or less number.

For every drove of hogs, six pence by the score, and so on in proportion.[52]

The technical view was that roads needed protecting. At the national level, the General Turnpike Act of 1773 contained 28 clauses relating to wheels, weights, and so forth.

Thomas Telford became involved with the technology of road construction in 1803 when Parliament considered the problem of roads in the highlands of Scotland. A development scheme was initiated: Parliament would pay one-half the costs if local politicians would propose schemes and pay the remaining costs. That's the first use of a financial matching scheme we know of. John L. McAdam obtained experience serving on trusts in the 1790s and 1800s, and he interested himself in both technology and administration. Figure 3.3 shows the first macadam road in the United States that was constructed on the Boonsborough Turnpike in 1823 outside Hagerstown, Maryland.

By the 1820s much of the road system had been improved to macadam standards. Extensive, organized coach service operated at 16 km/h (10 mph), there were mail wagons and mail service, and freight had emerged in two classes of service—fly wagons or vans for fast freight and heavy, slow wagons, about 10 tons.

FIGURE 3.3 First American Macadam Road (1823)

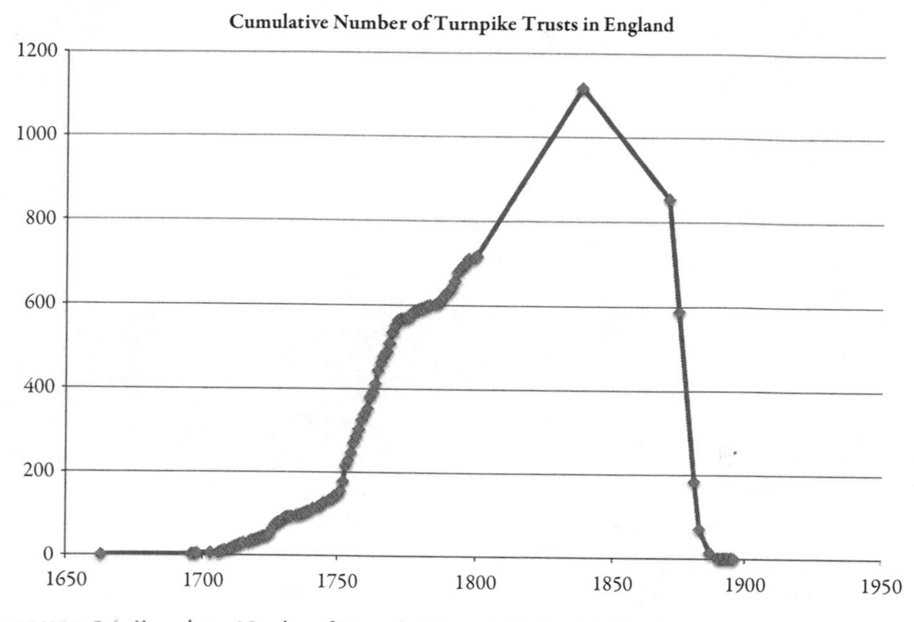

FIGURE 3.4 Cumulative Number of Turnpike Trusts in England, 1650–1800

Yet, the *Times of London* described toll collectors as: "men placed in a situation unfavourable to civilized manners, and who might be usefully employed in mending the roads which they now obstruct in a most disagreeable manner" in an 1816 editorial, giving evidence that tolls were not held in universal acclaim.[53]

Figure 3.4 shows how investment in turnpikes tapered even before investment in railroads increased. After the railroads, some turnpikes failed financially and were taken over by local governments, which converted them to free roads. Others simply had their franchises expire and not be renewed. In any case, turnpikes were merged with the parish road system as the trend toward a nationally controlled and financed system was strengthened.

3.8 Turnpike Companies

At the beginning of the 1800s, road transport in the United States was relatively expensive. "To transport a ton of good by wagon to a port city from thirty miles inland typically cost nine dollars; for the same price the goods could be shipped three thousand miles across the ocean."[54] Rural farmers could distill grain into spirits, one of the few goods that could justify transport by wagon. Whiskey production led to alcoholism. The means of transport profoundly affected the American diet.

Road (and canal) development in the United States lagged behind that in England. Instead of being "already ahead," the United States was "rushing to catch up." Also in contrast to France and England, the Unied States at the time was a federation of states with a weak central government. Hierarchical governments needed to learn roles. The main American turnpike experience differed from the British experience in several other ways. It began

TABLE 3.1

The Rise of Incorporation

Period	Incorporations
Colonial Era	6
1781–1785	11
1786–1790	22
1791–1795	114
1800–1817	1800

Note: Wood (2009) p. 462

later (in the 1780s) and ended later (the onset of the twentieth century). But the organizational form also differed. In the United States, private corporations held most turnpike charters, while in Britain, the turnpike trusts were institutions with more public control. The private turnpike companies were regulated, and often, especially in the early years, had some local government (public) investment, so we are speaking of degrees of public and private, not absolutes, when we talk about road networks.

Incorporations were rare in the colonial period, but became more common, as shown in Table 3.1.

The corporation was critical for the US model of turnpikes. The first authorized turnpike in the United States was from Alexandria to Berryville, Virginia, in 1785, though the road was a public tax-funded road. Maryland followed in 1787. However, the 100-km (62 mile) Lancaster Turnpike in Pennsylvania (with 9 toll gates), chartered in 1792, with land donated by the state, was the first notable turnpike company. As can be seen in Figure 3.5, 69 turnpikes were charted in the 1790s, rising to 398 by 1810.

The logic behind the turnpike boom is in many ways that of an arms race, though more productive. Turnpikes increased accessibility. With a turnpike (shown in Figure 3.6), Lancaster residents could now get to Philadelphia quickly on a broken stone and gravel road. This improved overall welfare (absolute accessibility). But perhaps more importantly, it improved Lancaster's position with respect to competing towns (its relative accessibility) who could only get to Philadelphia on slower dirt (and mud) trails. It was a mixed bag for Philadelphia, while it made the west more accessible helping spark the western migration, it brought Lancaster into Philadelphia's orbit and away from Baltimore's. Thus, it improved Philadelphia's position with respect to its rivals (Baltimore and New York). Moreover this turnpike was profitable, paying up to 15 percent annual dividends to shareholders.

A map of Lancaster turnpike is shown in Figure 3.7. Over time, Lancaster positioned itself at the center of a web of radial turnpikes, in addition to the route to Philadelphia: the Lancaster/Columbia Turnpike, the Lancaster-Marietta Turnpike, the Lancaster-Millersville Turnpike, the Rockville (now Rock Hill) Turnpike, and the Stumptown Turnpike (now Old Philadelphia Pike). As a side note, the Lancaster turnpike also helped birth the Conestoga wagon as a major vehicle type (which can be seen heading away from the inn, in Figure 3.6).

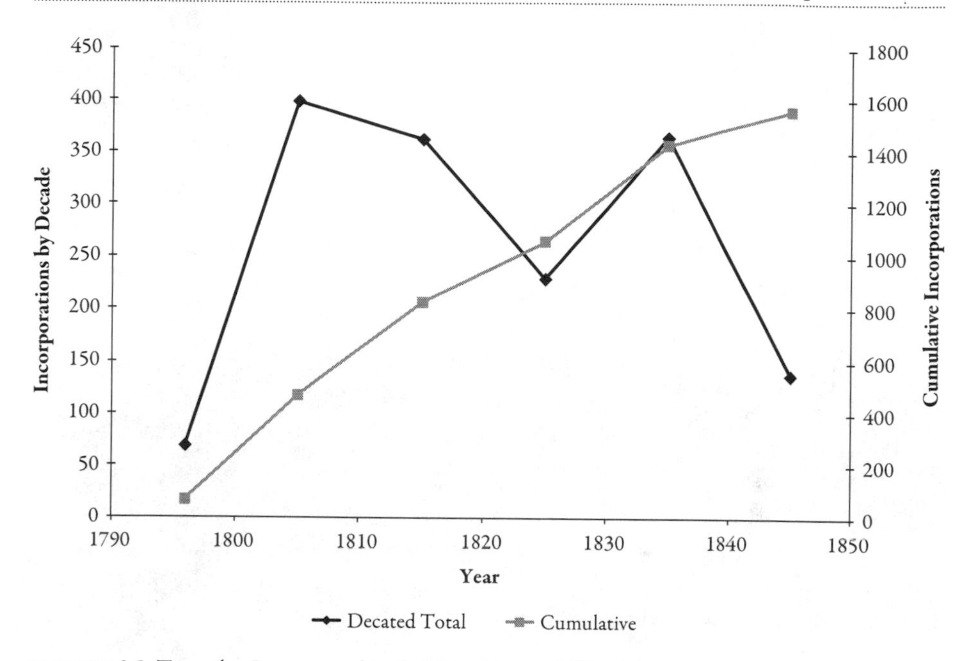

FIGURE 3.5 Turnpikes Incorporated in the United States, 1790–1845

Turnpikes were popular in the first two decades of the nineteenth century on the East Coast and Ohio. By the 1820s, the common stock in turnpikes exceeded the stock issued by banks. The corporation was still relatively new, and ownership was limited, so the owners would be required to continue to invest capital in the company even after the initial purchase of shares. While the largest shares were owned by the wealthiest in the community, shares were distributed widely. Governments also picked up shares; the Commonwealth of Pennsylvania, example, funded one-third of the turnpikes in that state. South Carolina and Indiana were bastions of state socialism in the turnpike sector.[55]

While there were turnpikes, county roads in the early nineteenth-century United States were in terrible condition. As in the United Kingdom, corvée statute labor road crews were untrained and unmotivated, farmers and farm-workers working on the roads for a few days in the off-season. They might clear stumps and do some leveling, but it was all temporary. In the snow-covered North, transport by sled would be much faster than road use in the warmer months. The South favored transport by water to animal power.[56]

In general, the turnpikes were disappointing, failing to reduce transport costs as much as hoped.[57] They often did not pay much if anything in the way of dividends and could not cover costs of collection plus operations and maintenance. Many turnpikes failed even before the emergence of canals and railroads. Further, shunpikes emerged, and teamsters waited until night to pass unobserved on the toll road. Unlike toll roads, toll-financed bridges were more likely to be profitable. There was greater demand and fewer alternatives. Capital sources were similar, local individuals along with governments. Some privately owned toll bridges lasted well into the twentieth century.

A US critic in the early 1800s might have said that the central government was destined to play only limited and rather ad hoc roles in transportation development. Secretary of the

FIGURE 3.6 The Philadelphia and Lancaster Turnpike Road (1795)

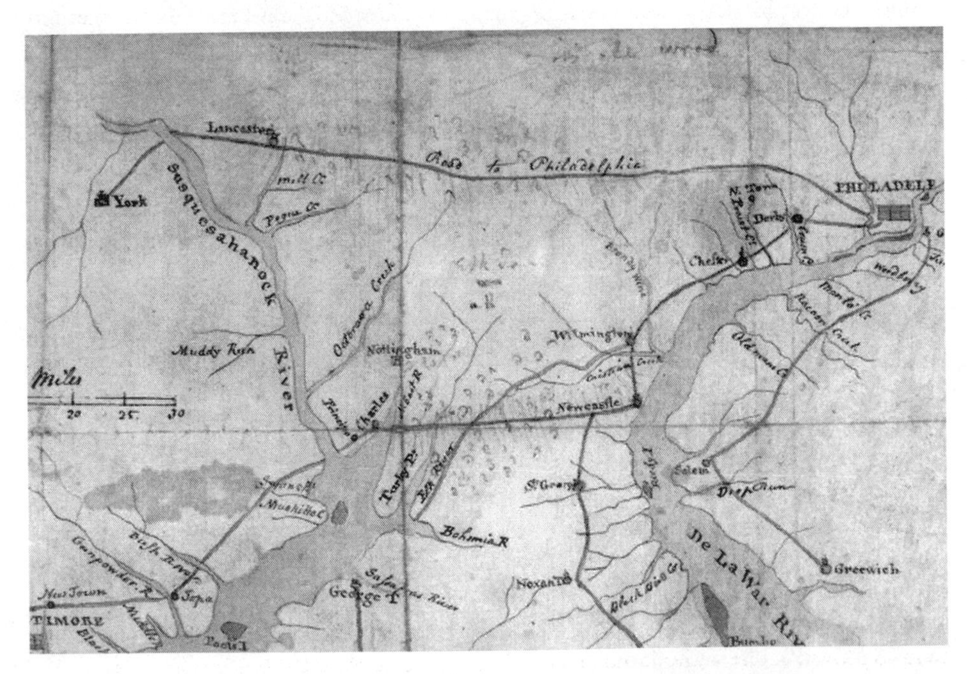

FIGURE 3.7 Map of Lancaster Turnpike (1796)

Treasury Albert Gallatin's report to Congress of 1808 had recommended an extensive set of canal and road improvements. Henry Clay also wanted the federal government to fund his American System of internal improvements. Transportation infrastructure was in the early nineteenth century justified as fostering "national grandeur and individual convenience."[58] But no action was taken. It was claimed that the Constitution did not provide for federal activities. Was that an excuse or reason for inaction? The real obstacle to action may well have been the regional jealousies of the times.

As a result (we think mainly of the problem of consensus about activities), federal government activities were small and scattered. The Ohio Statehood Act of 1803 set aside 2 percent (some sources say 5 percent) of the proceeds of public land sales for building the Cumberland Road, the first interstate highway, constructed by the US Army Corps of Engineers.[59] Those monies got the road built but did not provide for maintenance. An Act was passed in 1822 permitting the collection of tolls for maintenance, but was vetoed by James Monroe on the grounds that the federal government lacked the necessary jurisdictional power.

James Madison addressed Congress advocating "establishing throughout our country the roads and canals which can best be executed under national authority" but yet vetoed the Bonus Bill of 1817 (sponsored by Henry Clay and John C. Calhoun) providing federal funds for a highway linking the East and South to the West, from the profits of the Second Bank of the United States, as unconstitutional.[60] Nevertheless, some federal funding for some specific roads increased from $702,000 annually under John Quincy Adams to $1,323,000 under Andrew Jackson. Congressional support was weak for such improvements because of regional rivalries, New England with relatively good roads had little reason to support improvements in new emerging areas (e.g., the Midwest). Not many years later, in 1830, Andrew Jackson vetoed a bill funding the Maysville Turnpike in Kentucky on the grounds that the support of small local projects (akin to what we now call "earmarking") created "disharmony among the states."

Opponents to the turnpikes were in part Easterners, who objected to increasing the accessibility of the West (and thus the migration to the West), and those who opposed any federal involvement. Rival modes (canals and later railroads) also opposed government support for their competitors. Aside from the Cumberland (or National) Road (today's US 40), federal involvement in road building was limited. The National Road was not trivial, though, garnering $6.8 million from Congress until the 1840s when it was abandoned by the federal government.[61]

While the federal government was in general out of the picture, many states participated in private turnpike construction. Virginia, for instance, saw a network of toll roads emanating from Alexandria, sponsored by local merchants. Alexandria is just across the Potomac River from Washington, and was part of the District of Columbia from 1800 to 1846 before being retroceded back to Virginia. Local rivalries were intense;[62] sometimes opposition was, too, when local and town interest conflicted.[63] Beginning in the 1840s, states started to limit public funding via stock purchases or subscriptions, having been financially burned in the depression of the 1830s. Charters for mixed (public-private) corporations were becoming rarer, though municipalities and the private sector continued to cooperate to fund infrastructure without much federal or, now, state support.[64]

Bridges, with a stronger monopoly position due to their expense and scarcity, also sought exclusive rights. These were addressed by the United States Supreme Court in the Charles

River Bridge case of 1837, which concluded that a monopoly on a river crossing granted by the state had to be narrowly construed, allowing competition, which was deemed good for the public. The nature of charters was still being established, as charters were considered inviolable, adding risk on the part of the state when granting a charter, as it could not easily be dismissed.

3.9 Plank Roads

Turnpike incorporations slowed with the beginnings of the railroad in the 1830s. Aside from the plank road boom, turnpikes grew less and less important as an element of the American transportation network. By 1910, almost all turnpikes were assumed by state governments.

The plank road boom deserves some discussion. Wood planks provide a much smoother surface than gravel, and prior to concrete and asphalt, would enable the fastest off-rail transportation by wheeled vehicles. In the 1850s, especially in the timber-rich states, a boom in plank road construction took place, as shown in Table 3.2.

Table 3.3 shows a regression of the plank road data. The greater the rank in lumber production (i.e., the less lumber produced), the fewer plank roads built. Similarly, the greater the population, more roads built. So construction was a function of both material supply and travel demand.

TABLE 3.2

Plank Road Data

State	Rank in lumber production	1850 population	Plank roads
New York	1	3,097,394	335
Pennsylvania	2	2,311,786	315
Ohio	4	1,980,329	205
Wisconsin	11	305,391	130
Michigan	5	397,654	122
Illinois	10	851,470	88
North Corolina	16	869,039	54
Missouri	9	682,004	49
New Jersey	13	489,555	25
Georgia	18	906,185	16
Iowa	23	192,214	14
Vermont	20	314,120	14
Maryland	21	583,034	13
Connecticut	22	370,792	7
Massachusetts	7	994,514	1
Rhode Island	26	147,545	0
Maine	3	583,169	0

TABLE 3.3

Regression Results of Simple Plank Road Model

Variables	Coefficients	*t* Stat	*P*-Value
Intercept	133.91	2.90	0.01
Lumber rank	−7.19	−2.88	0.01
1850 population	0.000040	1.90	0.08
Adjusted R^2	0.51		
Standard error	75.61		
Observations	17		

The plank road boom ended as the first plank roads deteriorated. Wood planks typically last eight years in road use. As the planks went bad, they either needed to be replaced (which was costly) or abandoned, either of which would raise the cost of operation, and if passed through as tolls, the cost of travel, and make plank roads much less advantageous than they had seemed eight years earlier.

3.10 Mail and the Gospel of Speed

> Neither rain, nor snow, nor gloom of night stays these couriers from the swift completion of their appointed rounds
>
> —Herodotus, inscription on US Post Office in New York City[65]

The US Constitution made special provision for the Post Office and Post Roads. By tracking the change in communications, we can see some of the spread of transportation (see Figures 3.8 and 3.9).

The largest part of the US government, the Post Office was widely distributed as customers came to it to get their mail; there was not yet delivery. Alexis de Tocqueville called the Post Office "a great link between minds" that penetrated "the heart of the wilderness."[66] The time of news transmission between Pittsburgh and Philadelphia dropped from 30 days to 10 days between 1790 and 1794. Similarly, from Portland to Savannah it dropped from 40 to 27 days between 1790 and 1810.[67] As shown in Figure 3.8, postal officers were the largest share of federal officials, and in the early nineteenth century, both were increasing rapidly, along with total letters and newspapers delivered.

While communications speeds steadily increased, until the pervasiveness of the electric telegraph, speculation based on differential information was a major problem. The Post Office attempted to reduce this by speeding communications, and prohibiting contractors (who delivered the mail for the Post Office) from delivering non–post office messages faster. The express runs from New York to New Orleans achieved 13 to 16 km/h (8 to 10 miles per hour). To further increase speed, the Post Office considered opening an optical telegraph network, but this was never authorized by Congress. The federal government did invest in Morse's first telegraph line, but this interest was sold (see Chapter 9). The Post Office also served major public purposes by subsidizing the delivery of newspapers.

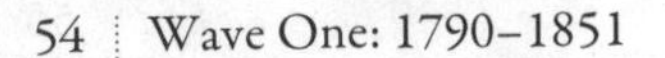

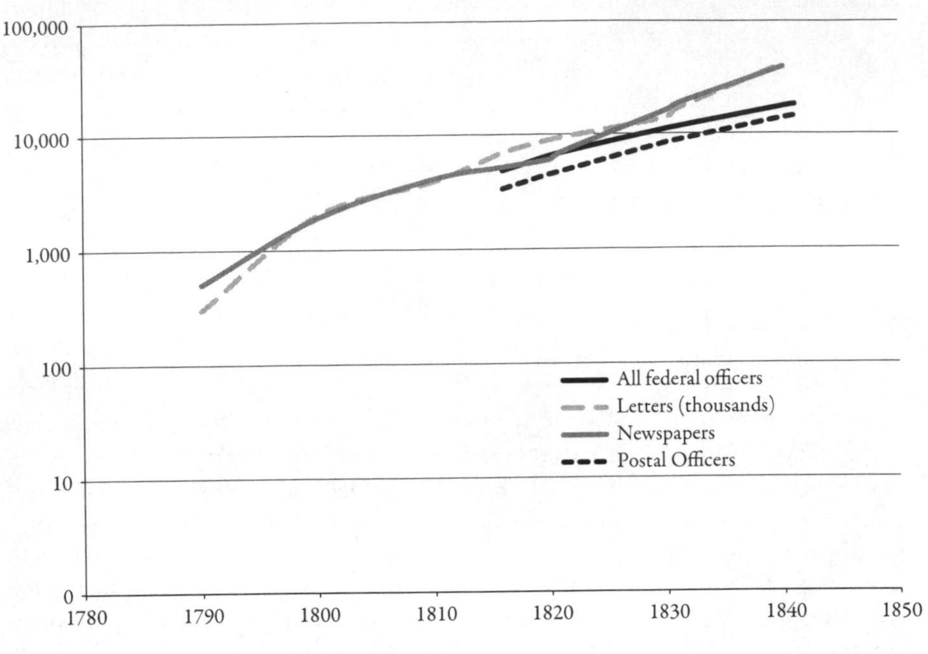

FIGURE 3.8 The Rise of the US Mail, 1790–1840

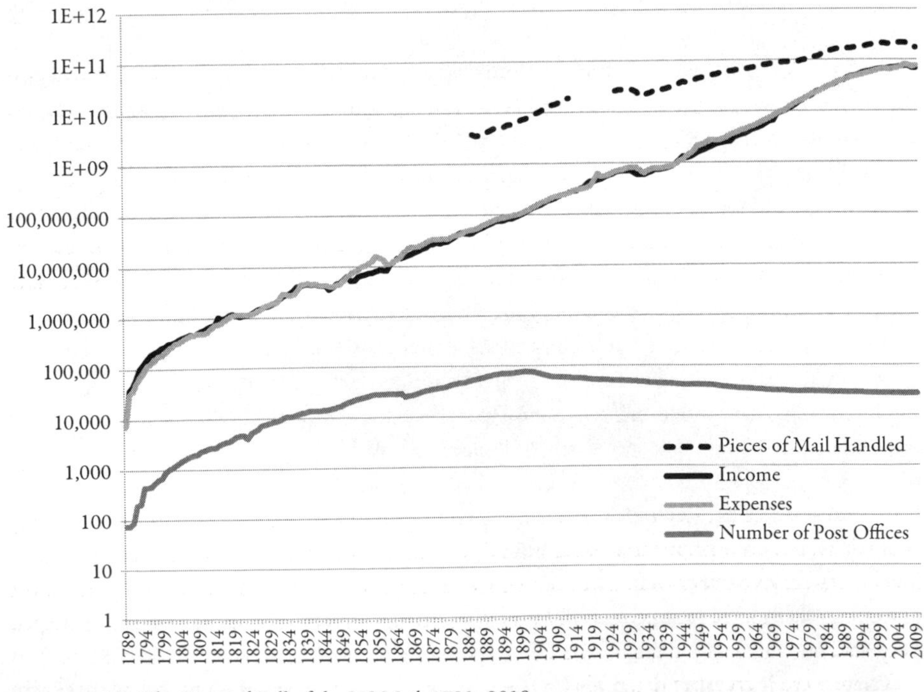

FIGURE 3.9 The Rise and Fall of the US Mail, 1790–2010

From the days of Postmaster Benjamin Franklin, the Post Office contracted with post riders as the primary means to distribute mail between cities. However, scheduled stagecoaches were also used, and on occasion steamboats. Obtaining a postal delivery contract gave enough steady business to enable some stagecoach firms to proceed to offer passenger service, putting the cart before the horse, so to speak. This was particularly important in the thin markets of the South and West. Later, trains garnered a large share of this market, and then airplanes and trucks. Mail-carrying stagecoaches were much more timely than mail-free coaches, so passengers with a choice preferred to follow the mail.

The US Post Office began its distribution pattern as a point-to-point network, with each post office routing mail to each other post office. This system broke down as the volume of mail and number of post offices increased. The system was redesigned as a hub-and-spoke operation by Postmaster General Joseph Habersham in 1800. This reduced the labor required and reduced the possibilities of tampering, as fewer hands would touch each piece of mail. This architecture lasted until the US Civil War, when continuous sorting on railway trains became more common. An important result of this change to hub-and-spoke is that the bottleneck was no longer link speeds, but instead sorting efficiency at nodes.[68]

The Post Office was the largest branch of the federal government, and its only representative in many small towns. It became a huge opportunity for patronage since the Jackson administration, though it was more professional before and has become more professionalized since. The Post Office was also profitable for many years, returning funds to the US Treasury in the first half of the nineteenth century. Under Postmaster McLean, it reinvested its profits. It also wanted to be the organization that provided a federal network of roads, for instance proposing a 1,750 km (1,100 mile) post road from Washington to New Orleans, which was never built, as Congress, and later President Jackson, resisted the most ambitious internal road improvements.

Through the beginning of the twenty first century, the US Postal Service is a government-owned corporation, despite some policy arguments.[69] In other countries, the posts have been privatized and deregulated. The provision and ownership of mail services will inevitably be a policy debate, as the US Post Office continues to require public subsidy. The attempt to save costs by shutting post offices and reducing services has been politically contentious. Its legacy retirement and health plans will also weigh on the service. Nevertheless, Figure 3.9 shows declining income, expenses, and mail handled in recent years (and declining number of post offices since 1900). Though it is hard to see on the figure, expenses now exceed income (though the objective is for them to be equal).

3.11 Fin de Siècle

Our discussion of the decline of turnpikes contemporary with the railroads' rise may imply that regional roads disappeared as railroads were deployed. That is too simple a view. Many local roads survived very well. They served feeder functions to railroads. Local governments provided local roads, and property taxes (or contribution of labor in lieu of taxes) helped keep that part of the road system viable.

There were increasing demands on the local road system as cities grew and as rural economic growth proceeded. With city growth, there were local markets for products of farms, and as grain, fiber, and meat production increased, specialized products were shipped from the farm via railroads to distant markets. These demands emerged in different ways in different farming communities. For example, the demands on the local road system for daily movement of fresh milk to the urban market were quite different from the demands by crops such as grain or cotton.

Improving the rural local road system began to be a matter of state concern during the latter part of the nineteenth century. The State of New Jersey, for example, passed legislation permitting counties to issue bonds for improvement of the local road system. A plausible explanation was the rapid growth of Philadelphia and New York, providing for market-garden type agriculture in New Jersey. Emulation may have played a role. References were made at that time to the lot of the European farmer with his somewhat better government-supplied road system. The spread of bicycle technology also played a role. Growing, wealthy urban areas demanded improved roads. Facility improvements like improved drainage, sidewalks, bridges, and pavements were regarded as highly desirable.

3.12 Discussion

There were two styles for road planning. On the Continent, and especially in France, central governments took responsibility for road development. In England (and places under English influence) road development was a local affair. As traffic between towns increased, enabling legislation was sought so that private organizations could improve roads and fund improvements from tolls. Mostly, short stretches of roads were developed, and the network evolved as links were improved end-to-end. To the extent that there were plans, they were business plans formed for the development or maintenance of an existing link.

The use of wagons and coaches required facilities quite different from precursor facilities. The requirement to reduce grades often called for new route locations. Pavements and drainage were used; bridges and inns and warehouses were constructed.[70] Requirements for investment and maintenance placed new demands on governments and for the development of capital markets, construction technologies, and management capabilities. These developments set the stage for the subsequent creation of canal and railroad companies.

Turnpikes emerged as a financing mechanism in the presence of free rider problems facing the existing road network. There were several free rides being taken. Laborers in the corvée/statute labor system were not putting in a full effort—they were shirking their taxes. Users were riding on non-local roads without compensating the locals who were responsible for them. A toll system remedied both of these problems. The remedies came at a cost, frequent stopping to pay tolls and high costs to collect tolls. Nevertheless, the system worked for almost 200 years before collapsing with the onslaught of intercity railroads, which were faster, better organized, and cheaper. Railroads avoided the free-rider problem from the start. Toll roads reorganized in the United States as feeders to the railroads, rather than competitors; but the long-distance roads collapsed into bankruptcy. By the end of the nineteenth century, and just before the widespread adoption of the automobile, those routes, as they had in Britain, were taken over by local government and dis-turnpiked.

The roads story continues in Chapter 7.

2 Phase I of the Life-cycle

You can't find anything unless you are looking for it.

—GEOLOGISTS' SAYING

4 Inventing and Innovating

THE START-UP, initiation, or birthing experiences have been told for three modes of travel. What do those stories tell us? Image a brainstorming session. To follow are some points we might hear. Think of them as characterizing Phase 1 of the larger story of the Life-cycle of a system.

The preceding chapters have emphasized events—actors doing things at places and times. These were creative events, things that mattered in the emergence of modern transportation systems. Chapters to follow will have similar emphases.

Events may be interesting, but they stand alone unless knitted into systematic ideas, and we strive for systematic ideas about the behaviors of actors and systems. The present chapter will begin our emphasis on systematics. Using definitions, assertions, and speculative remarks, it will ask the reader to muse with us about the nature of invention and innovation. We will repeat this pattern in subsequent waves of the book as *The Transportation Experience* unfolds.

4.1 There Are Multiple Models for Innovation and Invention

Robert Fulton did not invent the steamboat. The credit for invention should go to William Symington or others who worked earlier, example, James Rumsey or John Fitch. Fulton innovated a rather crude design suited to a market niche. Social innovation followed.

We have used the words "invention" and "innovation" and should remark on them and use them carefully. Though the words are often used synonymously, the difference between invention and innovation is slippery, but important.

An *invention*, from the Latin for "to find," is a new process; usability is not at issue. (But usability must be in the back of the inventor's mind because inventors are social beings.) A major new technology, or invention, requires the use of a new principle.[71]

An *innovation*, from the word for "introducing something new," assembles processes to work in markets. An innovation has both soft and hard technology aspects.

Some say that invention differs from innovation because the latter is clearly derivative.[72] We prefer to think of them as creating building blocks (invention) and bundling building blocks (innovation) into a larger assembly. Of course, today's bundled building blocks serve as components in tomorrow's even more complex systems.

Both invention and innovation can be contrasted with "discovery." A *discovery* finds something that existed but was previously unknown, whereas both invention and innovation create things that did not previously exist.

Many innovation paradigms exist for basic and applied research, and for individuals and large organizations.

Historically, the dominant paradigm was what we now call the "innovator in the garage" view. A lone worker (or small team) working in an appropriate space (a garage, a shed, etc.) develops something new, or adapts something to a new use. Steamboats, steam engines, even steam cars followed this view. It has carried forward to modern times as the founding stories of Hewlett-Packard and Apple. This is a view that things "just happen" without direct intervention. Of course there are incentives, both private and public. Patents, monopolies, and prizes are evidence of that. But direct government support or policy was not necessarily required, except to the extent of providing a legal framework and ensuring a civil society.

This paradigm evolved, especially with the emergence of complex systems. The innovator and assistants in the garage became a large team in a laboratory. Innovation became corporate. This is the model we associate with Thomas Edison and Menlo Park.

The end-driven paradigm, such as that used by the National Aeronautics and Space Administration (NASA), and the US Department of Defense (USDOD) states the desired end, and applies resources to get there (including the resources to perform research and development). Examples include the US space program, which was tasked with putting a man on the moon before the end of the decade (of the 1960s). In more conventional transportation, the Federal Highway Administrations (FHWA) Automated Highway System (AHS) program (described in Chapter 30) adopted this paradigm.

In contrast, the means-driven paradigm, followed by government science agencies like the National Science Foundation (NSF) and the National Institutes of Health (NIH), says encourage and support research in promising areas, with a nominal goal, but don't worry too much about the goal, and social advances will follow. This is a widely held view by individuals in universities and industry laboratories. The Defense Advanced Research Programs Administration (DARPA) used this when funding the development of what became the Internet.

We can think of two "motifs" for the invention of technologies: Darwinian and Kuhnian.[73] The motifs involve technologies adapting by adjusting internal parts so that novel structures arise from recombinations.[74] The Darwinian mechanism involves new technologies (like species) arising from selected variations on the old. The Kuhnian mechanism involves paradigm shifts as old technologies (like scientific theories) meet anomalies and are replaced by new ones. The Darwinian, incrementalist motif fails in a number of cases. Schumpeter writes "Add successively as many mail coaches as you please, you will never get a

TABLE 4.1

Typology of Innovation Paradigms, with Examples

	End-Driven	Means-Driven	Un-driven
Basic	Superconducting-Supercollider, Human Genome Project	Grants to Researchers (NSF)	Discovery of Gravity (Newton)
Applied	Automated Highway System (FHWA), Apollo Program (NASA), Telephone (Bell), Steamboat (Various)	Internet (DARPA), Transistor (Bell Labs)	Development of the automobile (Ford)

railway thereby."[75] Nevertheless, there is an evolution of sorts, and a selection process, even if there is also planning and design on the part of the designers, in contrast with biological evolution.

Another view is to think of the innovation within the supply chain of suppliers, makers, and users. Innovations can come from any of these parties, and in some segments come from users more than makers.[76]

Experience says that these and other views are correct. *There are multiple models for innovation and invention.* It also tells us some things that many don't want to read. The first is that "in certain situations and for certain sorts of outcomes" ought to preface statements of principles. Second, that technology is after all a creature of society (just as society depends on technology), and improving on the success of an innovation is very much a desired social activity. Therefore, successful innovations stimulate research to improve them.

4.2 Essential Knowledge May Follow Innovation

The people orientation in the previous chapters was deliberate because we would like to know what kinds of activities and actors yield consequential change. What's the conclusion? In what ways does rational, linear thinking (e.g., Smeaton's) occasion progress? What about more integrative thinking, such as that by Stephenson and Pease (discussed in Chapter 2)? Should we conclude that rational analysis has great strengths if the task is to improve an existing system, and integrative analysis has its advantages elsewhere? We hear that some things can't be done because we lack basic knowledge, and that there is a process: *Innovation follows essential knowledge* that works this way:

1. Inquire—Acquire knowledge;
2. Invent—Apply knowledge to new technology;
3. Innovate—Adapt technology to market niches.

But is this consistent with observation? The basic knowledge for the steam engine is thermodynamics, and it had not been acquired. Yet practical men were working with the laws of thermodynamics.

Essential knowledge may follow innovation. The First Law of Thermodynamics has to do with conservation of energy. Total energy was the sum of work plus loss. Efficiency was the percent of energy devoted to work. Inventors like Watt and Smeaton were working to convert lost energy into work. The Second Law of Thermodynamics says efficiency rises by increasing working temperature and pressure, which Trevithick (see Section 2.2) did, illustrating that what practical men did was suggestive to scientists and did not depend upon them. The idea of the electron didn't appear until after innovators figured out a lot of things about power transmission, transformers, and multiphase power. We are not saying that research and engineering don't contribute to building blocks for innovation. We object to the principle that research must be done first and argue that innovation may suggest directions for useful research.[77]

4.3 Technology Progresses with Building Blocks

Technology progresses with building blocks. But it does not progress top-down according to plan. Neither the ancient Greeks, nor Romans, nor Victorians had a plan for the modern jet engine. The early tinkerers with steam engines did not even have a steam locomotive in mind. Rather, they built something that suited them, and others took what they added to the community commons and extended it, to create new (more complex) building blocks. There has been a lot of recent study of how knowledge is created. The idea of *adjacencies* or the *adjacent possible* is valuable.[78] Most ideas derive from existing nearby ideas, ideas in similar fields or approaching similar problems. This process is sometimes given the name *bricolage*, which is construction from a wide range of available objects. Another recent book notes that unlike people, technologies (and the ideas that embody them) no longer die; everything ever invented can still be found.[79] The layers of technology involved in the modern world are too numerous to enumerate, but include fire, magnets, steam engine, electricity, wires, insulation, transistors, semi-conductors, computers, software, network protocols, markup languages, and so on.

4.4 Patents May Constrain Innovation

At the beginning of the book, inventors of steamboats competed for patents. These patents were developed to encourage invention, by giving inventors of intangible property (the idea underlying the invention) an exclusive monopoly on making that invention. The length of patents has increased across the board. It is widely disputed whether that in general promotes or stifles innovation.[80] It encourages innovation by rewarding expensive invention, which might not have otherwise taken place. It discourages it by limiting the amount of subsequent innovation, tying people in legal knots to bypass workable mechanisms to avoid losing all profits to the patent holder. In particular, patents can serve to gridlock the economy, creating a series of vertical monopolies, much like toll roads in series (or tolls on the

River Rhine).[81] The problem with vertical monopolies is that by taking as much profit from the customer as possible, they raise the price for the customer, who then lowers his demand not only for the good or service provided by the monopolist itself, but also for the upstream and downstream markets. We see this when toll roads are in series, but it is also important in patents. For example, imagine party *A* owns a patent on water navigation, party *B* owns a patent on steam power, party *C* owns a patent on boats and flotation. By diminishing demand so much, the value for the system as a whole is reduced.

There are strategies for this, where the common good requires cooperation rather than competition and short-term profit maximizing.

4.5 Innovation Requires an Adequate Design Serving the Right Market Niche

While building blocks are necessary, they are not sufficient for innovation. The perfection of the new building block (steam engine) was not key to the innovation. This is analogous to Simon's "satisficing" paradigm.[82] Fulton's key action was getting the market niche right.

> [Prime Minister] Pitt was the greatest fool that ever existed, to encourage a mode of warfare which they who commanded the seas did not want, and, which if successful, would deprive them of it.
>
> —The English Admiral Earl St.Vincent commenting on the Fulton submarine[83]

Fulton spent two decades outside the United States working on inventions such as the submarine and torpedo for naval warfare.

Small, tub-sized canals interested Fulton. He developed and wrote about them, and patented his ideas in France in 1798. Fulton developed the idea of submerged tunnels, and he also developed an idea for a submarine. Unable to sell the submarine idea in England, then at war with France, he went to France and got support for development and testing there. The submarine was to be used to break English blockades of Channel ports, but development was not brought to that stage. While in France, Fulton developed a steamboat and held trials on the Seine at Paris. However, his boat was burned by bargemen and never saw service. Returning to England, Fulton was able to interest the English in submarine actions against the French fleet, but that scheme also went only part way. Before leaving England and returning to the United States, Fulton ordered a Bolton-Watt engine, and this was the engine used on the Hudson River. The Hudson River service demonstrated feasibility and the existence of a market, and by 1820 there were about 300 steamboats in the United States Service was available on the Ohio-Mississippi system (see Section 1.5.1).

4.6 Policies May Be Forged to Aid Infant Industries

The submarine was the archetypical *infant industry*. Fulton is quoted as describing the submarine, which he hoped would lead to what we might now call "Mutual Assured

Destruction" and therefore end naval warfare and ensure freedom of the seas, "an Infant Hercules which at one grasp will Strangle the Serpents which poison and convulse the American Constitution."[84]

Policies may be forged to aid infant industries. Often, however, "infant industry" is an excuse for subsidy rather than a reason. Once established, subsidies are hard to get rid of.

Yet someone has to subsidize new industries, to make the capital investment in research and development, and to spend the capital costs needed for initial construction and manufacturing, which takes place before deployment results in funds returning to the proprietor. Whether that subsidy is public or private has historically varied by industry, and there is no one universal answer.

4.7 The Potential for Improvements as the Predominant Technology Emerges Is Critical

The steam engine wasn't successfully applied to wagons because, as the technology began to be deployed, it failed to improve in ways suited to steam cars. The main point in presenting this comment is to counter the argument that although the technology isn't so good now, if we can just get it started, costs will go down and quality will go up. That's the basis of infant industry claims for subsidy.

4.8 An Innovation Has to Be Consistent with Market (Client) Values

The Royal Navy was a prospective client for the submarine, and as the earlier quote from Earl St. Vincent suggests, its use wasn't consistent with the paradigms the Navy held about the correct ways to do battle. Perhaps if Fulton could have interested Napoleon and built on his wish to eliminate the English blockade, he would have been successful. It would await World War I and especially World War II before the submarine would become central to naval warfare. Note that the steamboat on the Hudson wasn't consistent with the values held by sailboat operators. That didn't matter, for the passengers were the clients.

Fulton moved to Pittsburgh and built the *Washington* for service on the Ohio-Mississippi Rivers, where there was a growing market. A low-pressure steamboat, the *Washington*, was a failure—insufficiently powerful to steam upriver. But current wasn't the Ohio-Mississippi problem, for more powerful steamboats were within the state-of-the-art and were soon in service. The problem was the state of the rivers: low water on the Ohio during much of the steaming season, lack of navigation aids, logjams blocking rivers, and snags (water-soaked logs, often not visible) that would damage boats. It took many years and innovation by people like Captain Shreve to remove snags and improve the Ohio-Mississippi system.

While going from New Orleans to Pittsburgh took about 120 days before steamboats, the application of steamboats reduced that time to 25 days, an 80 percent reduction.

4.9 For a System to Work, All Components Have to Function Appropriately

In the case of steamboats, this implies the steam engine, the mechanism for converting steam to power, the vehicle, and the water system on which the steamboat is running. Steam railroads were also made to work. Steam cars did not.

4.10 Innovative People Abound

> The reasonable man adapts himself to the world; the unreasonable one persists in trying to adapt the world to himself. Therefore, all progress depends on the unreasonable man.
>
> —George Barnard Shaw

This book profiles many of the innovative people who changed transportation. The list of names mentioned is long. The list of names unmentioned is longer. As the Shaw quote at the beginning of this section notes, progress depends on unreasonableness, changing the environment in which we act, bending space and time, instead of meekly accepting what the world deals. Innovators, who are "unreasonable men" in Shaw's now-sexist words, do not take the situation as given: for example, Trevithick did not allow himself to be bound by the constraints posed by Watt's patents.

4.11 Innovations Must Finesse Existing Constraints

> The painter does not fit the paints to the world. He fits himself to the paint.
>
> —Paul Klee[85]

> The visionary starts with a clean sheet of paper, and re-imagines the world. The tweaker inherits things as they are, and has to push and pull them toward some more nearly perfect solution. That is not a lesser task.
>
> —Malcolm Gladwell[86]

Progress depends on the unreasonable, but also depends on tweakers, those who finesse the system. They may be less remembered by history, but their collective work is just as important.

Innovations must finesse existing constraints. Existing constraints (policies, regulations, standards, etc.) were generally put in place for good, though perhaps obsolete, purposes. An innovation must adapt to those constraints or bend the constraints to adapt to the innovation. Brunel (see 5.7) poses a good case study, facing constraints in both rail and shipping.

4.12 Innovative People Cooperate

> A gift always looks for recompense.[87]

Informal know-how trading between rivals is another source of innovation. Why would competitors share information? If the market has many competitors, sharing with one of many competitors may have few costs, especially if the favor is returned. Reciprocity is required.

4.13 Excuses for Inaction Abound

Technical, regulatory, and social reasons can always be found for why things can't be done. The quotations at the beginning of many chapters in this book provide examples of incorrect negativity. Of course, there are many things that are proposed that do not work for one reason or another, and we note a few of those as well. But we do not need to determine success or failure in advance; we should think of it as probabilistic. Investors in new technology should rightly take a lot of low-probability, high-payoff, risks, as well as high-probability of success, low-payoff risks (the best are high-probability of success, high-payoff of course, but these are usually not available). A tree may release thousands of seeds every year; it is lucky if one takes root and grows to adulthood. We don't decry the wasted seeds, we celebrate the full-grown tree. Innovators must overcome the excuses, not only disbelieving them, but persuading others (sources of capital especially) of their falsity.

4.14 Innovation Can Be Innovated

Can we innovate (automate) the process of innovation?

Can we innovate (automate) the process of innovating the process of innovation?

If innovation is to accelerate, we need to innovate faster. Testing one hypothesis at a time is good, testing multiple hypotheses in parallel is better. Factories for innovation, as the classic Edison labs in Menlo Park, New Jersey, aimed to make innovation more efficient. While we do not see much of that in the first half of the nineteenth century, it becomes more and more widespread by the second half of the nineteenth century, when Edison got to work, and especially in the twentieth century. Certainly "normal innovation," working within the bounds of a well-known field, is easier to automate than more profound innovations. If it is simply a question of filling out a matrix (testing 100 drugs against 10,000 known diseases), there are mechanisms for automating that which can be developed.

4.15 Transportation Development Is Chancy

The transportation experience is full of what-might-have-beens, such as the what-might-have-been discussed for the steam carriage. We should pay attention to these and the processes of selection and choice that affected them.

Fulton first tried his steamboat in Paris. It steamed very well the first day, but was burned that night by bargemen concerned about their investments and their jobs. With the extensive canal system in place in France, success by Fulton might have set off technology development suited to canals. If the Duke of Bridgewater had lived to purchase and use Symington's designs, similar developments might have happened in England. Instead, the first developments were riverboats. We see many what-might-have-beens as we trace the transportation experience.

Often, chance is discounted in favor of choice. Choices were made driven by the tooth and claw of competition. "Best choices" are assumed to have been made. Even if one accepts that the tooth and claw of competition were less than perfect, "what's done is done." Development is an irreversible process, and our concern should be with today's outcomes. We disagree.

Phase 2 is discussed in Chapter 10.

3 Wave Two: 1844–1896

> . . . even if the [screw] propeller had the power of propelling a vessel, it would be found altogether useless in practice, because the power being applied in the stern it would be absolutely impossible to make the vessel steer.
>
> —SIR WILLIAM SYMONDS,
> Surveyor of the British Navy, 1837[88]

5 The Modern Maritime Modes Emerge

5.1 Beginnings

BENJAMIN FRANKLIN DID not discover the Gulf Stream; that is credited to Juan Ponce de Leon, but Franklin did document and map it.[89] The Gulf Stream is an Atlantic Ocean current running from the Gulf of Mexico past Florida, along the eastern seaboard of the United States to Delaware, before turning right across the ocean. Containing more water than all the rivers of the world, and moving faster (at 2.5 ms^{-1}, or about 216 km/day) than the surrounding water, it is a boon to trans-Atlantic navigation, helping shipments along in the eastward direction from New England to England (though hindering movements westward). Its importance in the era of sailing ships cannot be underestimated. The Gulf Stream is an Ocean-river provided to man by nature. Why nature provides it is beyond the scope of this book.

Before, during, and after Franklin's time, European colonial powers sought to profit from sailing to and from their colonies. Innovations such as mapping the ocean currents were one important aspect. But there were others. England, for example, established a set of navigation laws in the 1660s restricting trade with the colonies to English ships. This same urge, carry our trade in our ships, continues today. Actually, the colonies fared well from this policy on one dimension. A key resource for the building of ships was hardwood (oak, especially), and masts required tall pines. The United States had those resources and labor skilled in ship construction. So the American shipbuilding industry developed early, and by the time of the Revolutionary War, nearly one-third of all the ships in Britain registered as English were colonial built.[90] That was quite an industry, for the wood ships of the times lasted 10 years at best. The replacement market was coupled with market expansion.

Just after the revolution, the United States initiated the mercantilist policy to "carry our trade in our ships." In 1789, Congress gave a 10 percent tariff advantage to goods hauled in American bottoms. It also discriminated in port dues—they were lower for American ships, and such ships were only charged once a year. The latter assured that the coastal trade would move in American ships.

Wars between the United Kingdom and France in the 1790s and 1800s provided both a risk and opportunity for American shipping. As a neutral, American firms dominated what was called the "re-export" trade and could ship from French colonies in the West Indies to France when the French could not (due to British blockades). A few famous fortunes emerged in this period, including those of John Jacob Astor and Stephen Girard of Philadelphia. During this period (1793–1807) American trade (imports plus exports) rose from $43 million to $246 million.[91]

American ships suffered at the hands of the Barbary pirates in the 1780s and 1790s, as the United States had not yet paid them off or defeated them in battle. The British supported the Barbary pirates as a useful nuisance against competing naval trading countries (the US), keeping them out of the Mediterranean markets. The United States ultimately responded by building a strong navy to fight piracy (and protect shipping interests). By 1805 the British began seizing US merchant ships for trading with the enemy (France). This, coupled with systematic impressment of American seaman into the Royal Navy, were major grievances leading to the War of 1812.

Later, the Navigation Act of 1817 invented cabotage restrictions to assure control of coastal trade. In 1828, the United States passed a Reciprocity Act to deal one-on-one with other nations' trade restrictions. Today's situation is a modern version of these early matters.

5.2 Trading Companies

> Whosoever commands the sea commands trade; whosoever commands the trade of the world commands the riches of the world, and consequently the world itself.
>
> —Sir Walter Raleigh

European powers established a number of companies for trading with the West Indies (the Caribbean) and the East Indies, as well as other parts of the world.[92] In England, they arose as trade guilds: goldsmiths (1327), mercers (1393), haberdashers (1407), fishmongers (1433), vintners (1437), merchant taylors (1466). The Hanseatic League, dating from 1159, is also an early version of a trading organization. The large long-distance trading companies were instruments of European governments as part of their efforts at both trade and colonization.

- Muscovy Company, 1555–1917 (still operating as a charity in Russia)
- Eastland Company, 1579–1689 (trading in the Baltic)
- Levant Company, 1581–1825 (trading in the Middle East and Turkey)
- English East India Company, 1600–1874
- Dutch East India Company, *Vereenigde Oost-Indische Compagnie*, 1599–1798
- Danish East India Company, *Dansk Østindisk Kompagni*, 1616–1729, 1730–1779

- Dutch West India Company, *Geoctroyeerde Westindische Compagnie*, 1621–1792
- Portuguese East India Company, *Companhia do commércio da Índia*, 1628–1633
- Danish West India Company, *Vestindisk kompagni*, 1659–1776
- French East India Company, *Compagnie des Indes Orientales*, 1664–1769, 1785–1794
- French West India Company, *Compagnie des Indes Occidentales*, 1664–1674
- Swedish East India Company, *Svenska Ostindiska Companiet*, 1731–1813
- Swedish West India Company, *Svenska Västindiska Kompaniet*, 1787–1805

Early English trading companies aimed not only at the import of goods from abroad, but also export of English goods, mainly woolens. This was obviously more fruitful in northern climes than the East Indies, where "The Company of Merchants of London trading into the East Indies," perhaps the most important of these companies to emerge, was chartered to trade.[93]

Trade in condiments from east to west had taken place since Roman times. Nutmeg acted as preservative, pepper covered the odor of bad meat. The English East India Company initially sought to capture the spice trade in what is now Indonesia. It established its first trading post in the Kingdom of Aceh on the northwest tip of Sumatra, a place called the Gateway to Mecca by Indonesians looking westward. Aceh was devastated by the tsunami of December 26, 2004.

Importantly, this was a global company, one of the first. Unlike modern companies, which depend on decisions of headquarters, here the captains and crews were distant agents who could not be directly coordinated because the communications time entailed up to an 8-month voyage. It wouldn't be until widespread deployment of the telegraph (see Chapter 9) that instantaneous global communication would enable centralized decision making.

The English East India Company faced competitors from Portugal and the Netherlands, reflecting national rivalries. The English and the Dutch fought four wars between 1652 and 1784, and had many skirmishes in the colonies. In one case, battles with the Dutch continued even after a treaty had been signed in Europe between the English and the Dutch because the news did not travel quickly to the Islands halfway across the world. This phenomenon repeated itself at the end of the War of 1812.[94]

The English East India Company's main competitor, the Dutch East India Company, or VOC, was granted a royal charter at about the same time, in 1599. Lancaster's first voyage to the East Indies left from Woolwich in 1601 with four ships and 480 men; 105 men died *en voyage*.

As with any network, there was some hierarchical organization. The country trade comprised satellite or subsidiary trips in smaller ships (pinnaces and galleys, prahus and junks) from various islands and trading posts to factories (trading establishments) to concentrate commodities on a few large "tall ship" voyages back to Europe.

Many early voyages were not fully successful. Even when they were successful, there was no guarantee of markets. On an early voyage, once pepper shipment made it back to London, the price crashed. The company paid stockholder dividends in pepper as it was not allowed to sell its shipment (in competition with the horde somehow acquired by King James I).

The Portuguese and Spanish had established trading routes around Africa (Cape of Good Hope) and South America (Cape Horn). The English sought a Northwest Passage and sent a number of explorers, including Hudson, Frobisher, Davis, etc., to try to find it) unsuccessfully, as well as a Northeast Passage (explored by Willoughby and Chancellor) above Norway and the Barents Sea.

These trading companies were more like regulatory networks that governed the activities of its members and enabled them to form individual partnerships or syndicates to raise funds and trade for themselves. The trading company was not a company in the modern sense, but more like a guild. Unlike the Levant Company with which it shared officers and offices (in the home of their shared governor, Thomas Smythe) in the early years, the East India Company was both regulator and operator, and so risks were pooled into a joint and united stock. Initially (early 1600s) anyone could buy in, but later only existing shareholders could subscribe to new stock. At first, each voyage was a separate company, which paid all profits out in dividends, and each new voyage required selling new stock. Thus until the first voyage was complete, no one would risk on the second voyage. Even after completion of the first, the organization required all investors in the first voyage to subscribe to the second. As the venture proved itself, it became easier to sell stock for later voyages.

These were the first Limited Liability Companies (LLC). The East India Company not only coordinated voyages, but it also managed domestic politics in England, ensuring there were a sufficient number of Members of Parliament who benefited from its actions.

The East India Company policies were not without controversy. "It's worth keeping in mind that the patriots who threw English tea into Boston Harbor in 1773 were not revolting against higher taxes (the Tea Act, in fact, lowered the price of tea legally imported in America) but against the privileges granted to the British East India Co."[95]

The East India Company supported the British Empire. It invented the term "civil servant" and created its own college, Haileybury, to train future company men. It might be considered a state-backed company. Though the government did not directly own shares, individuals (politicians and courtiers) in government did. Further, at its peak, it controlled an army of 200,000. In the end, the East India Company's role was assumed by the British government. The government took control of the Board of Directors in 1784. It lost its trade monopoly in 1813. The Sepoy rebellion in 1857 was a major blow (and reiterating the points that critics had been making for decades, back to at least Adam Smith), and it was ultimately disbanded in 1874.

5.3 A Port in a Storm

The story of port growth parallels that of the ships that sail to them. As the ships grew, so did the ports. Sea and land transportation surged with the beginnings of modern commerce and industry, and the need for port facilities followed. Two styles emerged for development. On the European Continent and overseas where continental nations exerted power, central governments planned and developed ports. Forts protected space for housing, government functions, and material handling. Canals provided ship access to docks. The central government funded the ports using customs and rent receipts.[96] There were variations in physical layouts, of course, depending on local physical situations.

In contrast, in the early days, the central government in England did not exert power in the Continental style, and port development in England and in places where it influenced was *laissez-faire*, local initiative in style.[97] Although institutional and funding arrangements required central government permission, plans were developed and executed locally, often by a company established for development purposes. Defense facilities often were not constructed; port development varied greatly depending on local commercial and physical situations.

Ports increased in size to accommodate demands, and investments were made where increased throughput permitted the effective use of capital. If traffic were limited, cargo handling was managed by "lightering." A 100-ton ship would anchor in a river, and lighters of rowboat size or small barges would move cargo to and from shore. Landing stages accommodated lighters or small barges at the riverside.

Cargo handling was "by hand," with simple mechanical aids. Small ships had to stuff and unstuff, and quantities of throughput were small. In the 1700s a dock might have a manual capstan (a cylindrical machine for raising heavy weights by coiling a cable) used for warping (moving) ships, but that's all. Steam engines were available in the 1700s, but were not used on the docks. They were stationary and could not be well employed in loading or unloading of small quantities.

Some ports used manual cranes. A few specialized docks had walking cranes (1750–1850). Laborers walking around a 5-m (16 ft) drum geared to an external boom and hoist powered the crane. Powered hoists were not used until the mid-1800s. A central steam engine pumped water to a large storage tower, and the pressure head drove hydraulic crane motors (or water was pumped to an accumulator under pressure).

Increased movement of goods warranted investment in quays. One reference refers to 460 m (1,500 ft) of quays in London by the year 1700 and remarks on their crowding and needs for additional facilities.

Investment in Liverpool commenced in 1709 when an enclosed dock was constructed. The enclosed dock was built to deal with relatively high tides; the alternative would have been very deep quays. The facility was built using a small estuarial pool (Liver Pool), and it provided 5.5 m (18 ft) of water. Occupying 1.6 hectares (4 acres), it could accommodate 100 ships. Pools or basins began to be constructed in London some years later. The West India Docks opened in London in 1802 and enclosed some 12 hectares (30 acres). (The redevelopment of the Docklands (Canary Wharf) is discussed in Section 23.5.)

Where the topography was favorable, gravity flow of bulk commodities emerged. On the River Tees, for instance, Port Darlington was constructed downriver from Stockton so that railcars could be dumped directly into coastal vessels.

Other cases differed, depending on trade growth and local circumstance. In Hamburg, for instance, there was very early use of warehouses on small shallow canals. But as ship sizes increased, river anchoring and the use of lighters became common. Eventually, ships anchored to large barges for cargo handling and storage, and lighters worked the barges. That system continued until the 1880s.

Following these early beginnings, ports increased in size over the years. Facilities were improved to accommodate larger, deeper draft ships, and extended well beyond the early fort-like ports. Mechanical devices were developed and deployed for material handling. There was a burst of development at about 1850 as steam liners were placed in service. In spite of these great changes, we judge them as more growth than development in character, with development (change of form to accommodate new functions) awaiting the emergence of container liners and large cargo ships.

Ports grew, enabling economies of scale. One factor limiting the exploitation of economies of scale was ship size, which was not increasing very rapidly.

Local circumstance conditioned the story for each port. Growing ports increased capacity. Companies or local governments provided financing and exerted control, subject to enabling actions by central governments and to customs control and other operational matters of interest to central governments. Navigational aids were generally provided by local organizations. The City of London, for example, provided dredging and channel markings.

The theme in England was local government, quasi-government, or private institutions financing, constructing, and operating facilities. That theme continues for air and water ports in many parts of the world. On the Continent, central governments generally took the lead in port development. The position was that trade was essential to national development. (The crown could make money.) Defense was also to be considered.

5.4 Cargo Ships

This chapter is about emergence. Metal ships could be built significantly larger than their wooden counterparts. This upward scalability found applications of steam for cargo ships and ocean liners.

For centuries, general cargo ships were measured on register tonnage, with one register ton representing 2.8 m^3 (100 ft^3) of space. Net register tonnage space is tonnage so measured minus space used for machinery and crew quarters. Columbus's Santa Maria, the largest of his fleet, was a 300 register ton ship. Large sailing ships at the end of the sailing era were running 600–700 tons.

The register ton seems to have first been used in the Baltic trades, where 100 cubic feet = one ton of Russian wheat and the ship was full and down. Convention for construction changed, and by 1800 a ship in the Atlantic trades was down when its load was about 1.5 times register tonnage. A 400-ton ship in the Atlantic trades might have a cargo of 600 tons. If wheat, that load could be moved in 6 modern rail hopper cars. In the 1400s in Spain and Portugal, tons were measured in another way: two large kegs equaled one ton. By 1900, steel ships in the Atlantic trades were running 20,000 tons.[98]

Steel-hulled, steel-masted "grain racers" would carry more than ten times as much around the beginning of the twentieth century.[99] The economy of scale for ships just wasn't there. As ships upscaled, the number of sails and requirements for crew increased in an almost linear way, as did the expense and difficulty of construction. Rule of thumb called for three crew members per sail, though grain-racers had crews on the order of 30. Because of slow cargo handling, the "dead time" in port was long, disadvantaging large, expensive ships, the kind constructed by Brunel.

5.5 Ocean Liners

Several precursor service developments were "ready" to be incorporated into the liner system. Freedom of the seas had been established as principle; excepting coastal trade, anyone could sail in any service and anywhere. Beginning about 1812 and, in part, as a consequence of the War of 1812, a system of independent ship operators emerged. Previously, large shippers, like the East India Company, had tended to carry their goods on their own ships. Packet (scheduled) passenger and high priority goods and mail services had emerged in the New York–Europe trade.

The sea-lanes required no development. Excepting the bar that had to be dredged and where dredging had already started, New York was a deep-water port. Deep-water ports were already available on the South and West Coasts of England. A large shipyard with special tools had been constructed in Bristol.

After the first crossing of the *SS Great Western*, to be discussed shortly, the British offered mail contracts to support service. This led to the development of the Cunard Line, which soon dominated the trade. In 1840, the Peninsular and Orient (P & O) Line was given a contract to carry mail to Egypt and India. Many operational aids were in place or coming into place. Insurance systems improved and the British (soon followed by others) developed ship register and safety ratings systems. To illustrate the speeds at the time, the sailing ship *Rainbow* took 75 days to travel from Canton to New York in 1846.[100]

The growth cycle of passenger and high priority freight ran from the 1820s and 1830s through the end of World War II. At that time, container ships and aircraft displaced the service.

The takeover of liner services by steam-powered vessels was very quick. We do not have data on that; however, we do have data on steam versus sail for all services. Figure 5.1 displays a typical predator (steam) prey (sail) relationship. Sail tonnage grew by a factor of 3.5 between 1800 and 1880, and the tonnage in 1920 was about the same as it was in 1800. Actually, a better sense of the displacement of sail would be given if the tonnage of steam was inflated by a factor of about 4; the working capacity of a steamer was about 4 times that of a sailing ship. Figure 5.2 shows the improvement in transatlantic crossing times, which was rapid during the early days of steam, but began to level off by the middle of the twentieth century, when air travel took off.

FIGURE 5.1 British Fleets of Steamships and Sailing Ships, 1820–1920

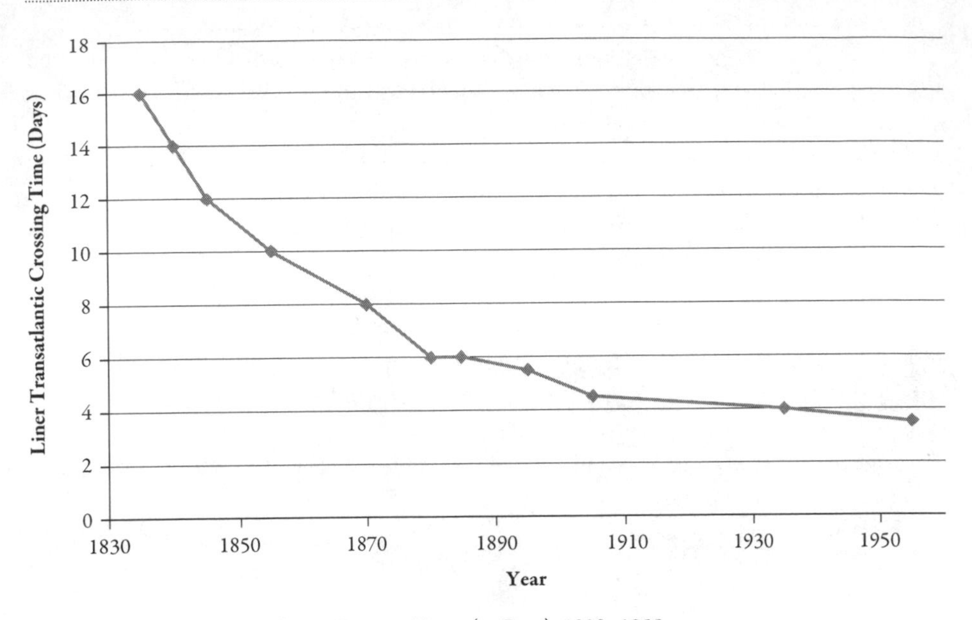

FIGURE 5.2 Liner Transatlantic Crossing Times (in Days): 1838–1955

The consequences of the rapid increase in ocean travel affected society in surprising ways. The eruption of cholera led to the development of the International Classification of Diseases. Its spread was facilitated by the rise of the steamship and rail, allowing sufferers to travel quickly enough to spread the disease widely before death.[101] The use of quarantines also spread widely as a response. Several international conventions on the subject were held in the second half of the nineteenth century, leading to more uniformity of application only to ships with known cases, not to healthy ships. The term "quarantine" itself comes from what is now Dubrovnik, Croatia, where it was used during the Black Death (1348–1359) to prevent spread of the plague by entering travelers, the term a word in Venetian Italian dialect contracted from *quaranta giorni*, meaning a forty-day period. Originally the quarantine was a trentine (thirty days).

5.6 The *SS Great Eastern*

The *SS Great Eastern* (Figure 5.3) was the largest ship of its era. Its tonnage anticipated the tonnages appearing about 40 years later, around 1900, although its speed was consistent with ships of its time. For a size comparison, the *SS Great Eastern* was 207 m (680 ft) in length; the *SS Kaiser Wilhelm II*, constructed in 1903 and the largest ship at that time, was 206 m (676 ft) in length.

Brunel took advantage of several things. The *SS Great Eastern* was of iron construction, and the limitations of wood could be set aside. The main disadvantage of wood was that connections were difficult, and, for the loads to be carried, structural pieces were large relative to iron.

FIGURE 5.3 *SS Great Eastern*

The form of the wooden ship was also limiting. Obtaining thrust from sails, forces had to be carried through the masts, rigging, and ship structure to overcome the resistance of water to ship movement. Ships had to be broad if they were to be stiff under heavy press of sail. Brunel's use of propellers and/or paddle wheels overcame that problem.

Brunel also took advantage of knowledge. Newton had studied the resistance of bodies moving in a fluid and had proposed the "principle of similitude." William Froude, working for Brunel, used the principle (for the first time in ship studies) to apply the results of physical model studies to predict the roll behavior of the *Great Eastern*. Froude (best known for the Froude number, a dimensionless parameter defining turbulence that relates the speed of open channel flow to the square root of gravity times depth) had some ideas on resistance/propulsion relations, yet the *Great Eastern* was underpowered.

This knowledge was the base from which Brunel acted. Relative to earlier ships, the *SS Great Eastern* was long and narrow—its profile was followed by subsequent liners. It had a double hull, and high bulkheads with no doors between them: 15 traverse and 2 longitudinal through the boiler and engine rooms. The *Great Eastern* was a nearly unsinkable ship.

The *Titanic* was the largest ship in the world in 1912, about 40,000 tons. She, too, had 15 traverse bulkheads, but they were carried only 3 m (10 ft) above the waterline (in contrast, the *Great Eastern*'s were over 9 m [30 ft]). The 30 to 40 doors in the bulkheads had to be closed by hand. The iceberg the Titanic met opened the first five bulkheads. As compartments filled, water overtopped the remaining bulkheads.

Brunel took advantage of the technology in the search for scale, yet the *Great Eastern* was not a commercial success. He should have achieved lower unit cost (cost to serve a passenger) than smaller vessels, though he didn't achieve quality (speed) advantages. Perhaps the size of the market prevented realization of lower unit cost.

5.7 Profile: Marc and Isambard Kingdom Brunel

Along with his father Marc Brunel (1769–1849), Isambard Kingdom Brunel (1806–1859) places near the top of any list of great engineers of the nineteenth century. In a recent BBC television poll, Isambard Kingdom Brunel (hereafter: Brunel) finished second (to Winston Churchill) in a ranking of "100 Greatest Britons."[102]

Though a UK native, Brunel studied during his high school years in France, as his father was French and his mother, Sophia Kingdom, had studied there. Though he applied to École Polytechnique, he was denied entry as a foreigner (English), and instead apprenticed with Abraham-Louis Breguet, a Swiss-born watchmaker.

His first assignment was working for his father on the Thames Tunnel, the first under a navigable river, from 1825 (before the era of railways). This was an especially challenging tunnel as it was under water and waterlogged land. The tunneling shield his father designed was necessary, but not perfected. In 1828 six workers were killed and the younger Brunel was injured for six months. The tunnel construction resumed after 1834 under the direction of Marc Brunel, and was completed in 1843. While financially unsuccessful (costing £454,000 to dig and £180,000 to equip) as a pedestrian tunnel, since 1865 it has served the East London Line, now part of the London Overground railway system.

The younger Brunel is more widely recognized for his great works: the Great Western Railway, tunnels, bridges like the Clifton Suspension Bridge in Bristol, and large steamships. Thomas Telford chaired the design competition for the Clifton Bridge, and rejected Brunel's first four designs in favor of his own. Public opposition forced the committee to recompete the bridge, which Brunel this time won. He was the chief engineer and a major promoter of the Great Western when it was organized in 1835. The route from Paddington Station in London to Bristol was opened in 1841. Afterward, Brunel constructed or purchased links and built a system to serve the west.

Brunel was sensitive to the economies of scale achieved by building larger equipment and suitable facilities. The Great Western Railway was built with mainly double track at a wide 2.1 m (7 ft) gauge. (Inverted) U-shaped rails were used. Where there was overlap with standard gauge railroads, as on the Oxford-Birmingham route, three rails were used.[103]

Passenger cars ran 24.4 m (80 ft) in length, and drawings of freight and passenger cars suggest that they were two to three times the size of cars on standard gauge railroads. At first, Brunel placed the body of the car within the gauge, that is, the wheels, for safety and ride quality reasons. Later, the bodies were raised above the wheels and made wider, enabling passengers to sit four on each side of the aisle.

A debate over gauge developed, perhaps because other railroads feared the expansion of the Great Western. In 1845 a Royal Commission on Railway Gauges was established. It looked into the gauge problem and concluded that the wide gauge was superior to Stephenson's standard 1.435 m gauge (4 ft 8.5 in).[104] But it established Stephenson's gauge as the standard because it already dominated the mileage.[105] The Great Western got along fine for a while using its gauge, but gradually shifted over to standard gauge, primarily by adding a third rail to accommodate both gauges, and completed the conversion in a final push over one weekend in 1892.

At the time of the start of the railroad, Brunel went into the steamship business. Brunel designed the *SS Great Western*, which entered service in 1837, steaming to New York in 19 days. The *SS Great Western*, a wooden, paddle wheel steamer with sails, was not the first of its type in service. But sized at 1340 gross tons,[106] it was two or more times as large as the few previous paddle steamers. Unlike previous steamers, its size permitted it to carry sufficient coal for full-time steaming and thus keep a schedule.[107]

The *SS Great Britain* followed in 1843. Although continuing the convention of sails, this was an iron ship with a screw propeller, sized-up to 3270 gross tons. (A somewhat similar ship, the *Rainbow* at 500 gross tons, was constructed in 1836 and had been successful in the European trades.) The *SS Great Britain* ran aground in 1846. Refloated and re-engined, it was used in the Australian trade until 1886.

The *SS Great Western* and the *SS Great Britain* were not new, strictly speaking, yet they triggered the development of the Atlantic liner trades (and some other liner trades). They were the right designs at the right time and place.

The story of the water modes continues in Chapter 11. The discussion of the process of growth appears in Chapter 10.

The rapid movement of trains must inevitable generate in travellers a brain disease, a special variety known as the *Delirium Furiosum*. If travelers are nevertheless determined to brave this fearful danger, the state must at least protect the onlookers, for otherwise these will be affected with the same brain disease at the sight of the rapidly running steam wagon. It is therefore necessary to enclose the railway on both sides with a high, tight board fence.

—BAVARIAN COLLEGE OF PHYSICIANS, 1832

6 Railroads Deployed: Learning from Experience

6.1 Trials and Errors

MANCHESTER, ENGLAND, has been called "the first industrial city,"[108] and the industrial revolution transformed this region more than others, with a population that doubled between 1801 and 1831. By 1824 Manchester had 30,000 power looms and was importing 400,000 bales of cotton. The cotton mills led to an engineering industry that first built textile machinery, and branched out, a chemical industry starting with dyes and then expanding, and a financial services industry that funded those and other sectors. Just down the Mersey River, Liverpool served as the primary port connecting England with the Americas.

The inland waterways of canals and rivers that connected Manchester with Liverpool were strained, and Manchester businessmen sought congestion relief. A Liverpool to Manchester cableway was proposed, and William James had begun surveys in 1822 (a few years before the Stockton & Darlington opened).

George Stephenson replaced William James on the Liverpool to Manchester line in 1824, and his plans went to Parliament in 1825. After overcoming much opposition from landowners and road and canal interests, an act was passed in 1826, and construction began.

Cable haulage was sure to work, and it had to be used in a tunnel connecting the station in Liverpool with the docks due to the steep grade and lack of ventilation. But there was concern about its use on the main line. It would provide excess capacity if traffic was light in the early days. Cables were not scalable (both upward and downward) in the same way locomotives were. Locomotives could be provided to scale with the growth of traffic. But would they be reliable and have the necessary power? The Rainhill trials proved they could.

The Rainhill Trials were conducted in 1829 to test the promise of locomotive engines to be used for the new Liverpool & Manchester Railway. Locomotives were to run at not less than 16 km/hr (10 mph), haul cars of three times their weight, adhere to a weight limit, and run the route twenty times to compare roughly with the distance from Liverpool to Manchester (50 km, 31 mi), without adding fuel or water.

Stephenson designed *The Rocket* to compete; it was another, even faster locomotive engine. *The Rocket*, shown in Figure 2.3, was innovative for its use of a multi-tubular boiler, as well as using a chimney to exhaust steam and bring in fresh air for the fire. Because all other competitors failed to finish the race, *The Rocket* won, but that should not diminish its accomplishment.

In 1828, the sole surviving signer of the Declaration of Independence, Charles Carroll, said at the ground-breaking of the Baltimore & Ohio (B&O) Railroad, "I consider this among the most important acts of my life, second, only to my signing the Declaration of Independence, if even it be second to that."[109] It is worth noting that a "railroad" is just the American version of the English word "railway."

The US experience was similar to the United Kingdom; new technologies had to prove themselves. In 1830, Peter Cooper's *Tom Thumb* lost to a horse in the famous race on the B&O, due to a minor mechanical failure when a belt slipped off a pulley. But the iron horse so impressed executives at the new B&O that it was adopted. While the *Tom Thumb* was proof-of-concept and not intended for regular service, other engines were soon perfected.

As a sidebar, Cooper himself had an interest in the success of the B&O, as he speculatively purchased land in the Canton neighborhood of Baltimore. Cooper, while at this point wealthy, became wealthier. One of his early holdings was a slaughterhouse, of which a by-product was gelatin, on which he obtained a patent for its use as a dessert. This was later commercialized successfully by others as Jell-O. Denis Papin, whom we met earlier, deserves some credit for gelatin as well, as an output of his digester. Another by-product of the slaughterhouse was glue. The animals that were slaughtered included horses, which his technologies helped make jobless. Cooper also founded one of the premier engineering schools, the Cooper Union, in New York.

Competitions have a long history in advancing technology. In addition to the Rainhill Trials and the Race of the Tom Thumb, we have DARPA's urban challenge (see Section 30.4), solar car competitions, or the story of the measurement of longitude,[110] which greatly aided navigation.

On September 15, 1830, the opening ceremonies for the Liverpool & Manchester Railway were held. The Prime Minster (the Duke of Wellington), Cabinet members, Members of Parliament, and other assorted dignitaries were present. Among those were an MP from Liverpool, and a 60-year-old former Leader of the House of Commons and Cabinet member, William Huskisson. The dignitaries had been riding on a train pulled by one of Stephenson's *Rockets*. Reports differ, but Lady Wilton, an observer on the same train wrote to Fanny Kimble:

> The engine had stopped to take a supply of water, and several of the gentlemen in the directors' carriage had jumped out to look about them. Lord Wilton, Count Bathany, Count Matuscenitz and Mr. Huskisson among the rest were standing talking in the

> middle of the road, when an engine on the other line, which was parading up and down merely to show its speed, was seen coming down upon them like lightening. The most active of those in peril sprang back into their seats; Lord Wilton saved his life only by rushing behind the Duke's carriage, and Count Matuscenitz had but just leaped into it, with the engine all but touching his heels as he did so; while poor Mr. Huskisson, less active from the effects of age and ill-health, bewildered, too, by the frantic cries of "Stop the engine! Clear the track!" that resounded on all sides, completely lost his head, looked helplessly to the right and left, and was instantaneously prostrated by the fatal machine, which dashed down like a thunderbolt upon him, and passed over his leg, smashing and mangling it in the most horrible way.

Stephenson personally helped Huskisson onto a locomotive and traversed 15 miles in 25 minutes (57.9 km/h) to receive medical attention in the nearby town of Eccles. But it was for nought. Huskisson amended his will and died within the hour.[111]

This was not the first death by steam locomotive, it was at least the third, but it was still the most notable.[112]

Despite this inauspicious beginning, both passengers and freight services (the latter opened in 1831) were immediate successes.

The Liverpool & Manchester rediscovered the passenger market. The route was planned as a freight route, and the amount of passenger demand came as a surprise, even though the Stockton & Darlington had shown latent passenger demand. Indeed, coach operators saw no threat to their business, and, unlike the canal interests, did not oppose the construction of the Liverpool & Manchester.

The Liverpool & Manchester also altered the common carrier concept. Managing locomotives and schedules required that the railroad own and operate equipment. If brought to the stations, freight of all kinds (fak) was accepted for shipment, with charges at set rates.

While the railroad attempted to arrange passenger service provision by a coach operator and freight service by a canal operator, the railroad's desire for both control of car training and train schedules and revenue, and tepid support from private operators, led the railroad to elect to operate services, a major reinterpretation of the common carrier tradition championed by canals.

Freight cars were built from tram and wagon experiences. First class passenger cars were, essentially, three road coach bodies mounted on a flat car. Second class cars were open sided cars with roofs. If a traveler wanted his own road-coach, it was mounted on a flat car.

This chapter watches the rise of the railroads, from the emulation of the first system to learning about networks, technology, passenger travel, freight transport, and embedded policies.

6.2 Emulation

> Andrew Jackson arrived in Washington in 1829 in a carriage and left eight years later on a train.
>
> —D.W. Howe, 2007[113]

Success followed success, and by the 1850s a good bit of the fabric of the world's rail routes were in operation.

- 1825: The Stockton & Darlington opened;
- 1827: The Baltimore & Ohio obtained a charter;
- 1829: The Baltimore & Ohio opened for service;
- 1830: The Liverpool & Manchester opened;
- 1834: The Baltimore & Ohio extended to Harpers Ferry (now in West Virginia);
- 1831: The Pontchartrain Railroad (New Orleans, Louisiana) opened;
- 1833: The Charleston & Hamburg (South Carolina), which for a short time was the world's longest railroad, opened;
- 1834: 14 short railroads had opened in the United States;
- 1837: The London & Birmingham opened;
- 1842: The Boston & Albany opened.

The London & Birmingham (L & BR) and the Baltimore & Ohio (B & O) brought learning further. Each required the establishment of a company of considerable size and the raising of capital in capital markets. (The City of Baltimore put up much of the funding for the B & O; the State of Maryland waived taxes.) The problems of locating terminals in large cities and managing large-scale engineering works were confronted and managed. Experiments with the technology continued. Except for learning about organizational structures and distributed management, which occurred later on the Erie and Pennsylvania Railroads, the very early railroads pretty much defined the railroad, how it developed, and what it could do.

Given the answer to the questions of what railroads would be and what they could do, deployment was very rapid. Figure 6.1 shows the growth of the network in England. As the pace of growth suggests, the system had largely climbed its S-curve by 1880.[114]

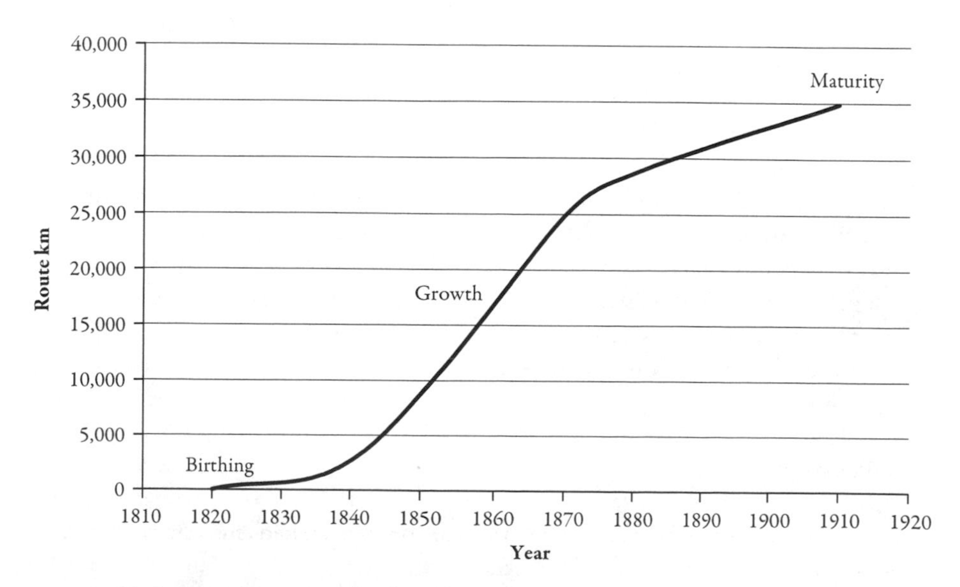

FIGURE 6.1 S-Curve of Route km of Railway in England

It should be noted that the overnight success of the railroad was not so overnight from the perspective of the time. It took the B & O three years before ridership exceeded 300 passengers per day, and it wasn't until it was open and operating that much emulation took place.[115]

Because the railroads were the first of the modern modes, there was much interesting learning-from-experience as they dealt with their growth and development problems. Later modes emulated much of what the railroads learned. For this reason, learning will be the operative word in our organization of the discussion.

6.3 Learning about Networks: The Legrand Star Plan

The Legrand Star plan, centering railroad service on Paris, was produced about 1830 by the French Corps des Ponts et Chaussées. Louis Navier, leader of the Corps, saw speed as the advantage of rail, and railroads were to move passengers and priority freight in a fashion complementary to canals. He felt that others, especially the Americans and Germans, were not building to high enough physical standards, and the plan called for limited curvature and grades, foreshadowing high-speed rail.

But there was a problem. In spite of taxing power, public capital wasn't sufficient to operate high quality railroads. The compromise was that the government would create the fixed facilities and that private companies would provide financing for equipment, stations, and so on. There would be private operations for 99 years, at which time the properties would revert to the State. In practice, there was some compromising of standards to reduce facility costs. Twenty-eight companies were created and eventually consolidated into six regional monopolies. To meet the requirements of the plan, main routes cross-subsidized the operation of routes in small markets.

State engineers planned the routes, and there were complaints from places not well served. State financing was partial, as mentioned. State engineers also established tariffs and fares (and to an extent, service).

Private companies could and did build feeder lines, as was the case with toll roads built in the regions. The point is that the central government took actions in an absolutist way in the spirit of Colbert.

These accomplishments were not made without great debate, and there were periods when anti-statist, anti-elite, liberal Adam Smith–like forces held power. Anti-statists argued that the time value of money made high quality facilities inefficient and that marginal cost ideas should take priority over state-determined prices and cross subsidies. The authority of government engineers versus engineers in the private sector was also debated. "Cheaper and better" argued for private sector engineers. But even with these debates, absolutism continued. This was in spite of the Revolution and Napoleon. Indeed, the Napoleonic Legal Code (Roman-based), in contrast with English common law, may have eased the implementation of absolutism.

Most critics say that France over-expanded its rail facilities and invested in canals long after they ceased to be competitive with rail. High standards for canals resulted in expensive facilities; they did not fit instances where water and/or traffic was in short supply.

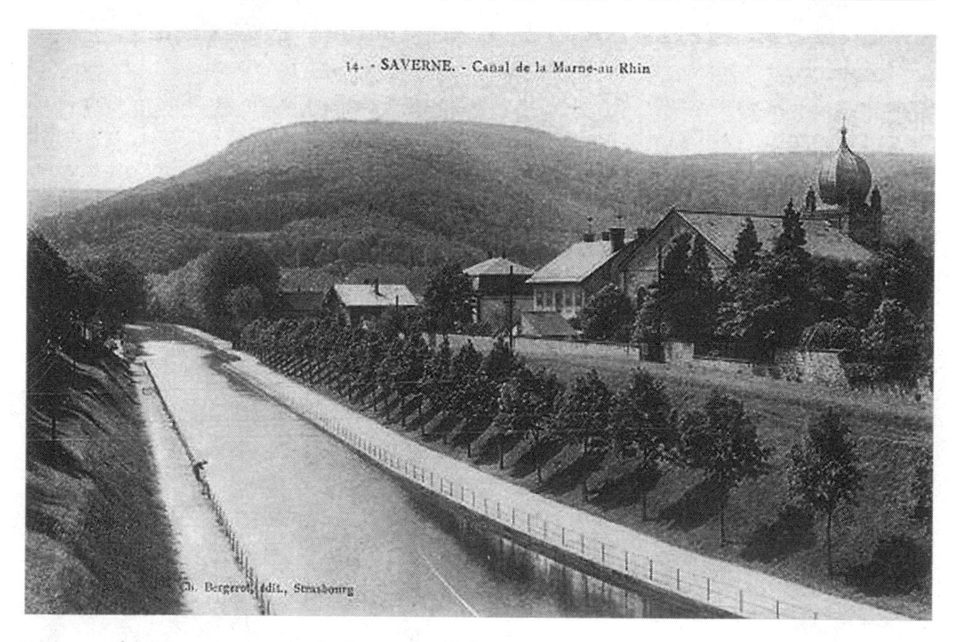

FIGURE 6.2 Transportation in the Zorn River Valley

Bismarck's invasion of France in 1870 brought the Paris-centered rail system into question, for the French could not move troops to the front as quickly as the Germans could. As a result, a grand plan was produced in 1880 resulting in the northeast rail corridor from Dunkirk to Nancy.[116]

Other critics say that France developed very fine systems as a result of the professionalism of the central government and the engineers' uses of science. There is no question that, from a strictly physical engineering view, the French developed superior facilities using superior knowledge.

Société Nationale des Chemins de fer Français was created in 1938, and another round of superior engineering was seen as the railroads were electrified after World War II (25,000 volt, AC transmission converted to DC by locomotive rectifiers). The Train à Grande Vitesse and today's expressway system followed.

Figure 6.2 illustrates inter-modal complementarity; in the River Zorn valley, side by side are the Rhine-Marne canal, the River Zorn, the road leading to Lutzelbourg, and the Paris-Strasbourg railway.

6.4 Learning about Technology

Railroad development began in the United States at a time when the predominant technology was still in a hardening phase, and the technology was specialized to the US situation. A key actor was Robert Livingston Stevens (1787–1859), son of Robert Stevens, the competitor of Robert Fulton of steamboat fame.

Trained in construction and steam technologies by his father, R. L. Stevens was first president and chief engineer of the Camden & Amboy Railroad (later, part of the Pennsylvania system), incorporated in 1830. Stevens journeyed to England in 1830 to meet Stephenson, and he purchased a planet series locomotive—the *John Bull*, which arrived in the United States in 1831. Stevens designed the "T" rail, which was known as the Stevens or American rail and is now universally used. He also developed the rail-spike-cross-tie system. The bogie truck and methods of wood preservation were developed on the Camden & Amboy. Also on that railroad was Issac Dripps, the innovator of the cowcatcher, which not only redirected lost livestock, but also (with additional wheels) helped to keep the locomotive on track.

Stevens was America's Stephenson, and the Camden & Amboy was its Stockton & Darlington.

The hardening of the technology America-style also yielded the "American" locomotive. The US terrain was unsuited to the British locomotive;[117] rigid British locomotives did not ride the poorly aligned, light American rails very well, nor did they serve well on the sharp turns of mountain passes. They were also designed for coal burning, while wood was abundant in eastern North America. American technology was still imported in the mid-nineteenth century; by 1840, of the 450 locomotives in US, 333 were made in the United States; there were 3,200 miles (5,150 km) of track (as many as canals) and as many miles as Europe. Much credit for this goes to Stevens—someone had to develop the American version.

To meet US needs, Henry R. Campbell of Philadelphia developed and patented the "American" 4-4-0 locomotive in 1836 (see Figure 6.3). The numbers assigned to the American trains are based on Frederick Methvan Whyte's system of classification. The first number is the number of leading wheels, the middle number is the number of driving wheels, and the last is the number of trailing wheels. Thus for the 4-4-0 there were two leading axles (four wheels), four driving wheels, and no trailing wheels and axles. The chief feature of the American locomotive was a 4-wheel, two-axle pivoting truck at the front, which would turn and guide the locomotive around curves.

John B. Jervis produced the first versions of the American design; his locomotive was a 4-2-0 and weighed about 7–10 tons. But within a year or so, the Campbell design evolved to 4-4-0s weighing about 15–25 tons.

FIGURE 6.3 American Locomotive from Northern Pacific Railroad

The wood versus coal question was touch-and-go. Initially, coal was expensive in the United States, but rail service soon began to make anthracite available. Just at that time, though, the demand for coal was affected by rapid growth of "hot blast" iron and steel production, which put pressure on anthracite prices. Less expensive wood became the fuel for most railroads, but a few railroads used anthracite.

As the technology hardened in a suitable form, railroad construction boomed (starting about 1845). The American locomotive dominated American practice through 1900. The weight and power of the locomotive increased by at least a factor of three, yet the real price remained steady, and even fell during some decades. There was a shift from wood to coal fuel at about 1880. There were many American-type locomotives still on the properties when diesel replaced steam in the 1950s. Quite a number were exported, some to England. Later, many 4-8-0 locomotives used in Europe were based on American designs.

One reason for the quick demise of canals was the unusually cold winters of the middle 1840s—there were short shipping seasons just at the time new canals were opening and facing competition from railroads.

The United States had some major systems integration issues, despite being able to follow the lead of England. Because engines were imported from England, many US railroads adopted standard gauge of 4′8.5″ (1,435 mm). Some Southern RRs used 5′ (1,525 mm), the Erie RR used 6′ (1,830 mm) to avoid diversion of traffic, but that was self-defeating. At the beginning of the US Civil War there were 11 different gauges in the North and an unknown number in the South. A train from Philadelphia to Charleston would have to change gauge 8 times.[118] A few larger railroads were built by single companies, example the Baltimore & Ohio, the Erie, and the Illinois Central, but the vast majority of lines were short and local. Longer trips would require using the lines of many different organizations.

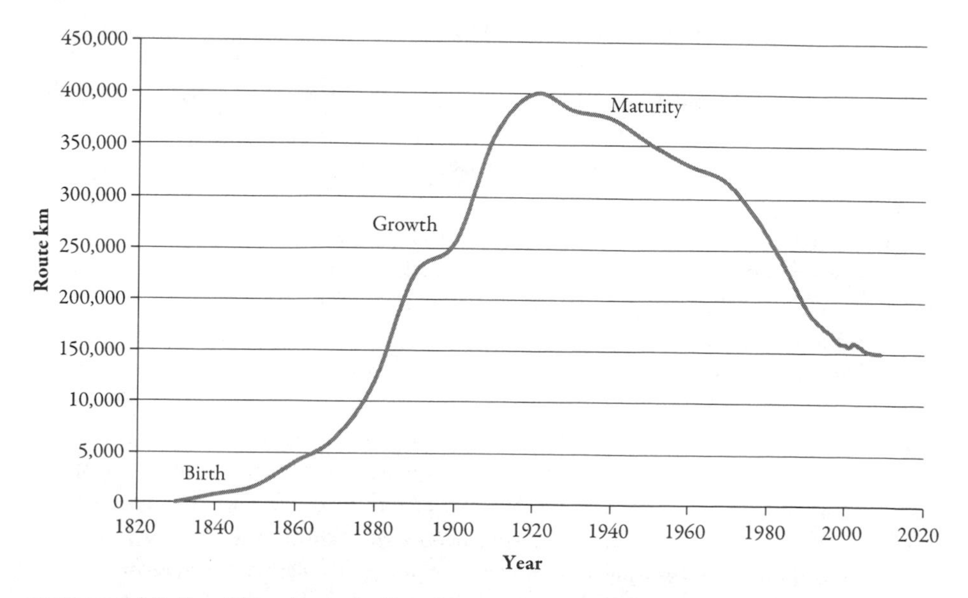

FIGURE 6.4 Railroad Route km in the United States, 1830–Present

The railroads were aided by grants of public land. At least as significant was the financial support from governments. In 1838, state debts totaling nearly $43 million were attributable to railroads.[119] William Ripley argues that local support was at least as great, and by 1870 was 20 percent of total construction expenses to date.[120] Virginia subscribed to 60 percent of the shares, provided the other 40 percent was taken up by the private sector.

A so-called debt-repudiation depression began at about 1839 and continued to about 1847. The conventional reason given for that depression is the write-down of investments made in canals and some changes in currency and banking policy. While not dismissing those factors, Santini (1988) links the depression to the clash between the new and old technology and, in particular, to the displacements the uses of the technology occasioned.

From the 40 km of railroad in the United States in 1830, in about nine decades the railroad grew to about 400,000 km. Subsequently, the length has declined to about what it was in 1890.

Another American innovation occurred in 1836, when the sleeping car emerged in the Cumberland Valley RR. This would prove to be important in the American context, as the trips were significantly longer than in England or Europe.

George Pullman essentially invented the hotel on wheels. He attained early notoriety for physically moving the four-story Tremont Hotel in Chicago, raising the building with jacks to match the road. He brought on the sleeping carriage in 1859 on the Chicago & Alton RR. Sleeping carriages were heavy in addition to being luxurious (not so much for the beds, but for the wooden furniture), and so railroads were required to improve their track and supports to accommodate them. They gained more fame when used on Abraham Lincoln's funeral. Pullman's labor practices led to one of the most violent strikes in US history.[121]

6.5 Learning about Passenger Service Standards

Charters for the Liverpool & Manchester and the London & Birmingham were obtained after debates about their diversion of freight from other modes. But diversion of freight was slow and selective. The latent demand for passenger transportation, however, came as a surprise.

Recognizing a role in passenger service, the Act of 1844 responded to public claims for rights to travel and set a minimum level of service availability. Known as Parliamentary trains, service was to be provided in each direction each weekday. Stops were to be made at each station. A maximum fare was set and running speeds were to be not less than 20 km/h (12 mph). This was third class service.

First and second class service reflected the coach, fly wagon, post horse, freight wagon services available on roads and fly boat service on canals. The classes of passenger service offered by the railroads mirrored canal and road services. They substituted for canal and road services and pretty much captured the market. The same economic rationale continues today. It yields, for example, first, business, coach, and a collection of lower cost special fares in air transportation. Jules Dupuit considered a situation where there was a fixed cost of providing service. Using concepts of consumer surplus and utility differentiated

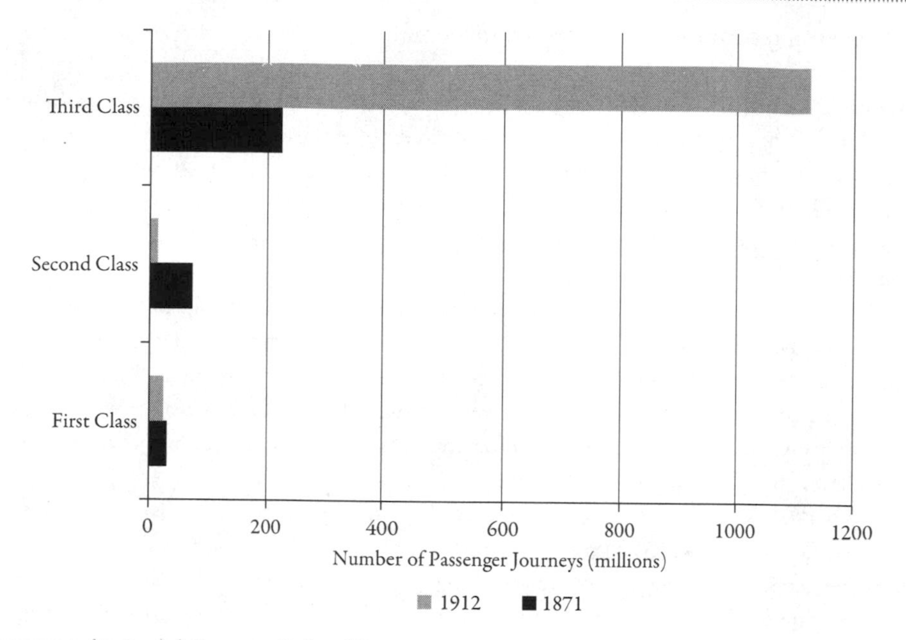

FIGURE 6.5 English Passenger Railroad Data

among users, he concluded the wisdom of differentiated tolls for differentiated passengers—the idea of price discrimination, referred to today as Ramsey or inverse-elasticity prices.

Railroads found demand and costs such that they could lower fares and upgrade service. Although Figure 6.5 does not reflect early experiences, it shows the consequences of railroad actions—low fare services grew very rapidly, while in absolute terms, high fare services withered.

Things were improving, and there was no need for central government action, with one exception. The railroads found the obligations of the 1844 Act onerous, especially the requirement to stop at every station. They pressed for relief and obtained it in an 1866 Act. However, there was a trade-off. As railroads asked for relief, some were required to operate Workmen's Trains in the vicinity of large cities (more on that later). We have seen:

- Policy stated as a minimal level of service requirement. A standardized requirement, one policy fits all situations.
- Requirements for minimal level of service policy are mooted if service performance improves, as it does in the early days of a system.
- Policy revision and the use of trade-offs to obtain revision.
- Experience said offer differentiated services, and custom and tradition have caused differentiated services to continue.

More generally, what had been learned in the pre-rail experience was incorporated in the rail experience and modified in a pragmatic way, and the rail experience led to additional modifications.

6.6 Learning about Freight Rate-Making

In the case of roads, the local justices of the peace were required to set maximum freight rates and service conditions (1691 Act). This was early in the development of wagon trade, and the concern was that wagoners were setting rates at excessive levels. It seems likely that they were. One had to have considerable (for the times) capital to enter the business, and competition must have been limited. The wagoners had mutual interests, and price rigging must have taken place off and on.

The experience of setting rates and service conditions was reflected in canal and turnpike charters: the canal firms and turnpike trusts were not allowed to monopolize the boat, wagon, and carriage business. Maximum tolls were set, and later refined. Road tolls, for example, had some relation to cost. They were computed on the basis of wagon wheel width and tons. Flyboats paid higher tolls than freight boats. In river navigation projects, there was a tendency toward complex tolls that charged according to value.

From a policy point of view, the Stockton & Darlington was a canal. Anyone with equipment could use the track. Maximum tolls were set. These conditions were carried over to the Manchester & Liverpool Charter. But the "anyone can use" policy was never implemented. The London & Birmingham charter simply required the railroad to carry all freight brought to its docks.

At this point in time, the system took off, and by 1850 there were some 215 railroad companies in England. Additional problems emerged, begging policy attention. One was the amount of time the legislature was spending on the private acts, one for each firm. Could generic legislation be stated and action delegated to some creature of the legislature?

Debate continued on this prior to the 1844 Act and a proposal was developed, but there was no full response to this problem until 1872 when the Railroad and Canal Traffic Act created a railroad commission and gave it powers. (This was about 15 years before the Interstate Commerce Commission was created in the US.) An interesting aspect of the 1844 debate treated government ownership. Using the turnpike model of reverting to government ownership, it was recommended that newly chartered railroads become the property of the government after 21 years (compensating their owners at reasonable prices).

Another problem was the growing complexity of commodities to be moved. Classification was the name of the game, as was the basing of rates on the value of the product. Traders pressed the railroads hard for advantages in order to attract traffic, and classifications and rates began to be complex. About 300 articles were recognized by 1850, and by 1871 the number had grown to 2,753. Classes were recognized, articles were placed within classes, and charges for them were related to class rates.

The situation was more complex than the increase of articles and classes would suggest. With the growth of the economy and trade, more and more shipments moved over multiple properties. By the 1880s there were 40,000 stations between which shipments could move. Some system-wide referee standardizing rates had to be established. Would government do this?

Not at first, although government did pass lots of acts relating to rates. The railroads created a clearinghouse to do the rate work and to articulate shipments among roads. Government action mainly responded to claims that the firms were making too much money—maximum rates should be reduced. Government tried to protect the public from monopolies by authorizing competitive lines and controlling amalgamations. It gave running powers to railroads that were locked out of markets.

Most researchers say that government actions were not very effective. Government control of rates was ineffective partly because control was aimed at line haul charges. Carriers could set terminal rates and collector/distributor rates.

The classification-rate problem was settled by mutual government-industry action. The clearinghouse had enforcement problems, and government had its problems. The clearinghouse worked with Parliament's Board of Trade.

The Act did more. It extended to terminal charges and it introduced tapered rates. In this small way, the Act, for the first time, related charges to costs. Previous Acts (1868 and 1873) had required that rates be published and that charges be itemized.

Government got involved in setting rates because, while the railroads could get some agreements on rates, an authority was needed to enforce agreements. Once government got involved, it did begin to exert some social policy influence. The requirement for the publishing of rates reduced the problem of asymmetric information between users and service suppliers.

The above discussion treats a lot of borrowing and learning. First, the railroads and government borrowed from previous experience. Next, they responded in pragmatic ways to the rail experience. The overarching challenge was to deal with complexity. Government worked with the railroads to create institutions for that purpose.

6.7 Learning about Embedded Policies: The Org Chart

The discussions to this point have stressed how railroads began by borrowing policies from previous modes and, then, modified policies based on experiences. A similar statement may be made about government. Passenger and freight services were examined. Now, the discussion turns to how policies became embedded within the structure of the mode.

By the late 1800s, many small rail starts had been absorbed by larger systems, and strong large systems had emerged. Within-railroad embedded policies evolved to manage these large systems, namely: functional and departmentalized administrative structures for day-to-day routine operating decisions with precise accountability for expenses and use of human resources. The telegraph provided operational control—on an hourly basis for trains and on a daily basis for (1) expenditures, (2) work performed by men and machines, and (3) freight and passenger movements. These innovations are attributed to D. C. McCallum of the Erie Railroad beginning in 1854. They were widely publicized and served as a model for other railroads and for industry generally.[122]

J. Edgar Thomson of the Pennsylvania Railroad, who, after adopting McCallum's scheme, divided the Pennsylvania into operating divisions in 1857 took the next step. Each division had a superintendent in charge of three functions: transportation, way, and equipment. The headquarters superintendent and the division superintendents were delegated operating authority over men and machinery. Headquarters was primarily involved with finance and planning and rules and procedures.

Other railroads in the United States quickly adopted this organizational system. Adoption elsewhere seems to have been slower, mainly because systems were smaller.

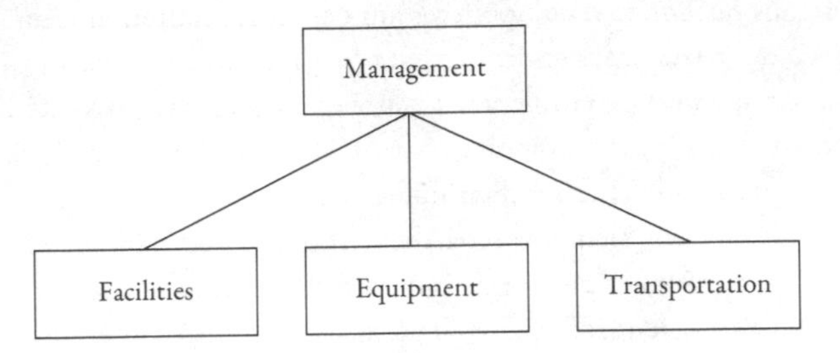

FIGURE 6.6 Illustrative Organization Chart of Railroads

As noted in the above, transportation systems have a three-component, triad structure.

- Fixed facilities created and managed by civil engineers (through, e.g., a public works department or the US Army Corps of Engineers);
- Equipment designed and manufactured by mechanical engineers (through, e.g., General Motors or North America Car);
- Vehicles operated by pilots, train engineers, or drivers.

Mirroring that structure, railroads are organized as shown in Figure 6.6.

The answer to the question of why railroads are organized that way requires little discussion: that was the division of talent and problems at the time. For the other modes, "modeled on railroads" is a good first approximation to the answer to the question of why they are organized the way they are.

But this structure yields considerable dysfunctions. The question is that of why modes have continued to be organized in triad structures in light of the dysfunctions. That's especially a question for railroads. As corporate entities, they should be able to adopt restructuring policies more easily than the other modes.

We may term one dysfunction the inability to make markets. A conversation with a railroad person would be misleading. Most of them can tell all about the goods being hauled and why. Those who have to know have deep knowledge of the trading terms and transportation costs interrelations in the capturing of markets. They know that a cent here and there or a quality of service attribute here can change the terms of transport. But the level of freight movement knowing by railroad people has an important constraint on it. Railroad people do not have sharp knowledge of the cost to the railroad when making a market.

The reason is structural. Each division of the railroad claims a budget to carry forward its task, and each tries to minimize costs and/or maximize output from given inputs. At the level of a geographical division, for example, the transportation (operations) manager sees certain tons to be moved. Given the equipment and fixed facility situation, a good manager keeps the transportation inputs required to a minimum. Individual shipments move across divisions and impose costs on way, equipment, and operations. Shipments are handled jointly with other shipments in different combinations at different places and times. So it is very difficult to identify the cost of a particular shipment.

Railroad people know this very well, and they give a lot of attention to costing. Indeed, Wellington's *The Economic Theory of the Location of Railways.* Original publication is 1887.

Subsequent editions published later, centered on cost information and its use.[123] Where special services are offered, such as unit trains on heavy haul routes, cost information may be pretty good. Some railroad marketing departments are organized by commodities and try to match costs and services as best they can.

Even so, there is a long-standing dysfunction locked in by structure. The situation is worse in other modes where the structure is not within mode-wide organizations. In air transportation, for example, we have authorities that operate airports, a Federal Aviation Administration that manages the airways, and airlines and private operators that move passengers. Individual firms such as trucking companies and ship builders know their costs and markets, of course. System costs are another matter. No one is in charge. In the economist's pure atomistic world, if there were perfect competition that is of no great concern. But structural interdependencies in transportation are strong, spatial monopolies exist, and competition is far from ideal.

The second dysfunction is associated with patterns of dynamic decision-making. The operations group on the railroad measures costs on a car basis, so they press the equipment division to buy larger cars. That's acceptable to equipment managers because it lowers their costs. But it has been found that large, heavy cars impose lots of costs on track structure. What has happened is that incremental decisions were made on the basis of component criteria, subject only to the conditions that it fit the system in a technical way. Examples of this kind of behavior proliferate throughout the modes. For instance, truckers buy high pressure, radial tires and ignore pavement damage questions. Sometimes this component sub-optimization begs remedial policy, as was the case when drivers started to use snow tires with metal studs. Often, policy is faulty because it fails to consider the system environment. In rail, there is policy about the condition of track and speed limits that pays no attention, as it should, to the type of equipment ordered.

One term for this pattern of behavior is *disjoint incrementalism*.[124] Incrementalism allows exploration of options in cases where wants are unknown, enables reversible and low-risk decisions, stabilizes organizations, permits compromise among goals and stakeholders, and has other desirable attributes.

Perhaps stasis from constrained behavior gives an additional flavor to the idea we strive to communicate. Actors in a component of a mode are constrained to do things that fit the structure of the mode, and actors' goals, values, and priorities are constrained to their component.[125]

As a consequence of the constrained behavior of components, systems have a highly constrained path of system development. The options that can be explored are very, very limited, for example. Actions may be costly because of impacts on others.

We think the limitations on the exploration of options are the major cost of disjoint incrementalism. When is this failure costly? It is clearly costly when systems age. Inability to change is one reason they sail along in the sunshine of their obsolescence. It can be costly early on, because it can lock in dysfunctions. Policy bearing on Brunel's wide gauge railways (see Section 5.7) is an example.

One dysfunction following from the organization of systems is the "environmental fallacy."[126] Centralizing actions are taken under a false pretext of crisis and urgency without considering the environment and feedbacks from environmental adjustments. That thought could be used at the level of how components interrelate with each other and at larger scope.

(This is distinct from the "ecological fallacy," in which inferences about the individual are made from aggregates of which they are a part).

6.8 Learning about Rules: The Code of Operations

As would be expected from our discussion of technology and as seen, the railroads borrowed policies from previous experience. For example, the first charters for railroads were modeled on those for canals, as already mentioned. Temporal organization was also borrowed.

With respect to embedded policies, management and operations policies were borrowed from military and industrial experiences. Military borrowing resulted in style or corporate cultural features. Such borrowing was necessary because of the geographical span of railroads and requirements that systems operate in a reliable fashion—in order to take an action, the actions of others elsewhere must be highly predictable. The railroads borrowed from the military, and especially from industry, the policy of operations by rules. Clear, carefully stated rules were necessary for predictability, for the training of labor, and for assuring task responsibility.

Workers who had lived for years with nature's clock and contracting for independent tasks (I will bring in your crop this fall) had to be trained to do coordinated things at specific times. To accomplish this, *The Law Book of the Crowley Iron Works* (late 1700s) contained 100,000 words.[127] Modeled on such rules, especially Josiah Wedgwood's chinaware factory rules, the Liverpool & Manchester opened for business with a well-developed *Rules and Regulations* book. The *American Standard Code of Operation Regulations* of 1887 illustrates the heavy hand of early beginnings.

6.9 Learning about Time: The Rise of the Time Zone

We of course have always had time, but its importance has changed over the centuries. Before attention to time was imposed by the industrial revolution, and standard time invented by the railroads,[128] lives ran on time measured by the days and seasons and benchmarked by church and state holidays, seasonal fairs, and other events. Work was organized by the job, say, contracting to bring in the hay, for a two-year trip on an India voyage, or for a mid-summer (between planting and harvesting season) raid on neighboring kingdoms.

Factory work and the spread of precursor and modern transportation systems restructured the daily, weekly, and annual activity paradigm. Employers imposed time contracts (much like the corvée, which required peasants to give 6 days a year, or more, to the government to maintain the roads, although nominally these contracts were voluntary). These contracts said: be there at ... stay until.... A weekly rhythm emerged, and people had to learn to work in tandem with others at specific times and places.

For transportation, the operation of scheduled services as on rail required time coordination to ensure reliability and safety; just as ships had "watches." Rail service also learned from the military, which heeded time's beat to beat the enemy.

Transportation was no mere user of time, the transportation sector fashioned tools for time. While the clock and calendar are old, transportation required precision timekeeping and coordination. Measurement of longitude required precise portable timekeeping devices.[129] The issue was so important that the British government established the Longitude Act in 1714, and a Board of Longitude, which would award a prize of £20,000, a nice sum today, but huge for the time, to anyone who could develop an accurate and practical measure of time at sea. The timekeeping devices developed by John Harrison would evolve into the portable watches required by train conductors.

Standardized railroad and air services need Greenwich Mean Time. Of course, such time was an arbitrary invention, established as standard by the Prime Meridian Conference in 1884. That Conference followed on the heels of Sunday morning, November 18, 1883, which was dubbed the day of "two noons" in the United States, as those in the eastern parts of the four time zones had a second noon to comport with the western part of those zones. Today's Global Positioning System (GPS) provides sufficiently accurate time and location that people are considering using it, or at least the next generation of GPS, for vehicle control, for instance keeping cars in their lanes.

This mode of living—activity scheduled by time and requiring specialized inputs, extensive use of capital (machinery), and requirements for tandem or sequential activities—emerged in the 1800s and now dominates life and thinking. Consider, for example, the public school system, which teaches students to live and work in the mode established by the industrial revolution.

So transportation (and the industrial revolution) framed the modern perception of time, and thus our daily schedules.

6.10 Learning about Traveler Information

Today, with GPS satellite navigation systems built into cars and ubiquitous cell phones, with on-demand access to the Internet, with information about everything at one's fingertips, it is hard to imagine a time with so much less. As a complement to the new transportation systems arose new guides that would enable tourists, relatively rare until the railway, to travel far beyond their experience with information about where to go and how to get there.

Many guidebooks arose. George Bradshaw developed what came to be known as *Bradshaw's Guide*. At first they were railway timetables, which of itself was quite important, since there were more than 150 different operators with many routes and schedules each, which changed frequently. Bradshaw also produced a *Bradshaw's Tourist Guide*. Michael Portillo, a retired British politician (a former cabinet officer and "Future Prime Minister," who was expected to ascend the political ladder, but didn't quite make it.) used that guide to revisit Victorian England in a television series *Great British Railway Journeys*. In England and its colonies, the term "Bradshaw" was genericized to be used for any railway guide. Other famous guides of the railway era included the German published *Baedeker's* guide.

In France, the *Guide Michelin* was sponsored from the early twentieth century by the eponymous tire company to aid travelers in the choice of hotels and most famously restaurants. This emerged with the automobile, rather than the railway, which brought about a further revolution in how people traveled.

Travelers needed information not only about their trip, but the world. Newsstands such as W. H. Smith (1848) in England and Hachette (1852) in France arose at railway stations to keep passengers entertained and informed. These complementary goods, along with other facilities at train stations, restaurants, hotels, booking offices, and waiting rooms, presage the rise of the roadside rest areas in the highway era.

6.11 Learning about Right-of-Way: The Conflict between Land for Access and Land for Activity

Consider early roads, trails, ways, and river or beach landings. Services were ubiquitous and suited the traffic. A "way" was a zone along which the public has right of movement, the right of way. It was a path that, perhaps, shifted seasonally and was suitable for people and their pack animals, as well as other animals walking to market. The term "way" describes many routes.[130,131]

A "way" is important, but as property rights are established, one needs a "right of way" in order to build a transportation system. What was government involvement in obtaining land, what did government learn, and what policy evolved?

Property conflicts occurred as landowners sought to fence property for field crops and restricted amount of way available. In many places, the abutting property owner had an obligation, but little incentive, to maintain the road. In England, the enclosure movement made establishing public ways more difficult.

The idea of a way emerged in the United States when the Midwest and Plains states were being settled. The government retained title to mile-wide (1.6 km) strips (sections). These sections would be used as ways for long-distance movement, an important matter when animals walked to market. TV Westerns often featured conflicts between cattle drives and settlers in their plots.

Turnpikes and canals required land taking by trusts and firms. These were typically small undertakings. The enabling acts passed by central governments endorsed local agreements and contained details on the land to be taken and prices to be paid. Large landowners typically were actors in the promoting organizations, and would not use land ownership to hold up or block development.

We have some evidence that the tramway land-taking problem was more difficult, perhaps because mine and transport operators were often not part of the landed gentry. For example, the first charter for the Stockton & Darlington Railway (the tramway charter) was requested in 1819. The Duke of Cleveland's concerns about disruption to his fox dens (and the consequences for his hunt) stalled construction of the railway for about two years, so the charter was not obtained until 1821.

The London & Birmingham Railway posed a new problem. It was not a local undertaking: Birmingham businessmen largely promoted it. It was strongly opposed by the owners of canals serving that market and points between. Its enabling Act could not be an endorsement of proposed deals. It said that market value would be paid for land and treated the matter of partial takings. This protected investors as well as those whose property was taken.

That is what the enabling Act said, but there is a good bit of evidence that acts were obtained by "buying off" opposition. In an address to the British Institute of Civil Engineers, Stephenson remarked (with perhaps some rhetorical flourish) that the costs of obtaining the enabling Act for the London & Birmingham was not much less than the cost of constructing the entire line. He remarked, "Great was the ingenuity of the agent who discovered the use of the word SEVERANCE," and went on to call it a system of "spoliation, permitted by Parliament."

If the opposition could not be bought off, the railroad had to avoid the property. That requirement caused the building of long tunnels on the London & Birmingham and building the Liverpool & Manchester through the Chad Moss. (Moss, short for Morass, is described as a "wet and dreary waste, composed of peat,"[132] and is named for St. Chad.)

Two things seem to have muted the conflicts between rural landowners and the railroads. First, with experience, rural landowners found that railroad construction was to their advantage, and began to try to attract routes. For instance, Lords Derby and Sefton, who, by their opposition, had forced the Liverpool & Manchester through the Chad Moss, became patrons for a rival line, on the condition that it would pass through their property. Chad Moss, which had been worthless, suddenly became valuable.

Second, the railroads offered productivity improvements. They could afford to pay "through the nose," if needed. The profits were there.

The construction of railroads into cities posed very difficult property taking problems. The railroad map of London (see Section 8.2), for example, shows that railroads were unable to penetrate into the urban core and that the railroads were more successful in taking land from the poor and powerless than from the rich. Workmen's trains were one consequence. The poor had to move to the suburbs, and were given low-cost commuter trains.

The situation yesterday was no different from today. Many of today's urban transportation problems are conflicts in the location of fixed facilities. For example, debates about roads, docks, and airports stress undesirable visual, noise, and air pollution; congestion spillover to adjoining areas; destruction of neighborhoods, and so on. What did we learn when the conflicts were between people and their possessions and activities and the new-kid-on-the-block, the railroad? Not much!

Every city has its own story, and each story is idiosyncratic. The influence of city morphology differed because of site differences between cities, and there is a cut between those cities well established prior to the coming of the railroad and those where the turf was not so occupied. The conflicts differ by facility type: passenger and freight terminals, yards, routes, facilities such as shops, and so on. Finally, the situation changed with time. Some railroads invaded cities on public street property. Electrification was pushed by the need to use tunnels and the problems with urban environments, as in Baltimore and Paris and, later, New York and London. The lack of "union" stations (the "rail terminal problem") is a long-standing consequence of this situation. In London, many stations, several nearly adjacent, competed for business. The clock on the King's Cross tower disagreed with the one on the next door St. Pancras.[133] Victoria Station housed two railway companies, separated by a wall.

As a result of differing temporal and morphological situations, the literature tends to be city specific. One can find multiple books on London and Chicago. Chapter-length discussions often appear in rail or city histories. Many cities have studied location problems, have penned reports, and have sometimes taken action. Simmons (1986) treats places large and small in England. To compress his remarks on London:

- Railroad passenger facilities penetrated into the city in varying degrees. During the 1830s, only three of six went well into the built-up area, and only one placed its terminal within the Corporation of London, the City proper.
- Construction was costly, and additional costs were incurred as traffic grew. Fragmented land ownership was especially a problem.
- Many small properties were demolished and "... the process ... was generally thought salutary in clearing away noisome dwellings that fostered disease and crime" (p. 32).
- Viaducts and cuttings became barriers.
- Replacements for housing demolished became a requirement in 1874, as did compensation to those forced to move. However, these requirements were not enforced very well.

Facilities were needed for car storage and sorting, trans-shipment of less-than-car-load lot (LCL) traffic, teaming yards, and access to large shippers. Although very demanding of land, little is said in the literature about freight transport conflicts in cities. In many cities, long viaducts were built across yards. These were a precursor experience for limited access highways. Pressure was exerted to elevate or tunnel heavily used routes. This was a force for electrification of railways. Many cities wanted "union stations." In the absence of union stations, interline transfers were arranged in difficult ways, example through the use of the Underground in London and through circumferential rail routes in other cities. Yet the rail experience did not influence other urban facility policy, such as freeway policy, as strongly as it should have.

Opportunities to improve facilities and take advantage of improved and larger vehicles emerged. A critical point is the changing relationship between fixed facility costs and operating costs. Costly but more efficient and higher quality facilities emerged—seaports and airports, limited access roads, and carefully engineered rail facilities, subways, and elevated transit. Now, the rule is to achieve lower door-to-door travel times, carry heavier loads, go safely with less stress, and save money.

Traffic has been concentrated, and accommodating that traffic has required land—land for container and bulk shipment ports, for multilane freeways, and for airports and railroad yards. Graceful abandonment and reuse of land has generally not been achieved.

Concentration and growth of traffic have also deepened issues of externalities: noise, air pollution, and so forth. In principle, one can tunnel for highways just as one tunnels for deep subways. Disruptive cut-and-cover is not required. French highway planning is taking advantage of tunnels (segregated to separate cars and trucks, and thereby allowing more vehicles in less volume) linking districts of Paris, for instance on the Paris beltway (orbital expressway A86).[134] A similar suggestion for the widespread use of small tunnels was discussed by Garrison (1974).

Externalities, especially those generated by traffic, make transportation land use undesirable for neighbors; construction of facilities has downsides, too. Dust, mud, noise, and disruption of traffic have their costs. The construction gangs and others associated with the building of the Erie Canal were not good neighbors.[135]

We need to improve our ability to remember and use experiences. The rail experiences in urban areas continued through the decades. Problems did, too, and many remain today. New construction is not so much at issue today. But there remain local issues of routes around cities, consolidation of terminals and lines, noise, and congestion at yards and terminals, grade separation, and the provision of workmen's trains (today's commuter service). Still, with respect to land taking and environmental insult, the highway designers failed to learn from the rail experience.

Some difficulties in transportation modes obtaining right-of-way results from the resulting negative externalities. An externality is an effect that occurs to a third party. While the railroad and its passengers engage in an economic transaction, an externality (positive or negative) is what happens as a result of that transaction which befalls other individuals. The gains in accessibility must offset the costs, including environmental costs.

The discussions above imply some conflicts between transportation and the environment, and a few instances were mentioned. Trains were dirty and noisy compared to previous modes, and they were broadly disruptive of previous ways of life. Highways bring construction, which disturbs parks, and cars, which spew filth into our skies.

As something new, railroads were viewed as an environmental insult.[136] Later, they were accepted as part of the landscape. Landowners opposed canals, "the thought of having their land cut through and separated by canals was far from appealing to them." Siddall (1974) goes on to say, "In general, though, the canal was soon appreciated as an attractive addition to the landscape . . . these stately roads' as Wordsworth . . . called them."[137]

With respect to railroads, there was "torrent of opposition in England." It is not clear how much of this opposition was disruption-of-the-natural-environment related and how much was addressed to the broad changes induced by the railroad. Charles Dickens, for example, was very anti-railroad (see Section 6.15). Elsewhere, he complained about smoke and noise. Complaints gradually declined, and the railroad was gradually accepted as part of the rural landscape.

Conservatives (not in the American political sense, but in the maintenance of the present world sense) oppose change. New facilities, especially new technologies, bring change. Eventually the change is brought and the new facility becomes old, a part of the environment. Opposition before evolves to acceptance after. That is not saying that given time we should sweep environmental problems "under the rug." It's what we learn from the rail and previous experiences.

6.12 Learning about Alliances

The organization chart eased the problem of within-railroad embedded policies (but not the making-markets or profit-cost centers problem). Developing between-railroad embedded policies proved tougher.

In the post–Civil War decade there were four major systems connecting the East Coast with terminals just west of the Allegheny Mountains, and those trunk lines sought alliances with western railroads and steamship lines to curb rate wars. Some purchased stock in the western roads, and all made agreements on through rates and services. But this did not go

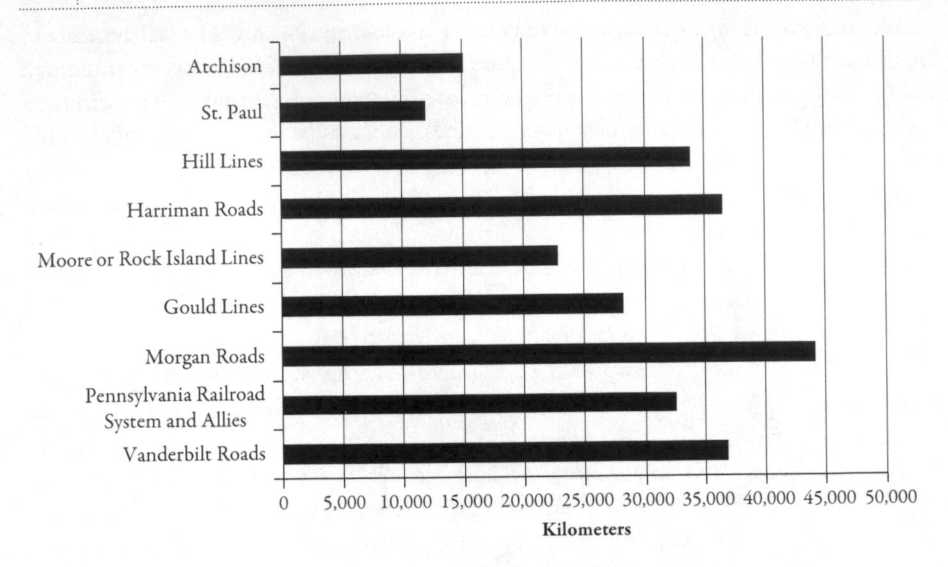

FIGURE 6.7 Nine Leading US Railroad Groups and Systems in 1908

smoothly because individual roads could gain by seeking favorable changes in the status quo. (Jay Gould sought such changes when he gained control of the Erie in 1867.)

A variety of supplemental techniques were attempted. The Vanderbilts, for example, sought to federate the New York Central with lines to the west. The western railroads, which were generally weaker than the eastern trunk lines, sought to pool receipts in order to reduce motives for price wars.

These efforts were unsuccessful, and in 1877 the eastern railroads formed the Eastern Trunk Line Association headed by Albert Fink. This was soon joined by a similar Midwestern and New England organizations, with the several organizations unified by a common executive committee. The regional organizations determined rates and classifications, which were then adopted for the areas covered.

Things seemed to have worked smoothly for about a decade until Gould and the Vanderbilts became aggressors in this non-zero sum game. Figure 6.7 indicates the size of the properties that we have been discussing (at a date a little later than the period we have been discussing).

Alliances proved to be important in the maritime (see Section 19.3) and aviation (see Section 20.5.1) sectors as well, and are used within many cities to organize the private provision of taxi, jitney, and bus services.

6.13 Profile: Cornelius Vanderbilt

According to legend, in 1810, at the age of 16, Cornelius Vanderbilt (1794–1877) borrowed $100 from his mother to purchase a periauger (a type of boat) and began to ferry passengers from Staten Island to Manhattan. His business expanded, and he soon was also the business manager for Thomas Gibbons's steam-powered ferry between New Jersey and New York. Gibbons was challenging the monopoly that had been granted by New York State to Robert

Fulton and Robert Livingston, and which had been sold to Aaron Ogden. Ultimately in the US Supreme Court case *Gibbons v. Ogden*, that monopoly on Hudson River steamboat traffic was broken, as it was interstate travel and could not be regulated by the states. In 1829 Vanderbilt left Gibbons's firm (now owned by Thomas Gibbons's son William) and started expanding his own ferry busines.

Cornelius Vanderbilt, along with former President John Quincy Adams, were aboard a train of the Camden & Amboy Railroad on November 8, 1833, on the first US train crash involving the death of passengers (similar to the first UK crash discussed in Section 6.1). Vanderbilt broke a leg and swore off train travel, a vow he would eventually break.[138]

Vanderbilt acquired other lines on the Hudson from his brother Jacob, and lines on Long Island Sound. On the Hudson, he faced competition from Daniel Drew, who entered the steamship business in 1831. Drew forced Vanderbilt to buy him out. Drew and Vanderbilt became secret partners for thirty years to avoid competition.

Vanderbilt's "The People's Line" served the Hudson from Albany to New York City against an entrenched monopoly, and was eventually bought out itself, so he moved most of his operation to Long Island Sound. He soon extended his transportation networks from water to land, acquiring feeder railroads. One of his strategies was to acquire competing lines to his target, lower rates on his own line, drive down the share price of his competitor, and acquire it at a discount. He used this to acquire the New York, Providence, & Boston RR, called the Stonington. He also took over the Staten Island Ferry in 1838.

He sought to build a canal across Nicaragua, and while that was not to be, he did ferry passengers across Lake Nicaragua and the San Juan River on his Accessory Transit Company, combined with road travel. This served the market of migrants to California, which was exploding with the Gold Rush.

While more than half of migrants chose an overland (through the United States) route to California in this period at a cost of $200 per person, this was only one-third the cost of the water journey around the Cape Horn (taking five to eight months), or going to Central America by boat and through Central America by mule (taking five to eight weeks), as promoted by Vanderbilt.[139]

Vanderbilt lost control of the Accessory Transit Company once American "filibuster" (unauthorized military expeditionary) William Walker gained control of Nicaragua. Vanderbilt helped to drive Walker out in the 1850s, but could not get control of his line back from the subsequent government, and so started a new transportation line in Panama, and split the market with the incumbent operator Pacific Mail. Uncommonly, Panama had a railway before a canal. Ferdinand de Lesseps, who built the Suez Canal, attempted a canal across Panama, and acquired the Panama railway, but the project failed as Panama is difficult to work in and was not amenable to a lockless canal. The Americans later completed the canal.[140]

6.14 Learning about Finance: The Erie War

He who sells what isn't his'n, must buy it back or go to pris'n.

—Daniel Drew (attributed)

From 1863, Cornelius Vanderbilt acquired more railroads. Beginning with the New York & Harlem Railroad, he started accumulating a network to connect New York with Chicago, which by 1870 comprised the New York Central & Hudson Railroad, then valued at $100 million, the most successful of the Northeast lines, which was able to pay dividends through the Panic of 1873. His methods were harsh. He cut off feeder traffic to force a downstream company into his control. He cornered the market in stocks to punish short-selling public officials.

Short-selling is selling stocks you don't own with the hope of buying them back later at a lower price, and is in contrast with going long on a stock, or buying it with the expectation it will rise in price. Clearly, if the price does not fall, a short-seller will lose money. If someone can acquire a majority of the outstanding stock, and an insufficient amount of stock is available for short-sellers to purchase, their losses can be astronomical. In one of his few failures, Vanderbilt got into a bidding war over the Erie Railroad.

The Erie War brought together a cast of some of America's most important financiers and railroad men in a battle over a second-tier railroad.

Daniel Drew (1797–1879) was once Vanderbilt's partner. He became a stockbroker in 1844, and joined the Board of the Erie RR in 1857. He shorted the stock of New York & Harlem RR, and lost a fortune in 1864. After the Erie War with Vanderbilt, in 1870 Fisk and Gould then played Drew, causing him to lose $1.5 million. The Panic of 1873 was no aid, and Drew filed for bankruptcy and died penniless.

James (Diamond Jim) Fisk (1835–1872) made his first fortune dealing in Army contracts in the US Civil War. While he lost this wealth in speculation, after the war he worked for Daniel Drew's brokerage. The famous Black Friday of 1869 resulted from Fisk and Gould's failed attempt to corner the gold market. After the Erie War, Fisk had a scandalous affair with showgirl Josie Mansfield, which ultimately broke off when Mansfield took up with Fisk associate Edward Stokes, who then attempted to blackmail Fisk for his illegal doings. Fisk had no part of that, and Stokes killed Fisk in 1872 (and went to prison for 4 years). Fisk was remembered as a populist loathed by high society.

Jay Gould (1836–1892), about the same age as Fisk, but much younger than Drew, was a surveyor and historian, and then formed a tanning business. It was the Panic of 1857 which moved him to high finance, when he bought out his partners' properties for himself. As with any good, on-the-edge capitalist, this led to some violent disturbances, but Gould profited, and soon used his profits to invest in the Rutland & Washington Railroad. He acquired a reputation of being able to move markets by cornering the market in gold in 1869, culminating in Black Friday, when the price of gold collapsed. After the Erie War, and then being forced out of that railroad, he acquired the Missouri Pacific, Union Pacific, and Western Union, and transit routes in New York City. among other properties. At one time he held 15 percent of all US rail mileage. When crushing the 1886 Great Southwest Railroad Strike, he is reported to have said "I can hire one-half of the working class to kill the other half."

These three—Daniel Drew, one-time partner of Cornelius Vanderbilt, James Fisk, and Jay Gould—illegally issued "watered-down" stock in the Erie Railroad, much of which was purchased by Cornelius Vanderbilt, who was aiming to get control. Watered-down stock entails the issuing of additional stock in a company, increasing the company's par value. Suppose a company had issued 10,000 shares of stock initially in exchange for $10,000 of capital.

The stock would be watered if the company acquired, say, $1,000 of additional real assets in exchange for $2,000 in stock. The value of the other stock would in reality be worth less (since the total company assets were now $11,000 but there were 12,000 shares outstanding, each share was only worth $0.91 instead of $1.00). This procedure is no longer done (as such) since par value is now nominal on companies, and the last court case involving watered stock was in 1956.

The more control Vanderbilt wanted, the more stock Drew, Fisk, and Gould issued, costing Vanderbilt $7 million between 1866 and 1868. While much of that was repaid to Vanderbilt, Gould himself could not retain the Erie. This was due to a strange immigrant.

Lord Gordon-Gordon (a.k.a. Lord Glencairn, Hon. Mr. Herbert Hamilton, George Herbert Gordon, George Gordon, George Hubert Smith, and John Herbert Charles Gordon) migrated from Britain to North America in 1870. He was not a Lord, as many Americans and Canadians later learned, but was merely impersonating a Scottish peer to borrow money to buy land. He landed in Minnesota in 1871, and deposited £20,000 in a local bank, establishing legitimacy. He promised to invest $5 million to help resettle 100 Scottish families on land managed for the Northern Pacific, Railroad. Col. Loomis, the land commissioner for the Northern Pacific spent $45,000 touring with Lord Gordon-Gordon through rural Minnesota.

While still leading Minnesota on, Lord Gordon-Gordon, using letters of introduction from Col. Loomis, soon moved to New York. On the train ride east, he befriended the wife of James Fisk. In 1872, he convinced Gould that he and his European friends already owned some 60,000 shares of the Erie RR and he could help Gould acquire control of the Board of Directors of the Erie Railroad, in exchange for $1 million in stock as part of a "pooling of interest." Upon receiving the stock, Gordon-Gordon promptly sold it, worsening Gould's position. Gould sued, and Gould's friends in New York City Hall (then under the reign of their ally and Erie RR fellow Board member, Boss Tweed) had Gordon-Gordon arrested. But Gordon-Gordon made bail based on the reputation of his purported European friends, and promptly fled to Manitoba, before such information could be confirmed, and before his history caught up with him.

> 600 shares of Erie, some 1,900 of corporations affiliated with Erie, and 4,722 of the Oil Creek and Allegheny Valley Railroad, twenty-one thousand dollar bonds of the Nyack and Northern Railroad, and $160,000 in currency. The careful recipient of these securities and cash presently found an error of forty thousand dollars in the footing of Gould's memorandum and sent word of the shortage. Gould did not think there was such an error, but under the circumstances he would not dispute the point and came back with an additional forty thousand dollars in cash. To a modest request for a memorandum receipt, his lordship replied with exceeding dignity that his word of honor ought to be receipt enough, and handed the bundle back to Gould. Gould took it, went as far as the door, returned, laid it down, and departed in faith that his property was in safe hands. It must have been sheer sport in playing a fish which had taken his hook so greedily that led Gordon to demand that Gould separate himself from the old directorate. On March 9 Gould delivered to him his resignation as director and president of the Erie Railroad Company, to take effect upon the appointment of his successor. The great covenant was complete.[141]

Gordon-Gordon offered to buy large parts of Manitoba from the government, which appealed to locals there, but soon his American enemies found out and sent a posse of bounty-hunters to bring him back to the United States. This posse included two future governors of Minnesota and three future members of Congress. They successfully kidnapped Gordon-Gordon but were stopped by the Mounties in Winnipeg and put in prison themselves. The Governor of Minnesota put the state militia on alert and President Grant authorized sending an army into Manitoba, and a major international incident between Canada and the United States was threatened. To avoid conflict, Canada released the posse, but Gordon-Gordon was already freed. Then his European enemies asked for his extradition on similar charges of swindling a jeweler of £25,000, and Canada agreed. Making his escape again, he was again arrested, and again released on bail. But before his extradition in 1874, he held a party, gave gifts to his local friends, and then shot himself.

Gould's loss of $1 million in stock may have been sufficient to cost him the Erie Railroad.

The Erie War was, of course, not the only instance of shenanigans in rail finance. Two other examples follow. The United Kingdom had a scandal involving the Hammersmith and City line, wherein the Chairman of the Board and another director acquired land on the proposed route before the company did, to earn great profits.[142] The Union Pacific (UP) Railway was famous for the Crédit Mobilier Scandal, in which the UP officials ensured that the Crédit Mobilier of America company, in which they held an interest, was issued construction contracts to great profit. The ringleader was Congressman Oakes Ames of Massachusetts, who held off Congressional inquiry by buying off public officials with discounted shares in Crédit Mobilier. It is not clear how many similar cases were not caught.

6.15 Comments by Social Critics

> We do not ride on the railroad; it rides upon us
>
> —Henry David Thoreau, *Walden*

The railroad has been an important symbol in literature, and a subject of social critics. Charles Dickens's writings fall in the same period as the rise of the railway and he had much to say about its emergence and growth, and his views changed over time.[143] In *Dombey and Son*, Dickens portrayed the railroad as symbol of power and ruthlessness. In the *Uncommercial Traveler* he said:

> I left Dullborough in the days when there were no railroads in the land. I left it in a stage-coach. I was cavalierly shunted back into Dullborough the other day by train, and the first discovery I made was that the station had swallowed up the playing field. It was gone. The beautiful hawthorn-trees, the hedge, the turf, and all those buttercups and daisies, had given place to the stoniest of jolting roads: while, beyond the Station, an ugly dark monster of a tunnel kept its jaws open, as if it had swallowed them and were ravenous for more destruction.

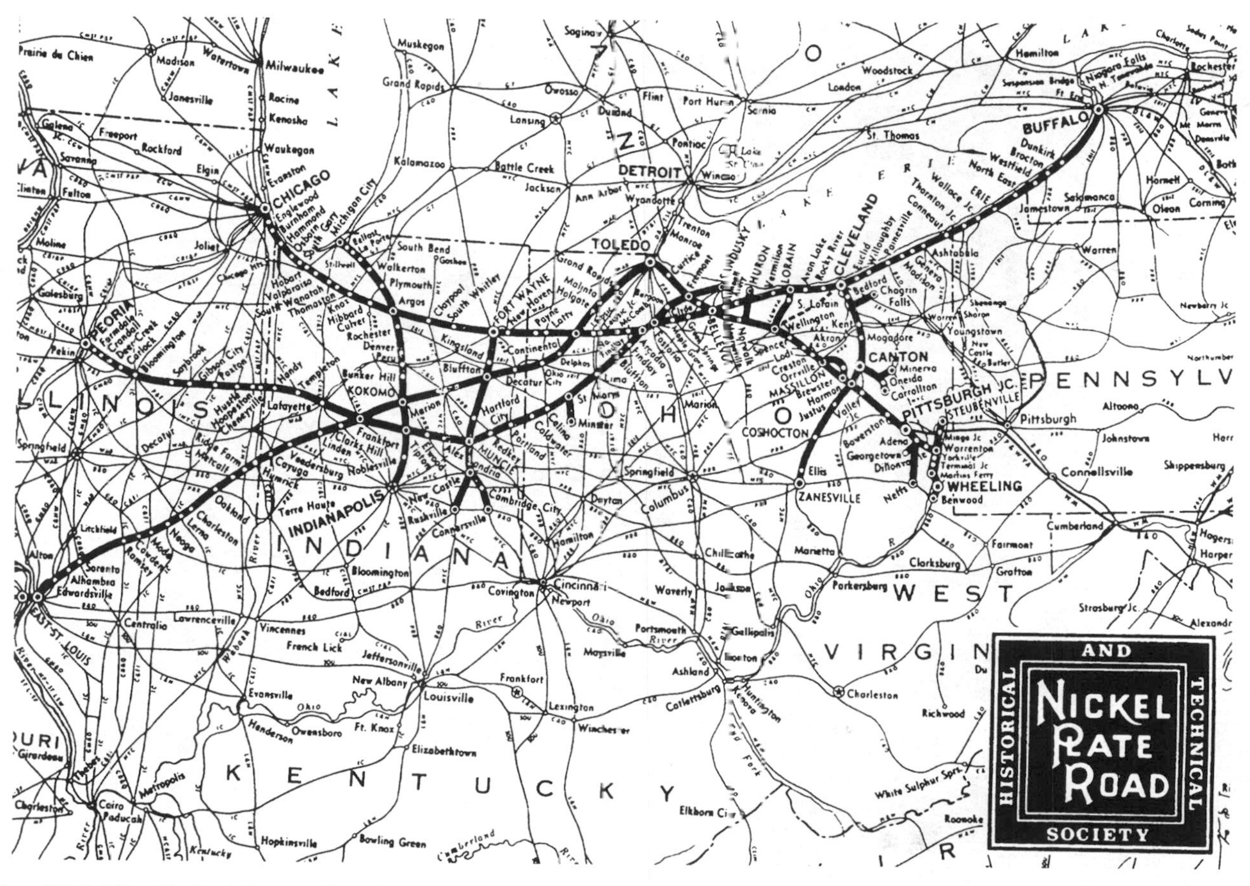

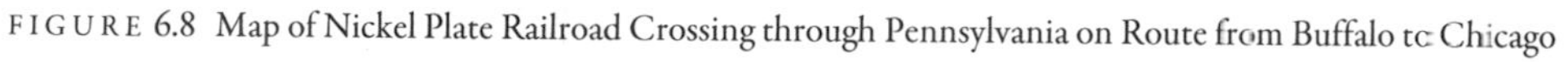
FIGURE 6.8 Map of Nickel Plate Railroad Crossing through Pennsylvania on Route from Buffalo to Chicago

Nathaniel Hawthorne indicated considerable resentment of railroads in his *American Notebook*. Thoreau's *Walden* was kinder. Commenting on hearing a train, he said: "It seems as if the earth has now got a race worthy to inhabit it."

Finally, the press also had fun. *Indiana and Indianans*[144] said:

> We renew an idea which we have propounded before, but which has been lost sight of by Railroad managers and by our citizens—namely that an unnecessary number of trains are run on our railroads to accommodate travelers from distant States, and that on train daily would be all the passenger business that is now done on any of the roads. This extravagance is wrong and ought not to be continued. I have no doubt that to change it would add 8 to 10 percent to the dividends to suit these railroads. They should arrange their running and time to suit the citizens of Indiana, and not those of Massachusetts or Texas. All trains should arrive in this city between 11 and 12 A.M., and leave from it at 1 or 2 o'clock P.M.
>
> The Roads of New York sought to pass through a portion of Pennsylvania without conferring any advantage upon the State—in fact, drawing away trade from her own metropolis and conferring no local advantage even upon Erie, unless there was a break of gauge. If the foreign companies succeed in their schemes, there is one great monopoly of railroad interest from Albany to Chicago, rich and powerful enough to buy out or trample upon any rival interest, combining a moneyed power which is without a parallel in the history of the country. We propose making at Erie a break in this interest, and it is this in reality which the railroad fears.

The reference is to Erie, Pennsylvania, and its situation is shown in the partial map of the Nickel Plate Railroad (New York, Chicago & St. Louis Railroad Company) in Figure 6.8. Interestingly, Erie is today's location of much of General Electric's transportation equipment manufacturing, suggesting perhaps that the railroad did offer some advantage.

The anxiety associated with the railroad finds parallels today with the concern of the all-consuming "sprawling" city and the shrinking countryside.

The story of the rail modes continues in Chapter 13.

I often hear now-a-days, the automobile instigated good roads; that the automobile is the parent of good roads. Well, the truth is, the bicycle is the father of the good roads movement in this country.

—HORATIO EARLE,

"Father of Good Roads" (1929)[145]

7 Good Roads, Bicycle Mechanics, and Horseless Carriages

7.1 Bicycles as Building Blocks

THE *VELOCIPEDE* DEVELOPED in France in the 1850s and 1860s[146] might be considered the first true bicycle. The front high-wheeler, or ordinary, evolved into the more modern-looking safety bicycle with two equal-sized wheels, making bicycling more popular.[147] In addition to being a significant mode in its own right, and driving ladies' fashion toward the bloomer, the bicycle enabled the automobile.

The bicycle gave the automobile a head start on pneumatic tires, ball bearings, paved roads, developing a set of manufacturers and mechanics who could adapt their skills to the more complex automobile, and "democratizing both touring and personal transport in cities."[148]

One of those manufacturers was Colonel Albert Pope (1843–1909), who acquired a number of European patents. His Columbia bicycles were manufactured in Hartford and distributed nationally (via the US's new railroad system and marketed via new popular magazines distributed by the post office) beginning in the late 1870s. Though prices dropped to $100 in nominal terms, they were still quite expensive. Pope convened thirty one independent cycling clubs and helped organize the League of American Wheelmen, which advocated for "good roads" on which to ride bicycles.

Pope's firms in 1897 exploited their bicycle knowledge of tubular steel to build the first mass-produced automobile, the Columbia Electric Phaeton,[149] Mark III, a relatively light carriage (800 kg [1800 lbs] of which 385 kg [850 lbs] were battery), and also adapted metal spoked wheels and pneumatic tires from bicycles.

The Duryea brothers, Charles (1861–1938) and Frank (1869–1967), were also bicycle mechanics interested in the automobile, developing the first US gasoline engine to run well in Springfield, Massachusetts. Frank won the first auto race with one of their cars, and Duryeas were the first commercially produced motor vehicles in the United States, turning out thirteen in 1896. Ironically, perhaps, the first auto crash involved a Duryea in New York City (driven by a Henry Wells, not the founder of Wells Fargo) hitting a Columbia bicycle (ridden by Evelyn Thomas) on May 31, 1896.

The most well-known bicycle mechanics to turn their attention elsewhere were of course the Wright Brothers (see Chapter 12).

This chapter examines the proto automobile/truck-highway system around the turn of the twentieth century. The bicycle was not the only antecedent to the modern automobile; the development of gasoline, steam, and electric engines (see 7.2) are quite important. But the emergence of good public roads (see 7.4) first came about due to avid bicyclists seeking a network, who were followed by even more avid auto users. By the turn of the twentieth century, there was a built-out, but mostly unpaved system of roads that were building blocks for the auto/truck/highway system that followed.

7.2 From Horse to Horseless

> Young man, that's the thing; you have it. Keep at it! Electric cars must keep near to power stations. The storage battery is too heavy. Steam cars won't do either, for they have to have a boiler and a fire. Your car is self-contained—carries its own power plant—no fire, no boiler, no smoke, and no steam. You have the thing. Keep at it.
> Thomas Edison to Henry Ford in 1896[150]

The automobile had long been forecast. Chapter 1 noted some of the early (often amusing) attempts with steam cars. During the late nineteenth century, the steam engine had been adapted from railroads to farm equipment, including a self-propelled steam tractor. Yet the gestation period for the automobile lasted over a century. In the 1890s, Indianapolis had one horse for every 14 people, while Kansas City one horse for every 7.4 people.[151] The peak year for horses in England was 1901, which saw 3.25 million, more than a century after the onset of the industrial revolution, and 75 years after the iron horse.[152] By 1924 there were fewer than 2 million. The cost of maintaining a horse, combined with its lesser efficiency, doomed this mode to be replaced by motorization.

The electric cars, too, had early prototypes:

- In 1835 Thomas Davenport of Vermont built, the first rotary electric motor and found many uses, including a toy train, later pulling 31–36 kg (70–80) carriages at 5 km/h (3 mph).
- In the late 1830s Robert Davidson of Scotland built a carriage powered by batteries and a motor, and later an electric coach, the Galvani, running on rail tracks.
- In 1851, Charles Page built an electric locomotive reaching a speed of 30 km/h (19 mph) on a run outside Washington, DC.

All these were important, but false, starts. The electric car was technically feasible but economically impractical given the relative costs of energy generation, transmission, and storage.

The electric grid, developed by Edison and others, was necessary for practical electrical transportation. Electricity was first applied in transportation to the streetcar. The year 1879 saw Siemens and Halske build a 2.6 km (1.6 mile) line in Berlin. Battery trolleys were tested in the early 1880s in places like the Leland Avenue Railway in Philadelphia. By 1887, a New York financial syndicate funded Frank Julian Sprague (and his company Sprague Electric Railroad and Motor Company) to build a 19.2 km (12 mile) line in Richmond, Virginia, for somewhat more than the $110,000 that was budgeted (a line that would be on the order of $1 billion in modern terms). Trolleys exploded across US cities, as is detailed in Chapter 8. The electric streetcar and other electric railways transmitted power to the vehicle via a cable. While tracks were not necessary (as later demonstrated by trackless, rubber-tired, electric trolley buses), batteries would be the key to enable electric transportation off the grid.

Gaston Planté developed the first workable storage batteries in 1859, which were initially used for telegraphy and for scientific instruments. Camille Faure later thought to use it for load leveling of electric dynamos, and then filed a patent for a battery-powered vehicle. Faure's battery comprised pure lead plates coated with a paste of sulfuric acid and lead. Technical problems emerged when the paste separated. In 1888 Faure was issued a US patent for recharging a battery when going downhill.

1893 World's Columbian Exposition displayed six automobiles. The only one from the United States, by William Morrison of Iowa, was electric. Yet the energy density of the battery remained the principal constraint on the electric vehicle's market share. By the turn of the century, range and the energy per unit weight of battery compared with gasoline engines were already defined as key weaknesses by the best engineering talent of the time.

Battery-powered vehicles have more limited range (distance before recharging/refueling) than gasoline-powered vehicles due to energy density. The limits to battery technology result from battery weight. Each additional battery reduces the effectiveness of all the others, as they must spend some of their stored energy moving around other batteries instead of the rest of the car and passenger. Diminishing returns set in quickly.

While longer distance touring was a relatively small market, people consider the extreme use for the vehicle they buy, not the average. A vehicle must be usable in a maximal number of conditions. People imagined traveling more than 100 miles in a day, something an electric could not do. Other problems were the underdeveloped electric grid (as late as 1900, only 5 percent of factory power was electric) and lack of charging stations, especially at homes. However, public lighting was a complement to the use of streets (and thus automobiles) at night.[153]

At home, electrification in the late 1800s and early 1900s was limited. The first rationale was to replace gas lighting. Hence sockets were the standard outlet, and appliances had to plug into ceiling lighting sockets rather than the wall outlet that is common today.

Electricity's famous "War of the Standards" was fought pitting Westinghouse and Tesla's alternating current (AC) against Edison's direct current (DC) in the 1890s. While AC eventually won for electric distribution, DC was simpler for charging cars. The 110 V DC system could be plugged into a rheostat (to adjust voltage) and then used to connect to the car battery. Overcharging was a risk, and so it was not quite a turnkey operation. In contrast, AC was more difficult, coupling an AC motor with a DC generator, along with a rheostat to match voltage. This was finally made available by Westinghouse as a consumer good in 1900. In any case, an enormous amount of power was lost as heat in this process. The plug to connect the car battery to the wall was not developed until 1901; prior batteries had to be removed from vehicles, no trivial task.

Electric utilities implemented time of day pricing to better balance peak loads.[154] Some electric utilities (e.g., Rockford, Illinois, which was DC based) encouraged electric vehicles (EVs) and helped charge and maintain them at central stations, and did locally have an effect on EV sales. Most electric utilities saw these customers as nuisances rather than a source of business. Range (c. 1901) was about 4 hours, so charging was a frequent event. Fast charging (one or two hours) deteriorated the batteries. The industry magazine *Electric World* predicted,[155] "the time will come—and not very far in the future—when electric automobiles will furnish one of the most important sources of income to the electric lighting companies."

A third strategy, a charging hydrant, dubbed an "electrant," located every few blocks would allow travelers to pull over and pay for a metered amount of electricity. This, however, was never deployed. These ideas have been regurgitated in the 1990s and 2000s as people seek to solve the same problems with electrics. Again, the number of charging stations remains quite limited, as no one wants to invest in a network of charging stations until there are many plug-in electrics requiring charges, and few will buy plug-in electrics if the cost and convenience do not match the technological competitors.

Another concept, developed by L. R. Wallis in 1900 was to have a parent battery company, from which batteries would be leased, and then swapped out when needing recharging for already charged batteries. This idea was revived with the company Better Place in the 2000s, which hoped to develop a network of battery exchange centers, and deployed some in Israel, its home country, to service specially designed Renault vehicles.

Similarly, electric garages, modeled on livery stables (for horses) were established to limit the owner's need to deal with the difficulties of charging and maintaining the car.[156]

By 1899 there were a variety of automobile types on the road at different sizes, aimed at replacing markets for buggies, wagons, and carriages. Body styles were kept from horse-powered vehicles to avoid too much shock to potential buyers. Many different names were applied: Runabouts for smaller open-air cars, Broughams for larger enclosed vehicles, among others.

In 1900, the United States produced 4,192 automobiles, 28 percent of which were electric. The dominant competing technologies at the time were steam, gasoline, and electric. Observers believed each would find niches, a "sphere of action" they would dominate, electrics for the ladies, who needed to travel in town, gasoline for the longer distance trips, and so on. The automobile was not yet mass produced, and it was emerging in an environment where horse-drawn coaches, carriages, and cabs dominated the pre-automobile personal transportation scene. An emerging middle class of urban professionals, managers, and white-collar workers formed a market for a new type of transportation.

The primary advantages of electric at the time had to do with user interface. Charles Kettering had yet to develop the self-starter, so gasoline engines required the user to get out and crank. This was a non-starter for upper-income women, who thus preferred electric vehicles. EVs were often marketed to women, but this feminizing of the product may have discouraged men.

Mass-produced automobiles, like the Oldsmobile, sold 425 vehicles in 1900. The market was still minuscule, but growing exponentially. Detroit in 1900 was much like Silicon

Valley in the 1970s, with its HomeBrew Computing Club that begat Apple Computer and Microsoft. In 1903, the first year of Ford's Model A ($750), Ford sold 1,708 vehicles. Olds sold 4,000. By 1905 Ford rolled out a Model B ($2,000), C ($950), F ($1,200), and a Doctor's Car ($850). In 1906, it introduced a model N ($600) and K ($2800). The company was trying to occupy a variety of market niches. But Ford soon developed what might be dubbed Henry's Law, "Thou shalt make only one model," and that model was the Model T, and for Ford and his customers, it was good.[157] By 1912, Model T sales reached 82,388, in 1914: 200,000, in 1915: 400,000. Though the Model T was originally available in other colors, to drive down costs, black became standard. Despite Edison's encouragement of Ford's gasoline-powered car, as noted in the opening quote of this section, later Edison and Ford worked together in a failed attempt to bring about an electric car that was competitive with gasoline-powered vehicles.

In 1905, the 1,200 electrics sold were fewer than 10 percent of all vehicle sales. Ultimately EVs fell further and further behind as economies of scale drove down the relative cost of its competitors, attracting a greater and greater share of consumers. Like internal combustion engines (ICEs), EVs were rising in sales, but at a much more modest pace, growing to only 6,000 vehicles in 1912.[158]

Electric trucks (ETs) had a better time than electric cars, as trucks had the advantage in many cases of fleet operation, and thus centralized maintenance and management. Electric trucks lasted longer, but that can be a disadvantage in a rapidly advancing field, as they would become functionally obsolete as newer, more modern gasoline- or diesel-powered trucks took the road and made the ETs look and feel antiquated in comparison. Compare a cell phone from today with one from five or ten years ago to get a sense of the rapidity of change.

Edison and others experimented with alternative battery materials, as lead-acid batteries were heavy, and had a low energy density. Other materials were technically better, but more expensive. Finding something less expensive with a higher energy density would be the key to solving the range problem.

Because of the difficulty consumers had with charging, Salom and Morris of the Electric Storage Battery (ESB) Company proposed a fleet of rental cars (an antecedent to car sharing), where professionals could do charging and maintenance. Individuals would still rent or lease a particular car. However, this failed to get critical mass, and required picking up the car, rather than storing it at home. In the end this became a fleet of cabs, where instead of recharging batteries in the vehicle, they would be swapped in and out, and charged (slowly) out of the vehicle.

Owner of New York's Metropolitan Street Railway Company, Henry Melville Whitney, consolidated the electric vehicle industry beginning in 1898, acquiring ESB, combining with Pope, and absorbing the Riker company, with the aim of establishing a fleet of 15,000 electric cabs to serve urban America.[159] Whitney also purchased the Selden patent on gasoline-powered cars. This "Lead Cab Trust" began to fail when the batteries, designed for smoother running streetcars or stationary operations, did not do well on bumpy road surfaces and the frequent charging and discharging use of cab service, rather than the more sedate private ownership. Batteries deteriorated with use more than simply age.

The Edison Storage Battery Company aimed to develop a nickel-iron alkaline battery to replace the lead-acid battery. Edison's competitor, ESB, tried to perfect the lead acid battery.

The New York Electric Vehicle Transportation Company, (EVTC) part of EVC (the Lead Cab Trust) was probably the largest consumer of such batteries. It also developed its own central station and substation, and started running electric buses on Fifth Avenue as well as other routes. Other subsidiaries of the Trust fared less well, the New England and Illinois branches of EVTC folded in 1901. Edison hyped his battery for years, but it was not widely used once it came to market, as the cost-energy density trade-off never worked favorably.

The internal combustion engine (ICE) was patented in 1860 by Belgian engineer Jean Joseph Étienne Lenoir, who applied a coal-gas and air burning version to his three-wheeled *Hippomobile*. Nikolaus Otto developed his engine in the 1870s and Karl Benz used Otto's engine to power a 600-watt (0.8 horsepower) three-wheel carriage in 1885.

At the 1876 Centennial Exhibition in Philadelphia, George B. Selden saw George Brayton's internal combustion engine and applied for a patent on an ICE powered automobile in 1879. The Selden patent, issued sixteen years after filing, in 1895, was used as a hammer against competing markers of gasoline powered autos, who were required to license his patent. An Association of Licensed Automobile Manufacturers, who were legally authorized to produce gasoline-powered cars, was established. (And licensees did not want unlicensed autos about, as they would drive down prices.)

Selden, widely criticized in histories of the automobile, has his defenders, example Byers (1940), who argue that Selden in fact did develop the gasoline-powered automobile in the United States before anyone else. To quote the decision sustaining the patent against Henry Ford and others who did not want to pay the license "There was no such [horseless carriage] industry; the art existed only in talk and hope." A later court ruling did not overturn the patent, but ruled that Ford's vehicle was not in violation. This came as the patent was winding down, approaching its sixteenth year (in 1911), so had little effect.

Charles Kettering developed the electric starter, which temporarily overloaded the motor. Interestingly, Kettering modeled his innovation on the self-starter with his work on motorizing the cash register when he was an engineer at National Cash Register in Dayton, Ohio. Kettering later founded Dayton Engineering Laboratories Company (DELCO), soon acquired by General Motors. The self-starter eliminated the disease of "Ford's fracture," a broken arm resulting from cranking accidents.[160]

After Kettering, the automobile become an electric system in miniature: its generator (with the battery) was the central station, which distributed current through a network to uses like starting the car, but also for headlights, and later radios and other purposes. Surprisingly, battery makers boomed not from selling batteries to makers of EVs but from selling to makers of gasoline-powered cars containing an electric self-starter.[161]

Endosymbiosis in biology refers to the idea that organelles of eukaryotic cells (like mitochondria and chlorplasts) were originally free-living micro-organisms that combined symbiotically (to mutual benefit). The internal combustion engine adopted the battery as a self-starter, and is a technological version of this biological process. Hybrid vehicles, which ramp up the battery so that the vehicle can travel on either electric or gasoline power, are another version of this.

Ferdinand Porsche built a hybrid in 1901. In 1905 Fischer Motor Vehicle Company developed an early gas-electric hybrid omnibus with an electric starter; the battery helped the gasoline engine in overload conditions.[162] There were other early attempts, but nothing persisted in the marketplace. It wasn't until 1969 that General Motors began experiments with hybrids. Other automakers similarly experimented. In 1997 the Toyota Prius went on sale in Japan, and in 2000 in the United States, while the Honda Insight started sales in the United States in 1999. Electric vehicles would see a revival in the early 2000s (see Section 27.5.2).

7.3 Road Trips and Races

> The Long Island Motor Parkway will supply an uninterrupted route across the Island that, owing to its proximity to the metropolis, is destined to be the home of millions with business and social interests in New York City. Someday the state will supply such motorways.
>
> — *Automobile* magazine (1908)

Cars had several key properties: they could go farther than other private modes of transport (bikes and horses), and they could go faster. The way to demonstrate these properties was through various stunts. Long road trips, especially on pre-paved rural American byways, was one such class of venture. Road and off-road races were another. Some notable early American road trips are listed below:

- 1897: Alexander Winton, a bicycle dealer and automaker, drove from Cleveland to New York in his own creation.
- 1899: John and Louise Hitchcock Davis tried to go from San Francisco to New York in a Duryea, but only made it to Chicago after three months and gave up.
- 1901: Alexander Winton tried to go from San Francisco to Chicago, but gave up in Nevada.
- 1901: Roy Chapin took an Oldsmobile from Detroit to New York in just over a week.
- 1903: Horatio Nelson Jackson and Sewall K. Crocker from Burlington, Vermont, traveled from San Francisco to New York in 63 days, completing the first successful transcontinental car trip. This has been chronicled in the book and documentary *Horatio's Drive*.[163]
- 1915: Transcontinental film convoy: a four-month trip to the Panama-Pacific International Exposition to complete a film (The amount of film taken was three miles long), sponsored by the Lincoln Highway Association.
- 1919: Motor Transport Corps convoy: a truck train from Washington, D.C., to Oakland, California, which encountered 230 incidents (breakdowns of various sorts), and broke and repaired 88 wooden bridges. Lt. Col. Dwight David Eisenhower participated. The speed of travel was 5.67 mph (9.1 km/h) for 573.5 hours. This illustrated the poor quality of US roads.
- 1920: Motor Transport Corps convoy: a truck train from Washington to San Diego on the Bankhead Highway, which departed June 14, 1920, and arrived October 2. This illustrated the especially poor quality of Southern US roads.

After North America was conquered, promoting round-the-world races became a popular stunt. This was generally infeasible (and not just because of several oceans in the way).

Early car demonstrations had races for speed as well as endurance. In 1896 at a Providence Horseless Carriage race, Riker Electric Motors won the prize with an electric that covered a mile in 2:13. Second place went to an Electrobat running the fastest five miles in 11:27.[164] Racing was important for a variety of reasons. It was a public test of the ability of the vehicle. It was a thorough technological challenge; if a car could survive the stresses of the race, it would do better in real driving conditions than one which could not.

Henry Ford took his car with the Detroit Automobile Company to a 1901 race at Grosse Pointe, and driving it himself, won the race against experienced racer Alexander Winton of Cleveland. Barney Oldfield, inspired by the 1900 New York Auto Show at the old Madison Square Garden, which he attended with his Indianapolis friend Carl Fisher,[165] became the first famous race driver.

Carl Graham Fisher (1874–1939) was an auto dealer of multiple–brands (Packard, Oldsmobile, REO), but he was more than that. Fisher was a promoter and racer (later in life he would help develop Miami Beach). In a throwback to the race of the *Tom Thumb*, he would race his car against horses (for money), and win. Car races were increasingly popular in the 1910s, and lethal to both drivers and fans. Oldfield wrecked his car, resulting in both injuries to him and death to a fan. The same day, a car of Fisher's also killed an onlooker.[166] Fisher co-founded Prest-O-Lite, using compressed acetylene for headlights. These soon became standard from the automakers, making Fisher, his friend James Allison (co-founder of what became Allison Transmissions, later part of General Motors), and inventor Percy Avery wealthy. Prest-O-Lite was purchased by Union Carbide in 1917.

In 1908, with Allison and others, Fisher incorporated the Indianapolis Motor Speedway. Its tar-macadam surface was dangerous under the high speeds of racing, and the American Automobile Association threatened to withdraw its sanctioning of races. Still, by its third race, the racetrack had claimed three lives and many injuries. The tar surface was replaced with bricks and concrete walls, giving the track its Brickyard nickname. Within two years, the first Indianapolis 500 was held.

The Long Island Motor Parkway, also called the Vanderbilt Parkway, was a private road built by William Kissam Vanderbilt, Commodore Vanderbilt's great grandson. It was built as a turnpike, and was the first limited-access roadway, running from Queens to Lake Ronkonkoma on Long Island, opening in part in 1908 and completed in 1911. The route was intended to be used for road races—the Vanderbilt Cup (since using existing roads for races was deadly), which was run from 1908 to 1910. As the first limited access route, it was built to standards that would soon be considered inferior. Robert Moses (see Section 22.3.4) constructed the Northern State Parkway in a competing right-of-way, and the state purchased the parkway in 1938 in lieu of back taxes, and it was shuttered. Parts have been converted to bicycle trails, electric transmission rights-of-way, and other parkways.

7.4 Object Lessons

The Good Roads movement arose out of the efforts of the League of American Wheelmen, an association of state bicycle clubs formed by Hartford-based manufacturer Colonel Albert Pope.[167] The League was the AAA (or Cyclopath) of its day, developing maps and recommending cycling routes for the recreational cyclist, with user ratings. It had over 25,000 members by 1891.[168] Pope used his funds to help the League establish the magazine *Good Roads*, which advocated for pavements and endowed a highway engineering course at the Massachusetts Institute of Technology (just as Studebaker later endowed traffic safety programs at Harvard and Yale). The League opposed the corvée (statute labor) system of working out the road tax (see Section 3.3), instead demanding professionally built, government funded good roads. It encouraged allies from all areas, including farmers, engineers, manufacturers, and merchants.

Farmers constituted an important part of the Good Roads coalition, and they exerted pressure to improve the rural road system once they began to purchase and use automobiles and trucks. Cars and roads sharply imprinted rural activities as a new spatial arrangement of activities (market, recreation, religious, school, etc.) emerged. For instance, small, local hamlets withered as commercial and social activities increased at larger centers.[169]

Rural residents claimed a right to an improved road system. Necessary governmental institutions were in place (township, county, and, later, state agencies) and technology (knowledge of how to construct roads and bridges) was available and/or could be evolved. Authority and power were in place, too, lending political interest and support; at the time, rural interests dominated the state and federal governments. However, money was a problem.

Pope went on to start a lobbying organization, the National League for Good Roads. These efforts were soon followed at the federal level with the establishment in 1893 of an Office of Road Inquiry (ORI) in the US Department of Agriculture, led by general Roy Stone, a Civil War veteran. Note that there was not yet a Department of Commerce (where the later Bureau of Public Roads would long be housed) or a Department of Transportation (where the Federal Highway Administration would find itself), and the role of roads was seen as largely rural to support agriculture and intercity commerce. The ORI funded demonstration projects, dubbed "object-lesson" roads, to illustrate the benefits of improvements. Ultimately 410 such were object-lesson roads were constructed.[170]

The best "object-lesson" road was privately funded, by Coleman duPont, heir to the eponymous chemical company. An MIT-trained engineer, in 1911 he designed and funded a "Grand Boulevard" (much of which is now US 113) through his home state of Delaware. Notably it included 61 m (200 ft) rights-of-way, passing lanes, bypasses around towns, and curves designed for the motor age. He donated the road to the state.

Connecticut, for example, by the late 1890s was paving 130 km (80 miles) of road per year,[171] and by 1905 had designated a network of 1,600 km (1,000 miles) of state trunk highways. These new good roads made the automobile, which was much more sensitive to the vagaries of off-road conditions, feasible. Pope, the Duryea brothers, and others advocated for more good roads to help promote the motor vehicle.

One issue debated in the early 1900s was the emphasis to be given to business roads versus touring roads. Before the Model-T, autos were expensive and were used mainly by the upper classes. Touring was an important leisure time activity for the idle rich and those aspiring to that job, so they asked for touring roads, as did the auto manufacturers who thought touring roads would increase the market for vehicles. As less expensive vehicles came along, there was a great touring boom. Along with the development of National Parks and as interest in the great outdoors increased, there was a strong basis for a touring road emphasis. The debate lasted about ten years. The pragmatic business road concept won the day, although the business promoted was often tourism.[172]

Not only bicyclists and automakers were advocates for good roads. Perhaps the most famous advocate was the son of a road overseer for Washington Township, Missouri. Harry S Truman (1884–1972) assumed his father's role upon the latter's death in 1914 and served for another six months until politics turned him out of office. Harry Truman served as a Jackson County, Missouri, judge (an administrative rather than judicial position, like a county commissioner) between 1922 and 1924, and again between 1927 and 1934. He ran on a platform of good roads and good management. In his Jackson County interregnum, he served as a salesman for the Kansas City Auto Club and as president of the National Old Trails Road Association, one of many Road Associations (see Section 15.2), like the Lincoln Highway Association, advocating for good roads. He is credited with bringing Jackson County out of the mud, and establishing one of the best county road systems in the United States.[173] Yet despite his experiences with road-building, it was not under President Truman that the Interstate Act was fully funded. While the plan was established by BPR engineers during his tenure as vice president, and ratified by Congress in the Federal-Aid Highway Act of 1944, almost no funds came with it. His presidency saw limited increases in highway funding, constrained both by the Korean War and by the fact that states could not spend the money that was already being raised.

7.5 Discussion

The bicycle was one of many important building blocks in the automobile highway system. As a technology, it trained engineers who went on to more complex machinery. The metal tubing was an important input, as were the tires. The demands from bicyclists for good roads saw the creation of an infrastructure that was critical for the emergence of the automobile. The irony is that the combination of good roads with adequate automobiles made life difficult for bicyclists. Not just the on-road part, where there were natural conflicts between large motor vehicles and the unprotected bicyclists, but also the expanded distances enabled by the automobile, making not having an auto that much more difficult.

The engine was another critical aspect. The same debates at the turn of the twentieth century about batteries versus petroleum are replaying at the beginning of the twenty-first century. The same trade-offs are apparent. Although both technologies have advanced, nothing so much better has emerged to obviate the question.

The roads story continues in Chapter 15.

After being tried for eighteen months a street railroad in Richmond, Va., which has been operated by electricity has been declared a failure, and the company will go back to horse or mule power.

—*The Railway Age*, 1889

8 Transit

8.1 Omnes Omnibus

FROM 1662 TO 1675, mathematician Blaise Pascal secured a royal monopoly from King Louis XIV to operate Carosses à Cinq Sous (five-penny coaches), seven horse-drawn vehicles carrying six to eight passengers, along five regular routes in Paris.[174] Pascal's routes, serving the wealthy classes but not soldiers or peasants, did not last much more than a decade. They were expensive; the working class would take in roughly 8 sous per day, so one ride was more than half the daily income.

Fixed-route transit services followed upon the more common taxi services. In 1617, Nicolas Sauvage established a business of hired cars serving Hotel de St. Fiacre. As the location of the first taxi-stand, St. Fiacre became the patron saint of taxis. Early services included the Oystermouth Railway in Wales, which began passenger service in 1807 (and is also the first fare-paying passenger rail service) and Mexico City, which had public transit as early as 1820.[175]

In the 1820s, Stanislas Baudry, proprietor of a local public baths, opened a transport service in Nantes to serve his customers. In the event, people got off along the route, and fixed-route transit was reborn. One of the first bus stops there was in front of a hatter's shop owned by a man named Omnés. The Latin word *omnibus* means "for all," and the slogan "Omnes Omnibus" (all for all) was used. The name was eventually clipped to omnibus, and then bus. Baudry soon expanded to Paris by 1828, taking the name "Omnes Omnibus" with him. Competition soon followed. Ultimately, motorized omnibuses in France began to be called "autobus." Sadly, Baudry committed suicide in 1830.

By 1829, the Omnibus idea expanded to London, where George Shillibeer, a coach maker, started service.

FIGURE 8.1 Streetcars in United States History

The first horse-drawn omnibus in the United States, a 12 seater named *Accommodation*, plied New York City's Broadway from Battery to Bleecker Street for 24 US cents[176] starting in 1827. It was organized by Abraham Brower, a stagecoach operator, who also helped start the New York Fire Department. Others soon followed, and then a horse-drawn street railroad was launched in 1832 from Fourth Avenue at Fourteenth Street to the Bowery at Prince Street. By 1850, 593 omnibuses ran on 27 routes in New York, though no new horsecar tracks were laid.[177] Figure 8.1 illustrates the history of streetcars in the United States. In the upper left corner of the figure is a tram (or horse) car (using rope or cable haulage), the John Mason, placed in service in New York City in 1832. Note both its similarity to road coaches and how later designs evolved. Omnibuses could be found in other cities as well. In Washington, D.C., an omnibus ran from Georgetown to the Navy Yard as early as 1830, following an earlier streetcar.[178]

Horse trams (on rails) had been used in special circumstances, such as mining, for a long period of time, and they appeared in cities as passenger traffic grew enough to warrant investment in rails. These trams were quite competitive with the omnibus because of higher productivity—larger vehicles and increased speeds. Because rail investment was required, firms provided the service. New York had over thirty tram roads by 1860.

8.2 Going Underground[179]

Prior to the advent of the steam railway, London was a metropolis of just over 1 million people.[180] It was well served by both canals and turnpikes connecting to other parts of Great Britain.[181] Internally, there were omnibus services. The London & Greenwich Railway was the first of many railways to reach London, with the first section opening in 1836 and being completed in 1838, making it possible to reach Greenwich in twelve minutes instead of the hour required by horse-drawn omnibus or steamboat. Famously built on a viaduct, the route was initially paralleled by a tree-lined boulevard that operated as a toll road, serving those unwilling to pay rail fares. However, the toll road was disbanded when the viaduct was widened to enable more frequent services to the densely populated urban core, ultimately growing from two tracks to eleven.[182]

Soon many other railways sought to connect to London. To avoid disruption in the core, a Royal Commission on Railway Termini, appointed in 1846, drew a box around central London and decreed that no line shall enter the cordon. (This box resembles the congestion charging zone adopted in the early twenty first century, which aimed to reduce cars, rather than prohibit trains.) The result was railway terminals locating on the edges of the central region. London, like many cities, has no unified railway station (see Section 8.4), as the North, South, East, and West lines have no common intersection.[183] The problem is worse, though, in London, as even lines from the north run by different organizations would be build adjacent (St. Pancras/Kings Cross), or nearly adjacent (Euston), stations without convenient interchange. Later (between 1858–1860) some penetrations of the box were permitted by Parliament, but most of the City of London (the original walled city where the financial district still lies) remained untouched.[184] While preventing railways from severing the most densely populated part of the city, which would have been expensive for both the railways and the city, it created a need for a connection between the termini to allow transfers. The Metropolitan Railway, a private concern like all railways of the era but with some support from the Corporation of the City of London, was approved by Parliament in 1854. It aimed to connect the northern termini (Paddington, Euston, St. Pancras, King's Cross, and Farringdon, which was later added to the plan) to ease movement for through travelers.

The trends in the City of London were quite different from the rest of London. The City of London has seen a long trend of depopulation from 1851 (prior to the first Underground line) and for many years saw increasing employment, lending support to the notion that the railways, especially the Underground, enabled decentralization of residences and concentration of employment.

The Metropolitan Railway opened in January 1863 and was extremely successful. Clearly the market was much larger than inter-line transfers. The firm paid dividends throughout its life.[185] Emulation is the proof of success. Many new railway lines were proposed, the 219 London-area railway bills brought before Parliament during the period 1860–1869 totaled 1,420 km (882 miles) (see Figure 8.2).[186]

Some of those lines were proposed prior to the opening of the Metropolitan, indicating the smell of success was in the air, though the peak years were between 1863 and 1866, following closely on the heels of the Metropolitan's opening. The most important of these was the Metropolitan District Railway (later called the District line), which ran just north of the River Thames, but south of the Metropolitan, connecting a number of the southern railway termini (Victoria, Charing Cross, Blackfriars, Cannon Street). Proposals for what became the Circle Line service linking the Metropolitan and District (roughly inscribing the box described above) were quickly proposed, but the two lines were not connected on both ends until 1884. Both the Metropolitan and District lines were constructed using cut

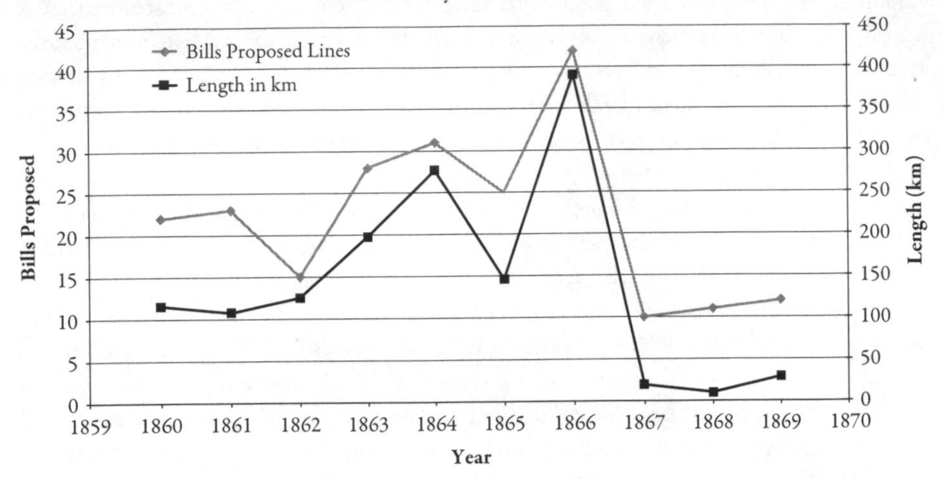

FIGURE 8.2 Railway Bills in London in the 1860s

and cover techniques. Later lines, from the City and South London Railway (first section opened in 1890) onward, generally used deep-level tunneling techniques to avoid disruption of city streets, existing railway lines, and public utilities when they needed to be below grade. Outside the Circle Line, however, the railways could emerge above ground and competed fiercely in some markets, while operating unfettered in others, to provide suburban services. In some cases this involved building new lines, in others it involved acquiring running rights on (or ownership of) existing lines. The development of suburbs was a way to develop traffic for lines that in the city, though profitable, were operating below maximum capacity, and thus maximum profitability.

8.3 Above and Below New York

The London Underground opened in 1863. By 1870 other cities tried to copy. In New York, the publisher of *Scientific American*, Alfred Beach, constructed in secret a short pneumatic tube railroad under Broadway. That it was constructed in secret (at night) is surprising to modern eyes, and was done because Beach did not have the approval of the Boss Tweed ring then governing New York City. Ultimately Tweed killed this nascent technological path though his influence over the governor, who instead approved charters for elevated railroads. Though Beach tried to lower costs by switching from shield-tunneling to cut-and-cover, and Tweed soon went to jail, he could never get enough financial support to proceed. New York was condemned to elevated railroads rather than subways for the next 34 years until a new, non-secret, subway was opened.[187]

Elevated Railways (Els) were constructed in Manhattan beginning in 1870. The initial foray using cable technology was soon replaced with steam engines, basically a railroad in miniature (though the gauge was standard). Els were eventually found on Nineth, Sixth, Third, and Second Avenues, and the latter two of these routes were ultimately extended to the Bronx. Manhattan Railways, their operator, was controlled by famed rail financier Jay Gould along with partner Russell Sage. These were electrified between 1900 and 1903, adopting the Multiple Unit Control system developed Frank Sprague, which was also applied to streetcars.

Despite New York's vastly greater population, Boston preceded New York in operating a successful subway line, opening in 1897 with trolley cars operating in a subway adjacent to Boston Common. Electricity enabled deep-bore subways (which steam made infeasible for anything but cut-and-cover technology). For instance, the Pennsylvania Railroad, which previously served New York City via ferry from New Jersey, now could tunnel under the Hudson and open up a station on the island of Manhattan.

In 1894 municipal voters under the leadership of New York Mayor Abram Hewitt approved the Rapid Transit Act, authorizing a new Rapid Transit Commission to contract with a private firm to construct and operate for 50 years a subway line. The Rapid Transit Construction Company (later the Interborough Rapid Transit Company) was formed to bid on this contract, led by John B. McDonald and August Belmont. After being awarded the contract, it acquired the Manhattan Railway company, operator of the Els, so that it could offer integrated service. Fares were capped at a nickel. The initial route was dubbed Contract One, and its extension Contract Two.

Advantages of shallow excavation (cut-and-cover) over deep bore tunneling included easier access to the tunnel from the ground and, once operating, shorter distances for travelers, so that elevators would not be required. The major downside was the expense of utility relocation and shoring existing buildings.

The new cars were eventually built of steel rather than wood. Though there was some concern about the greater difficulty of rescue given a crash with steel construction (axes have a hard time breaking through steel), the greater protection in event of crash proved to be a more important consideration.

In the first year, the New York Subway attracted 106 million passengers.[188]

Advertising on the subway platforms was an early issue. The franchisee for advertising was Artemas Ward (descended from the eponymous US Revolutionary War general). Opponents were not happy that advertising obscured the then new and nice tilework in the subway stations. Proponents argued that advertising provided useful information for potential customers. This tension would last for decades. The Washington Metro famously limited in-station advertising, while other systems (less graced by federal largesse) embrace it more widely.

Belmont's IRT ultimately took over the Metropolitan Street Railway Company, consolidating control over transit. However, the Brooklyn Rapid Transit Company (operator of the Els in Brooklyn) was given authority to build subway lines into Manhattan. The BRT entered Manhattan in 1908, taking its elevated trains across the Williamsburg Bridge into a Manhattan subway.

The control of transit became one of many fronts in local newspaper rivalries. The *New York Times* supported Belmont and continued private control of the subway, suggesting the test of a subway was a "reasonable certainty of profit," while William Randolph Hearst's newspapers supported municipalization.[189]

The 1908 Elsberg Law shortened the length of contracts, which made it more difficult for potential competitors to enter the subway market (as they would have less time to amortized fixed capital facilities like power stations), and was passed over the opposition of the Rapid Transit Commission.

The Dual System was established in 1913, locking in the BRT and the IRT as the dual private subway providers. The Dual System established the network for each, providing competition in Manhattan, while the IRT dominated the Bronx and BRT was the primary provider in Brooklyn and Queens. Initially IRT and BRT supported the five cent fare because it provided a minimum floor that could not be violated. It later turned into a difficult ceiling for them.

BRT went into receivership in 1918 (after a strike and the resulting tragic Malbone Street Wreck, among other events). It was reorganized as the Brooklyn Manhattan Transit Company (BMT). The IRT narrowly averted the same fate.

In 1924, a new Independent City Owned System (IND) was established (opening in 1932) to compete with the private BMT and IRT, with a line from the Bronx through Manhattan to Brooklyn. While fares were fixed, competition was hoped to improve quality of service, and the new line would add needed capacity on a system now handling over 700 million riders per year.

The capacity on the different services varied, as they were constructed at different times and used slightly different technologies. At the time of the IND opening, the IND could move 90,160 persons per hour per track, the BMT 73,680, and the IRT 59,400. Later technological improvements in signaling improved capacity on BMT and IRT.[190] Turnstiles also were innovated to improve flow entering congested stations and the accuracy of revenue collection.

The downturn in the US economy was felt in New York. The IRT went into receivership in 1932, like the BRT before it. The private lines were municipalized in a process called Unification that was complete in 1940. Only $19 million from the IRT (and none from the BMT) was recovered to repay the funds laid out by the city as part of the Dual Contracts.

New York also has a number of commuter railroads, notably Metro North and the Long Island RR, a separately operated PATH (Port Authority Trans Hudson) service into lower Manhattan from New Jersey, and Staten Island Rapid Transit, which does not directly connect with the rest of New York's transit.

Unlike the Underground of London, or the streetcars of Minneapolis–Saint Paul, the subways in New York largely followed existing populations, rather than being built into greenfields,[191] and acted as an agent of decentralization from Lower Manhattan.

8.4 Transit Surfaces

Early railroads penetrated cities along lakefronts, along rivers, or through lands that had not already been co-opted for urban purposes. But in cities like London and New York, where the urban area was large and dense, railroads were unable to penetrate all the way to the heart of the city.

The passenger market was a big part of the rail business, and railroads established passenger stations as close to the centers of urban markets as practical. For competitive and geographic reasons, each railroad established its own stations. "Union stations" serving multiple railroads came later in a few cities (notably in North America: Chattanooga,

Cincinnati, Denver, El Paso, Indianapolis, Kansas City, Los Angeles, Nashville, St. Louis, St. Paul, Toronto, Washington, DC., and Winnipeg). To accommodate non-bulk freight, teaming yards were also built near downtown. Because of the requirements for car storage as well as room to work teams of horses, the yards involved considerable amounts of land.

As cities grew, additional teaming yards and considerable industrial track were constructed. Today, the parts of cities built during the period we are describing have considerable interlacing of rail facilities within them.

With continuing urban growth, higher capacity modes became practical, even in cities not as primate as London and New York. The reliance on horse (both for personal and mass transportation) was problematic, not just for the cost, and the pollution that resulted (manure), but also the risk of disease from interacting with animals on a regular basis. The shift from feed for animals to fossil fuels for energy became more and more justified as energy conversion processes were improved and favored fossil fuels. Along routes where traffic warranted, steam trains were run. (These evolved into subway/elevated systems, commuter railroads, or some combination of the two, as noted in the previous sections.) The superior technology for many surface streets next became the cable car, and cable car systems were installed in many cities.[192]

Electric utilities promoted and often owned electrically powered transportation properties. The markets that the properties provided aided the utilities by providing for base load, aiding in the achievement of scale economies in production and distribution of power, and also aiding the diffusion of service throughout urban areas.

Rail-like services pushed aside other services, so the predominant technology of transit shares much with railroads, as does the embedded policy. Public policy was railroad policy writ small for the scale of the city.

Systems were given charters (franchised) by the cities and policy treated safety, service, and prices. A flat fare policy was widely adopted—usually five-cents. (The suburban railroads used distance-graded fares.) The five-cent fare was not a burden in the early years when traffic was increasing and economies of scale were achieved. It became a burden as inflation (especially at about the time of World War I) decreased buying power and the transit worker unions demanded higher wages and better working conditions.

Electric railways (streetcars) were deployed rapidly between 1895 and 1915. As is true of lots of new developments, some false starts were made. However, once established, streetcars enjoyed a sharp cost advantage over precursor systems. Numbers we have seen suggest that the cable car gave a cost saving over horse cars of about 20 percent. At first, cost savings were not so marked for streetcars because of the voltage drops on early DC street car lines. It wasn't until AC electric power was more widely distributed that converters could be used to supply DC to outlying lines. Car size and service frequency adjusted to markets; streetcars quickly won over cable car markets, except in very hilly terrain. Cable car mileage peaked in 1893. The cost comparisons tell only part of the story because electric cars were larger than precursor vehicles, and they offered service advantages such as higher speeds (see Table 8.1).

These electric railways (and steam trains where they had been deployed previously) had considerable impact on the forms of cities. Suburban development increased and downtowns could draw on larger areas for labor and sales and service activities. In short, the centralizing/decentralizing forces that we usually associate with the automobile were quite strong before the automobile entered the city.

TABLE 8.1

Cost of Alternative Technologies (1893, 1894)

Mode	Average Cost per Car per km
Horse	$0.06
Cable	$0.0525
Electric	$0.0625

Source: Cable and Horse data for 1894 from Hubert, Philip G. Jr The Cable Street-Railway Scribner's Magazine Volume 15, Issue 3, March, 1894 http://www.cable-car-guy.com/html/ccscrib.html Electric Data for 1893 from "St. Charles Avenue Streetcar Line, 1835" National Historic Mechanical Engineering Landmark brochure, American Society of Mechanical Engineers and Regional Transit Authority http://www.asme.org/history/brochures/h101.pdf.

Streetcar service was highly desired. Cities were quick to issue franchises. There was over-building of lines, and some lines were not well connected to others. The opportunity, need, and ability to consolidate rail properties seem to have varied greatly from city to city. In Los Angeles some thirteen systems became the Los Angeles Railway Corporation. By 1890 there were four properties. Consolidation was complete by the time automobiles began to appear in numbers, 1910 and onward. Washington, D.C., had twenty seven distinct street railways chartered between 1862 and 1902, and also consolidated.[193] These railways were tied to real estate development.

At about the time rail transit services became available, cities were growing very rapidly; the highest growth rates for cities occurred at the turn of the twentieth century. Making land available for many types of development, the transit services had massive off-system effects. The physical realizations of the impacts included streetcar suburbs and the decentralization of shopping from the central business district (CBD) to outlying retail centers.

The systems suffered from the ills we will be discussing for the railroads (rail-itis). The problem of the five-cent fare has been mentioned. Route and service rationalization was needed due to duplicating properties. Desirable network connections were needed because lack of physical connections and/or transfer rights thwarted end-to-end services. Congestion occurred during rush hours, yet low demand characterized some routes and times of the day. Importantly, the systems had lost much public support. In the 1890s, transit was regarded as highly desirable from a social point of view. Transit-based developments reduced residential crowding and made job sites available. Yet by 1920, systems were regarded as poorly managed "rip-offs".[194]

Problems became acute by the 1910s and 1920s when the car and truck began to provide competitive services and other factors affected the transit market. The truck eased the logistics problems of stores in decentralized locations that reduced CBD shopping activities. With auto mobility, such activities no longer needed to be on transit lines. The truck also enabled the decentralization of many kinds of manufacturing; especially the activities that had previously found their most desirable locations near the downtown rail freight yards and where labor could travel to work using CBD oriented transit.

With activities at decentralized locations, the automobile offered an option for the journey to work.

The adoption of electric motors and other factors began to favor single store manufacturing establishments. Land requirements influenced the building of new plants at the outskirts of cities.

Actually, the early impact of the auto on transit was modest. At first, the automobile was used for recreational travel and was considered a rich man's toy. Only later did it begin to be used for the journey to work and for shopping as the sites of those activities began to shift. The competitive impact of the automobile differed depending on city morphology and type of transit. Today, for example, the CBD of Chicago continues to be served by transit in its historical forms (which Charles Yerkes helped establish)—commuter rail, subway-elevated, and bus (which substituted for the streetcar).

Other factors affecting the demand for transit included growth in real incomes and the shortening of the workweek from six to five days. As a result of the factors at play, transit demand, which had been roughly evenly distributed seven days of the week, was reduced on Saturdays and Sundays. The morning and evening peaks of traffic became relatively steeper as off-peak CBD shopping and travel for warehousing and manufacturing work in the vicinity of the CBD decreased.

All of the developments required planning. Most was project planning in style and scope, and its first client was the city government because franchises were needed before work could go forward. The topic of how the system was to be operated (fares, schedules) was treated because it was of great interest to governments. Consolidation planning (or rationalization) had a different style. It extended to the best ways to combine properties and the appropriate treatment of stakeholders in the properties. These plans were developed by (relatively) famous consultants from out of town, and were usually sponsored by city governments.

8.5 Metro-Land

The suburban extensions of what came to be known as the London Underground were much more speculative than those built in the city.[195] Though at-grade suburban lines were less expensive to construct, they also had a lower expectation of revenue. While in the city the demand was present through the high density of existing development, the suburban lines in many cases went through greenfields. In contrast with the main line railways, which had a long-distance market and could add a station on an existing line to test a new short-distance market at minimal cost, an Underground extension required both the line and the station in order to provide a new service to (rather than through) an initially sparsely developed area, what one might dub a "railway to nowhere."

The suburban development along the surface railway lines began early. In 1853, the London & North Western railway advertised in the *Illustrated London News*

> To persons intending to build houses of a suitable character and of a value of not less than £50 annual rent within two miles of the following stations viz. Harrow, Pinner, Bushey, Watford, Kings Langley, Boxmoor [now Hemel Hemstead] and Tring. A free

annual first class pass to one resident of each such house for the following periods—Harrow 11 years, Pinner 13 years, Bushey 16 years, Watford 17 years and the other stations mentioned 21 years.[196]

The London & North Western Railway provided an enormous subsidy to the wealthy to build on land the railway did not own. The reasons for this can only be surmised: was this to prime the pump, since the amount of revenue coming from these houses with an eleven- to twenty-one-year free pass, discounted back to the present, would be quite small; was it to encourage non-work travel, especially by other household members; or did the railway, or its executives, have a hidden interest in development somehow?

Two decades later, the suburbs had begun to take shape. "About the Barnet station has sprung up within the last few years one of those new half-finished railway villages which we have come to look on as almost a necessary adjunct to every station within a moderate distance of London."[197] The network in the northwest quadrant of London was the least dense with surface railway lines. Not surprisingly, those were the unharvested suburban pastures that attracted the most attention from the Underground railways.

The powers to develop land varied. The Land Clauses Consolidation Act of 1845 required that within ten years, railways must dispose of land not required for the projects identified in the Parliamentary Act describing and authorizing the railway. The Metropolitan Railway Acts of 1868 and 1873, however, allowed the Metropolitan to hold on to such lands for a longer period of time, putting the Metropolitan into the development business in a way that other railways were not permitted.[198] The lack of effort to get similar provisions inserted into the Act of other railways suggests that those railways did not wish to enter the development business.

The Metropolitan Railway acquired surplus land as it acquired right-of-way. In large part, this was the acquisition of whole parcels when sellers did not want their property bisected by the railway. Unlike other railways in the United Kingdom, however, the Metropolitan had Parliamentary authority to develop that land, and did not sell surplus land as quickly as possible. The consequence of this was the development of Metro-land, and the ability to pay higher dividends due to its real estate division than other Underground railways were able to. That said, other railways, along with the Metropolitan, did work with developers to obtain subsidies for building stations near their new developments and promising to run services.

Charles Yerkes (1837–1905), a Philadelphia bond salesman, was perhaps the leading entrepreneur of the rail transit era. He moved to Chicago in 1881, and earned a shady reputation due to some poor economic transactions associated with the Chicago Fire. Nevertheless, Yerkes was able to buy respectability, which was for sale in the Chicago of the 1890s, and was able to acquire most of Chicago's public transit system through bribery of pubic officials. Eventually he was driven out of town when the political winds changed and he was unable to obtain a fare increase (provoking riots in the attempt), and wound up in London, where he helped build the loop and consolidate the Underground. Yerkes' life was fictionalized as the character Frank Cowperwood by novelist Theodore Dreiser in three novels, *The Financier*, *The Titan*, and the posthumous *The Stoic*.

When Charles Yerkes, who acquired a number of the deep-level tube lines and the Metropolitan District Railway, as well as other transport properties, was investigating whether he should invest in the proposed Hampstead (now Northern) line in 1900, the following came to pass:

> When they came to Golders Green, Lauderbeck [agent for Charles Yerkes] stopped and told Dalrymple-Hay [British civil engineer] that here was the proper site for the terminus, meeting protests about the absence of houses by pointing out that in the USA, railroads were built and the people followed. After visiting the site himself, Yerkes asked, "Where's London?" and on being shown, turned to his companion with the words "Davis, I'll make this railway."[199]

White writes:

> The [Hampstead, later Northern] line emerged into daylight to terminate in the fields bordering the Finchley Road. Most people were unimpressed by the wisdom of this move, but it is said a syndicate had already been formed to buy up the turnip fields before the announcement of the new line had affected land values. Thus at Golders Green first began the typical pattern of twentieth-century suburban development, the arrival of an electric railway in some untouched rural area, soaring land values, semi-detached villas and chain stores.[200]

Despite this activity, the sum of Metro-land developments created directly by the Metropolitan Railway and its subsidiaries amounts to only about 15,000 houses on about 2,200 acres.[201] But when considering the accompanying private developments that took advantage of the accessibility created by the railways, and thus gave traffic to the railways, the change was enormous.[202] The change is illustrated in Figure 8.3, which shows the development of both the railway networks and the population density of London from 1850 to 2000 at fifty-year intervals.

Ultimately, the co-development of suburbs and Underground lines came to a halt with the 1948 designation of a Green Belt around London.[203] This resulted in the cancellation of proposed line extensions, and hemmed in the Underground-served suburbs. Later suburban developments jumped the Green Belt, but these were to be served by automobile, bus, or surface rail.

8.6 Discussion: The Land Value Metric

Transportation makes it easier to reach places. A transportation improvement may increase the value of a site for production (decrease the costs of consumption), that is, increase its economic rent. Thus we expect that land value reflects the benefit with a higher market price.

There is long-standing interest in measuring development impacts by observing land value changes. Speaking of a road opened in 1756, Rev. Young observed, "It was no sooner completed than rents rose from 7s to 11s, per acre; nor is there a gentlemen in the country who does not acknowledge and date the prosperity of the country to this road. . . ."[204] Today, improved analysis techniques are aiding estimates of land value changes, and value capture to aid infrastructure financing is an attractive option to public agencies short of funds. Value

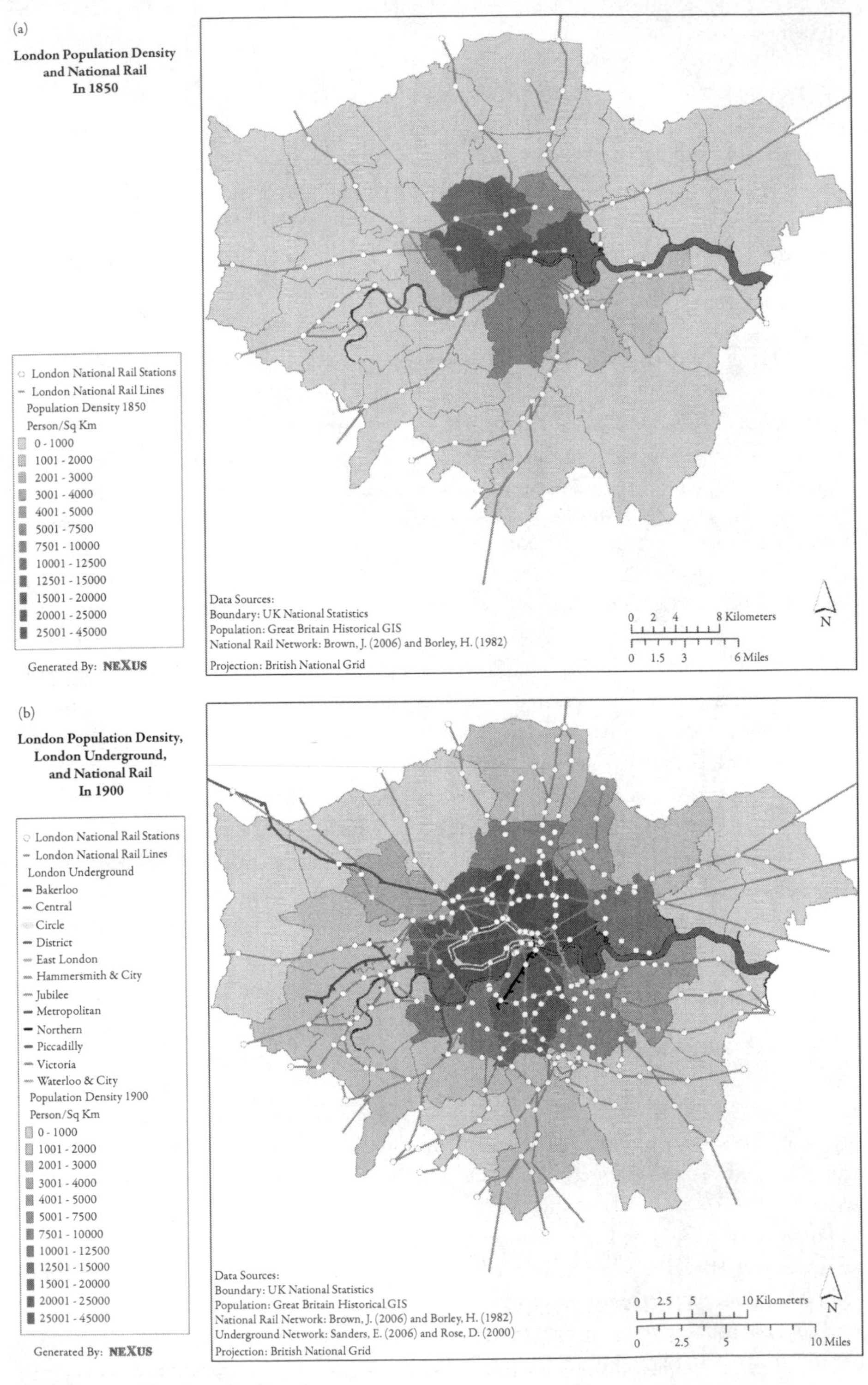

FIGURE 8.3 The Growth of London

(c)

London Population Density, London Underground, and National Rail In 1950

London National Rail Stations
London National Rail Lines
London Underground
Bakerloo
Central
Circle
District
East London
Hammersmith & City
Jubilee
Metropolitan
Northern
Piccadilly
Victoria
Waterloo & City
Population Density 1950
Person/Sq Km
0 - 1000
1001 - 2000
2001 - 3000
3001 - 4000
4001 - 5000
5001 - 7500
7501 - 10000
10001 - 12500
12501 - 15000
15001 - 20000
20001 - 25000
25001 - 45000

Generated By: **NEXUS**

(d)

London Population Density, London Underground, National Rail, and Docklands Light Railway In 2000

London National Rail Stations
London National Rail Lines
London Underground
Bakerloo
Central
Circle
District
East London
Hammersmith & City
Jubilee
Metropolitan
Northern
Piccadilly
Victoria
Waterloo & City
Docklands Light Railway Stations
Docklands Light Railway Lines
Population Density 2000
Person/Sq Km
0 - 1000
1001 - 2000
2001 - 3000
3001 - 4000
4001 - 5000
5001 - 7500
7501 - 10000
10001 - 12500
12501 - 15000
15001 - 20000
20001 - 25000
25001 - 45000

Generated By: **NEXUS**

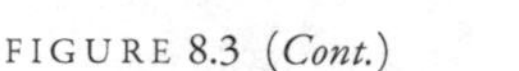

FIGURE 8.3 (*Cont.*)

capture dates at least from port financing efforts when French royalty first strived to develop colonies and expand maritime trade.[205]

The theory is simple in outline—all other things being equal, land derives value from its location. A change in transportation costs thus has a direct affect on the location-derived value of land. That simple mechanism is attractive to analysts. It fits very well with the view that benefits may be measured as user cost reductions.

Analysts armed with benefit-cost or similar ideas are cautious about claiming more than user cost savings. That is partly because of the experience as the interstate highway system was constructed. Many studies were made under the rubric "highway impact." They were motivated by the notion that increments in land values might be captured to aid in financing the interstate (US Congress, 1961). While impacts were often indicated, these tended to be modest, and there were double counting issues—counting both user cost savings and increments in transportation rents. Critics of study designs and workers concerned with theory stressed that what was mainly observed was the shifting of development from here to there. That is, development, which would have happened somewhere, was focused on areas proximate to newly constructed facilities.[206]

The same story holds for urban transit development. Analysts have long known that as deployment winds down, development impacts are modest. Edwin Spengler found in his 1930 study in New York City that trends overrode impacts. If an area was declining, it continued to decline as transit service was improved; stagnant areas remained stagnant; and growing areas continued to grow. Spengler's findings appear almost word-for-word in Walmsley and Perrett's review of the situation.[207] Other studies tracing transit development impacts have managed to find impacts, yet they are small.[208] To magnify impacts, cities are encouraged to use zoning and other tools to concentrate development at transit stations.[209]

As stated, the transportation rent-land value notion is simple, but in real world contexts there are the workings of land markets, agglomeration economies, and historic path dependence. It is partly for these reasons that very little can be said "... unless a great deal is known about ... (households and) ... the production processes of business firms."[210]

Yet, we can show that the value of land is intimately tied to its accessibility to other pieces of land and their associated activities. If a transportation or land use change enables people to reach activities that are more desirable in less time, then the value of their land increases. Hedonic theory suggests that individuals do not purchase goods, but rather the bundle of attributes composing the good. Someone does not buy a house, but rather the qualities of that house: location (accessibility), size, type of construction, appliances, noise from nearby roads, and so forth. Every house combines the various attributes slightly differently. Hedonic models are used to pull apart these attributes and develop demand curves for the various attributes (goods or bads). However, these attributes are interrelated; houses with high accessibility will be more expensive, which will lead to more investment in other attributes, leading to better maintenance and more frequent remodeling.

The hypothesis of the standard urban economic residential location model, that individuals trade housing costs for transport costs, has been tested in numerous studies. All else equal, housing prices decline as one retreats from the center of the region (the location with highest accessibility).[211]

Land rents influence land uses because the market asks that land be put to its highest and best use—uses have to pay the rents. From this observation we derive the notion that transportation improvements may change patterns of land uses. Some years ago Garrison and colleagues engaged in extensive analyses of this proposition.[212] Economies of agglomeration, central place theory, and some of the other processes bearing on land use patterns were considered. The general conclusions were that (1) patterns did change and (2) benefits were present (termed reorganization or structural change benefits), although benefits were not easy to measure.

Although the term "reorganization benefits" did not come into wide use, the ideas that one pattern of land use may be better than another and that transportation improvements might change land use patterns remains popular. We hear that transit investment will yield "better" land use patterns than will highway investment. Of course, better is in the mind of the beholder. We have some difficulty with the notion that the station-focused accessibility of transit improvements would produce a better stage (as opposed to a different stage) for the running of consumer sovereignty than the more diffused theater given by highway investments.

Reasoning from the dynamics of transportation system development, the expansion of existing transportation infrastructures using conventional technology in a mature market would not be expected to have much impact—the developments triggered by investments have already run their course. To put the matter at the extreme and use an overworked metaphor—observing land value changes on a mature network is similar to observing the deck chairs sliding on the sinking *Titanic* and measuring changes in their comparative (locational) advantages.

The story of transit continues in Chapter 14.

What hath God wrought
—SAMUEL F. B. MORSE (1844)

9 Telegraph

THE ELECTRIC TELEGRAPH grew up with the railroad, and it made long distance railroading possible. But the first "telegraph" (writing from afar) was built in 1791 by Claude and René Chappe in France. It used sound along with times on a clock face, rather than electrical impulses, to send messages, and the revolutionary government soon adopted it. This telegraph simply enabled communicating messages to afar without physically moving a person or document. Abraham Niclas Edelcrantz developed an optical telegraph in Sweden soon after hearing of the Chappe invention in 1794. By 1795 the British had also devised a telegraph system using shutters (that passed or didn't pass light) to transmit messages visually. Different combinations of open and closed shutters indicated different letters or words. By the 1830s, such systems were common throughout Europe. Typically, relay stations (or nodes) were located on hills for maximum visibility, hence the term "telegraph hill." An optical telegraph line operated between Philadelphia and New York in 1840, for the benefit of its owners.[213]

We don't have networks of optical telegraphs dotting the landscape today, as the technology was made obsolete by the electric telegraph. Some difficulties in the technology had to be overcome, including transmission of electric signals over long distances and efficient coding. Samuel F. B. Morse, a famous American artist, apparently unaware of the difficulties others had had (few advertise their failures widely), proceeded to develop some of the core technologies that made up the telegraph system. He is most famous for the signaling code of dots and dashes named for him. He also developed a mechanism for recording the signals on papers, creating a transcript of the messages. Joseph Henry, an American scientist (and first head of the Smithsonian Institution), developed the means for transmitting electric signals over long distances by using small batteries in series rather than a single large battery.

In parallel with American efforts, Cooke and Wheatstone were developing a different electric telegraph in England. The first telegraph line was 2.1 km (1.25 miles) between the Euston and Camden Town stations on the London & Birmingham Railway in 1837. The first line had five wires (because of the complicated code they used). A later deployment on the Blackwell railway did as well, but as some of the wires broke, operators improvised with a two wire system, developing a more parsimonious code. Realizing fewer wires were needed, future deployments were simplified.

In 1844, after ten years of lobbying, using a federal appropriation of $30,000 (a not inconsiderable sum), Samuel Morse built a line between Baltimore and Washington along the right-of-way of a skeptical Baltimore & Ohio Railroad, which had concerns that a telegraph might substitute and eliminate much passenger travel.[214] A demonstration transmitting the passenger list on a train ahead of the passengers gave proof of the utility of the telegraph. Morse also took in a financing partner, Alfred Vail. Other applications of the Baltimore-Washington line include apprehending criminals and verifying checks. By 1845 the Magnetic Telegraph Company was formed to exploit the invention, building lines between major East Coast cities (the first opened in 1846). The first year the line was still running a loss, despite charging one cent for four characters. The federal government allowed Vail and his partners to acquire the federally funded Baltimore-Washington line in 1847.[215] But soon telegraphy turned profitable and development exploded (see Figure 9.1).[216]

By the 1850s, consideration was given to laying a trans-Atlantic cable, despite huge technical obstacles. By 1858, the Atlantic Telegraph Company had laid the first transatlantic cable, and messages were transmitted. However the cable soon broke. This "Thread Across the Ocean," as goes the title of one book on the subject,[217] was the premier engineering accomplishment of its day, employing the labors of Isambard Kingdom Brunel and the future Lord Kelvin, a work that failed four times before achieving success. The effort was

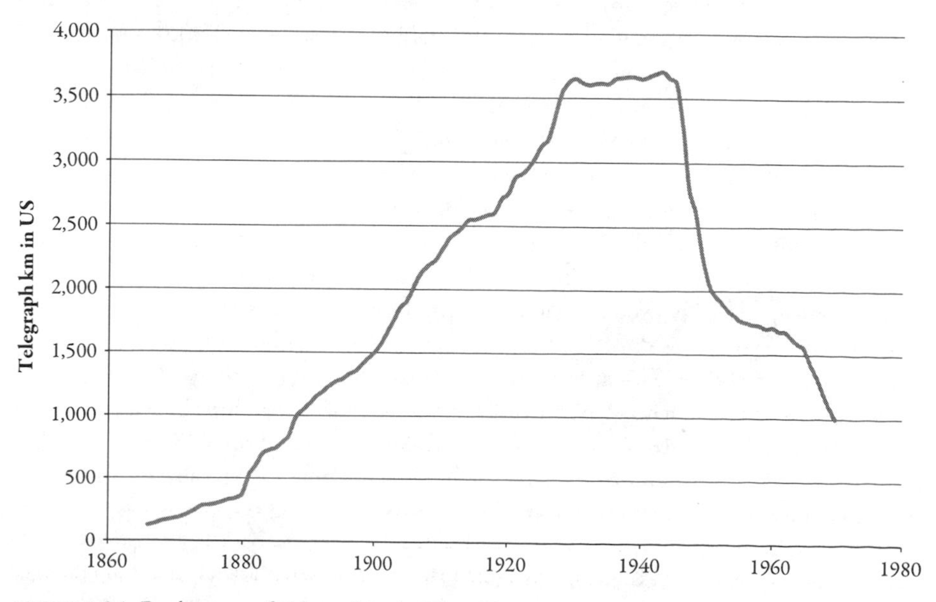

FIGURE 9.1 Deployment of Telegraph in the United States

led by Cyrus Field, an American entrepreneur and visionary who, despite failing, kept raising funds and kept hope alive that the Transatlantic telegraph cable was not a fool's errand, but a wise investment that would return profits to investors. Each failure, resulting from cables snapping or going dead, led to design improvements, stronger more ductile cables that would be more robust in the difficult design conditions of the cold Atlantic seafloor. But stronger heavier cables could not be handled by the conventional ships of the age; it required the world's largest ship, Isambard Kingdom Brunel's *The Great Eastern* (see Section 5.7), which was used on the fourth and the ultimately successful fifth voyages laying cables. Constructing the greatest communications network at the time, an endeavor that would substitute for traditional transportation-reliant material communications (mail) required the greatest transportation vehicle at the time to lay.

The learning involved was not simply building stronger cable, but how to lay cable so it wouldn't break, how to fetch "dead" and broken cable from the ocean floor, how to send electric signals thousands of kilometers without fading despite the cable being submerged inside an electrical conductor (seawater), and how to manage a transatlantic enterprise. Soon after the ultimate success in 1866, the fourth failed cable was resurrected and reconnected, giving a second, redundant link. Its owners enjoyed a brief monopoly, but soon brought prices down to increase demand, and shortly thereafter faced competition. (The competing cables were also laid by *The Great Eastern*.) By 1903 there were more than 15 distinct transatlantic cables, which meant that communications would not be severed even during two world wars.

By 1861, a North American transcontinental line was opened by Western Union, forcing the Pony Express out of business. While the Pony Express was fast (10 days to travel 2,900 km [1,800 miles]), the telegraph was almost instantaneous. Western Union, which eventually acquired a monopoly in telegraph services, was founded in 1851 in Rochester, New York, as the New York and Mississippi Valley Printing Telegraph Company, and took its current name in 1856 before opening up the transcontinental line. By 1866, the Western Union had acquired another 340 telegraph companies. It established relationships with most United States railroads, and carried some 80 percent of telegraph messages by the 1880s.

The telegraph proved very important to the operation of long distance train services. Imagine you have one rail line between two points, but you want to operate trains in two directions. In the absence of instantaneous communications, you can set up a very precise schedule; but trains sometimes get delayed, or you can set up a very coarse schedule (eastbound in the AM, westbound in the PM), but then trains will needlessly sit idle for hours. Or you could have a rule whereby inferior trains would wait until superior trains passed. With the telegraph, you can set up a dispatch and signaling system, and know when trains are coming from one direction to put a "red light" in the other, delaying trains for the minimal amount of time. This system also reduces the amount of track you need to build, lowering costs (wire is much less expensive than track). Train dispatching by telegraph took place in Great Britain by 1839, and was deployed along the New York and Erie railroad in 1849. In the 1850s telegraphic control of trains was commonplace. Typically, clerks and stationmasters at train stations were telegraph operators, suggesting another synergy to more effectively use available labor.

The telegraph was also responsible for establishing standard railway time, which became standard time shaping everyone's life. The railways adopted time zones so that a train leaving

New York at 11 AM was also recorded as leaving at 11 AM by observers in Philadelphia or Washington (both of which are west of New York, and had been offset by a few minutes). The Western Union time ball would drop daily from their headquarters (the tallest building in the United States at the time) at noon in New York, triggered by an operator at the National Observatory in Washington, D.C., from 1883 to 1913. Western Union came under the control of financier (and railroad magnate) Jay Gould.

In the two decades after the telegraph was developed, it had become the most important communications technology since the printing press, obviating distance as a barrier to exchange of information, so long as wires connected the two ends. Where previously it may have taken days or weeks for messages to be received, with the telegraph it was reduced to seconds. Yet not all ends were connected with wires, and though information could travel instantaneously within Europe, or within North America, it could not travel the same distance between the continents but aboard a ship. Bringing the Old World and New World together required laying a telegraph cable across the Atlantic Ocean.

That cable would complete the missing link in the global communications system, melding the separate networks into one. But not only would the communications networks now operate faster, the implications for commerce would be enormous—the world could operate as a single financial market. A live wire between Europe and its overseas colonies meant that military crises could be defused, or managed, or exacerbated without the delay imposed by requiring communications to ride upon the fastest means of transportation.

The telegraph improved logistics. The railroads used it to coordinate the movement of people. One could, for instance buy a ticket on the Louisville & Nashville Railroad with connections to the Milwaukee Road and Great Northern, a multi-firm transaction enabled by electronic information. Firms such as Ford used the telegraph to monitor and control inputs and outputs.

Telegraphy peaked between 1930 when there were about 16,000 US telegraph messengers total, and 1945, when the number of telegrams sent in the United States peaked. Western Union sent 200 million telegrams globally at its peak in 1929.[218] The number has declined since then, being insignificant after 1950.[219]

The term "Battle of the Networks" applied to the competition of alternating and direct electrical current. It could just as easily apply to the many transatlantic telegraph cables, the railroad wars, and events such as Gould's largely legal threat to break the Union Pacific's spatial monopoly on traffic in the middle part of the country to extort a payoff. Consolidating monopolies is important to control costs and ensure profits in network industries, and the railroads, with over 150 years of consolidation behind them, have done that. AT&T did that for 70 years. It tried to do that again in the cable television industry. Other firms are fighting a losing battle in trying to gain monopoly status. Airlines are attempting to establish alliances with other airlines to provide better services, and further entrench customers within their system through frequent flyer systems, thereby making them slightly less price-sensitive.

It appears that the structure, performance, and conduct of transportation systems are replicated in other systems. Transportation shares characteristics with other systems that:

- Are networked and highly standardized;
- Serve as universal input industries and are everywhere available;
- Have strong economies of scale;

- Affect the public interest so much that they have been publicly supplied or closely controlled.

While water supply and electricity come to mind, there are also systems where the predominant technology is soft, for instance, the public school system.

In addition to structural similarities, we see functional similarities between transportation and communications. Each is an organizing system: they organize the structure of production and consumption. The "What does it do that is worth doing?" question is similar for each, and answers tend to be rather hidden.

The story of communications continues in Chapter 17.

4 Phase 2 of the Life-cycle

Always remember that someone, somewhere is making a product that will make yours obsolete.

—G. F. DORIOT,

quoted in *Time*, 15 May 1987

10 The Magic Bullet

10.1 Growth

CHAPTER 4 MADE A number of comments about inventing and innovating, which we identified as Phase 1 of the Life-cycle process. Phase 2 of the Life-cycle process involves growing the system. Systems grow because bigger has advantages over smaller. These economies of scale comprise what we call a "magic bullet" when they work to expand a system. In this chapter we comment on the magic bullet, with references to cases we have seen to date.

The modes discussed in previous chapters started small and successfully scaled upward as markets were expanded and grew. Networks expanded, and on-network activities increased, resulting in heavy haul railroads, high throughput transit systems, and large barge tows. Filling in networks and expanding them at the margins sometimes requires downscaling where success was mixed. The discussions here (as well as in subsequent chapters) consider scalability. Scalability has a "with respect to what" character. Larger is advantageous if it enables faster, not just line haul speed, but also end-to-end speed, considering access costs, waiting and transfer delays, and so on. Networks enable economies of both speed and frequency.

10.2 Growth Takes Off Due to Magic Bullets

The magic bullet describes how the feedback amoung economies of scale; service quality, demand, and cost drive the growth of systems.

"Economies of scale" is the property that average costs decrease as throughput (satisfied demand) increases. (Economists give attention to the economy of scale in relation to firm size; we refer both to that and to economies resulting from the concentration of traffic on routes, sometimes called "economies of density.") Cost reductions may be kept as profits by the service provider in an uncompetitive situation.[220] In a competitive market, economies of scale are returned to users either as price reductions or service quality improvements (allowing users to upgrade). On a freight rail link, for example, the more the traffic, the less the cost of movement and the better the service, at least until congestion sets in.

For freight and passenger services, "network externalities" may also emerge, wherein the system is more valuable the more people who use it. If 500 people want to ride a train each day from Minneapolis to Chicago, there will be one train. If 5,000 people want to ride the train, there will be ten, meaning the trains will be more frequent (once every 2.4 hours instead of once every 24), so that my value of the trip increases as I have less schedule delay (the desired arrival time is closer to the actual arrival time).

For any technology, learning, adaptation, and borrowing drive down costs and improve quality, as new techniques are developed, perfected, and applied. These learning curves are not strictly (in the economist's sense) scale economies, but they operate the same way; especially in the early years of deployment, learning increases with output, as there are more and more opportunities to perfect the system.

There are lots of "ifs" about that generalization. Networks, especially fixed networks, can respond only in the long run, so sudden changes are hard to deal with. For example, a sharp rise in traffic on a rail route will create congestion and impair service quality. That's been the case in the United States in those years when exports of farm products and coal soared. But where there are gradual increases in demand, adjustments to hard and soft technology can be made more easily, and the links work better and better.

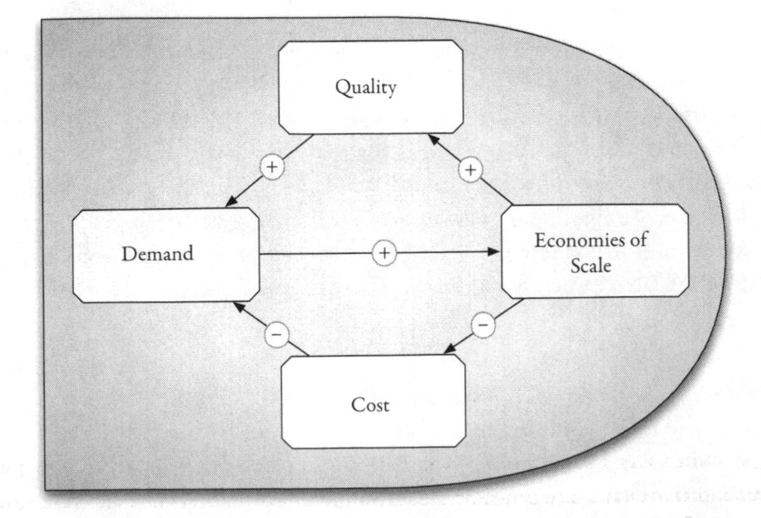

FIGURE 10.1 The Magic Bullet

So there are opportunities to combine new technologies with investments that yield scale economies, as shown in Figure 10.1.

- Increased use yields economies of scale. (+)
- Economies of scale allow:

  The use of product quality–improving technologies (+), and
  A lower per user cost of existing technologies (−).

- The improved quality yields increased system use (+).
- Higher (lower) costs would decrease (increase) system use (−).

10.3 Standardization Enables Magic Bullets, Thwarts Innovation

Christensen's (2012) model suggests that there are three types of innovations that capitalists may invest in: empowering, which take elite products and transform them to the mass market; sustaining, which replace old products with new; and efficiency, which reduce costs, and drive what we call the magic bullet. As a system transitions from youth to maturity, both product and process-of-production technologies standardize. Standards tend to thwart improvements that can be obtained from hard and soft technological changes, and product improvements come largely from returns to scale. Improvements also result from large-market-derived opportunities to provide variations on services or products tailored to market niches (specialization to market niches).

Today's transportation systems are standardized and pretty well deployed. To compete, an organization may locate and satisfy market niches by tailoring products, so long as it can be done with a more-or-less standardized technology. The many models of automobiles are examples of market niche tailoring.

There is another option: strive for a big marginal gain (reduction in cost or improvement in product quality) by combining scale economies with quality-enhancing "new" technology.

Combining is the important word, for standing alone, neither scale nor quality-enhancing technology has much to offer. With respect to scale, reductions in costs can be achieved by serving more customers, but many systems have gone that route, and as systems reach maturity there is not much room for more cost reduction. The market is saturated. (Though saturated, markets are still increasing with wealth and population, and there are thus some options to achieve more economies of scale.) Quality-enhancing technology has been strived for, too. But often only marginal things can be done absent the spreading of costs over larger markets.

10.4 Purported Magic Bullets Are Sometimes Tragic Bullets

The magic bullet has been framed as combining quality-enhancing technologies with economies of scale. That's one point on a continuum of options. At one extreme, improved quality can result just from scale—the quality improvement from more frequent service is an example. At another extreme, improved quality can be derived from technology alone. We have stressed the combining case because that seems to be the usual case. The extreme cases represent options already pretty much taken up by mature systems, as mentioned.

The combination of quality-enhancing technologies can be targeted on a market niche. We are not the first ones to find that magic bullet. Wonderful things could be done if there were just enough system use.

We have had many policy experiences where promoters put forth policies to steer users to systems so that wonderful things could be done to improve their quality. For example, we have heard many times that if policy would restrict air travel along the Northeast Corridor (Boston-Washington), then wonderful train service would evolve. Service would improve because of higher schedule frequencies and from investments in quality-enhancing technologies.

Magic bullet proposals are very risky. To make a difference that's more than marginal, significant capital must be invested. If the technology doesn't work just right, quality improvements aren't there, costs are underestimated, and if ridership is overestimated, there can be a disaster. The planner must be very careful.

The high risk of blatant disasters is a feature of magic bullet efforts, and that point has been made. Yet there is a larger risk, the risk of disasters that aren't blatant. We now attempt to develop this idea. This requires considerations of the development paths created by magic bullet solutions, the hidden character of some disasters, and the effects of concentrating system use.

Consider the returns from system improvements that history tells us about. Early in the life of a new system, it may be two times as efficient as previous systems. As it grows, takes on new functions, and achieves economies of scale, its improvement runs an order of magnitude, and there are large off-system gains.

Even so, consideration of pathways for development raises the question of whether magic bullet–type actions are a desirable way to go. Even if an endeavor is successful, it is on a path that mines out opportunities in a mature system rather than opening a new path of much greater potential. It's not getting the factor of two improvements over previous services, and it doesn't hold out promise for order of magnitude improvements and large off-system gains.

If there are economies of scale to a point, this does not imply economies of scale at all points. Unlike Fulton's steamboat, Brunel seems to have overshot, so to speak, when sizing the *Great Eastern*. We don't know exactly why it wasn't a success—not enough passengers, too long a turn-around time in port? Yet for some reason, it didn't fit the market situation. Even so, the kinds of ships imagined by Brunel were successful in the liner trades—the *Great Eastern* was "too soon." What about his 2.1 m (7 ft) gauge railroad? It was "too late" to establish the standard and obtain economies of scale. By logic copying, English railroads had already committed to standard gauge—a bandwagon effect—and networking requirements forced Brunel to join.

Other examples support the claim. The clipper ship era is interesting. Long a leader in maritime service, US operators bet on the clipper ship over emerging steam and steel hull technology. Clipper ships offered fast service compared to previous sailing ships, and their use was targeted on markets where they soon dominated. The clipper ship was a magic bullet in the sense that we have used the term, and it was a rousing success, for a short period of time.

But British operators bet on steam propulsion and metal hulls and the soft technologies that matched them. They got on a take-off path that displaced the clipper ships. US operators gave up clipper ships and got on the steam path too late. US maritime services have never regained the momentum they lost.

This raises another issue about magic bullets. They run the risk of technological obsolescence, and are very much at risk for loss of scale economies, which may turn a seeming success into a massive failure. The general rule was given in the chapter's opening quote.

We have discussed the clipper ship's sailing in the sunshine of its obsolescence. There are many other examples, such as the quick obsolescence of high tech, fast passenger trains of the 1950s that were pushed aside by air service.

This wrong-path discussion is very value laden. Benefits from magic bullet actions accrue in a short time frame. Their costs are in some unknown future. Our view that we would be better to privilege actions that open up new and rewarding paths stems from our ethical stance. We highly value creating options for the future.

There is another value-oriented thought. Capturing scale economies usually concentrates service and off-system benefits and costs. Air, ocean liner, and rail operators have captured scale economies by concentrating service on particular routes and terminals. Those who are in the right places have many advantages from such concentrated services; those elsewhere are relatively disadvantaged. There is a horizontal (spatial) equity problem. It is one thing to observe that in the aggregate everyone is better off; the distribution of gains is another matter.

There can be another downside. Externalities are concentrated, as those who live close to high-density coal routes, highways, or large airports are quick to remark.

To summarize, it's certainly true that many of the improvements in systems are scale-derived. That's especially the case as technologies are standardized and grow to maturity. The points made above stress that there is a downside, too. In today's situation we are dealing with mature systems, and there may be traps when economy of scale–based improvements are sought.

10.5 The Trajectory of Magic Bullets Is Difficult to Alter

10.5.1 INTERVENTION CAN SLOW DOWN OR SPEED UP THE TEMPORAL PACE OR REALIZATION OF SYSTEM DEVELOPMENT

Interventions to affect the rate of deployment are consequential, for impacts running for decades may be involved. This is particularly true for imitators (evidence is the second country deploys faster than first country; because a lot of questions are answered and the path is clear, public sector may be able to gather the capital and resources to accelerate this).

The interventions (actions going against the tide, so to speak) to be treated here are by no means exceptional. We can find similar efforts from time to time in all the modes. In the transit case, for example, there was extensive planning and investment in subway improvements serving central business districts (CBDs) during the 1930s. Its purpose was

TABLE 10.1

Typology of Interventions, with Examples

	Accelerants	Brakes
Growth	Chinese Twenty-first Century Infrastructure Investment	European Automobile Tax Rates
Decline	US Fuel Economy Standards (Oil consumption)	US Twentieth-Century Transit Investments

to reverse the erosion of transit ridership associated with the decentralization of activities from the CBD. National airport planning and investment programs have attempted to steer the development of the US transportation system by decentralizing service.

Further, countries can slow deployment of unwanted technologies. During the early decades of the twentieth century, European nations and Japan attempted to intervene in the growth of automobilization using taxation schemes and by restricting the development of infrastructure. Many countries today still impose taxes and regulations on cars, at least nominally to address environmental and congestion effects, though surely some revenue transfer is involved.

Interventions are summarized in Table 10.1. Accelerating growth makes it go faster, braking growth makes it go slower. Braking decline aims to arrest or reverse direction. Accelerating decline aims to put it out of its misery sooner.

To extend the metaphor, we think of these interventions as changing the medium through which the magic bullet moves. Bullets slow when traveling through media with greater drag than air, like water.[221]

10.5.2 INTERVENTIONS CAN COORDINATE BETWEEN DISJOINT ACTORS

We think of two situations where intervening in within-system decision making can be successful.

- Systems behave disjointly. Intervention has forced joint action by components, actions that might not have occurred if matters had been left to run their course. Left to equipment owners and operators, for example, weight limits on trucks would not have been planned and implemented.
- Systems generate externalities. There has been successful intervention to reduce noise emissions from aircraft and trucks, for instance.

10.5.3 HARDENING IMPLIES INTERVENTIONS MAY BE INCONSEQUENTIAL

Intervention can counter system decision-making behavior in the accomplishment of things that aren't very consequential. This statement needs tempering. The record seems to say that it becomes more and more difficult to counter system behavior as systems age. Early on, systems are clay-like, and more consequential things can be done. As time goes by, their features harden; they become more like bricks.

Notably, policy and regulation are reactive systems, and so tend to intervene in the later stages of markets, limiting the consequences of their interventions.

10.6 Speed enables specialization

Users of systems choose based on better, faster, cheaper service attributes, and how such attributes change as size increases/decreases is at issue. Speed is one of the three most important attributes. As we learned in algebra class, $D = S \cdot T$, Distance equals Speed times Time. If I increase speed, I either lower my travel time or increase my travel distance.

Speeds of technologies have risen significantly over the past centuries, since at least the industrial revolution. Speeds of modes have been getting successively faster, and while a particular technology may have an upper limit, soon (it seems to happen every 50 years or so, and by the graph in Figure 10.2 we are due) a new faster technology will come along. This is illustrated in Figure 10.2, which primarily considers long-distance modes.

We can draw an envelope (which may be a macroscopic S-curve for all transportation modes), which is either continuing to rise (if rocket or other high-speed transportation takes off for transportation rather than just exploration), or leveling off, if the jet plane is the practical upper limit (and the Concorde story told in Chapter 12 would seem to indicate that air transport is not getting faster any time soon). The changes in short-distance transportation are not as stark (we can take out the fast air and slow maritime modes). Yet clearly the auto

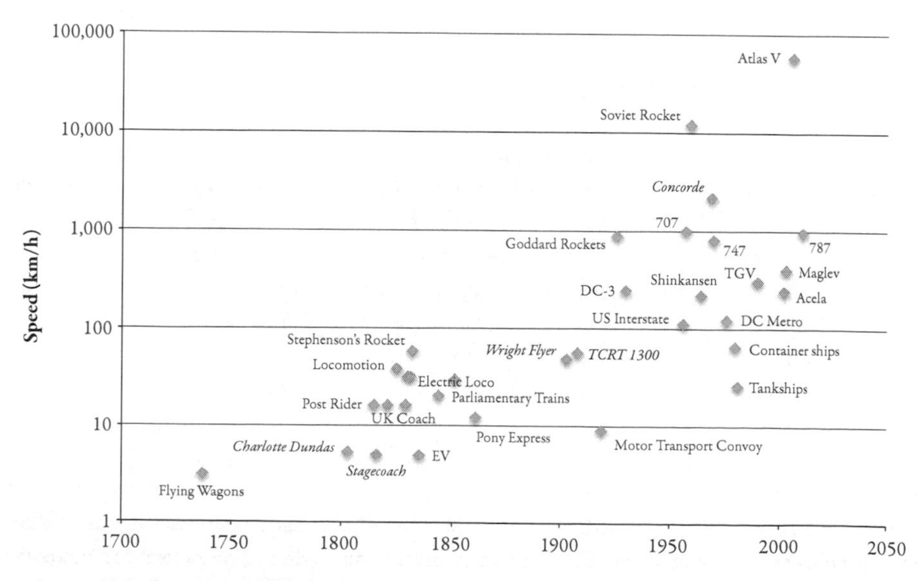

FIGURE 10.2 Operating Speeds for Major Transport Modes, 1750–2000

is faster than the horse-drawn carriage, which was faster than the pedestrian. Similarly, the DC Metro (subway) is faster than the Twin Cities Rapid Transit streetcar.

We observe that the speed gains over time are leading primarily not to time savings but rather to distance increases. If there is not quite a travel time budget, there is still a tendency that should be noted. We discuss this further in Section 24.5.1. The ability to cover a longer distance enables greater access to "the country" (parks and farms) and larger yards. This increase in speed has been reshaping urban regions for over a century, starting first with transit (see Chapter 8).

The changes in speed are also among the dominant factors in mode choice. Chapter 14 notes that in the United States, passengers for transit modes declined steadily from 1920 until near the end of the twentieth century, and have been flat since. One reason is that rising income enables people to buy cars, but the reason people want to buy cars is that they are faster. At best, rail transit and bus rapid transit (bus on HOV lanes) are faster in the very specific corridor they serve, but unless travelers have their origin and destination on top of the corridor, it is likely that the costs to access and egress the transit system will be high. In contrast, cars are parked at home, and for the vast majority of US residents at least, can be parked very close to the final destination. Time, which includes time in-vehicle as well as access, egress, and waiting time, is the main measure of convenience, and is central to all travel choices. It explains some of the difficulties with getting people to carpool. If individuals have different home or work locations, a carpool requires at least one of the parties to go out of his or her way to pick up or drop off another. The time savings on the long-haul part of the trip must be significant to overcome this inconvenience.

The simplest explanation one can give for "need for speed" is that value of time is proportional to income. As incomes rise, the disutility of time expended in travel is greater. Travelers need speed to decrease the time wasted in travel. Similar reasons apply to freight transportation: higher valued goods, just-in-time inventory systems, and higher real wages to vehicle operators are manifestations of the issue.

Such explanations are quite satisfactory to many. They are simple, and that's a virtue. They also allow the analyst to go one step backward in the reasoning. In the case of the value of the individual's time, for example, one can cite the relations between the wage rate and the value of time, reasons for higher real wages, and so on.

Yet speed is being consumed not in time savings, but in distance. The trend is toward lower residential densities within cities, at regional scales, and at sites of activities. We see more space consumption per person both at home and at work.

Perhaps the better explanation is that specialization of activities and places is enabled by higher speeds. Specialization is a source of higher incomes. So there is a magic bullet–like positive feedback loop: specialization increases income leading to higher speeds enabling more specialization, as seen in Figure 10.3.

This loop, while positive, is not exponential; rather, it may have limits. Consider the marginal velocity gain that early automobiles gave over horse-drawn vehicles, say, from 8 to 24 km/h (5 to 15 mph). That 16 km/h (10 mph) added to, say, 88 km/h (55 mph) hardly compares in the magnitude of relatives. But we also observe that there is a geometric relationship at work. In a simple form, we observe the increased quantity of space accessible as the radius of a circle is increased. Here, we would think of the radius defined by velocity. How far can we travel in some amount of time? As (Euclidean) speed tripled from 8 to 24

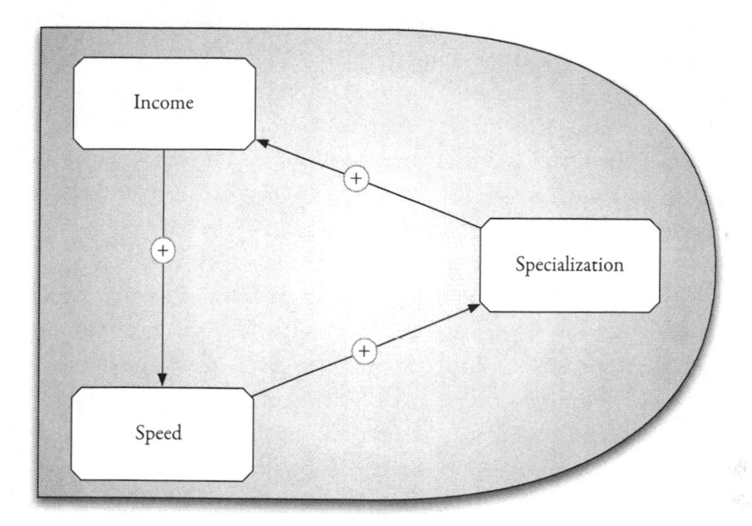

FIGURE 10.3 Accelerator

km/h, time spent travelling 1 km dropped 5 minutes from 7.5 min to 2.5 min. Alternatively, in 7.5 minutes when we triple speed, we move from being able to reach 1π km^2 to 9π km^2. If densities are uniform, we can reach nine times as many things when we triple speed. This enables huge specializations, and allows metropolitan areas to serve many more people in a reasonable amount of time.

As speed increases 16 km/h from 88 km/h to 104 km/h, the time drops for our 1 km trip from 41 seconds to 35 seconds—a mere 6.3 seconds—which hardly seems worth discussing. Though a time savings of 6.3 seconds per vehicle sounds small, it still adds up to 175 person hours a day (or 7.3 person days/day) on a busy road with 100,000 cars per day. The lives of 7.3 persons per day, at a value of life of about $3 million suggests the speed hike is saving in economic terms $21,875,000. Alternatively, at a $10/hour value of time, the speed increase saves $1,750/day, or $638,750 per year, or $19,157,100 over thirty years with 0 percent interest rates. Those numbers might be worth talking about, but whether small values of time are additive in such a fashion is a controversial question in transportation economics, though most practice does add them. In brief, since time is likely the dominant benefit, if the project to save 6.3 seconds per vehicle costs much less than $19 million, it is probably worthwhile; but if it costs much more, it probably isn't. Ezra Hauer has an interesting paper comparing value of time and value of life "Can One Estimate the Value of Life: Or Is It Better to Be Dead Than Stuck in Traffic?"[222] There are diminishing returns to our positive feedback loop, and they may be at work at the urban scale.

Now, consider the problem of structuring causality. We could just as soon say that higher velocities result from the desire to use more space as to say that they result from higher incomes. Furthermore, we could tie the specialization trend to either of the bases for causal reasoning. What is the bottom line from this discussion? We need to engage in empirical work exploring causal models. That work would make a difference in practical affairs and could guide planning.

Suppose we want to engage in research tied to the higher velocity trend. We have some alternative technologies in mind, for example, using high-speed rail as one building block in a system or using high-velocity automobiles as a building block for another system.

What to do? Unless we can be "wise" about causes of the velocity trend, there is little to guide us. If it is more space that is at issue, then perhaps line-haul velocities are not as important as collector distributor ones. (That is, we want to make our areas that can be accessed work more like circles, reducing the circuity so the ratio of network distance to Euclidean distance is closer to 1.0.[223]) If the issue is the higher value of time, then we might want to look at some places where velocities are low and where marginal changes will make a big difference. (A difference like the one that the Model-T made over horse systems.) Eliminating waiting in the system, places where the speed is zero, is a good place to start. Making waiting or in-vehicle time productive, with the use of example mobile computing and transit or self-driving vehicles, is another potential direction, so the value of time lost while traveling is diminished, since time in motion can be almost as productive as time at the destination.

Deregulation, increased specialization of transportation services, and developments in communications and control are enabling changes in procurement and distribution to which we have attached words such as "just-in-time manufacturing." Deregulation has been important because shippers and customers can now bargain with transportation carriers and suppliers. Logistics managers consider inventory costs, size of shipment, and many other things. As with any transportation changes, there are two impacts. The first is the reduced logistics cost (doing the same things faster, cheaper). The second is the improved quality, resulting from an increased variety of goods and services offered by both being able to reach new markets and being able to do new things. The explosion of logistics activities and the reshaping they occasion serves as a model for changes in the ways people live and work.

Our discussion above described schedules in the late twentieth and early twenty-first centuries. The status and future of such scheduled living is at question because of the changing nature of work, improved communication, and other things. The requirement to work (play, learn) at an appointed place and time lessens as intellectual capital replaces physical capital as the standard. The web and the wireless are much less cruel taskmasters than the factory and the field. As life in the advanced countries becomes more flexible, the rigor of the weekday becomes weakened and the schedule flexibility of the weekend becomes the daily norm. Freedom is achieved when citizens regain control of their schedules. There is an important corollary: freedom is lost when schedules are tightened.

10.7 Discussion

Frequency dependence (network externalities) appears when a new technology succeeds because it has been adopted by a large number of users. The classic example is the railway gauge. Standards ensure the ability to communicate and inter-operate, but this makes change and innovation more and more difficult. Social learning occurs as individuals short-circuit the long painful development process by copying and help to reinforce standards. Learning-by-doing has average costs declining with cumulative output. All of this learning is good, as it drives the magic bullet. However, it has a cost. As we climb up one mountain, we are further and further away from the peak of another mountain, we get locked in to one way of doing things, and foreclose opportunities.[224]

Yet not every potential market is realized. There are many might-have-beens. Would small aircraft like the Ford Flivver have become popular after World War II with sufficient development, and overtaken commercial aviation? Could the steam engine have been revived in the automobile with enough concerted investment by auto-makers?[225]

Scalability and associated economies, speed, safety, and economy increases have been treated in a system growth/deployment frame. Another scale is introduced when we think of changes to an existing system: How well can an existing service respond to increases or decreases in demand?

Phase 3, the next phase of the Life-cycle, is discussed in Chapter 18.

5 Wave Three: 1890–1950

You don't understand! I could've had class. I could have been a contender. I could have been somebody, instead of a bum, which is what I am.

—TERRY MALLOY IN MOVIE *On The Waterfront*

11 American Shipping

LIKE LONGSHOREMAN TERRY Malloy in *On The Waterfront*, American shipping has become a bum instead of a contender. This happened over many years, for many reasons, and policy has failed to reverse the trends. The American shipbuilding advantage was first strained in the 1830s when the British began to use their advantages in propulsion power know-how and iron and then steel production to build large, fast steamships. The British also started a mail contract program in 1839, a program emulated by the United States in 1845. Recall Brunel's construction of the *Great Western*. With the Brunel "seeded" advantage, the British began to dominate the trades.

The merchant marine developed two types of ships serving different markets: transients (tramps) and regulars, the analogy here is to taxis serving various origins and destinations, and fixed route services, going back and forth on pre-specified routes. There were also a variety of vessels. Sloops (one mast, fore and aft rig, 25 tons) and schooners (two mast, fore and aft rig, 75 tons) were sailing vessels important on the coastal trade. Square rigged "brigs" (2 mast, 150 tons) and "ships" (3 mast, 300 tons) conducted ocean trade.

The clipper ship was the US response to the British steam and iron hull advantage, a response that kept important market niches for the United States for some years. However, the US Civil War destroyed the US fleet, and by the 1870s, the United States had little presence in maritime trade. Steps were taken to reverse this in the late 1800s. To build up the inputs needed for the shipbuilding industry, the Morrill Act of 1884 placed tariffs on iron and steel plates and on marine engines. To build up the market, it restricted the US registry to ships built in the United States. Many say that the nation was preoccupied with internal development during this period, and this was the reason for decline in the shipping industry. To us, that sounds more like an excuse than a reason. The United States was not competitive. The Morrill Act did not make the United States competitive, by 1900 US tonnage was at the 1807 level, and ships of US registry moved only 9 percent of US foreign trade.

By the turn of the twentieth century, the United States acquired some colonial interests and a taste for world power. A symbol was the around the world trip of the Great White Fleet of 1908, which brought the US maritime situation in focus. For logistics, the fleet had to be followed by a ragtag collection of foreign-flag trampers. The Navy had long been concerned about logistics dependence on foreign flag vessels.

The Cargo Preference Act of 1904 had already begun to respond to that concern; it required that military goods shipped overseas be carried on US ships. To help enlarge the fleet, the Panama Canal Act of 1912 allowed the registry of foreign-built vessels, and it enlarged the market by extending cabotage (the exclusive right to navigate in coastal waters; see Section 20.3) to Hawaii and the Philippines. The Seaman's Act of 1915 required US crews on ships of US registry. These were efforts to enlarge and control a fleet of vessels, but their impact was not great.

The United States Shipping Board was established in 1916 at the advent of World War I to coordinate the logistics of the US Merchant Marine. When the United States entered the war, the Board acted with vigor. The main result was a large US fleet at the end of a war which saw 470 ships completed. Another 1,300 were delivered in the four subsequent years. All flew the US flag, and the US fleet followed only the merchant fleet of the United Kingdom in size. The act also endorsed the participation of US carriers in open conferences to set rates (i.e., participate in the maritime rate setting in the style of rail tariff bureaus). This was a murky area before. The European carriers had established conferences in the Atlantic at about 1875. Because antitrust laws were established and enforced in the United States, the participation by US carriers was at question. An open conference is one that publishes rates and permits anyone to join. The Act also allowed transfer of Shipping Board ships to foreign registry—Panama at first, and then Honduras. The reason was to encourage neutral flag carriers during wartime.

The Jones Act of 1920 required US flagged vessels built in the United States for internal US commerce, and disposed of the Shipping Board and its ships. The ships were sold at near scrap prices. This law thus endorsed the principle of cabotage. With the disposal of the fleet, the US maritime activities slumped. The slump set off half-hearted debates, and it yielded the 1928 Act, which reestablished the mail subsidy and made available construction loans for shipbuilding. However, for international trade, US flagged vessels need not be built in United States. There was division of domestic shipping into intercoastal trade (from East to West coasts), and coastwise trade (along either the Atlantic or Pacific coasts, but not between them. Clearly domestic US shipping dropped significantly over the twentieth century as rail and especially trucking soared. Intercoastal vessels declined from 165 in 1939 to 57 in 1954, coastwise vessels from 543 to 283 in the same period. Since rarely were ports the final destination, shorter distance trips by trucks could reduce handling costs (even in the more efficient container era). Rail improved to handle most longer distance shipments. Petroleum tankers were the vast majority of coastwise trade by 1954.

The Great Depression reduced international trade, and things drifted along until the 1936 Merchant Marine Act established the US Maritime Commission (USMC) to develop standardized specifications for new ships.[226] As with the PCC Streetcar (Section 14.2), consumers demanded standardized vehicles from producers to lower costs. The USMC also established essential trade routes, and assigning services, and establishing wage and manning requirements for ships. The Act canceled mail subsidies and established construction-differential subsidies (CDS) and operation-differential subsidies (ODS). The CDS ranged from 35 to 50 percent and made low cost loans available. The subsidy was only available if approved by the US Navy, and the ships constructed were in the essential naval reserve. The ODS had no limit placed on them, except that ships had to be less than 20 years of age and operated in the foreign trades. These subsidies began to rebuild the US fleet modestly, and by 1941 the US fleet had about 16 percent of world tonnage.

The United States emerged from World War II with 60 percent of the world's tonnage. The 1948 Merchant Ship Sales Act was passed to dispose of the tonnage, and by 1950, the US fleet was reduced to 36 percent of world tonnage. The industry was heavily regulated. After a lot of paperwork and time, the Interstate Commerce Commission (ICC) issued "certificates of convenience" before service could be provided. Only trade internal to companies, for example an oil company with its own fleet, would have flexibility on pricing and service.

Policy at the beginning of World War II was to build easily assembled, inexpensive, and rather slow sailing ships, example, Liberty Ships. But some analysis soon showed payoffs from improved technology tankers (also termed tankships) and other new models that were placed in production and began accumulating in the fleet. Many older ships were scrapped, and these newer ships, purchased especially by US and European operators, formed the basis of the post–World War II fleets.

Although not generally recognized at first, there was still room for technological change. In part this was scale-up in form, especially for tankers. Tankers ranged from 16,000 dwts T-2 to upward of 600,000 dwts.[227] Tankers larger than 160,000 dwts are termed very large crude carriers (VLCCS). Most are in the size range of 250,000–450,000 dwts. Modern designs have the bridge aft, and use a single screw and rudder.

While Figure 11.1 shows the domestic fleet, some say the real fleet is the effective US control (EUSC) fleet. Ships registered with Liberia, Panama, Honduras, the Bahamas, and the Marshall Islands are often considered part of the EUSC fleet, and simply registered with these "flags of convenience."[228] These ships are generally owned by US citizens or corporations, and are subject to requisition.[229] The EUSC concept originated prior to World War II to avoid the Neutrality Act as a way to provide matériel to Great Britain. Most of the ships are crewed by non-US citizens, and thus seizing the ships will not result in gaining the crews (the principle that sailors were not to be pressed into service was established by the War of 1812).

The EUSC fleet defined as above (7,927 ships) is about 18 times the size of the flag fleet (443 ships), and constitutes about 28 percent of the world's fleet (28,296 ships) (see Figure 11.2). The Maritime Administration encouraged the development of the EUSC in the late 1940s by allowing World War II ships to be transferred to foreign flags in order to encourage new work for American shipyards. These ships were transferred with the agreement of the Defense Department; they could be recalled, if needed. Some effective control was maintained then, but effective control of today's EUSC fleet is a fiction. Many of the ships are tank-ships, and are too large to call at US ports. Yet, during the Gulf War of 1990–1991, many tankers switched to a US flag to ensure the protection of the US military.

As we will see in Chapter 19, the adoption of sensors, communications, and computers, along with containerization, has yielded sharp reductions in the labor required, as well as

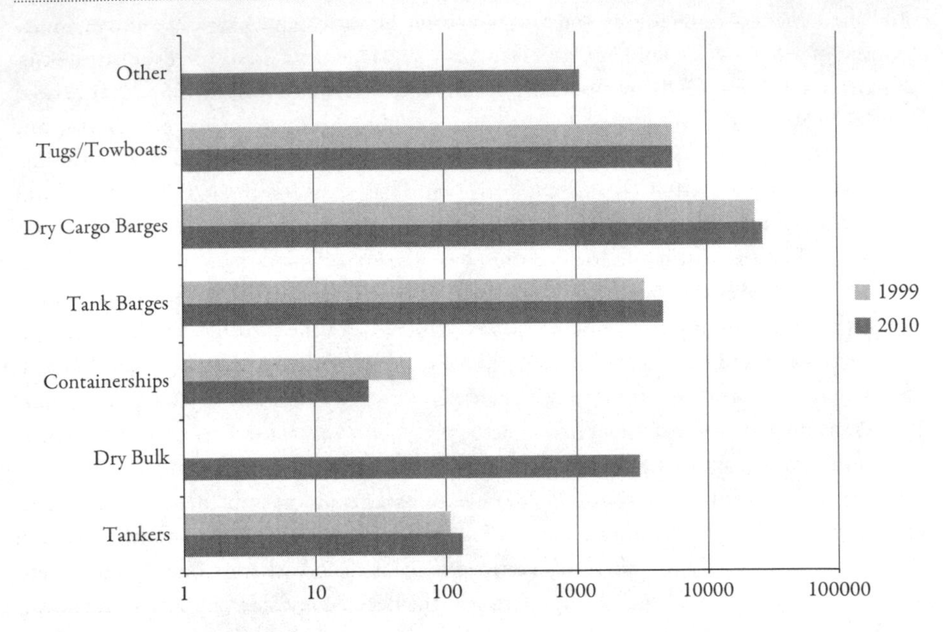

FIGURE 11.1 US Domestic Fleet

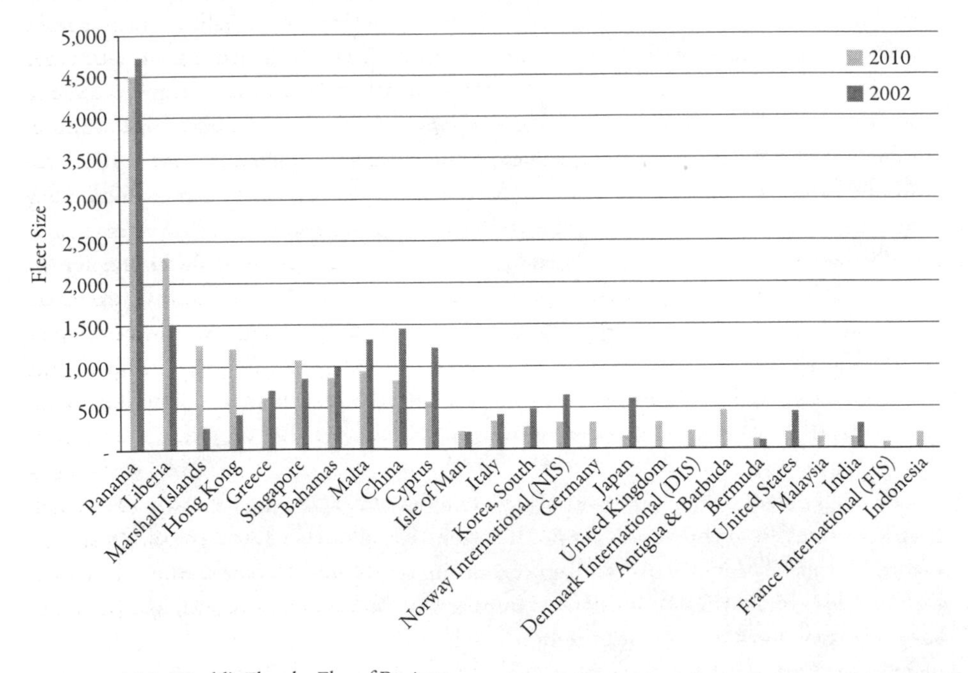

FIGURE 11.2 World's Fleet by Flag of Registry

advances in ship routing and scheduling. Armed with technical know-how, one would think the US fleet gained from these developments. One would think that the US fleet could have been a contender for world trade. Instead, the US has, on world comparison, a small fleet.

We do not have a good answer to the question of the lack of US comparative advantage. Discussions we hear stress the low cost of buying and using World War II ships—that was a bargain with the devil for Western European and American operators. Other nations without those ships adopted new technology forms to produce competitive ships, and that eventually gave them a lead. The tankships were the first ships to scale-up, and shipyards building those gained technology experience. We also hear that the US shipbuilding industry is too Navy-oriented, both in technology and contracting style.

The debate continues about the fiscal aids that government gives sea trade (not Jones Act) shipbuilders and operators, and it is argued that these aids are counterproductive. Construction-differential subsidies through its tie to naval interests overemphasized the break-bulk cargo liner, and the operation-differential subsidies blunted operators' searches for efficient ships and ship operations. To avoid the special US conditions on registry, some large operators have extensively used flags of convenience. Today, the Liberian flag is widely used.

To try to turn this situation around, 1970 legislation extended the construction and operating subsidies to non-liner ships, and the CDS was increased to 50 percent. Legislation in 1984 allowed liner service to inland points using contracts with other modes. That advantage is available to all liner operators, of course. But one cannot blame the situation entirely on subsidies.

Issues of regulation and protectionism remain within the global ocean freight industry (as in international air travel). While it is one thing for a government to let domestic firms compete for domestic trade unfettered (at least all of the gains stay national), it is another to allow unregulated international flows. Ships from most countries can carry US trade to other countries. In that sense, trade is already deregulated. But there are periodic pressures (to support the domestic shipping or shipbuilding industries) to impose some sort of protectionist measures (eg., cargo reservation), to reserve a certain fraction of cargo for US flag ships. Other countries do similar things to a greater or lesser extent.

With respect to the vitality of the US fleet and construction industry, one impression we have is that we ought to let things run their course. It has been said that US bottoms are more expensive because of mandated safety features, and they are more expensive to operate because of high labor costs. But as other nations develop, labor costs and requirements for safety go up. One can make similar comments about shipbuilding. The problem with that view is that cost differences are bound to be around for a long time, and the US fleet and shipbuilding would disappear if there were no construction and operating costs subsidies. At best, we would end up with a US-owned foreign flag fleet, as we have in the bulk trades. But that's a risk. The defense sector wouldn't accept that risk, because they wouldn't have an effective US controlled (EUSC) resource. They wouldn't have the shipyards either. An option would be to continue policies now in force, make them stronger, and put more money behind them.

The story of the water modes continues in Chapter 19.

Airplanes will eventually be fast, they will be used in sport, but they are not to be thought of as commercial carriers. To say nothing of the danger, the sizes must remain small and the passengers few, because the weight will, for the same design, increase as the cube of the dimensions, while the supporting surfaces will only increase as the square.

—OCTAVE CHANUTE, AMERICAN AVIATION PIONEER (1904)

12 Taking Flight

12.1 Profile: Juan Trippe

TO PROVIDE MASS air transportation for the average man at rates he can afford to pay.

– Juan Trippe's stated goal for Pan Am

The Trippe family were bankers descended from English seafarers who had settled in Maryland in the 1600s. Born in Sea Bright, New Jersey, Juan Trippe (1899–1981) attended Yale, where he played football and served as a Navy flier during World War I. Trippe formed his first airline, Long Island Airways, to provide taxi service in the New York area to other wealthy gadabouts.

While Long Island Airways failed, he and classmates from Yale, William A. Rockefeller and Cornelius Vanderbilt Whitney (great great grandson of Commodore Cornelius Vanderbilt, and son of Harry Payne Whitney [and thus a second cousin, once removed to Henry Melville Whitney of the New York Metropolitan Street Railway Company]), joined Colonial Air Transport in 1922, which was a beneficiary of the federal airmail contracts, serving the Boston to New York market. It added passenger service in 1927.

While not Hispanic, Trippe used his name as a way to enter the Latin American markets (the source of his name "Juan" is disputed). After clashing with Colonial over expansion, which Trippe favored but the stockholders opposed, Trippe formed a new company (Aviation Corporation of America) in 1927, and using his name as an aid, obtained Cuban landing rights from Cuban President Garardo Machado. However, when the airline won the landing rights, it had no airplanes, no money, and no contract. One competitor had the airmail contract, but no landing rights. A third company, Atlantic, Gulf, and Caribbean

Airways, had access to Wall Street financing. These three companies (none of which were operating planes at the time) merged and adopted the name Pan Am, from one of those newly merged companies (run by Henry "Hap" Arnold, who went on to run the Army Air Force during World War II). In 1927, the new company under Trippe's leadership was able to provide airmail and, by 1928, passenger service between Havana, Cuba, and Key West, Florida.

Trippe successfully used the levers of government (both US and non-US) to further private ends, making Pan Am effectively the US's flag carrier. With airmail contracts lined up in Latin American markets, the State Department would work on his behalf to ensure it a monopoly position where possible. Trippe then used Pan Am's dominant position in the air carrier business to play airframe manufacturers such as Boeing and Douglas against each other. The Clipper services establishing long-distance over-water services (first Caribbean, then Atlantic and Pacific routes) were essential for Pan Am to ensure dominance.

While Trippe started out providing luxury services with his Long Island Airways, he was the first to expand the market to the middle classes by establishing "tourist class" fares between New York and London ($275 in 1945).[230] This drew the ire of competing flag carriers and the IATA, the International Air Transport Association (originally International Air Traffic Association), and Britain closed its airports to Pan Am, forcing the airline to fly to Shannon Airport in Ireland.

Trippe introduced Jet aircraft, the Boeing 707, in 1958, which were faster, higher, and carried twice the load. Trippe created the "jet set" with the trans-Atlantic jet services, but then brought the "jet set" within reason by continuously lowering costs and prices. While the 707 had a cost of $0.041 per seat km ($0.066 per seat mile), he wanted to lower that further, which could only be achieved with greater economies of scale.[231]

A new class of still larger jets was needed. To that end, he told Boeing, "If you build it, I'll buy it," to which Boeing's CEO William Allen replied, "If you buy it, I'll build it." The supplier-customer relationship was not simply a market transaction, but involved significant negotiation. Aircraft were far from a perfectly competitive market, and at the time, the same was true of airlines. Pan Am was intimately involved in the design of the 747, including ensuring that part of the upper deck would be for passengers.

While the 747 was ultimately a very successful aircraft, still in widespread use, it had such high fixed costs that it almost bankrupted both Pan Am and Boeing. Pan Am was hit hard by rising oil prices in the 1970s, which forced it to raise fares and reduced demand. Trippe died in 1981 and Pan Am merged with National Airlines, a Miami-based domestic carrier. The airline failed in 1991 for a variety of reasons, among them the December 1988 Lockerbie bombing sponsored by Libya.

12.2 System Evolution

Like many technologies, there were many false starts. The first manned flight may have been in 559, when the Emporer's son, Yuan Huangtou of Ye, China, was forcibly strapped to a kite and set airborne from a tower. Yuan Huangtou was later executed, and this experiment did not lead to any follow-on. The French Montgolfier brothers designed and took off in a hot-air balloon in 1783 (with others), which is credited as the first manned free-flight. In

1891, German Otto Lilienthal flew in a glider. By 1898, internal combustion engines were powering airships (dirigibles) designed and flown by Brazilian Alberto Santos-Dumont. Numerous other pioneers made attempts.

Coming to the United States as an observer to the American Civil War representing Württemberg, Count Ferdinand von Zeppelin decided to explore the North American frontier.[232] While staying in St. Paul, Minnesota, he encountered a fellow German who had served for the Union inflating a hot-air balloon. It was here Count Zeppelin first went airborne in 1863. Almost half a century later (1909), the Zeppelin Company he founded was facing financial difficulties selling airships to the German military, and decided to start an airline (DELAG). While it was not at first as successful in organizing regular service, it did provide some (with logistical support from the Hamburg-Amerika steamship lines), marking the first commercial airline. By 1914, DELAG had made over 2,000 flights, totaling 100,000 miles (160,000 km), carrying 34,028 passengers. World War I changed the nature of airship use, and the German military's interest. By war's end, almost 100 airships had been used for the German army and navy, and more than half were lost, indicating lower success than hoped for, and significantly underperforming airplanes. Still, the Americans showed some interest, and after the war, the US Navy ordered some airships from the Zeppelin Company as part of reparations payments, to the anger of some Germans, who disapproved of the technology transfer, and an assassin skulked Dr. Hugo Eckener, the manager of the Zeppelin Company. Eckener almost stood for president of Germany against Hitler, but stepped aside when war hero Count von Hindenberg chose to run again. Prior to the 1938 *LZ 129 Hindenberg* (named for the Count) disaster in Lakehurst, New Jersey, no passenger had died due to a crash of a Zeppelin airship. Transatlantic airship travel was not doomed due to this one well-publicized crash, but rather to the rise of Pan American Airways flights, which were much faster, though less comfortable. Though there was an attempt to convert future airships (such as the *LZ 130 Graf Zeppelin II*) to helium rather than hydrogen, the US government withheld helium supplies under the 1937 Neutrality Act.

Lifting vehicles have been available for centuries (kites, gliders, balloons), and as propulsion systems came along there were efforts to apply them. An early assumption was that vehicles would emulate birds, and that assumption continued with respect to control of flight into the 1920s, even after the propeller was adopted. (And well after the Wright brothers flew with a fixed wing in 1903.) In hindsight, we know that the developers of the times took an overly simplistic view of the way bird flight is controlled.

The system began to evolve in the 1920s in over-water market niches. Important actors were Pan American Airlines and producers of "flying boats." Commercially viable planes were produced, corporate management and passenger interrelations were worked out, and there was important development of en-route control systems. Although, where national borders were crossed, firms had to make arrangements for rights to provide service and there was airmail subsidy, governments were not much involved in the 1920s and early 1930s.

Overland commercial systems began to evolve in the 1920s, but their birth is best marked at about 1935 by the development of the DC-3, a suitable aircraft for such services.[233] While Boeing's Model-247 was the innovative aircraft, combining the best of existing know-how, Douglas's DC-2 and DC-3 were larger and faster, and flew farther. In particular, the DC-3 "got it right" for conditions at the time (size, velocity, cost); it was the Model-T of aircraft, so to speak.[234]

At about that same time, suitable aircraft for other markets began to evolve, example, the Gypsy Moth and the Piper Cub for recreational flying, the Norseman for arctic flying, fighter and bomber aircraft, and so forth.

Evolution of the system includes more than equipment. Aids to en-route navigation evolved. The 1930s saw development of the AM radio range, suitable voice communication radio, tower and en-route traffic control, weather forecasting, and so on. Air traffic control improvements introduced after World War II included changing to FM beacons and the use of radar. The airmail subsidy played a major role (see Section 12.3).

Heavy government involvement in the 1930s included the development of the precursor to the Federal Aviation Administration (FAA) and government regulation along the railroad model. Management of labor disputes was handled by the Railroad Labor Relations Board. Costs were set for passenger travel using tapered per mile rates. Established during the 1930s, regulation shared an aim of trucking regulation: protect existing operators from new entrants. Policy was also concerned with service to small communities; there was the requirement that small community service be included in the route structures of firms and a minimum level of service provided.

Regulation kept new firms out of the business. Even so, old firms were able to increase service areas, and service was gradually increased in the larger markets. Complex classes of fares had been introduced, along with hubbing. Indeed, O'Hare Airport in Chicago was a hub for DC-6 service as early as the 1950s.

The situation was quite similar in other nations. However, during this period they were more affected by the winds of war. The increased internationalization of air travel after World War II drove an international policy agenda beginning in the late 1940s. As part of that, a debate in the United States continued from the 1930s about whether there should there be a recognized US flag carrier. The United Nations, established at the end of World War II, became involved in air transportation rather early on through its International Civil Aviation Organization (ICAO).

In the early days, services had mainly been from small, privately owned fields, many not much improved over pastures. All that was needed was a level grass field and a shack or so. Some people suggested that train stations, which already had passenger facilities and long straight stretches of land, could be adapted to serve as airports as well. That was not to be, as airplanes need runways pointing in specific alignments. Train station architecture was, however, adapted to serve airports.

As DC-3 service came along, longer runways and pavements were needed. Airports began to be developed by cities.[235] Airports provided a gateway to a city, and they enabled the city to participate in the air system. On the heels of World War II, many military fields adjoining cities were obtained for commercial use. Although a few airports have been constructed from scratch, in the main the former military fields have been expanded to accommodate traffic growth.

In metropolitan New York, Newark Airport was the first major facility with paved runways and runway lights. It was followed by LaGuardia. When Robert Moses (see Section 22.3.4) misstepped and proposed raising rates to Eddie Rickenbacker, the aviator running Eastern Airlines at the new Idlewild (later John F. Kennedy) Airport, Eastern left for Newark. To avoid such intra-metropolitan competition (which was good for the airlines, not so good for the airport owners), New York Mayor O'Dwyer transferred control of New York City's airports to the Port Authority of New York (and New Jersey), which soon controlled Newark Airport as well as Port Newark. The Port Authority was modeled on the Port of London Authority, and was established in response to inefficiencies during World War I mobilization.[236] Its territory encompassed all the area within a 35-mile radius of the Statue of Liberty.[237]

12.3 Air Mail

Mail (see Section 3.10) has been used to justify US federal involvement in transportation since the beginning of the Constitution. Mail subsidies aided the railroads. Similarly, airmail contracts were the savior for young airlines. Melville Kelly authored the Contract Air Mail Act of 1925 and the Foreign Air Mail Act of 1928 (the Kelly Acts), which used the post office to subsidize (and ensure monopolies) for given origin-destination pairs. The mail contracts covered the costs of the flights, enabling airlines to sell seats for a profit.

In 1930, the Watres-McNary Act enabled the Postmaster General to award competitive air mail contracts to the lowest responsible bidder, rather than simply the lowest bidder. This made the Postmaster General, Walter Folger Brown (who helped draft the Act), a virtual czar over the airlines, determining their fate. The main three east-west routes were given to United Airlines (controlled by Boeing at the time) (the northern route: New York–Chicago–San Francisco), Transcontinental and Western (TWA) (the middle route New York–St. Louis–Los Angeles), and American Airways (the southern route Washington–Dallas–Los Angeles). Eastern Airlines was awarded the main north-south route on the east coast (Boston–New York–Miami). Braniff was later awarded a north-south route connecting Chicago and Dallas, and Delta was given a route connecting Charleston with Dallas. These big six continued to dominate US domestic air travel until the era of deregulation, and their hubs (or hubs of successor airlines) are still located in cities for which they received contracts in 1930, showing the effects of lock-in.

Airmail was expected to become standard by many prognosticators. Roger W. Babson was an investor, serial college founder, and 1940 Prohibition Party candidate for president of the United States. He wrote: "Within the next fifty years practically all mail and most of the express will be carried by airplanes. When a standard commercial plane is developed, the depreciation charges can be reduced so that the cost of carrying passengers and light freight will be less by plane than by rail. When this comes at least one third of the railroad mileage will be scrapped."[238] It was not the plane that caused the railroad to scrap mileage, but moreso the truck, and general consolidation trends as discussed in Chapter 21. The prediction was generally accurate, though.

As an amusing side note, there have been several attempts to use rockets or missiles to deliver the mail. In 1931, Freidrich Scmiedl launched a V-7 rocket with 102 letters in Austria. In 1934, Gerhard Zucker used powder rockets in the United Kingdom. In 1934, Stephen Smith used rockets to deliver some mail in India. In 1936, rockets containing letters were launched and landed in Greenwood Lake, New Jersey.

The United States announced its "first official missile mail" (despite the obvious antecedents) in 1959, when letters launched from a submarine on a cruise missile hit a target in Mayport, Florida, and was then brought to Jacksonville, Florida, for sorting and delivery.

In the end, rocket mail has thus far been a technological dud.

12.4 Discussion

As can be seen in Figure 12.1, the median passenger-weighted airport among the top thirty as of 2006 was founded in the early 1940s. The era of new airports is over, and the rate of new important airports joining the world system is very slow. Large, hub-worthy cities have airports, and while new cities may still join the world system, all the best cities have been founded.

Hubs at these airports persist. Airports that are hub-worthy tend to get attached to specific airlines. The three large network carriers in the US systems (c. 2012), American, United, and Delta, and their largest recent merger partners (US Airways, Continental, and Northwest, respectively) are shown in Table 12.1. The early airmail routes from the Post Office, or passenger routes granted by the Civil Aeronautics Board, were sufficient to establish hub lock-in some 90 years later. Not all early carriers survived; Braniff, Eastern, Pan-Am, and TWA are some of the largest names not to have made it to the present day, though several were acquired rather than dissolved. Consolidation in aviation, as other transportation modes, is an important story. Students of business history can map the particulars, but for instance, American Airlines was the product of combination of 72 earlier companies, Northwest of 8, and so on. First mover advantage is strong, despite the mobile capital in the airline sector.[239]

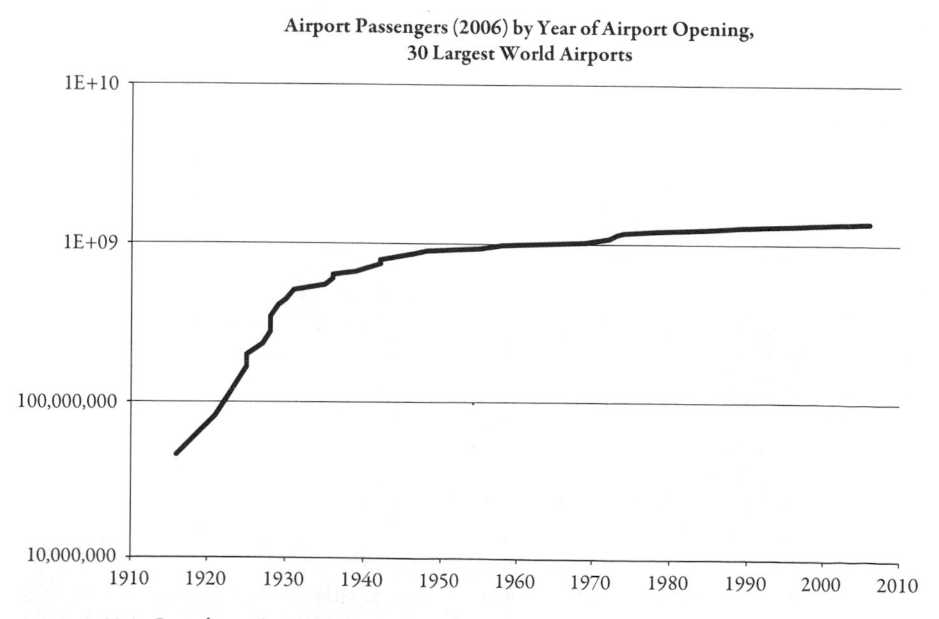

FIGURE 12.1 Cumulative Passengers at 30 Largest World Airports by Opening Year

TABLE 12.1

US Network Airline Hub Cities

Airline	Year	Hub Cities	First mail service	First passenger service
American Airlines	1930	**Dallas**, Miami, San Juan, **Chicago**, St. Louis	St. Louis, Chicago	Dallas, Chicago, Boston
+ US Airways	1939	*Charlotte*, Philadelphia, *Phoenix*, Las Vegas, **Pittsburgh**	Pittsburgh	Pittsburgh
United Airlines	1926	Chicago, Denver, Washington (IAD), San Francisco, Los Angeles	Boise, Pasco	Chicago, Kansas City, Dallas
+ Continental	1934	Houston, *Newark*, Cleveland		El Paso, Las Vegas, Albuquerque, Santa Fe, Pueblo
Delta Airlines	1924	Atlanta, Cincinnati, *Salt Lake City*, New York (JFK)	Fort Worth, Atlanta, Charleston	Dallas, Jackson
+ Northwest AIrlines	1926	**Minneapolis**, Detroit, Memphis, Tokyo, Amsterdam	Minneapolis, Chicago	Minneapolis, Chicago

Source: Airline websites

Note: **bold** indicates original airport served by airline. Other hubs were often served by acquired companies; *italics* indicates original airport served by an acquired company.

The aviation story continues in Chapter 20.

...(policy has been) transformed into a kind of witchcraft
—DARLING (1980)

13 Railroads Regulated

13.1 Federal Triangle

THE FEDERAL TRIANGLE, located in Washington, D.C., just south of Pennsylvania Avenue between the Capitol and the White House, was the largest building program the US government had undertaken by the time of its opening in 1934. The new buildings aimed to show the "dignity and power of the nation."[240] The Interstate Commerce Commission (ICC) occupied both the apex of the "iron triangle" (see Section 13.8) of regulators, the regulated, and the legislators, and one of those Federal Triangle buildings continuously from 1934 until its dissolution in 1995.

A late 1980s visit to the impressive ICC edifice contrasts with the interior. Behind the curtain, a walk down the green narrow halls with cracked linoleum floors, visits to offices, and attending an ICC hearing gave an impression of senescence. It was a sharp contrast to the impressions from the newer regulatory agencies, with modern fixtures and furnishings if not modern policies. Hence the replacement of the ICC by the Surface Transportation Board (housed elsewhere) came as little surprise—its old and run-down furnishings and fixtures symbolized its old and run-down ideas.

Like the ICC that regulated them, most of today's transportation systems are mature or nearly so. They are fully deployed, structures are fixed and inflexible, and the mix and variety of the services that can be offered are limited. Given that situation, how do systems respond to changes in their environments? We explore that question, using the US railroads as case at point.

Using our life-cycle scheme, we are in the maturity phase of the cycle, and the matters we treat generalize very well to the situation across the board in transportation today. It's fair to say that in the more developed nations, most systems are nearly or fully deployed, and service is available everywhere. Designs are fixed; unitary technologies are used. So while we are looking at the rail system, we see behavior, problems, and reactions to problems common across today's systems.

In the early days, as discussed in the UK context, the government role was largely enabling (chartering firms, e.g.) and by default, doing things in the interest of the railroads that, for one reason or another, the railroads couldn't build a consensus for action (e.g., Board of Trade intervention when railroads couldn't agree on running rights).

Tasks similar to these carried forward into the near and full deployment phase of development, the phase now examined. The government role was broadened and fine-tuned. In addition to this already established role, we will see some other roles emerge. In general, we can say these emerged because the "bloom was no longer on the rose." During the start-up and expansion phase, new services were being made available and existing services improved. Improvements tend to stagnate as the unitary technology is locked in and service becomes widely available. At this point, the debate over who claims the transportation rents becomes fiercer, absent improvement, ills are more visible, and problems covered up by improvements emerge for management.

In response, we see a wave of tightening regulation, with regulation steered in many ways by the "Who gets the rents?" question. We also see a notion of rationalization emerge and be modified and modified. The rationalization story is treated in Chapter 21. This chapter focuses on the emergence and maturity of modal regulation.

13.2 Correspondence and the Locus of Authority

Medieval society ran on the common will. Later, as commerce extended village to village, shops and inns and larger enterprises were subjected to medieval, common law attitudes. Shopkeepers and others derived their rights and powers from the community, and communities specified the terms of services and prices charged. For example, inns charged "fair" rates; they could not refuse service without good reason. Communities developing in the United States were clothed in that spirit. The villages and towns continued to control the activities within them. In addition to stating the terms in which individual entrepreneurs operated within their span of control, these communities undertook public works to serve the members of the community.

But railroads and canals were big enterprises, and they were more than villages and towns could handle from the standpoints of financing and geographic scope of activities. Extending the medieval tradition, rail and canal organizations were chartered by the states to serve the public. Essential elements of those charters were the notions that service to the public was to be without discrimination and favoritism; users in like circumstances were to be served the same; and government was entitled to regulate rates, terms of service, quality of service, and so on. Government delegated authority necessary to the purposes of the corporations—for example, the power of claiming eminent domain over property was delegated to the railroads.

In the early 1800s, public attitudes drifted from medieval logic. The spirit of Adam Smith and laissez-faire gripped the nation. (Actually, the spirit was behind many early settlements; they were commercial enterprises with certain rights and restrictions—the Virginia

Company and the Massachusetts Bay Company were examples.[241]) These enterprises are also worth comparing with the chartered trading companies of the same era, discussed in Section 5.2. Small farmers, merchants, and larger organizations began to do what they wanted to do. There had been a breakdown of the medieval village and, at any rate, many new activities were non-traditional and also did not fit traditional governing units. The nation was growing very rapidly; most people shared in that growth, and laissez-faire was seen as a touchstone for growth.

The within-railroad and between-railroad embedded policies and rules explain much about their structure and behavior. (A similar statement holds for all transportation systems.) However, governments were asked to develop policies when the railroads could not develop and enforce needed embedded policies. One case is when the domain of the policy extends beyond the domain of the mode; for example, policies needed to assist in acquiring right-of-way. Another case is when there are winners and losers within the modal community, and governments are asked to play the referee. The granting of track rights and criteria for the division of income from movements involving more than one firm are examples.

This ad hoc development of government policy seems especially true during the early days of systems. Such policies aim to help the system get deployed and work. Later on, government policies respond more to broad social consensus about systems. They intended to control the system for social purposes.

Regulatory acts and institutional and administrative actions relative to freight tariffs and passenger service occurred in Britain in 1868, 1873, and 1893. Those initiatives and the US federal initiatives of the late 1800s involved values, ideas, and concepts that were commonplace then and remain so today in the verbiage of common law. Yet, the imposition of those ideas on the railroad "turned the clock back."

There was reference to a mismatch in government structures, and we have used some ideas about that to aid in recognizing issues and policy responses, as shown in Table 13.1. Using that classification device, issues may be divided into two classes: on- and off-diagonal. The problems of dealing with off-diagonal issues are much discussed in the literature, especially by persons engaged in intergovernmental relations. The table and its implications are transparent and will not be further remarked on at this point.

Just as with the spatial correspondence problem discussed above, there are things which are political (subject to planning processes) that sometimes fall to the technicals (engineers) to decide. Things which are technical sometimes fall to the politicians to decide. These

TABLE 13.1

The Spatial Policy Correspondence Problem: Loci of Authority vs. Loci of Issues

		Loci of Issues		
		National	Regional	Local
Loci of Authority	National			
	Regional			
	Local			

TABLE 13.2

The Institutional Correspondence Problem: Loci of Authority vs. Loci of Issues

		Loci of Issues	
		Technical	Political
	Technical		
Loci of Authority	Political		

mismatches are the cause of major resource misallocations. We can frame this as in Table 13.2.

Things which are on diagonal might include technical authority and technical issues, such as striping of roadways, or political issues with political authority, such as the allocation of tax revenue. But we can see some mismatches. The original Interstate Highway System was laid out by engineers, but had an obvious political set of issues that were initially decided only within the technical value system. This led to conflict. The ICC used political decision making to set prices in potentially competitive markets. Today, we have political interference in what ought to be routine technical decisions, like highway maintenance on existing facilities, where the objectives and values are not seriously questioned.

13.3 Mighty Elevators of Grain

The social mood of the Gilded Age was that the public and individual interest was very much served by entrepreneurial freedom.[242] The railroads on some measures were providing better and better services, so the public should have been pleased by current service compared to previous and have expectations for even better service in the future. But there were problems, straining the "social contract," as described in Section 13.4.

So although the terms of reference in which the federal government exercised control over the railroads seem very familiar today, federal control was a turning of the clock back (from laissez-faire) so far as the evolution of the economy and the roles of individuals and corporations within it were concerned. Government began to exercise its traditional medieval rights in a manner that had not been exercised visibly and strongly for some decades, if not centuries.

The states attempted to manage by extending general laws so that they dealt with specific ills and by development of commissions—some advisory and some regulatory—to make the laws work.

Although individually owned, the grain elevators colluded in the setting of prices, and they controlled the movement of grain from much of the Midwest to markets in the East. In 1873, the State of Illinois passed a Warehouse Act to regulate the rates and terms of service in Chicago's grain elevators. This act was challenged by successors to the firm of Ira Munn and George Scott, whose elevator business failed during the Panic of 1873, in part because of corruption.

Munn v. Illinois (1877) worked its way up to the United States Supreme Court. Citing common law, the Supreme Court, in a decision by Chief Justice Morrison Waite, found that the activities of the grain elevators were a matter of great public consequence. He wrote, "Common carriers exercise a sort of public office, and have duties to perform in which the public is interested.... Their business is, therefore, 'affected with a public interest'". Countering the claim that the law destroyed private property, Waite argued that legislators, elected by the people, were the appropriate forum to establish regulation, and the courts should behave with self-restraint in the economic arena. Thus, Illinois had a clear right to regulate those grain elevators in the public interest.

Munn v. Illinois was a precedent-setting case, and it applied clearly to the railroads. This enabled the states to regulate the railroads and federal regulation of interstate commerce. Reasoning from *Munn v. Illinois,* findings in *Wabash, St. Louis, and Pacific Railroad v. Illinois* established the right of Congress to regulate interstate commerce.

The state actions were, however, ineffective, and debates escalated to the federal level. In 1872, President Grant appointed the Windom Committee to look into the matter. It recommended government ownership of railroads. However, legislation failed until the McCullom report, which led to the Act to Regulate Commerce of 1887, recommended an Interstate Commerce Commission to regulate rates. The Act was not much more than an appeal to common law, and it lacked enforcement means. Its first three parts required rates to be just and reasonable, forbade personal discrimination and rebates, and outlawed undue or unreasonable preferences. All of this is very much in common law tradition.

Other requirements limited long-haul and short-haul rates, prohibited pooling arrangements, and required the publication of rates and fares, as well as due notice of increases. The long-haul and short-haul limitations referred to equality of rates under "substantially similar circumstances."

All of the phraseology rolls smoothly on the tongue. We nod in agreement—that is how actors are supposed to behave. But the Act served up general prescriptions, and it was not exactly clear how the Interstate Commerce Commission (ICC) was to implement common law prescriptions. After all, the ICC was a small, new organization created to regulate a vast enterprise when communications, office record keeping, availability of skilled bureaucrats, and other attributes that are commonplace today were not available. The Interstate Commerce Act more than anything else strengthened the license of individuals and organizations to ask the courts to make decisions about the activities of railroads. However, this was unworkable due to the tremendous volume of case law involved. Moreover, it resulted in individual courts deciding on matters of preference, reasonable treatment, and other such factors that require guidelines and comparisons. The Courts would not expedite injunctions to enforce ICC orders. They took testimony and acted on their own. They limited ICC control over just and reasonable rates, and they interpreted the phrase "substantially similar circumstances and conditions" in quite a literal manner. Rebating continued.

Congress seemed satisfied with the situation for about a decade and a half and then moved to strengthen the ICC. The Elkins Act of 1903 amended the 1887 Act limiting rebates. Congress corrected this problem in the Hepburn Act of 1906, which removed the courts from the critical path. The Act further extended jurisdiction of ICC regulation to sleeping car companies, express companies, pipelines, and railroad terminals. It prohibited free railroad passes, permitted specification of maximum rates and prescription of through and joint rates, forbade the railroads' hauling goods (except lumber) they produced, and put commission orders into effect in thirty days without prior approval of the courts. Finally, the Mann-Elkins Act of 1910 reformatted the long- and short-haul clause by deleting the "substantially similar" phrasing, and extended ICC rule to telephone, telegraph, and cable companies (later to be regulated by the Federal Communications Commission).

Though there were additional acts affecting the ICC, the ICC as we have known it during the decades of the present century was mainly shaped by the Hepburn Act. To make a long story short, we may observe that laying down "fair" rules of behavior and giving the ICC enough clout to enforce them solved the problem of planning for the operations of railroads.

The legislation discussed above deals mainly with rate discrimination. The predatory behavior of firms, stock watering, traffic pooling, and monopoly practices were somewhat slower in being managed and began to be treated under anti-trust law rather than ICC law. Acts eventually tempered that treatment, while the ICC gained control over security issues and bankruptcy proceedings, and the Reed-Bullwinkle Act circumvented the effect of anti-trust on rate bureaus.

The above discussion doesn't begin to list all the details of legislation, but it is enough to identify trends and major matters at issue, especially rates. Because of the very large number of things carried, commodities were grouped into classes, and rates developed for classes. On account of fixed terminal costs, these rates incorporated a tapering distance scale. (This convention did not hold for intercity passenger rates.)

Rates reflected what-the-traffic-would-bear, and also considered the weight-volume relationships. (Actually, comparatively little traffic moved under class rates. Most moved under commodity rates, which were specific to commodities and points of origin and destination. Nonetheless, the rate class concepts were reflected in commodity rates.)

The Transportation Act of 1920 emphasized the responsibility of government to assure that adequate transportation was provided—very much a change of mood. These are important words; they stress the positivism of governments. It is a notion that has reappeared in bits and pieces of legislative rhetoric at all levels of government. Yet that positive role in strengthening activities is not in the present rail debate. It is a "get the government out of the way" debate.

13.4 Government's Proactive, Normative Rule

An issue often arises when it is recognized that some activity is dysfunctional; something isn't working correctly. One litmus test for dysfunctional activities is the logic of what governments do. Among other things, governments are expected to manage things that are regarded as wrong. If something is going on that counters the way governments are supposed to manage things, there is an issue.

The logic of what governments do has grown out of long experiences. Transportation roles in the United States are seen for governments when there are Constitution-based responsibilities: the economic integration of the states (interstate trade), safety (health), and provision of services (welfare). The presence of market failure, limits on expertise, control of illegal activities, non-transportation goals affecting transportation, and many other considerations over the years have broadened the bases from which roles are argued.

We have attempted to make a list of the reasons that governments have regulated (or nationalized) transportation systems:

- Fairness
  - Service is not available everywhere.[243]
  - Small and large customers (shippers) are not treated in the same way.
  - Gains from trade should be split differently between service providers and service users.
  - Transportation (or location) rents ought to be shared between capital and labor.
  - Monopsony customers and competition mean not all providers make adequate profits.[244]
- Competition
  - Dysfunctional market organization is a market failure needing repair.
  - Monopoly transportation providers abuse their economic power.
  - Natural monopolies cannot recover high fixed costs.[245]
- Progress
  - Infant industries need assistance. A variation on this is the notion that social overhead capital requires up-front investment.
  - Processes of innovation and technology development are not working in viable ways.
  - Productivity is falling (or improvements are diminishing) off-system and on-system.
  - Technology change begs new arrangements.
- Stability
  - Control of between- or within-system competition is necessary in order to achieve some desired result.[246]
  - Stability is a good thing, but transportation development (or deterioration) may upset an existing equilibrium.
  - A social contract begins to fail.
  - The value of capital stock to the value of the plant varies widely.[247]
- Off-System
  - Transportation may need to be limited, expanded, or coordinated to achieve some off-system goal.
  - The health and safety of the public must be ensured.
  - The health and safety of labor must be ensured.
  - Effective workings of governments, national economic systems, or defense systems must be ensured.

The list of rationales is rough, and we would not argue its perfection. Indeed, each time we use it we think of needed revisions. The point of the list is that it helps identify the nature of issues and their origins. As we see it, an issue arises when something is wrong. Wrong with respect to what? Rail (or air) service to small communities is seen as a fairness issue, for example. The railroads (airlines) are not providing services to all.

13.5 Regulating Labor Relations

Stakeholders in transportation enterprises include users, owners, managers, and labor, and the labor part of the equation is big—about one of every nine employed persons in the United States is employed in transportation or in a transportation-related activity such as gasoline service stations. Even so, one finds little explicit consideration of labor interests in the academic literature on policy. (Transit labor inputs are an exception.) Labor has its own special iron triangles. Why is labor treated apart from general debates?

Labor organizations evolved with modern factory and transportation systems. Prior to modern production and transportation, there were artisan labor organizations (guilds), and the organizational basis shifted from things produced to workplaces. Early organizational purposes included social and retirement arrangements, but soon extended to the notion that labor could organize to sell its product (labor) to the workplace. That notion, however, was in direct conflict with common law. Labor was seen as conspiring to take the property of others, and court-made law said that conspiring to strike was the taking of the rights of owners to use labor as they wished.

Work at home can be compared with the factory (work out of home) systems, basically as trade-offs in scale economies and transport costs. To the extent that scale economies are less achieved with people commuting out of the house, and because manufacturing requires less and less labor, shopping can be more effectively conducted virtually rather than in store, and telecommunications become better and better substitutes for in-person contact, we should expect the substitution to become the more dominant feature of the relationship between transportation and communication.

Common law's restrictions on labor organizing held until 1842 when the Supreme Judicial Court of Massachusetts broke the conspiracy theory, holding that labor organizations may exist for "public spirited" purposes. That was a means-ends doctrine. The end was lawful; therefore, the means were lawful. The case had to do with the refusal of union workers to work for employers using non-union employees.

Means had to be mild. Workers could resign and thus not work. But they could not take actions to prevent owners from operating their businesses. Beginning in the 1870s, injunctions prohibited picketing, boycotts, and trespassing. That situation holds today, though the notion of appropriate means has been vastly softened.

The railroads were big businesses and, during the period we are discussing, were the stage for much of labor strife. (Coal mines, steel mills, and packing plants were other large stages.) And they attracted national attention because of the wide effects of labor disruptions.

Carrying forward the guild tradition somewhat, rail labor was organized into craft unions, which reduced its effectiveness because of the indifference of other unions to a craft union's special interest. There was common interest when railroads asked for across-the-board pay reductions and in other common interest cases. But when strikes occurred, injunctions stopped them.

Unions had other problems. For instance, Eugene Debs and other radical union organizers found little support for their views in the general population.

The perceived and real abuses by big business resulted in the Sherman Antitrust Act of 1890. "Every contract, combination... or conspiracy, in the restraint of trade of commerce among the several states... is illegal...." Although unexpected by the drafter of the legislation, the Act was quickly used against unions. For years, indeed, its main use was against unions.

Strikes and strife continued, even so. The result was the ineffective Erdman Act of 1898 and the more effective Newlands Act of 1913 providing for arbitration panels. By this time, the stage was national, for the railroads grouped and worked together, as did labor.

But arbitration doesn't always work. The Newlands Act foundered on labor's desire to change the dual rate of pay from 10 hours or 100 miles, whichever came first, to 8 hours or 100 miles. Labor refused arbitration, and to avoid a crisis on the eve of World War I, Congress passed the Adamson Act of 1916 granting labor's demand. (The matter was again arbitrated, and in 1985 the 100-mile component was raised by two miles per year, so in 1988, 108 miles was the rule.)

In 1917 President Wilson seized the railroads, citing ineffective operations and labor difficulties. They remained in government control for twenty-six months, until March 1, 1920. Rail wages doubled during this period. The broadly important matter, however, was that government abruptly shifted from an adversary of organized labor to a supporter. It sought legislation to encourage union membership and to take labor relations out of the courts and avoid the injunction process.

A series of acts followed. The Transportation Act of 1920, the Rail Labor Act of 1926, and the Rail Labor Act of 1934 set up special provisions for dealing with the railroads (and the airlines in 1934). A parallel set of acts and, in particular, the National Labor Relations Act of 1935 extended to labor generally. These acts state the right to organize and bargain collectively without interference from employers. In the rail and air cases: "It is the duty of all carriers and employees to exert every reasonable effort to voluntary settle disputes."

For major disputes, the (rail, air) National Mediation Board establishes an Arbitration Board with subpoena powers over persons and documents. Awards are enforceable through the courts. If that process doesn't work, then the National Mediation Board notifies the president, who can establish an Emergency Board. Once the Board is appointed, parties must hold the status quo while data are gathered and for 30 days after a finding is made. Findings have only the force of the status of the president and public opinion. If the findings are not accepted, then the president may ask Congress for special legislation. So today, major issues are settled by Congress. The carriers are very unhappy about this, for labor often has more power in Congress than carriers do.

Our question is about the ways that labor considerations enter the policy debate. We said that labor considerations are a thing aside. Labor is a thing aside because it has its own story and traditions. There are explicit processes for labor considerations that are distinctive. Legislative and administrative considerations and the trail of law for labor questions are apart from those used in policy generally. We made one not-very-surprising finding: The railroad experience had general spinoffs. Railroad problem resolution served as a model for other activities. Much of labor law is based on railroad labor precedence. Why, from a labor perspective, are airlines treated as railroads? Mainly, we see that as an accident of timing. Airlines were beginning to be regulated at a time when there were major revisions in rail labor legislation. Unlike interstate trucking, the airlines were regulated prior to the development of general labor law.

13.6 Deregulation

Jumping to the railroad situation in the last two decades of the twentieth century, there were new ideas floating around, with the return to laissez-faire conditions via deregulation carrying the day. One reason for the deregulation thrust was the ICC's debilitating impact on the viability of the railroads. It has been said, and it is no doubt true, that regulation cost the nation a great deal in efficiency terms. Without regulation, the railroads would act more wisely, obtain efficiencies, and the nation would be better off. In general, it is accepted that competition is a good thing.

It was necessary to regulate the railroads at one time because they were monopolies. Also, conditions of entry were such that no reasonable organization would enter the market even if the actor already there misbehaved—the critical factor being the large investment in fixed plant required in order to enter the market. A related matter was that fixed costs were relatively high, so marginal costs decreased over the range of outputs feasible in markets. This phenomenon blocked marginal cost pricing guides, as well as the ability of competitive actors to enter the market.

The deregulation argument is that much has changed. In particular, trucks offered competition to rail (and barges and pipelines under certain circumstances). No actor can misbehave for long. The extent to which that is true has bothered Congress, and the ICC examined the question of captive shippers and found a few of them. It required that the railroads make the contents of contracts public.

Thinking about the impact of the ICC on the industry involves what we may call the internal dynamic of a bureaucratic agency. Political scientists and students of organizations have written a variety of things on this. "One way to summarize is to say that agencies go through a life cycle." They start out by having trouble getting their feet on the ground and learning how to walk, then they walk with great vigor, and finally they fall into fixed patterns of behavior and stasis. That notion fits the ICC and other regulatory agencies very well, and it is fair to say that in its last decades the agency had a certain rigor mortis.

A companion factor is that regulatory agencies become captives of those they regulate. That was the case for the ICC; it generally acquiesced to what the railroads wanted to do. Beyond requiring hearings and publications, it did not regulate rates much at all. Another observation stems from an interpretation of what the railroads collectively "wanted to do." There is a common denominator interpretation of what the railroads wanted to do. For instance, if a railroad wanted to do something very aggressive and other railroads didn't agree, the aggressive railroad would be stifled. This was especially the case in merger policy where third-party railroads would be very demanding. Deregulation is aimed at avoiding such least-common denominator actions.

The debate over deregulation concerned the stifling impact of a bureaucracy and the advantages of competitive behavior. The latter was enabled by arguments about the nonexistence of monopolies because of truck competition or the potential for service by competitors. Note that this debate is in essence quite different from the one in the late 1800s, which had more to do with how organizations should act under the rules of common law and the notion that big, important things are clothed in the public interest. The second observation is that the debate in the ICC arena was limited to mergers, rates, and terms of services. It overlooks the presence of much regulation elsewhere in government, and in organizations

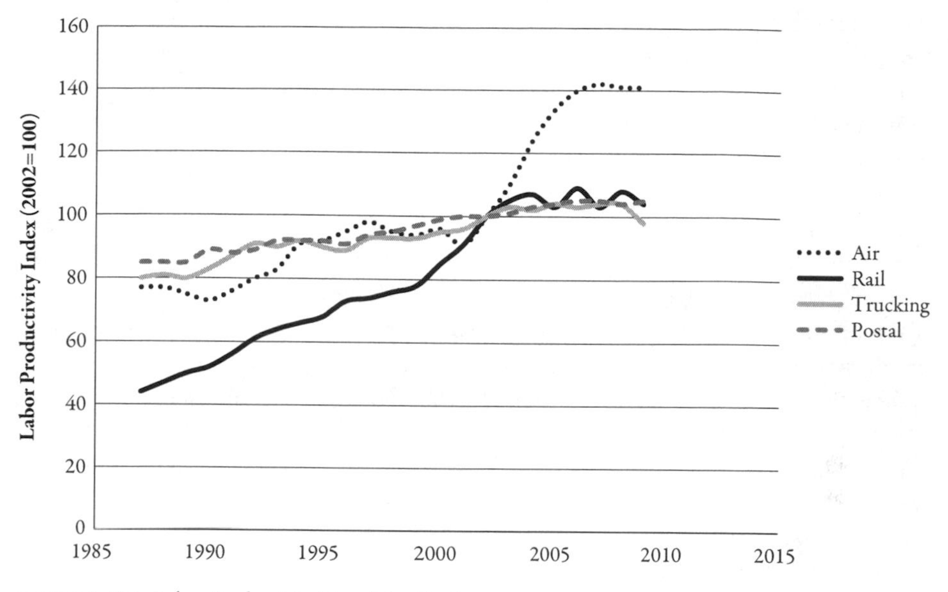

FIGURE 13.1 Labor Productivity Trends by Mode

created by the railroads themselves—rate bureaus and Association of American Railroads standards setting, in particular—as well as by organizations such as the National Industrial Transportation League (formed in 1907 to serve shippers before the ICC, but since deregulation has sought a broader role serving carriers as well). Rate bureaus are no more. (Rate bureaus never told the railroads what to do, anyway. They were a communications medium and a consensus-building arena).

By and large, the return on investment in railroads has never been very great. Many railroads were in trouble before the ICC was created (before 1887), during the period when the ICC gathered some strength (1877–1915), and later.

Much of the economic literature on rail cost and deregulation largely justified moving toward a deregulated system (Meyer et al. 1959, Keeler 1983, Friedlander and Spady 1980). The ICC was replaced by the Surface Transportation Board in 1995 after steadily losing importance due to the deregulation promoted by the Staggers Rail Act and the Motor Carrier Act, both passed in 1980.

The most recent data (see Figure 13.1) suggest significant productivity increases in rail in the 1980s, post-deregulation, and in aviation in the 2000s, post-bankruptcy and reorganization. The post office and long-distance trucking have seen only modest gains.

13.7 Comparisons of the Developed World

Sometimes when you have everything, you can't really tell what matters.

—Christina Onassis

Striving to summarize the English, French, and US policy experiences and anticipate the developing world's future experiences, some broad-brush remarks will be made. With the exception of France, experiences similar to those discussed for England held on the Continent.[248] Early on, France differed because of its strong central government and its traditions of scientific work, rationalization, and professionalism.[249]

We think the term *entrepreneurial capitalism* best captures English policy. English policy had a "hands off" character, and this seems to have flowed from the notion that the national interest was best served by competitive, independent firms creating jobs and wealth. Rate setting by agreements among firms, rates high enough to return profits, and so on, resulted in profitable firms. There was a bit of welfare tempering for example, workmens' trains, publication of rates and listing of charges, but not much. Also, Parliament's careful examination of requests for charters protected property owners, existing properties, and investors from poorly developed schemes.

The French policy experience may be labeled absolutism or statism; all decisions turned on what was taken to be good for the state and were made by the state. Plans were made by the state, instead of by firms as in England. The same was true of tariffs (rates, fares). There was state financing (partial) instead of private capital. Operations had heavy state input. There was to be eventual state ownership of facilities.

US policy doesn't summarize so easily. Early on, local and state mercantilism makes a good label, and there was also flirtation with aspects of federal absolutism (when railroads linking the West Coast with the Midwest were debated and when regulation was considered). After some bad experiences, policy evolved into what may be termed a US version of entrepreneurial capitalism: capitalism constrained by antitrust and rail-specific regulation. That continues even though the Interstate Commerce Commission has been eliminated. Let's call it constrained entrepreneurial capitalism.

An example may aid in contrasting the differing national approaches to policy. Early trains had no brakes, and as trains became larger, faster, and heavier, hand-applied mechanical brakes began to be used. However, poor braking was the cause of many accidents. Late in the 1800s, George Westinghouse developed the air brake. The French adopted it by fiat: use it. The English took the attitude that when the properties saw that air brakes made good sense they would adopt them, no government action was needed. Indeed, the English had a hands-off policy on safety generally. There were government inspectors that reviewed properties, and their reports were published, but government took no action beyond that. The US government negotiated with the rail properties and adopted a legislative requirement for air brakes. But the refinement and implementation of air brakes was pretty much left to the railroads and their decision-making arrangements.

We think of the summary just given as broad-brush. We would be remiss not to point out even broader-brush thoughts. Each national rail policy experience was rooted in situational attributes, as well as the ideas and attitudes that influenced the thinking about related government organizations and roles. All that can be termed political, national, ethic, or culture. That's the kind of thing Alexis de Tocqueville commented so skillfully on in his *Democracy in America* and in his *The Old Regime and the French Revolution*.[250]

As stated, the rail experience provides a "mother logic" for policy in other modes. We need to go beyond that to point out that rail-mother-logic was shaped by national culture; it was also a shaper of national culture, especially in the United States. In this respect, a very broad-brush view of rail policy provides a comment on national industrial policy. Note that the labels we have attached to English, French, and US rail policy apply to industrial policy in general; the French have elected to develop key industries, the British and the Americans have had rather different industrial policies. Other countries, including the less-developed ones, fall here or there between the extremes of absolutism and entrepreneurial capitalism.

Earlier we said that all nations are developed, given historic paths, resources, tools, and other attributes. We now invert that statement, for if we accept that change in transportation is possible, then new futures are available to all nations.

In some ways the transportation development problem is most urgent in the so-called developed nations. The old systems have clearly run their courses, and that's one reason for the slow-down in development and lack of investment opportunities.

But there are other urgent development objectives, many of a "reduce constraints" type (reduce government regulation, limit the number of product liability cases). Some objectives call for investments, such as in new technology and education. That is, there are a lot of urgent things to do that are more visible and closer than transportation to the mainstream of economic and political thought.

Wilfred Owen, a long-time student of economic development, now gives overriding attention to equity. He says that all less-developed nations should be provided with upgraded road systems. Without such systems, residents cannot participate in modern life, acquire health services, and so forth. Leaving aside the point that roads are not enough, Owen has an attractive thought: a basic level of transportation service may be prerequisite to the rest of civilization. Compared to other nations, the existing transportation systems of the United States are regarded as in pretty good shape, despite congestion and some disrepair.

13.8 Iron Triangles and Aluminum Rectangles

> The "Act to Regulate Commerce" goes into effect on Tuesday next.... The National Commission ... is now in session.... The railways as well as the public generally ... will give them every encouragement possible for the execution of their duties.
>
> —*The Railway Age*, April 1887

In the United States, the national government structure comprises three branches: the legislature (Congress), the executive (president and administrators), and the judiciary (courts). Informed by the executive and in other ways, Congress makes policy. Policy implementation takes place within independent regulatory agencies or executive branch departments. The courts interpret the intent of Congress, the relations among branches of government, and government powers. That's the balance of powers and triad that civics textbooks show.

Actually, policy is made by a different triad. There is negotiation between those who are affected and those who affect. The playground for the formation and implementation of transportation policy are regulatory agencies or units of departments, a subcommittee of the Congress, and those who are regulated, example, the railroads. The structure is so tight that the expression "iron triangle" has been used to describe it. It is a government within and partly outside government.

The ICC was imposed in a fashion satisfactory to interested publics—the stockholders of railroads, railroad management and labor, shippers, the states, and so on. The conventional view for the origin of the ICC is that actors in the so-called Granger states coalesced politically around an effort to achieve fairness of treatment from railroads. There were conflicts between the states and the federal government as well as the differing regional interests in the Granger area.[251] But there is the opposite view: the railroads wanted to stabilize their environments and dampen competition, which may explain why the railroads seemed to welcome regulation, as noted in this chapter's opening quote.[252]

This iron triangle substructure provides the context for many studies of the interrelations of structure and behavior. That model of a tight substructure described well the situation in previous decades for the trucking, maritime, air, intercity bus, and rail modes. However, now things are a little fuzzier. That is partly because the United States Department of Transportation (USDOT) has gained strength and has begun to advocate policy. Also, the USDOT has picked up safety regulation tasks, though the situation differs by mode. Moreover, independent environmental and consumer groups now keep a closer tab on transportation policy and try to influence, moving us from an iron triangle to what we might call an aluminum rectangle. As more players involve themselves in transportation policy, the rigid policy structure becomes far less rigid. This has become most pronounced in highway and transit policy since the 1991 Intermodal Surface Transportation Efficiency Act, which opened up transportation policy making to many more players.

The Federal Highway Administration (FHWA) has worked hard to define its turf, and it works hard to keep and expand what it has. Although the FHWA is only one component of the truck-highway-user system and it must adhere to system rules, it can take considerable independent action. It is a powerful creator and implementer of policy. On the matter of turf, the Surface Transportation Board (STB) has no difficulty interrelating with railroad fixed facility topics, such as route abandonment. It has not said a word about the FHWA's debate with truckers over facilities suitable for large trucks. It would not dare to, for highways are not part of the STB turf.

The FHWA and the truck situation is an example of an iron triangle (few outside groups have involved themselves effectively in truck regulation, causing us to keep only three corners). Other cases include the Army Corps of Engineers and its waterways and wetlands and the Federal Aviation Administration and its airways and airport systems. The Federal Transit Administration situation is a bit different because of scale. Transit regulation triangles historically have been local ones, with local public utility commissions or similar agencies having regulatory functions. Recent changes have reduced the strength of these arrangements, and have created national sets of relations.

One trend in the United States is increased interest of the DOT in policy making and implementation. The DOT and the OMB more and more exert their influence on policy and regulation, with varying results. The DOT had considerable influence on the Interstate Commerce Commission (ICC) de facto deregulation of the railroads.

Structure plays out differently in other nations, of course. Even so, other nations usually have a regulatory organization, and special interest groups (such as highway lobbies) have their say.[253] In some, there is the "white paper" style of policy analysis. In many other nations, systems are (or were) nationalized, especially railroads. The iron triangle concept doesn't have the inside government–outside government split found in the United States, that is, government is regulating inside-government railroads. Is there more "government versus those who are governed" conflict in the United States than in the nationalized system case? It would be hard to show that there is.

13.9 Discussion

Although rail construction continued into the 1900s, the fabric of rail networks was pretty much deployed in the United States and Western Europe by the last third of the 1800s. At this juncture a series of problems became more visible, and there is a striking parallel between those problems and the ones commonly listed for today's transportation problems.

Management of the system became the priority. That is, interest shifted from planning construction (tactics) to planning control (operations), and little attention was given to planning technology (strategy), which was largely thought too late to shift.

Government control was imposed to:

1. Manage the fiscal problems of the properties and assure their fair treatment (i.e., by other roads and large shippers);
2. Obtain fair treatment of individual users, large or small;
3. Stabilize the competitive positions of regions; and
4. Rationalize investment and disinvestment.

Many would use the term "regulation" to characterize ICC activities. We think it is useful to see regulation as the outcome of a style of planning. The goal was to maintain a viable physical system; the concentration was on soft instruments rather than investment.

Regulation was innovated. It was trial and error, especially with respect to needed legislation. Things that worked were adopted. The building blocks of the times were used. Such building blocks included the "rules" tradition, then accounting systems, legal styles of inquiry and management, and the rate-making agencies of the roads. The ICC agency culture and style resulted, a culture and style that was imitated as government began to regulate in other areas.

That's one result of railroad regulation: it taught governments how to regulate. What were other results?

It took a while, but regulation did repair most of the tears in the social contract. With respect to the embedded policy problems the railroads could not solve, regulation enforced rate and service agreements and stabilized competitive relationships. It protected weak roads from the strong.

Stabilization is perhaps a key word. Mergers and purchases were slowed down and price changes were tempered. At least to some extent, management became concerned with regulation rather than efficiency, and innovation suffered in some respects. Changes that would have occurred in a more open situation were slowed down.

The resulting lack of responsiveness began to be seen as a problem in the 1920s, and an acute problem was recognized as other modes, especially truck, inland water, and pipelines began to compete. The solution to that problem was to rationalize the situation, a matter to be discussed in chapter 21.

Before leaving this topic we should point out that government regulation was hardly a standoff, except for the public interest matter. This was especially true after ills in the social contract were repaired. Section 13.8 introduced the notion of the Iron Triangle. Regulatory agencies get co-opted by the organizations they are to regulate.[254] The Ralph Nader study group's *The Interstate Commerce Omission* put that in harsher terms, "... the ICC and the transport industries forged a corporate state that utilized public power for corporate purposes."[255]

The chapter began by looking back in time, looking all the way back to medieval society in order to interpret the regulation of the railroads in the late 1800s. Before regulation, laissez-faire ruled. But laissez-faire behaviors sow the seeds of regulation. Regulation controlled the behavior of railroads and brought them in line with common law norms.

However, just as free market excesses create the rationale for regulation, regulatory excesses of the ICC-type build in their own destruction. "There is a life cycle." In the early days, political interest is high, and young professionals have the political power and energy to aggressively pursue regulatory tasks. But regulators age, the low cost, big payoff things get done, political interest and power decrease, and agency culture and procedures harden. Special interest groups can exert more and more power as others pay less attention. Agencies become a liability to a democratic government, and are at risk of death.

The story of the rail modes continues in Chapter 21.

I see a place where people get on and off the freeway. On and off. On and Off. All day, all night. Soon where Toontown once stood will be a string of gas stations. Inexpensive motels, restaurants that serve rapidly prepared food, tire salons, automobile dealerships, and wonderful, wonderful billboards reaching as far as the eye can see. My God, it'll be beautiful.

—JUDGE DOOM in *Who Framed Roger Rabbit*

Nobody's gonna drive this lousy freeway when they can take the Red Car for a nickel.

—EDDIE VALIANT, *Judge Doom's nemesis*

14 Bustitution

14.1 Myths in Motion

In the 1920s, a General Motors (GM) subsidiary acquired streetcar systems in Springfield, Ohio, and Kalamazoo and Saginaw, Michigan. Contracts required the transit systems to buy only GM and Mack buses, Firestone tires, and fuels and lubricants from Standard Oil of California. In exchange, GM, Firestone, and Standard Oil provided the capital. In 1936, E. Roy Fitzgerald, a bus company operator who hauled miners in northern Minnesota, was made president of National City Lines (NCL), General Motor's transit holding company (controlled through GM bus units Yellow Coach and Greyhound). By 1946, NCL and its subsidiaries, American City Lines and Pacific City Lines, owned transit systems in 45 cities (including the Red Car lines in Los Angeles and the Key Lines in Oakland). That same year, industry trade publication Mass Transportation named Fitzgerald "Transportation Man of the Year." Fitzgerald also made the cover of the July 20, 1946, issue of *Business Week* as NCL became a publicly traded company on the New York Stock Exchange.

On April 10, 1947, United States Attorney General Tom Clark indicted nine corporations and seven individuals (including Fitzgerald) on antitrust charges of conspiracy in the sale of equipment to a "nationwide combine of city bus lines."[256] March 13, 1949, they were all convicted on one count of conspiring to monopolize a part of the trade and commerce of the United States. National City Lines and General Motors and the other companies were fined $5,000 apiece, while their managers were fined $1 each.

It is important to rail transit advocates that they tell the story that America's streetcar system was done in by a conspiracy, because if it were a conspiracy, rather than market forces or democratic preferences, it was a wrong to be righted by new investment in rail transit. Advocates tell this story in a number of forums, including the (relatively) widely seen documentary *Taken for a Ride*. A written work, *The Decline of Transit*,[257] a comparative study of the United States and Germany, posits a grand conspiracy to do away with transit by automobile and suburban development interests. At places, it states that the whole story had to do with finding employment for labor in a capitalistic society.[258] Similarly, in the early 1970s, Bradford C. Snell charged that General Motors (GM) had used NCL to destroy thriving street railways in the interest of selling buses and, in the longer run, automobiles.[259] Snell's *American Ground Transport* was first published as a House Committee Print. It was not limited to NCL. It charged GM with forcing the railroads to adopt its diesels and with aiding the Nazi government during World War II. Needless to say, Snell's charges garnered significant newspaper attention.[260]

Did a Judge Doom–like General Motors do in the streetcar, or was the streetcar already doomed due to the general rise of the automobile and the greater flexibility of the bus?

14.2 Deterioration

> Rule Three: Invest in companies which are not dependent on high tariffs and would not suffer from European or Asiatic competition in the years ahead. A thoroughly reorganized traction company in a large city would qualify, while a textile company would not.
>
> —Babson (1948)[261]

Subway and elevated transit, streetcars, passenger railroad service, and ferry service were deployed in large cities prior to the coming of the automobile, and by the late 1920s the widespread use of the automobile in urban areas either brought the growth of ridership to an abrupt halt or turned it downward. Streetcar service was impacted hardest. Views of the problem had several ingredients.[262] One was that suburban areas not served by transit required transit expansion. Otherwise, market shares could not be maintained.

Figure 14.1 shows long-term loss of market shares, though some recent modest gains after 30 years of concerted investments in rail.

Another view said that the automobile was a superior vehicle, so transit vehicles of higher quality were needed. There were two responses to these views. One response was to deploy the bus. It could serve newly developing residential areas and serve as an economical substitute for streetcars on thinly traveled lines. Transit operators liked the bus because a one-man crew could be used, the bus was somewhat more maneuverable than the streetcar in traffic, and interruptions of power, fires on streets, and so on could be detoured. Additionally, many cities required that streetcar operators maintain the roadway occupied by streetcars, an expensive endeavor and one disliked by transit operators because those facilities were used by automobiles.

The other response was to develop a superior technology. The Presidents' Conference Committee (PCC) streetcar resulted. The story of the PCC car is an interesting one. The

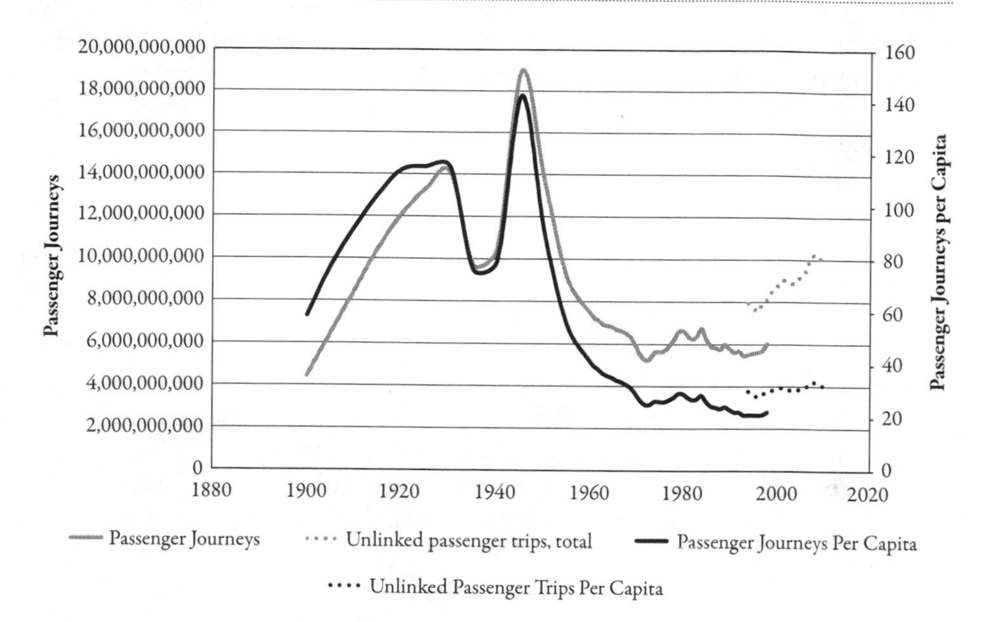

FIGURE 14.1 United States Transit Ridership Trends

Electric Railway Presidents' Conference Committee (PCC) was made up of the presidents of the large transit properties. It examined the situation and published reports and recommendations, most of which were ignored. Many of the ideas developed are topical today, for example, the idea of having downtown activities (land owners) subsidize CBD transit services. It was claimed that transit service resulted in a socially desirable urban form.

The PCC undertook the design of a modern electric streetcar (the PCC car), which was built in small numbers just before World War II and greater numbers just afterward. Several car builders were involved, and overseas builders were licensed to use the technology.[263] European designs deployed in the United States as light rail ventures, from San Diego (1980) forward, were second-generation PCC cars.[264] The PCC car was much improved over previous cars—better acceleration, quieter, better ventilated, better seats, stronger, and so forth. Yet it could not undo the reshaping of demand and competition from the automobile.

The third perception of the transit problem sums as "rationalization." In many cities there were several properties operating competing services, and services were often not coordinated. Because the transit properties operated on franchises, matters of fares and service had often been politicized. Many properties were operated by electric utilities. Tackling these problems was the third challenge to transit planning. The rationalization issue was not new, of course, for consolidation of properties had been undertaken previously.

The transit properties were mainly private properties, so often planning studies were undertaken by the property owners. But in many cases problems were so blatant that municipalities sponsored studies. Also, the franchise, fare, and service leverage exercised by cities on the properties gave them a considerable involvement in problem solving. Many studies were made in the "great men" style during the 1920s and 1930s. One actor was

Henry M. Brinkerhoff, who worked as an engineering manager in the transit industry in Chicago. During the 1910s, he was chief engineer of the Chicago Subway Commission and he developed Chicago's consolidation plan. He subsequently worked in Detroit, Cleveland, Philadelphia, and a number of other cities.[265]

Considerations of a more policy planning–strategic sort were treated by the American Electric Railway Association. Its publication *The Urban Transportation Problem* (1932) reads very much like today's problem discussions. For instance, special assessment districts in downtowns were proposed for transit funding.

Using the criterion for success "did the plans accomplish what they set out to do," transit planning worked pretty well. Buses replaced streetcars and offered service in suburban areas; the PCC car was quite a remarkable advance over previous technology. Rationalization was not as successful. In many places properties were not consolidated, and everywhere little progress was made on adjusting fares and service.

The transit context was the problem. The properties were either losing market share (subways and elevated lines) or losing in an absolute fashion (bus, streetcar). The problem was to manage stasis or graceful decline, and we have four observations on that problem.

First we may observe that changes pose no special problems for individuals and institutions if the physical and psychological magnitudes of those changes are slight. An example is the switch from the horse-drawn tramcar to the cable car, which involved a familiar technology and not much new knowledge. It did require some change in behavior on the part of institutions as the use of horses was phased out and new investments made and as new kinds of people were brought on board to operate the cable works. The change from the electric streetcar to the bus was also manageable. However, suppose that physical and psychological magnitudes are great. Consider the shift from the steam locomotive to the diesel. Traditional locomotive manufacturers had almost none of the relevant knowledge, and it was unthinkable that something so different could replace the iron horse that built the railroads. Those manufacturers are no longer in business.

The next three observations are also well-known. We will follow Simon's wording of them:[266]

- When an institution is not performing as well as it aspires to perform, search behavior is induced.
- The institution's level of aspiration drifts downward until it matches obtainable performance.
- But there is a catch. The level of aspiration may be so locked in by peer group values and traditions that it does not drift down easily. Search behavior may be fruitless or only marginally rewarding. The mechanisms causing performance to fall short of aspirations may continue in place, and declining aspirations chase ever-declining performance possibilities. (So there may be a fast variable, slow variable situation where things get out of synchronization.)

If the latter situation holds, then the possibility of catastrophe looms. In the private sector this may mean bankruptcy, depending upon the structure of competition. In any circumstance, one observes rational adaptive behavior being pushed aside by apathy and/or aggression.

The transit industry seems to have maintained its "cool." It searched for new options (e.g., bus, PCC car), lowered its aspirations, and acted in a rational fashion. Today, however, it has the problem that concerned publics have high aspirations for it—save the city and eliminate congestion, among others. But those aspirations have not permeated the industry very deeply.

14.3 Motorization

One by one, the properties began to substitute electric or motor buses for streetcars and inter-urban services gave way to bus services. Public policy actions enabled and often encouraged the shift from streetcars to buses.[267] Policy allowed for some rationalization of properties and services, as well as fare adjustments. The industry muddled through the Great Depression. World War II saw great increases in ridership as employment increased and automobile production and use was discouraged. Many properties were "flush" with cash at the end of the war.

But it wasn't long after the war until trends dating from the 1920s again took hold. The fiscal problems of private operators called out for public actions in the larger older cities where the loss of transit services could not be tolerated (San Francisco, Chicago, New Orleans, New York, Philadelphia, Pittsburgh, etc.) and the cities began to take over services. The acuteness of problems, steps taken, and success in managing problems varied by venue. Seattle and San Francisco took over private services early on; some cities took over step by step. Chicago, for example, took over subway-elevated services first and bus and streetcar services later.

In general, the cities undertook modest "downsizing," fare rationalization, and other efforts to improve the matching of services to markets. They did not experiment with deregulation in a full sense. However, in the 1950s and 1960s, quite a few services in the metropolitan areas continued to be privately operated, typically providing small market niche services.

Several key points can be drawn from our history that relate back to the conspiracy. First, while General Motors' National City Lines was a private venture that converted streetcars to buses, a number of other cities, in which National City Lines was not a player, did so as well. Some of these cities made conversions while the city government controlled the routes, others while they were in private hands. These other (non-NCL) cities did not always make quite as extensive a conversion, but they did a conversion nonetheless. (Evidence for this includes locales beyond the US, for instance London, which phased out [double-decker] electric streetcar and trolleybus lines for the ubiquitous double-decker motorbuses that are world famous).

Second, the PCC car was promoted by private companies as well; General Electric, Westinghouse, and St. Louis Car were not non-entities in the business. If there was a good business to be made in selling modern streetcars, there were capitalists willing to give it a go.

Third, the NCL organization was profitable for many years. General Motors entered the business to make money by selling buses, not to destroy transit to increase auto sales, which were on an independent trajectory and didn't need the assistance. While the business may not have been GM's most profitable, it was just an element of a large firm that diversified to spread risk and invest profits. Other GM subsidiaries at various times included appliances (Frigidaire) as well as bus, taxi, and rental car companies. The literature in the 1930s often refers to the saturation of the automobile market. Then as now, there was interest in the industry in growing through diversification.

Fourth, buses were at the time seen as the "new and improved" version of transit. Rational transportation planners comparing buses to streetcars would point to their flexibility, their ability to free ride on roads provided for other reasons (no maintenance of track or overhead electrical was necessary), the smoother ride, the ability to follow demand to the suburbs at relatively low cost, and so on.

At worst, the NCL conspiracy accelerated a trend already in place. We judge that the matter was not a big thing in the decline of transit. The market was the strong force at work. NCL may have yielded minor returns to the GM bus business, but as a diversification step, it was not a very good idea.

14.3.1 TWIN CITIES TRANSIT

While there is clearly some dispute as to the importance of a national conspiracy in the shift from streetcars to transit, one should not dismiss the existence of criminals in the streetcar companies of the era (like the rail era before them). A marked example is in the Twin Cities of Minneapolis and St. Paul, Minnesota. The streetcar lines in the Twin Cities were built by Tom Lowry in the nineteenth and early twentieth centuries and, as in many cities, were aimed in large part at land development. For the period between 1925 and 1948, fares held steady at $0.10, leading to capital shortfalls. The Twin Cities lines were publicly traded and most shareholders were non-local. The conversion from streetcars to buses took place after a series of events helped drain the company of even more resources. In 1949, Charles Green undertook a hostile takeover. He asked for a fare hike, fired 25 percent of the workforce, and canceled capital investment. He was employing a traditional "cash cow" model, wherein new owners milked the system of resources to pay for its own takeover.

A strange turn took place when Isadore Blumenfeld, a.k.a. Kid Cann (rumored to be a gangster and murderer) and Fred Osanna (known to be a lawyer) tried to take the system from Green. The State Railway Commission made an investigation of bribery, embezzlement, kickbacks, and death threats. Osanna and company did successfully take over the Twin Cities Rapid Transit in 1951, and sold off the streetcars and many of the rails. It is reported that the vehicles are still running in New Jersey and in Mexico City, though while the shells may still operate, whether the mechanics in the vehicles do is unclear. Osanna claimed "the fastest and most massive streetcar-to-bus conversion ever undertaken in any major US city." However, Osanna wound up in jail for fraud. The system was subsequently sold to Carl Pohlad (later owner of the Minnesota Twins), and was eventually sold to the public Metropolitan Transit Commission in 1970 for $7.9 million.

14.4 Angels and Devils

The conspiracy is clearly the most popular view of why transit failed. Milder "devil" theories have to do with tax subsidy for suburban family homes and subsidies for the automobile.

Sometimes it is said that mortgage insurance offered by the federal government influenced development, but such insurance was available for many kinds of residential development. One gets the impression that somehow the devil made otherwise good people make bad decisions about auto ownership and suburban development.[268] Devil theories go a little ways but end up as not very consequential.

Angel theories give the same impression: We would be in heaven if smarter, better trained managers were hired, if new equipment were purchased, if new systems were developed, and so forth.

A couple of cost-related angel theories are that transit is inherently less expensive than the auto alternative. Intuitive clarity takes over—moving a bunch of people in a vehicle is relatively inexpensive. That extends to the notion that the higher the capacity, the lower the cost: example, heavy urban rail transit systems are lower cost than light rail, and light rail is lower cost than bus. This may be true at capacity, if the costs of construction are not too different. In practice, the evidence is otherwise (see Section 23.2).

Another angel theory is that of avoided cost—providing transit is much less expensive than providing for increased automobile use. That is a very situation-specific assertion, and work we have seen says that it doesn't hold very widely. At any rate, it makes assumptions about market growth and about unchanging spatial patterns of demand that are not very tenable.

Somehow we went from this assertion to the conclusion that we ought to be concerned only with transit operating costs (fixed costs are of no matter because we would have had to spend at least that amount of money on auto facilities). This kind of thinking leads to the conclusion that BART is more efficient than bus because the fare box covers more of its operating costs and in general the nonsense that high fixed costs, low variable cost systems are always superior. Amtrak has taken such thinking one step further. At one time it treated its subsidy as income on its books, and claimed to make a profit.

The notion arises that highly desirable (angelic) urban lifestyles and urban forms will be achieved by transit development. Because of market segmentation, at some level the notion has to be true, and the issue becomes, "For how many?" If there were lots of people who find that to be true, then ridership growth would reflect it. Alternatively, changes in urban land uses and urban form would reflect it.

Urban transit was pretty well deployed by the 1920s, and subsequent further deployment has had marginal impact on urban land uses. As early as 1930, when transit services were provided, areas that were declining continued to decline; where there was little change, little change continued; and growing areas continued to grow.[269]

Even so, transit's favorable impact on urban development is the Holy Grail of transit advocates, and they continue to look for it. Federal transit agencies funded study after study, and the Transit Cooperative Research Program (TCRP) has begun work in the area, to no avail so far.

A variety of views exist. Some hold that transit-walking neighborhoods might be created with desirable social values. Demand is diverse, so there is certainly a market for such neighborhood developments. Issues are those of the size of the market and why private decision making hasn't cleared such markets. If there is a market and there is some dysfunction constraining it, there might be room for policy actions.

At a much broader view, there are those who advocate return to transit-like cities in order to improve the quality of life and, in particular, deal with energy use and sustainability problems. Newman and Kenworthy (1989) have pointed out that dense, transit dependent cities use less gasoline than "auto cities." The Newman-Kenworthy policy of returning to transit-served cities has intuitive clarity to many, but it is not sensible.

Why? As discussed, transit served the city of the year 1900 very well. Subsequent development of economic and social activities saw cities taking on more of an "auto" structure. The Newman-Kenworthy policy is much more than just a matter of substituting one mode for another; it has to do with reversing social and economic trends. They are tilting at windmills, trying to put Humpty-Dumpty back together again. History does not repeat itself.

This listing and debating of angel theories could go on and on. Important items to be included are reduced negative externalities (compared to autos), reduced space utilization, favorable social impacts of an "enabling the transportation disadvantaged to enter into societal activities" type, and expanding the energy efficiency discussion.

14.5 Symbolic Systems

The formation and use of myths are natural in human communication. Symbolic systems and images provide the tools used to simplify the complex world; the tools translate complexity to "common sense" and, then, provide a common sense basis for actions. What's wrong and what to do becomes simple and transparent. Richmond (1998) applies the extensive literature in semiotics and linguistics to transportation.

Richmond provides concise discussions of symbols, images, and metaphors. In brief, symbols are instruments of thought. They are partial information that can provide a gateway to a larger pattern or may be part of that larger pattern. For instance, a symbol on a map provides a gateway to the reader's knowing that, say, a park or a freeway is present; such a symbol may also be part of the larger pattern. It may show the extent of the freeway network or of a park. The point is that symbols allow one to conceive objects, so once we see the symbol for a freeway, we see beyond the symbol to the larger object.

A sign acts as an announcer: ENTER FREEWAY, CURVE AHEAD, RAILROAD CROSSING, and so on. It doesn't always lead the reader to conceive the subject. It's quite different from a symbol, although a sign may introduce a symbol. An image is intertwined with a symbol—it's how a symbol is understood. One abstracts from the pattern brought to mind by the symbol using a calculus of images to make sense out of the complexity that a symbol represents. A symbol on a map brings forward the pattern of a freeway. One then understands that pattern, using images that abstract from the larger whole. The image held by a pavement expert might differ from that held by a traffic engineer or an ordinary user. The Transportation Security Administration uniform brings the military and police to mind, even though their personnel lack the equivalent authority. The hope is, of course, that evoking the uniform of heroic officers will encourage airline passengers to comply.

One tool we use is a metaphor, one thing standing in place of another. A metaphor permits our drawing on a variety of experiences and applying them to a new situation. We might say, for example, that Brunel did a Stephenson (the developer of the first railroad) when he developed steamships for the packet trade. That's very useful. It is a heuristic fiction that names, frames, and connects what Brunel did. The birthing, growth and development, and maturity metaphor is useful.

Schön (1963) uses the notion of *generative metaphor* to describe the development of social policy. Participants in the policy process tell stories about what is wrong and how things can be fixed. These stories are metaphor based, and the naming, framing, and fixing of problems reflect the metaphors behind the stories. In policy discussions, metaphors serve to cut through and organize complexity; they also execute a normative leap. They pinpoint what is obviously wrong, and they make what to do also seem obvious.

Richmond also addresses the myth that rail transit can alleviate the transportation problems of Western cities, Los Angeles in particular. What is a myth? Symbols direct us to particular conceptions of objects, images form the impressions we draw about the objects, and metaphors frame and structure these. The result is a myth. Richmond's study involved interviews with political actors in Los Angeles, and he uses extensive quotes to show the images of trains: "powerful," go "whoosh," "straight through." He presents an aside on the sexual potency of the technological power of a train. A train is interesting because it is imaged as a "she" as well as a penis symbol. Inquiring into the way that symbols and images are organized in metaphors, Richmond refers to the organic metaphors so often used in transportation (arterials, free flow, etc.). He stresses the metaphor of the body in balance. The body and its transportation system aren't in balance or equilibrium because the circulation is congested. Trains are needed to get circulation in balance.

We think that it is helpful to consider metonymies because they may serve as a building block for images. A metonymy is the use of the name of one thing for the name of another, and that is not the same as a metaphor. It is contiguity/association based. For a simple example, we might refer to America's land resources although the land is actually owned by many people. Metonymies shape images. The rugged cowboy grins, sits on a horse in a beautiful Western landscape, and smokes a cigarette. The cigarette takes on the name of the landscape. Automobile advertisements show sports cars going fast on rural California roads; family cars are shown in the driveways of expensive houses. Sports cars have the meaning of going fast on rural roads, and family cars are the good suburban life. We think metonymies play a major role in shaping transportation images. Have photographs of BART with the Bay Area landscape in the background shaped mass transit policy? We think, yes, just as photographs of Shinkansen ("new train") bullet trains with Mt. Fuji in the background and flowers in the foreground have shaped interest in high-speed rail proposals.

We think that there are differences in the metonymies typically associated with the modes. A bus runs in a grubby place, streetcars and trains run (are imaged to run) in the country or in beautiful suburbs and malls connecting to vibrant, but not too crowded, downtowns.

Why do attitudes change? For example, cars used to be good things, now they are bad. Perhaps that is because the metonymies change. In the early days, cars were much used for touring; they were associated with the countryside and were imaged as carefree exploration of the countryside. Cars were viewed as much superior to the train for touring. Today, cars and congested freeways go together, at least in the debate on transportation problems of large cities. Some now argue that to be carefree you must be car-free.

Symbols direct us to particular conceptions of objects, images form the impressions we draw about the objects, and metaphors frame and structure these. Richmond says the result is a myth, but many political leaders and voters in Los Angeles say the contrary.

When Darling (1980) said "... (policy has been) transformed into a kind of witchcraft" it seems that he was commenting on processes involving Schön's generative metaphors, the metaphors used to name and frame policy problems. Darling saw naming and framing as witchcraft. The use of the word "witchcraft" itself tries to reframe what was happening to policy to evoke an irrational brew where the output has no relation to the input. This metaphor is far stronger than that relating sausage-making and bill-writing, which suggests you don't want to see it happen, because at least there is a rational (if unsightly) chain from beginning to end.

Because reasoning involves naming and framing, and that naming and framing is based on symbols, metaphors, and so on, the results of all reasoning must be myths. The results of reasoning are artificial constructs. That reasoning leads to the conclusion that all transportation knowledge and policy are steered by myth. While that conclusion must be true, it is destructive and immobilizing. If everything is a myth, why bother? Surely some myths are superior to others; while not generally agreed to, Richmond's and Darling's myths are superior to the wisdom generally held. Why? The obvious answer is the analytic content of Richmond's and Darling's knowledge. Analytic knowledge lets us see many myths: such as, railroad management tried to discourage passenger ridership, regulation was forced on the railroads, expansion of transit would be energy efficient, and so on.

Who generates myths? Who is using metonymies, constructing metaphors, and engaging in the policy debate? Who is debating policy? White-collar urban planners and politicians engage in service activities and commute to congested office districts. The nature of their transportation problems and their solutions thus follow.[270]

Politicians adjudicate the interest of differing publics so the balance metaphor may be especially appealing to them. Urban planners also have balance upfront; they are attracted to certain metonymies and images, such as the "right" kind of urban landscape. Now, urban transportation planners use the Urban Transportation Planning (or UTP) process; process and equilibrium metaphors are valued. If trained in engineering, transportation planners have a classical, Newtonian physical science worldview. Nature is orderly and exact; it doesn't change by leaps and bounds. Aristotle said about the year 300 BCE: "Nature does nothing in vain."

Because non-engineers engage in the policy process and their experiences are different, we suppose that there are classes of myths: economists' myths, urban planners' myths, environmentalists' myths, traffic engineers' myths, politicians' myths, transportation analysts' myths, and so on. There are different modes and these operate in different situations. We sense that these suppositions are true, but have explored their content only in a casual fashion.

An important point is that it seems that some actors' myths have more authority than myths held by other actors, with authority varying from situation to situation. This speculation contradicts the statement above that authority is gained from the analytic content of myths. The contradiction is easy to explain. Analytic content appeals to some, such as the authors. Others give authority on other grounds.

14.6 Discussion

> One should not increase, beyond what is necessary, the number of entities required to explain anything.
>
> —Sir William of Occam's Razor

The observation that government steps in and does what systems cannot do for themselves is perhaps the simplest explanation for the situation in mass transit. The observation applies worldwide. Straight-jacketed by high fixed costs, militant labor organizations, and regulatory constraints, by the 1920s systems were having difficulty adjusting to changes in competition and markets. So local and/or national governments began to acquire ownership to maintain the status quo.

In the United States, local actions gave way to federal, as will be discussed in Chapter 23. This follows from the federal taste for large programs acquired when water resource, agricultural, and defense programs increased in the 1930s and 1940s. With its monetary resources and the tilt to urban-based power in Congress, urban programs followed. With clamor from large city business, political, and transit labor interests, transit was there for the taking. Along with urban freeways and airports, it became one of several national programs leading to megaprojects. Analyses of such projects include reasoning about federal involvement similar to ours.[271]

So far, Sir William holds very well.

It is also easy to provide explanations for distorted project costs and ridership projections, the sometimes venomous anti-automobile, transit-first clamor, the willingness to use urban planning to (try to) control urban travel, and the use of "ends justify means" lines of pro-transit reasoning and analysis. (It might be worth noting that highway builder Robert Moses once said "If the ends don't justify the means, what does?") People value the status quo. Explanations that involve human avarice and efforts to impose planners' values on others are supplementary.

Mass (or public) transit receives a lot of attention from political actors and the press, and many professionals think transit "is where the action is" and will be in the future. This chapter traced transit's middle age, from maturity through decline.

The argument of the devil versus the angels is very much one of style versus substance. By demonizing their opponents, angelic transit supporters aim to seize the moral high ground. Yet their arguments and facts are as mythic as their styles. This raises an important issue about the ethics of advocacy planning. If the supporters of the conspiracy know that the conspiracy is not really to blame for the loss of the streetcar (and we would venture some of them do), then they as planners are behaving unethically in using that story as the cornerstone in their arguments for new rail transit investment.

The story of transit continues in Chapter 23.

Concrete costs less in the long run!

Concrete lasts longer with less need for maintenance and repair!

Concrete is quiet!

Concrete is safer!

Concrete is environmentally friendly!

Concrete is aesthetically pleasing!

—Concrete Paving Association of Minnesota at `http://www.concreteisbetter.com`.

Quick easy construction, easy design capabilities, outperforms competitive materials.

Doesn't your community deserve smooth, attractive, asphalt roadways? American's best communities choose asphalt, shouldn't it be your choice?

Don't be Rigid in Your PAVEMENT Thinking. Choose asphalt, the Versatile Paving Material.

Customer satisfaction, economical, speed of construction, durability, versatility, recyclability, smooth & quiet, maintenance application.

—Minnesota Asphalt Pavement Association at `http://www.asphaltisbest.com`.[272]

15 Public Roads and Private Cars

15.1 Whitetop and Blacktop

FLEXIBLE ASPHALT VERSUS rigid concrete—it seems a strange battle, even to those of us steeped in transportation. In many states, consolidated pavement manufacturers sell the appropriate material for the job. However, for whatever historical reason, in Minnesota, the producers remain small and their associations have chosen to compete over the fixed pie of pavement surfacing and resurfacing rather than coalesce to increase the amount of pavement demanded. They promote competing studies arguing in favor of their material. Historically, low-volume roads have gone asphalt and high-volume roads have gone concrete, and they compete at the margins. Since there are many more miles of low-volume roads, there is a great deal more asphalt on the ground.

Differences in pavements are, of course, but one aspect of highways, an aspect that makes little difference to the transportation itself, aside from minor adjustments to costs and delays during construction. Yet in the aggregate, pavements are a huge industry, and a large element of the transportation experience that we would be remiss to miss. The difference between asphalt and concrete is much less than the difference between paved and unpaved. Like the debate between Telford and McAdam (see Sections 3.5 and 3.4) in the early nineteenth century, the battle between asphalt and concrete is both slow, puerile, and futile. A combination of the two (e.g., concrete with asphalt overlay) often makes the best technique, though the sides are often too competitive to see it.

In response to pavement advocates, Francis Turner, onetime head of the US Bureau of Public Roads, would sometimes close the debate by jesting, "The Interstate is a balanced system, one-half of the mileage is concrete, and the other half is asphalt." Thomas MacDonald, his predecessor, had no prejudices for asphalt or concrete, but wanted the best for the job, which meant balancing costs and benefits, as overbuilding was as much a problem as underbuilding.[273]

In the United States, the modern auto-truck-paved highway system dates from the mass-produced automobile, example, the Ford Model T, around 1910. Of course, there had long been roads and personal transportation. So, one might say that steam, electric, and internal combustion engine vehicles triggered a new realization of old systems, though their birth dated from many centuries ago. The automobile and truck offered increased performance (higher horsepower per ton) and, eventually, lower costs compared to animal propulsion systems. This enabled a renewal or transformation of the precursor system or, if you wish, the birth of the modern system. The title of Barker and Gerhold's *The Rise and Rise of Road Transport* mirrors the renewal situation.[274]

Much has been accomplished since the early 1900s. It was decided what governments would do versus what private actors would do; funding responsibilities were set, learning about system capabilities occurred, and the system was deployed.

Some key matters in automobile development were the availability of high energy fuels, the shift from the steerable front axle to steerable front wheels, the use of specialized labor and Fordist mass production, and the adoption of interchangeable parts. Interestingly, it took until the 1930s for the predominant design of the automobile and truck to emerge. Process-of-production technology was also well developed by about that time. Development took off in the United States. It is said that circumstances were first ready in the United States, where incomes and preferences supported the development of mass markets and enabled mass production.

It is instructive to compare the time constants describing market penetration in various environments, say, the time required to go from 5 percent to 95 percent of market saturation. This required about seven decades in the United States, two to three decades in Western Europe, and a little over one decade in Japan. That partly has to do with the rise of incomes in those areas.

We might expect places that began to automobilize after the United States to automobilize faster than the United States did. Because the technology had been standardized, only emulation was required. But the notion that the environment was somehow ready in the United States and not in other places is harder to explain.

The time constants for vehicle market saturation compare in an interesting way with deployment of improved roads. Comparison of the S-shaped deployment curves for roads and vehicles (Figure 15.1) suggests that road improvements led and vehicle populations followed. In fact, there were many roads before motor vehicles. That observation raises an interesting question about causality in system deployment.

The automobile and truck appeared in environments where road improvements were underway. Motorized vehicles begged incremental changes, including pavements to keep down dust and protect the road surface, to reduce vehicle damage, and stronger pavements to carry loads, bridge improvements, increased capacity, and design changes to allow for higher velocities.

Many pre-automobilization design protocols were unchanged. For example, the higher horsepower per unit of weight of motorized vehicles would have enabled much steeper

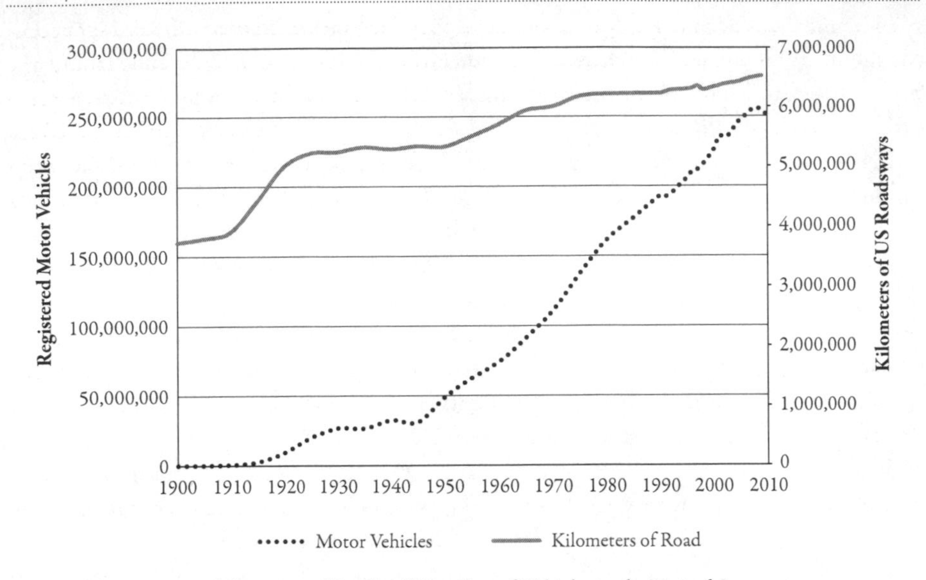

FIGURE 15.1 Figure Kilometers of Road and Number of Vehicles in the United States.

grades. (The critical road grade for a wagon was downhill). The then current brake technology allowed for a grade of about 2 percent, which was the maximum load horses or mules could hold back). Although special facilities for heavy vehicles were discussed in the early nineteenth century (the idea of iron roads), the protocol of limited specialization of the road to vehicles and operations continued. Specialized designs (truck-only; auto-only highways) were discussed in the 1920s and 1930s; but, except for a few parkways (such as those developed by Robert Moses in New York, see Section 22.3.4), those designs were never adopted (see Section 22.6.2 for a discussion of the topic).

The system characteristics of automobiles, trucks, and highway facilities were frozen when the predominant technology hardened in the 1930s. The roads, vehicles, and operations developed by the end of the 1930s have been deployed, and, today, only modest "polishing" changes are underway. That's true in the sweep of history in spite of, if not because of, the increasing number of regulations affecting vehicles and highway designs.

There is more to be said about the emergence of highways, and this chapter explores their policy, planning, deployment, and management. Although some aspects of state activities are treated in the discussion of the evolution of modern urban transportation planning system, the present discussion extends Chapter 7 aims to provide a coherent view of the emergence of modern, paved highways. We will see that three levels of policy and planning evolved: federal, state, and local.[275] In some states, there is a fair amount of articulation between county and state planning, in others there is not.

15.2 Auto Trails

To say that the country did not recognize the auto for what it was is to understate the case. The country recognized the auto as a rattling piece of machinery that could be counted on to break down every three or four miles.

— Norman Bel Geddes, *Magic Motorways*[276].

By 1912, Carl Fisher (auto dealer, land developer, racetrack entrepreneur, headlight company founder) launched a new project, a transcontinental rock highway, running from New York to California. While the idea was not original (The National Old Trails Road Association was already promoting a road from Washington, D.C., to Los Angeles, and the idea was "in the air"), Fisher was the necessary promoter to get it done. Fisher planned to build the road incrementally, piecing together existing segments and filling the voids with new construction, upgrading the whole thing over time. He wanted carmakers and auto dealers to contribute one-third of one percent of their gross revenue for three years to help pay for the route, which would be built by volunteers and governments.

Funds came from Goodyear Tire, cement magnate A. Y. Gowen, and automakers Hudson (now run by Roy Chapin), Willys (still led by founder John Willys), and Packard (run by Henry Joy), but Ford resisted, stating "as long as private interests are willing to build good roads for the general public, the general public will not be very interested in building good roads for itself."[277] Henry Joy became the leader of the group to build a "Coast-to-Coast Rock Highway," which was soon renamed the "Lincoln Highway." He predicted by 1915 coast-to-coast travels would be complete in as few as 11 days.

Fisher also promoted the Dixie Highway (Chicago and Sault Ste. Marie, Michigan, to Miami), which was actually one eastern and one western highway following parallel paths (as the compromise was easier to agree upon than a single route) with periodic connectors. The route upgraded nearly 4,000 miles (6,400 km) of dirt and poor quality asphalt trails to concrete and brick.

Other "Auto Trails" were established as well; each was designated with a distinct signage, each positioning itself as the most efficient, or the most scenic (they could not be both). The most famous beyond the Lincoln Highway were the National Old Trails Road (Baltimore to Los Angeles), Bankhead Highway (Washington, D.C. to San Diego), Atlantic Highway (Fort Kent, Maine, to Miami), Jefferson Highway (Winnipeg to New Orleans), and Pacific Highway (Port Angeles, Washington, to San Diego); though there were at least 250 US and Canadian highways and trails pumped by promoters and designated by distinct organizations.[278]

Many of the more important US routes were mapped onto the much more rational numbered US Highways systems, and, of course, many of the trails used common road segments. However, the named routes were not given a single number, instead being assigned to portions of multiple numbered routes, to the dismay of most of the route associations. The Lincoln Highway, the most important of the routes, was mostly designated US 30 from the East Coast to Salt Lake City. The Boy Scouts of America placed 3,000 concrete markers along the Lincoln Highway route to commemorate the 16th president, and then the Lincoln Highway Association disbanded in 1928.

The auto trails were seen by some as the start of something greater. Thomas MacDonald (1881–1957), later leader of the Bureau of Public Roads (BPR), was one such visionary. He called the Lincoln Highway "the first outlet for the road-building energies of this community."[279] New farm-to-market roads were valued as a way to reduce costs and increase productivity of farmers, recreational motorists wanted long-distance inter-city routes, while urban interests valued a different set of roads.

What MacDonald had in mind was federalizing the problem, rather than relying on a set of National Associations to build roads. The American Association of State Highway Organizations (AASHO), the organization of state highway agencies, developed national standards for numbering routes, the map, how signs would work, and so on. The US Highway numbering system, developed by Edwin James (even routes were east-west, odd routes were north-south, with the important routes on the 0s and 1s and 5s, starting numbering in the north and east), was later reconfigured for the Interstate Highway System (with important routes on the 0s and 5s, and the starting of the numbering in the south and west to avoid confusion).[280]

15.3 Safe Streets

The International Classification of Diseases defines roads (as a location where death occurs): "A public highway trafficway or street is the entire width between property lines or other boundary lines of every way or place, of which any part is open to the use of the public for purposes of vehicular traffic as a matter of right or custom. A roadway is that part of the public highway designed, improved, and ordinarily used for vehicular travel."[281] The classifiers have already conceded roads to the owners of vehicles, a transition that occurred in the early twentieth century. Traffic deaths, as of 2011 running at about 33,000 a year in the United States, exceeded 20,000 in 1924 and 32,000 a year in 1940, with far fewer vehicles and miles traveled.[282]

Who controls the streets? Historically, the pedestrian could use the entire street, cross at will, and there was no such thing as jaywalking. Only with the automobile, and the subsequent establishment of traffic control laws, was the pedestrian reigned in for safety and motorized mobility.[283]

Local Safety Councils advocated for safety in the new environment with lethal automobiles. Some of the local organizations were large; by 1922, St. Louis's Safety Council had over 25,000 members. Exhortation was one strategy the safety organizations used, with their iconic character "Otto Nobetter" who did everything unsafely. The Council established Safety Weeks and other strategies of moral suasion to improve driver behavior.

Juries were generally sympathetic to pedestrians in the early 1920s, blaming motorists for most crashes, even when fault was mixed or the pedestrian's. In part this was a class issue; the average motorist was wealthier than the average pedestrian, or jury-member. Similarly, two-thirds of police chiefs believed automobiles should have speed controls.

Over time, the safety movement ultimately but subtly shifted blame from careless motorists (who in the early 1920s were still widely blamed), who ought to drive more safely, to their victim the pedestrian. Clearly it takes two to have an externality, the polluter and the pollutee, each needs to take some responsibility. But in the end, someone has a right to establish the status quo.

Some blamed congestion on the automobile, others on the street. Too much demand and not enough supply are two sides of the same coin, but where you come down depends on whether you think there should be more cars and roads, or fewer. To the auto interests, road-builders, and prospective motorists, the answer was obvious.

William Phelps Eno's Traffic Safety Foundation established a set of rules of the road that became widely, though not universally or uniformly, adopted. Known as the "Father of Traffic Safety," Eno (1858–1945) designed traffic plans for some of the world's great cities, and wrote the first traffic code, for New York. He is credited with the stop sign, crosswalk, traffic circle, one-way street, taxi-stand, and pedestrian safety islands.

It was believed that application of scientific principles to traffic control could halve congestion.[284] Zoning, the control of land use, and traffic control were two prongs of the larger progressive movement aiming to reduce chaos and increase order in the city. Moving traffic control from the police, with their ad hoc ways, to a systematic engineering profession was the realization of this plan. But engineers needed to be trained. This was a new field.

In 1925, noted traffic engineer (and the first Ph.D. in the field) Miller McClintock (1894–1960) established the Bureau of Street Traffic Research at the University of California, Los Angeles (UCLA). After gaining a grant from Studebaker, McClintock set up a Bureau of Street Traffic Research at Harvard University within the Graduate School of Engineering. The Center was named for Studebaker's president, Albert Russel Erskine.[285] The Bureau of Highway Traffic, which started at Harvard, then moved to Yale in 1938, and then to Pennsylvania State University in 1968, for several decades trained the elites in US transportation engineering.[286] Since 2009, Miller McClintock's grandson Thomas Miller McClintock has served in the US House of Representatives from California as a Republican.

Engineers competed against others for authority. The existing streetcar systems had long managed their own ways, and could make a claim to manage roads, but as streetcars were themselves being decommissioned in favor of buses, this argument was losing. On the Twin Cities Line, Motorman Bill, a local iconic character, pled for consideration from motorists.[287]

Jaywalking was criminalized to control the pedestrian, reduce crashes, and improve the environment for the motorist. Sidewalks became the exclusive paths for pedestrians, and the road for vehicles. Conflicts would remain at intersections, but mid-block conflicts declined. At intersections, traffic lights and the Silent Policeman (small traffic circles) replaced the Cornerman and greatly improved traffic efficiency on urban streets. The Yellow Cab Company implemented traffic lights in Chicago's downtown Loop to aid pedestrians, at the price of annoying motorists.[288] Traffic signal timing on a grid, establishing the "green wave" to allow motorists traveling at the appropriate speed to catch multiple green lights in a row, was perfected by pioneering engineer Henry Barnes (1907–1968), who also developed what was dubbed the "Barnes Dance" or all-pedestrian phase during his time in Denver, which he brought to Baltimore when recruited by Mayor Tommy D'Alesandro Jr. (father of US House of Representatives leader Nancy Pelosi) and then to New York (when recruited by Mayor Robert F. Wagner) where he served as the Traffic Commissioner.[289]

Uniformity in traffic regulations was important. If each state, or worse, municipality, had separate rules of behavior, travelers would have a hard time traveling from place to place. A uniform vehicle code was proposed by many in the industry, and was largely adopted.

Were streets a public utility, to be regulated and controlled and for the public good, like street railways, water, or gas? Or were streets a market-place to serve the individual demand of private travelers? A public utility is often thought of as an enterprise where competition is infeasible. There is no competition for the supply of roads; there is, of course, competition for the use of roads, as for the use of the outputs of other public utilities.

Ultimately the Bureau of Public Roads started scientific investigations of driver and vehicle performance to improve road design and increase safety.

15.4 Transportation and Traffic Planning

Just about every city of some size undertook transportation and traffic planning during the 1910s or early 1920s to reconstitute streets for automobiles. Plans were sponsored by civic organizations, "great" men (such as Daniel Burnham) were imported to make pronouncements, and a fancy report followed. Typically, the first chapter said "what a fantastic city." Following chapters provided and discussed maps of arterials and land uses. Technical materials then provided street designs, including drainage, bridges, lighting, and so on. The plans also had fiscal and institutional recommendations. Local streets were to be paid for by special property assessments. Arterials were funded from citywide taxes. The city engineering office was to execute the plan.

These authoritative plans were very successful, welcomed by various types of clients. The plans dictated "what to do" and "how to do it." The imprint of these plans on cities can be seen today.

Plans met the expectations of social critics. The automobile was welcome for many reasons, among them: to reduce crowding and social tensions, to lower housing costs, to increase access to air and light.[290]

Finally, plans were consistent with the ways previous tasks had been undertaken: improved water supply, drainage, sewage works, streets, and the rebuilding of cities following fires. They were based on learning, and they took advantage of institutional resources already developed. Institutions and procedures developed to handle public works construction and operations formed the core resources for planning.[291]

A new activity emerged—traffic studies responding to the failure of capacity additions to solve capacity problems. The transportation and traffic studies of the 1910s and 1920s did lead to the construction of physical facilities with the assumption that the facilities would handle the traffic. They did not. So the cities began to commission a new type of study. The review of a typical study will be helpful.

Sponsored by the Chicago Association of Commerce, Miller McClintock's *Metropolitan Street Traffic Survey* for Chicago is an examplar.[292] It was done in the great man/powerful sponsor style. The Association of Commerce was the sponsor; McClintock, director of Street Traffic Research at Harvard, (whom we met in Section 15.3) was thus the great man.

The existing situation was surveyed:

- Street use and congestion;
- Factors complicating traffic flow;
- Costs of congestion;
- Traffic accidents and accident control;
- Parking and pedestrian problems;
- Traffic policy problems;
- Traffic signals and signs;
- Administration;
- Traffic courts.

The study recommended an ordinance that would:

- Create a division of traffic engineering in the department of public works;
- Reorganize the traffic division in the police department;

- Standardize signs and signals;
- Improve traffic law enforcement;
- Create a city traffic commission.

Studies of this type were common during the 1920s, and they continued with an intensified traffic safety flair into the 1930s. They didn't address financing, their recommendations were mainly operational in type, and beyond the passing of the ordinance (a city council matter), they split the responsibility for execution between two institutions: the police and public works departments.

Although traffic engineering had begun to take on an analytic stance during the period when these studies were done, there was a mismatch between the scientific fact-oriented views in the federal Bureau of Public Roads and the style of the traffic studies. The Bureau's developed facilities; the output of the traffic study was an ordinance. Traffic studies proceeded in the "great man" style; the Bureau's style was different. Where the Bureau had been involved in urban areas (very rare), the Bureau's study style was different. Whatever the reason, this round of traffic studies in the cities had little lasting impact, aside from forming a market for the development of traffic engineering.[293]

The plans and their execution by city engineering departments tamed the problem of the deployment of urban streets, but not traffic problems: the building of facilities did not take care of congestion, while traffic safety problems remained. Consequently, in later decades the traffic problem emerged and reemerged, and management of the traffic problem diverged somewhat from the management of the road infrastructure problem. We will refer again to the traffic debate, although it will not be pursued particularly in discussions to follow.

15.5 Free Curb Parking, or Who Controls the Roads as Commons?

Does prohibiting parking in neighborhoods respond to objections to the "outsiders using the roads we labored to provide" theme reported by Webb and Webb (1913) when they discussed outsiders using roads improved by local farmers—the situation when commerce was increasing and the toll road movement had not yet flowered. Is the concern that of protecting local communities from outside merchants and other influences, as reported for Rhode Island when toll roads were proposed?[294]

In *My Own Private Idaho*, S. Gofman used the subtitle *Staking Claims to the Public Streets*. For years public streets have been just that, public, and rights to use have been highly valued, and Gofman asks why the turnaround, why are local residents' now able to take public or communal property (streets) for their private purposes? He reviews court cases and highlights a 1977 Supreme Court decision that is taken to allow most any resident-only parking program on the theory that such programs reduce air pollution.

That's interesting, but is reduced air pollution the motive for residents promoting restricted parking? Does restricted parking increase transit use and decreased emissions from vehicles? If it does, why are residents generally able to purchase parking permits for their visitors? Does restricted parking exercise local residents' feelings against commercial or other traffic attractions near their neighborhoods?

Interested in property rights, Adler (1985) has made the case for streets as communal property. He observed that streets become semi-private when residents implement restricted parking programs, and they have a private aspect when driveways require curb cuts and take space that could otherwise be used for parking. There are variations on communal and semi-private when time limits on parking are applied and when parking rights are restricted to neighborhoods or are applied citywide.

Adler imagines a traffic destination (shopping, religious, school, or employment center) where there is parking spillover into neighborhood streets. As a matter of policy, is it in the public interest for those streets to be private, semi-private, or community property? Adler observes that welfare is maximized if streets are regarded as communal property and used by non-residents for parking. To engage in activities at the center, non-residents have driven some distance and that effort implies the values that they place on parking. Unrestricted parking allows society to realize those values.

Shoup (2005) argues that there is no such thing as "free parking" and asks for cost-based parking fees or "cash out" payments, say for employees who do not use the company parking lot. He suggests that by giving control of on-street parking to neighborhoods, revenue can be raised for many neighborhood public goods such as street landscaping.

15.6 The 7 Percent Solution

Roads were popular, and the popularity of road improvements was pushed even before the auto by needs to get farm products to railheads and by use of bicycles. In addition to state and county initiatives, national level legislators responded warmly to needs for funds, Table 15.1 gives a partial list of national legislation.

The legislation framed the context of road planning. The 1916 Act referred to post(al) roads as a way to designate routes to be improved. The Act contained an interesting equity-oriented statement of purpose, including the need to decrease the isolation of rural settlements. Rural facilities were poor relative to urban. World War I reduced the federal role, as resources were diverted to the war effort. Motorized military convoys (as noted in Section 7.3) made mincemeat of what paved roads were built, but nevertheless eventually were able to travel throughout the eastern United States.

The 1921 Act had concentrated resources and triggered the establishment of state highway departments where they had not already been established. It also gave those departments specific turf, the 7 percent of all rural highways that would be on the ABC system. "ABC" refers to a classification of state highways eligible for federal funding. Planning came along in 1934, following BPR experiments with state planning. The Federal Aid system ultimately totaled over 271,000 km (168,000 miles), and was far larger than the US Highway System, as it included many local roads serving county seats, but not interstate travel.

Why 7 percent? The number is arbitrary, of course, and may have resulted from the smallest number that would ensure senators in each western state could deliver both an East-West and a North-South highway, or based on MacDonald's experience in Iowa.[295]

State planning, initiated as per the 1934 Act, was not completely co-opted by needs and costs studies and by attention to construction of the interstate. Rather, it evolved into a fairly short-term routine leading to the placing of projects in the construction pipeline.

Though the scope of the state versus the county road systems and how federal money would be divided was debated, the 7 percent solution for the allocation of federal money in

TABLE 15.1

Timeline of Federal Road Legislation

Year	Action
1916	PL-64-355 First Federal Aid Road Act set a precedent by allocating $5 million in federal funds to be matched 1:1 by the states for construction of rural roads. The Bureau of Public Roads (BPR) administered federal aid to state highway departments
1921	PL-67-212 Federal Highway Act required state highway departments with BPR to designate a system of state ABC roads
1934	PL 73-393 Federal Aid Highway Act (Hayden-Cartwright Act) authorized funds for highway planning
1944	National Interregional Highway Committee recommends a 55,000 km (34,000 mile) system 1944 PL 78-521 Federal Aid Highway Act created a national system of interstate highways, not exceeding 64,000 km (40,000 miles), no funding authorized
1955	Clay Committee (President's Advisory Committee on a National Highway Program) recommended $22.5 billion bond-financed interstate highway program
1956	PL 84-627 Federal Aid to Highway/Interstate Highway Act, Highway Revenue Act creates Highway Trust Funda
1958	PL 85-625 Transportation Act
1965	PL 89-285 Highway Beautification Act
1966	PL 89-670 Department of Transportation established
1973	PL 93-87 Federal Aid Highway Act
1980	PL 96-296 Motor Carrier Act
1982	PL 97-424 Surface Transportation Assistance Act
1987	PL 100-17 Surface Transportation Act
1991	PL 102-240 Intermodal Surface Transportation Act
1998	PL 105-178 Transportation Equity Act for the 21st Century For a discussion of federal legislation close on the heels of the interstate program see Highway Assistance Programs: A Historical Perspective, Congressional Budget Office (1978).
2005	PL 109-59 Safe, Accountable, Flexible, Efficient Transportation Equity Act: A Legacy for Users
2012	PL 112-141 Moving Ahead for Progress in the 21st Century Act

the 1921 legislation co-opted some of the debate. Also, states worked out different solutions for allocating dollars between the counties and the state systems. The give and take of state politics managed the county-state issue.

During this period and into the 1940s, the rural emphasis of state programs continued. In 1928 there was an Act allowing expenditures in the sparsely settled parts of cities, but such areas were essentially the rural parts of cities (the Acts listed in Table 15.1 are not exhaustive). The 1934 Act extended federal aid to cities, a response to the depression and unemployment problems of the times. The 1932 Act had stressed the same reasons in its emphasis on farm-to-market roads.

The concentration on high-quality design can be traced as far back as the 1920s; further, the acts leading to the interstate reflect the interest in high-quality design.

Bonds had been proposed to fund the interstate, but the 1956 Act adopted the fee and fuel tax protocols used by the states. The subsequent evolution of state/federal protocols is complex.

An issue that should have been debated was the matter of rational shifts in the road pattern, given changes in the ratio of fixed to variable costs. In horse and mule days, variable costs were high relative to fixed costs when compared to the case once autos and trucks came along. Consequently, there was an opportunity to reduce road mileage while providing improved quality roads. The rural system would be improved by mileage reductions and concentrating investment on remaining roads.[296] Just as there was a rationalization of the railroads, a rationalization of rural roads would be desirable. This fits the case of counties that are road poor, in the sense that they have lots of mileage but little in the way of economic activity to support the road system. Those counties are engaged in minimum maintenance; they are allowing some paved roads to deteriorate to unpaved conditions. Though retroversion of pavement to gravel may be warranted, some planning that supported systematic reductions in the road network would be in order.

15.7 Financing Roads (c. 1920)

The booming automobile and the demand for roads led governments to seek a revenue source that would be "fair," one which would charge users for the benefits they received, but did not have the disadvantages of tolls. In the 1910s, attempts at instituting a federal gas tax were defeated, on the grounds that roads and streets were public, and should be paid for by the general public, not just users. By the late 1910s, vehicle technology had shaken out so that most cars used a gasoline-powered internal combustion engine. The gas tax was first established in Oregon in 1919 at one cent per gallon. By the 1920s, local legislators overcame resistance from motorists to gas tax legislation by linking the tax to road bills.[297] The idea was quickly emulated, and a decade later, the last state—New York—finally adopted a tax; then it was raised a bit higher, leading to total national collections of about $450 million.[298] Almost two decades after the gas tax began to pay for new roads, it was already time to start recapitalizing the system. Illinois, which had in 1938 more concrete roads than other states, found it would have to reconstruct the roads and had not at the time a way to finance this, as existing revenue sources were already dedicated to principal and interest payments.[299]

In 1932 the federal government followed the states, enacting a gasoline excise tax. Until 1956, with the passage of the Interstate Highway Act, funds raised by the gas tax were commingled with general revenue, though the amount of revenue was considered a benchmark for federal road spending.[300] Between 1956 and 1983, all of the revenue went to the highway account (although in the 1970s provisions allowed the money to be spent on transit in cities that could not complete their interstate highways); after 1983 one cent out of the nine-cent tax went to the mass transit account. From 1990–1996, some of the revenue was further diverted to general revenue to assist in deficit reduction. By 1997, of the 18.4 cent tax, 15.44 cents were spent on highways.

The US experience differs from that in other countries. Hypothecation, earmarking or dedicating fuel taxes for road use, is not nearly so common, since the revenue raised from fuel taxes far exceeds what is spent on roads. The practice is opposed by finance ministers, who want freedom to act, rather than having their hands tied. Great Britain effectively ended hypothecation in 1937[301] (though it maintained a fictional "road account" for a number of years) and Australia in 1959.

In the US, earmarking of transportation revenues for transportation services is natural and expected, and complaints about taxing transportation services for non-transportation uses arise immediately. In other countries, a fuel tax of four dollars per gallon is tolerated. Financing transportation with either user fees or general revenue and using transportation revenues to finance other services are highly charged topics.

US counties often levy a (in lieu of having property owners devote work days to road repair, the corvée system discussed in Section 3.3) to satisfy the demand for modern roads. With increases in the numbers of automobiles and trucks, the need for roads was seen as obvious, and in the early decades of the twentieth century increasing the property tax for roads presented no political problems. In a way, local governments solved the problem of building and maintaining the highway system in the same way the cities did—local taxes paid for local roads. As was the case in the cities, the property tax base was augmented by the development of vehicle fees and fuel taxes beginning in the late 1910s; these imposts were widely adopted by the 1920s. While the pattern of fees and taxes differed from state to state,[302] in almost every state, farm trucks received favored treatment in tax systems. Roads were popular, and the popularity of road improvements was pushed even before the auto, by the need to get farm products to railheads and by the use of bicycles. State and national legislators responded warmly to the need for funds.

Figure 15.2 shows the share of revenue spent on urban roads in 1920. Fees for vehicles ran about $10.70 for autos and $21.90 for trucks. The states had instituted gasoline taxes that ran about four cents per gallon. Imposts on trucks were running about 0.6 cents per ton-km (one cent per ton-mile). The situation varied from state to state, but in almost all states, the cities (sometimes via the counties) were receiving income from these taxes. For example, in California, urban counties were receiving about 50 percent of the taxes and fees paid by urban vehicles. At the national level, highway expenditures were running about $2 billion, and it was estimated that about $325 million were expended in cities. Special assessment districts were widely used, paying about one-half of the capital investment costs.

Issues at the time were the cross-subsidy of rural roads by urban vehicle taxpayers, urban congestion, and the large part of the urban highway bill being paid by general taxes. Although it was pointed out that urban vehicle owners used rural facilities for as much as 60 percent of their travel, it was thought that the urban to rural subsidy situation was unstable—example, an adjustment would be made.

With respect to the contribution of general taxes on property to road expenditures, it was pointed out that urban street facilities had purposes other than carrying traffic. They were used for walking and recreation and they provided light and air to adjoining property. They provided right-of-way for utilities. Expenditures would be necessary even if there were no motorized vehicles.

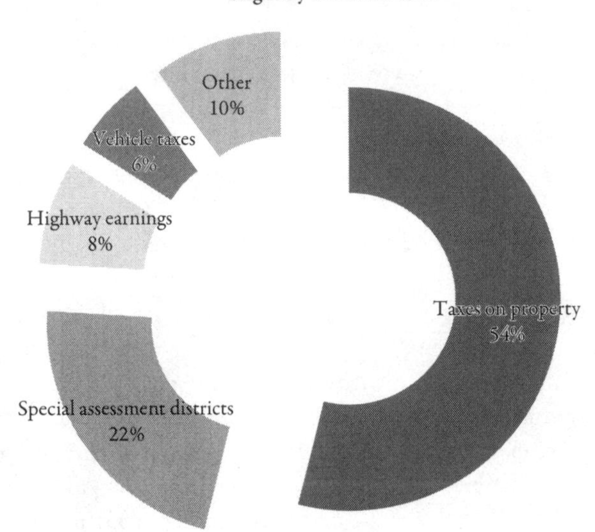

FIGURE 15.2 Urban Expenditures on Roads and Streets (1920)

But even in 1920, the contribution of general taxes to streets was on the decrease. In 1903, 28 percent of city expenditures were for streets, and this had declined to 17 percent by 1923.

Congestion was an issue in the 1920s, as it was before and as it is now. At that time its management was via improvements in the street system, although it was recognized that there were limits on what could be done because of existing properties and the political and fiscal costs of taking property. Zoning was held out as the solution; it would assure property development that would allow adequate street spaces and distribute traffic intensive land uses.

Jacob Viner, a well-known economist at the University of Chicago, discussed the control of congestion by "restricting the wasteful use of street space." He mentions systems of charges on parking and on traffic and restrictions on truck traffic. He predicted "there will not be as much restriction on traffic as the prevailing conditions require," but adds "the public will submit in time to the painful necessities of the situation."[303]

The astute reader will have noted that the issues in 1920 largely remain unresolved. Land use is still considered a possible solution, as are pricing parking and traffic. The source of funds has shifted somewhat over eighty years, and varies somewhat by region, but Figure 15.3, using data from the Twin Cities, is representative. A surprisingly large amount of road revenue comes from non-transportation sources. Students of federal level data will tend to miss spending on local roads. While it is arguable to what extent that road spending is required for the purposes of moving people, as opposed to accessing property, it is still spent on roads.

What changed in seventy six years? Property taxes are half as important. Special assessments are one-fourth as important. Gas taxes (highway earnings) from the state and federal government are about four times as important, and vehicle taxes are also four times as

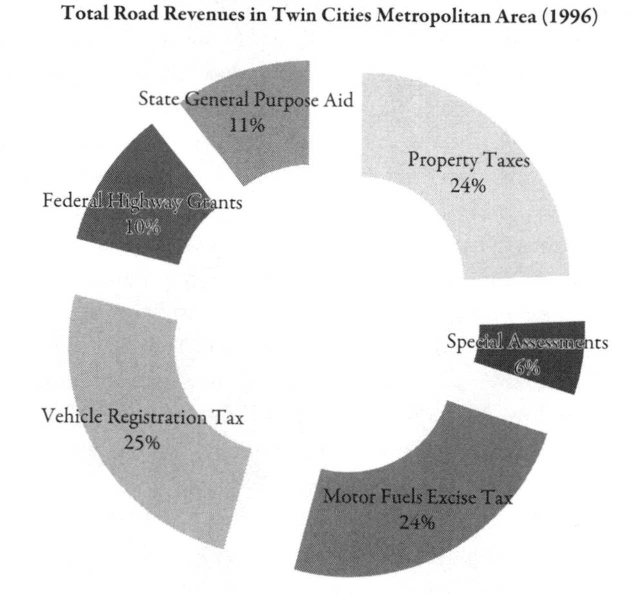

FIGURE 15.3 Expenditures on Roads and Streets (2000)

important. At this rate of change, in another century or so, road financing might actually be efficient.

15.8 Bureau of Public Roads

> Thomas Harris MacDonald earned his place in history less as a visionary than as a relentless refiner of the existing. He was an engineer's engineer, a man gifted at recognizing a problem and developing a methodical plan for fixing it. He did nothing 'on spec,' took no gambles; his decisions were founded on careful research, overlaid arguments, numbers and accompanied by charts, measurements, cost figures, traffic counts. All of which made him the perfect man for the nation's top roads job in 1919, because the American highway system was a chaos of overlapping auto trails, disconnected state highways, dead ends, and doglegs, and MacDonald was order personified.
>
> —Earl Swift, *The Big Roads* (2011)[304]

The BPR was established in the Department of Agriculture to "aid in the improvement and construction of public roads and to promote better agricultural engineering work" (Holt 1923). Originally founded as the Office of Road Inquiry in 1893, the function of the Office from its founding to 1912 was primarily educational, conducting studies and disseminating information through lecture and publications. From 1912 onward, the agency expanded its power to include the construction of post roads (as authorized by the Constitution) and the administrative responsibility for federal aid programs.

In the early part of this century, the Bureau of Public Roads encouraged states to develop statewide plans for the growing network of roads and highways. It is important to note the motive for these plans. Washington politicians made a commitment to assist highway development in the states, and significant funding became available. But there is never enough funding, and the BPR's response was to concentrate resources on through and important collector routes. Indeed, one of its first actions was the "7 percent solution": only 7 percent of the state mileage was to be eligible for federal assistance. As we will see, the strategy to concentrate resources permeated the subsequent BPR freeway programs.

In the early days, the BPR had a role limited to working out techniques and designs, demonstrating roads, and encouraging states to act. The result was a number of BPR-type roads. These roads failed badly under the traffic demands of World War I—indeed, so badly that there was a political backlash against the Bureau. In response, and adopting the "science is good for the country" thinking after World War I, the Bureau adopted a "rational" scientific method approach in the 1920s. This orientation was successfully applied first to soil mechanics.

In the 1920s and 1930s, the Bureau was finding that facilities were becoming obsolete before they wore out. With higher velocity traffic, designs needed upgrading. Also, traffic growth was pressing capacity. In response, the Bureau became involved in getting the travel and traffic facts to support planning and design. It began to go beyond recommending, and required that states develop state plans as a condition for federal aid funding.

The state surveys of transportation participated in by the Bureau were essentially identical in their organization, data analysis, and findings. State surveys from New Hampshire, Ohio, and Vermont illustrate the Bureau's early work.[305] The surveys aimed to provide not only a detailed description of the existing network, but also a plan for future highway development. Each survey contains a background description of the history of roadway development in the state, the organization of the individual state highway department, highway revenues and expenditures, and methods of state aid and supervision. There were minor variations among the plans, for instance a discussion of significant geographical features in the New Hampshire survey and a greater discussion of the urban needs and the state role in the Ohio survey.

Methods used in the surveys were stressed. Data presented include descriptions of existing roadways by roadway type, measures of traffic density, truck and bus traffic, traffic composition, and classification and forecasts of highway traffic. Traffic density was measured as the number of motor vehicles that passed a count station during a 24-hour period (ADT, average daily traffic), and were classified as light (less than 200 ADT), medium (200–499 ADT), and heavy (500 and more ADT). Truck traffic and density, motorbus routes, interstate travel, and trip length data were also presented. Forecasts of future traffic and future motor vehicle registration were also developed.

The final sections of the surveys proposed plans for future highway improvement (for example, five-year plans for Ohio and Vermont, a ten-year plan for New Hampshire). These plans outlined the extent of road needed to meet future demand, provided estimates of the cost for the proposed improvements, and discussed safety concerns, areas of growth, and special study needs.

There was nothing radical or breakthrough in these early surveys. They scaled up project type work and used the conventional engineering standards of the day.[306] They illustrate the emerging relations between the Bureau and the states: the investing of primary responsibility in the states for development and maintenance of the highway network, with the federal government serving to establish both uniform standards and financing. A kind of partnership emerged which is still a fundamental part of US road transportation administration. The roles of the states and the federal government have remained substantially the same.

The plans also indicate a change in the role of the roadway network. Then primarily established as roads to move agricultural products to rail facilities for long-distance travel, the surveys point to the development of a network that serves additional needs, including social, economic, national defense, and long distance transport. The increasing roles of the automobile and truck were recognized—the role could hardly have been overlooked at the time. But it is interesting that long-distance truck transport was not much emphasized. The plans are key documents for the time, indicating the change and how it came about.

In addition to the concentrate resources theme and the fact-based, partnership theme, the BPR was very much a product of progressive government, professional leadership thinking in the early decades of the century. The long-time chief of the Bureau, Thomas MacDonald, and Bureau staff personified progressiveness.[307] The progressive movement affected state and local government, of course, but its impact was concentrated in the Midwest and Western states. As discussed (in Chapter 16), engineering studies and management confronted the "city beautiful" movement.[308]

With the growth of motor vehicle use and the importance of the highway network, the role of the federal government expanded. The BPR, as part of its responsibility for administering federal aid for roads, required the states to submit plans for the existing road network, the organization of the state highway departments, and the method for securing the necessary finances for road construction and maintenance. The state surveys were further developed as part of this process. The federal government viewed the relationship with the states as a cooperative partnership, and while it provided some funding, did no road-building itself. At the request of MacDonald, the US Army prepared the 75,000 mile (120,000 km) *Pershing Map*, an early interstate highway map that reflected the wishes of the Army in providing roads that would connect important locations, and was developed based on the bad experiences of the Army in the Transcontinental Motor Convoys. Much of this map was eventually constructed as US or Interstate Highway.

The Bureau also began to make resources available for planning. The Federal-Aid Highway Act of 1934 authorized expenditures of 1.5 percent of federal monies made available to states for surveys, plans, engineering, and economic analyses of projects for future construction, and by 1940 all of the states had created highway planning surveys.[309]

The Bureau began to tiptoe into the urban areas. Urban area travel studies were not included in the highway planning surveys until 1944, because it was not until the 1944 Act that appreciable federal funds were made available to urban areas.[310] In anticipation of that Act, the Bureau had "... developed a method to give the needed information" and refer to origin and destination of trips, mode of travel, and trip purpose.[311]

The federal government prioritized rural and inter-city travel during the 1930s. As mentioned, there was the problem that early road designs were inadequate to the growth of traffic (number of vehicles, size of trucks, etc.), and a modernization and replacement program was needed. Local agencies were making claims on highway user taxes. Should trucks pay higher taxes? Finally, the state secondary road system, for which funds were beginning to become available, needed to be identified. These problems called for "facts," and the highway planning surveys responded. Study techniques were honed, and began to include financial topics and road-life analyses.

15.9 Discussion: Highway Needs

The word "needs" is a tough one to work with: How do we establish need for a highway improvement? Needs were operationalized by the 7 percent solution in the 1921 Act, and that was tempered in 1932 by legislation that said that states could increase their mileage by 1 percent per year when the original 7 percent was 90 percent complete and well maintained. In other words, there is no limit to needs; they grow 1 percent per year.

The 1934 legislation provided for state highway planning, and the states gradually built up planning capabilities. Although the BPR suggested procedures and the contents of plans, those plans varied somewhat from state to state and yielded no results in the rationalist facts style desired by the Bureau. At the end of World War II, the Automotive Safety Foundation undertook some needs studies for several states, and these yielded facts. The Bureau recommended the study strategy to the states, calling attention in particular to the Automotive Safety Foundation study in California.

At first, the needs studies that followed BPR urging were made on an episodic timing, and were sponsored by legislators. A study commission was established with advisory committees. A sizable staff, budget, and block of time were devoted to the work. Staff was largely engineering but included legal and fiscal specialists. The resulting needs were then considered by the legislature.

Currently, needs studies respond both to local interests and to Congress. Congress requires that the USDOT report to Congress on the "condition, performance, and future capital investment requirements of the Nation's highway and transit systems." This report is carried out every two to three years. The Highway condition information is drawn from the Highway Performance Monitoring System (HPMS), which is a federal database fed by state DOTs.

The cost to improve highways and bridges is developed from the FHWA's Highway Economic Requirement System (HERS) and the National Bridge Investment Analysis (NBIA). HERS determines investment requirements by using the HPMS data on pavement, geometry, traffic volumes, vehicle mix, and other characteristics at the road segment level and making changes to the road segment to evaluate the benefit-cost ratio. It chooses the improvements that produce the greatest benefit for the purpose of estimating recommended improvements. It aggregates the results over all the road segments and statistically adjusts the results to get a national estimate. It thus approximates local decision making (though of course each project still has its own much more detailed analysis before it actually gets funded).

Roads can be improved in a number of dimensions, including the capacity of the road, the strength of the pavement, and the strength of the bridges. For pavements, a serviceability index, which depends on cracking, variance of slope, and amount of patching is applied, and projected into the future. Sufficiency ratings are also used for pavements and bridges.

As stated, today the needs studies are made by the states in a style largely mandated by the federal government. Manuals of procedure, computer information systems, and analysis programs are extensively used—embedded in the Highway Performance Monitoring System (HPMS), which includes data on the extent, condition, performance, use, and operating characteristics of roads. For instance, the pavement condition is measured on an index from 0 (extremely poor) to 5 (extremely good). That measurement produces its PSI (pavement serviceability index). Typically, new roads have a PSI of 4.5, and roads are mended when the PSI drops to about 2.5.[312]

It's generally agreed that the needs studies do what they say they do.[313] Even so, there is much to criticize. For example, a need may be recognized to increase the width of an arterial in a residential area and the cost of that action included in dollar needs. But as a practical matter, that monetary cost should not be included because the political-social cost of the improvement makes it highly unlikely (alternatively put, the political and social costs plus the monetary costs outweigh the benefits, indicating that it has a net negative net present value, and is thus not a need). Too, there are diverse pavement or traffic management strategies that might be less costly than the one incorporated in the analysis system. Moreover, the studies are not very demand sensitive; engineering standards drive needs.

The situation is in considerable disarray. Can we get at needs in a better way? Can we change the point of view of cost studies? Are additional styles of planning needed?

The needs study is straightforward in concept, but it is hard to know what goes on inside the "black box" used when calculations are made. Seemingly simple changes in standards and/or calculation methods can change results. Everybody says we need more demand-oriented methods to estimate needs, and the studies have responded by including calculations based on traffic. But is that the answer? Recently, for example, the studies have been performing triage on low-volume rural roads: the needs studies find no need because of light traffic use.

The incremental cost method was established to assign costs of construction. Should we go to cost methods more responsive to the marginal costs of using the system?

An alternative way to allocate costs is to charge based on benefits received rather than costs occasioned. Thus those who benefit most would pay the most. This is more difficult to determine, especially in an unpriced system (vehicles do not pay for each segment at each time; rather, they currently pay for their share of the system).[314]

The story of roads continues in Chapter 22.

Make no little plans. They have no magic to stir men's blood and probably themselves will not be realized. Make big plans; aim high in hope and work, remembering that a noble, logical diagram once recorded will never die, but long after we are gone will be a living thing, asserting itself with ever-growing insistency. Remember that our sons and grandsons are going to do things that would stagger us. Let your watchword be order and your beacon beauty. Think big.

—DANIEL BURNHAM, CHICAGO ARCHITECT (1864–1912)

16 Urban Planning: Who Controls the Turf?

16.1 Introduction

WHAT ARE PLANS, and how do they differ from policies? To be brief, we think of policies as rules for behaving, and plans as the schematic of the desired outcome. Thus policies focus on the process and plans on the product. Planning itself doesn't create products; it only specifies how products should be created. The plan may say what should go where, and which should come first. A plan is thus a blueprint for building something and a schedule for building it. Obviously, process affects product, and products shape future processes. Nevertheless, they are distinct things that are too often confused (especially by those who do policy).

Plans are applied in all fields of human endeavor, but we are most concerned about plans for transportation systems and networks and plans for cities and other places.[315] Why are transportation plans and land use plans conducted by different professions: engineers and planners, respectively?

The chapter begins with some theories about how cities are organized, and some discussion about how the positive becomes normative, and vice versa. It then explores the questions of who are the engineers, who are the planners, and what is the turf over which they contend.

16.2 The Urban Wheel

Early on, cities serving government, military, church, and/or local market functions were scattered about. The commercial revolution pushed the growth of those suitably sited either for transfer of goods by water or to hinterlands served by roads. Industrialization again pushed growth of those cities, as well as cities favorably sited relative to industrial resources such as coal. However, lack of internal transportation affected urban structure.[316] Pittsburgh, Pennsylvania, for instance, was not in early days an integrated city as we observe today. Rather, it was a collection of interdigitated, but largely independent, mill towns and commercial centers.

About 1910, the automobile populations of cities began to grow very rapidly.[317] As a result, several urban transportation planning activities were put in place.

The stage through the 1920s for these activities was very rapid urban growth (see Figure 16.1), the centralization of income and wealth in cities, and the emergence of new concepts of planning and management, especially the progressive government movement. There was the grand notion that scientific principles could be used in planning and management to achieve efficiency.

By 1920, there were 6.7 million motorized vehicles in the Unite States (about 60 percent urban), with the number to grow to about 23 million by 1930. That was about four urbanites per vehicle.

While there has been a great centralization of people from the countryside to metropolitan areas, over most of the past century, the concern has been about suburbanization, movement from the core of the metropolis (the central city) to the suburbs. Suburbanization began early, notably with the streetcar, followed by the automobile. Central business district (CBD) interests about accessibility to the core still dominate urban transportation planning. Wheel-like radial routes and circumferential routes have been seen as the facility need. The concentric circle theory of urban form consistent with these notions affected urban freeway planning.

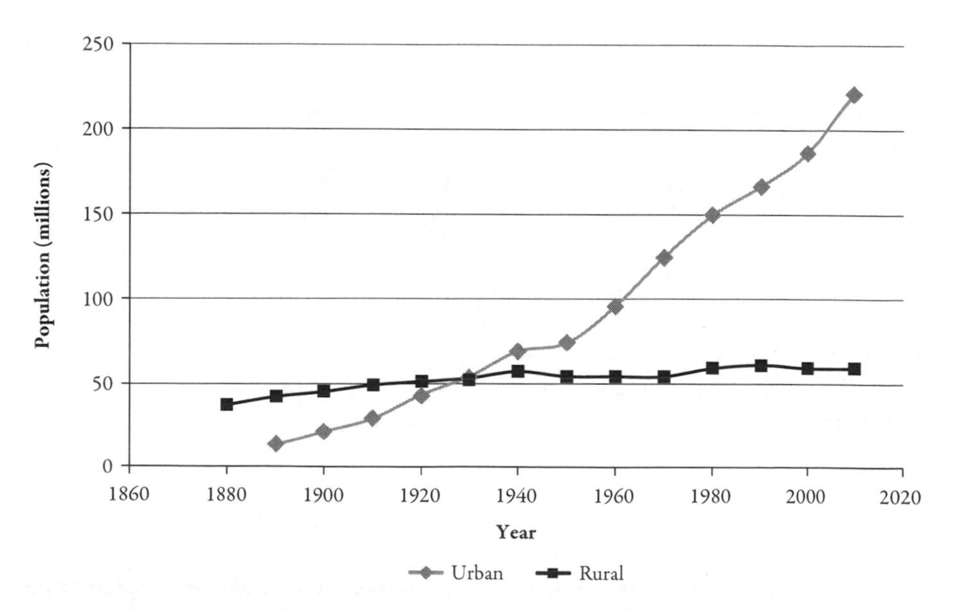

FIGURE 16.1 Population of the United States (in millions)

Wheel-like language described the streetcar city of 1920 very well: at the hub lies the CBD; arteries of transportation radiate to the rim where housing is found. But "what is" became "what ought to be"; that is, descriptive language became normative language. Plans began to prescribe radial routes and circumferential routes.

Architect and planner Daniel Burnham made the "normative wheel" the theme in his work, and seems to have been most responsible for the normative leap. In the general theory section of his 1905 study of San Francisco, he noted that the finest examples of cities of the Old World (Paris, Berlin, Vienna, Moscow, and London) consisted of concentric rings separated by boulevards, with the smallest ring (inner circle) around the civic center. From the inner boulevard ran diagonal arterials. Having presented his findings about the finest European cities, he said, "It is on this study that the proposed system of circulation for a larger and greater San Francisco is based."

Ten years later, Werner Hegemann, a leading German planner who at one time worked on Oakland and Berkeley, undertook a critique of Burnham's theory.[318] Burnham had given credit to architect Eugène Hénard's study of Paris.[319] Indeed, Burnham used Hénard's figures in his report. But Hénard was dealing with the deficiencies in the Paris circulation system; his was hardly a scheme to be emulated. He went on to say that concentric boulevards were the result of the elimination of city walls as medieval cities expanded; radial streets started out as paths from outside the city through city gates.

Hegemann didn't reject concentric boulevards completely. He saw an inner concentric boulevard as distributing traffic downtown. He also thought radials were desirable. "The cities of the future will resemble a wheel . . . With spokes radiating in all directions and the 'spokes' connected by a network of cross streets."[320] The radials would intersect the grid of the urban core. Hénard, who advocated traffic circles, pointed out that blocks formed between radial streets gradually shrink as radial streets approach the center of the city. This creates problems.

There is more to Hegemann and Hénard's works, and no one would say that either writer claimed a complete theory of streets and circulation. Burnham did, and he influenced US practice.

Burnham represented "city beautiful" thinking, and one might think that the rise of the "city practical" movement would have questioned Burnham's construct. It didn't. BPR chief and engineer Thomas MacDonald looked to Paris for inspiration, and referred in 1937 to competent French engineers having planned a circular highway enclosing the city and radial arterials; the underlying soundness seemed self-evident. Following MacDonald, the 1957 AASHO policy on arterial highways in urban areas referred to a wheel-like basic pattern.

By 1950 there were three widely known "theories" or generalizations about urban form. Sociologist Ernest Burgess's Concentric Zone Theory argued that cities grow outward in concentric circles through a process of invasion and replacement.[321] Jobs and other central city functions would be located in the CBD, and these rings were the product of a competitive economic process. The rings were: the commercial center, the zone of transition, working class residences, middle class residences, and the commuter zone.

In contrast, economist Homer Hoyt proposed Sector Theory in 1939.[322] Each sector, or wedge, of the city would have different economic activities. Hoyt made specific predictions of which type of activity would show up where in the city.

Geographers Chauncy Harris and Edward Ullman suggested the polycentric or Multiple Nuclei Theory.[323] They noted that cities do not always have a single center, but many mini-centers. Because of economies of agglomeration, similar activities group together in these mini-centers.

The wheel metaphor seems most compatible with the concentric circle theory. Examining freeway plans, civil engineer and planner Edgar Horwood compares three cities and their radial, circumferential, and enclosed cellular areas.[324] Horwood makes no normative claim for wheel-like designs.

Even in the twenty first century, US cities are building rail transit networks focused on central business districts that have not added jobs in decades, to serve central cities that are well off 1950 population highs. The power of the radial network is strong, as is the image of downtown.[325]

Today, there are two approaches to urban morphogenesis. One is an urban modeling approach in which rents and market activities steer toward a unique equilibrium pattern of activities. The other acknowledges increasing returns and the kinds of processes studied by George Polya in the 1930s.[326]

Just as positive (descriptive) models of how cities once worked become normative models of how cities should work, the increasing returns model imposes lock-in, and suggests that investments shaped by normative models of how cities should work will affect how cities actually work. For more on normative models and why they exist, see Section 24.4.5. As should be apparent, we align with the increasing returns models, and network morphology is discussed further in Section 25.4.

16.3 Civil Engineering

Who are the engineers? Engineers are those who use ingenuity to create new technologies. The first searching questions about the social role of engineers were in the non-civil engineering fields. Most of those engineers worked for private organizations. Were the engineers responsible just to the organizations that employed them, or, as professionals, did they have responsibilities to society as a whole? This debate re-emerged in the late nineteenth century when the social role of big business was at question, revisiting interests held by James Watt and other engineer-scientists of the times in social matters, interest dampened by the excesses of the French Revolution. This was the "trust busting" era.[327]

The civil engineers were not much involved in that debate. Civil engineering was, even then, an old field and had not much in common with the new fields, such as electrical engineering. There were about 2,000 civil engineers in the United States during the period of canal building. The ASCE was established in 1852. The mining, electrical, and mechanical organizations were not established until the 1880s, and most other engineering professional organizations were not established until the twentieth century.

Most civil engineers worked for government, construction, and consulting organizations. Civils were not "big business" payroll employees. Finally, the civils had a sense of social role, for they were heavily involved in the emerging conservation movement at the onset of the twentieth century.

The civil engineers did no formal joining with those debating. For instance, when Carnegie gave money for a headquarters building for the engineering societies in 1903, the civil engineers elected not to join others in a unified headquarters. This was an important matter, for the societies were seeking strength through unification.

The matter of just what social responsibility meant got messy when Fredrick W. Taylor (a mechanical engineer) began to study engineering efficiency in manufacturing, example, using time and motion studies. Frank Gilbreth (of *Cheaper by the Dozen* fame) and others rapidly implemented Taylor's notions. Why did things get messy? Taylor's notions gave engineers the tools and concepts so that they could be the best managers. The problem was that of reconciling the best manager role (scientific management) with democracy and obligations beyond the place of employment.

The debate continued strong into the 1930s. As late as 1933, Thorstein Veblen saw the engineers as the real social revolutionaries. But the debate began to peter out. Perhaps President Hoover's move from hero to villain in the public eye was a factor. Hoover, an engineer himself, had been a champion of the engineers' views. Perhaps the Great Depression found engineers void of usable ideas.

16.4 City Planning

Who are the planners? Planning emerged from the design professions (engineering, architecture, and landscape architecture). However, as the profession of planning has matured, it has been slowly captured by the process professions (law, policy analysis, management, economics).

Although city planning has ancient roots, it came into its own in the first decades of the twentieth century. Sometimes sponsored by city governments and sometimes sponsored by groups of concerned citizens, at the turn of the twentieth century, planning was highly aesthetic in orientation. The reasons seem numerous—the classical education of many social and economic leaders with interest in the architecture of Rome, Greece, and Egypt; the impact of the 1893 Chicago World's Fair with its emphasis on civic ideals; and the view that beautiful environments would uplift man's souls. So cities planned and built great regional parks, parkways, ornate civic centers, and museums. Frederick Law Olmsted, Daniel Burnham, and Arnold Brunner were leaders in that "city beautiful" movement.[328]

The city beautiful movement was badly damaged, although not dead, by the end of the first decade of the century. What seemed to have stopped it was the realization that in spite of monuments, haphazard growth was leading to many thorny infrastructure problems and that lofty ideals and monuments were not as relevant as were plans and designs to efficiently control growth and deploy needed facilities. Actually, the conflict between the practical and the aesthetic in design was long-standing. Pierre L'Enfant (1754–1825), for example, found his plan for radial streets in Washington, D.C. (reflecting a quasi-organic baroque city notion) in conflict with Thomas Jefferson's gridiron pattern of roads (an ordered city). L'Enfant was later fired over an environmental conflict (trying to run New Jersey Avenue through the newly built house of Daniel Carroll, a signer of the Declaration of Independence). Matters such as these are discussed in several histories of city planning.[329]

The "city beautiful" versus the "city practical" (or "city efficient") debate of the early 1900s saw advocates of ideas debating and pushing their worldviews. Frederick Law Olmsted was one of the more successful of the former. As a landscape consultant, he produced designs for dozens of cities, often working with others. In addition to producing plans that advocated his concepts, he debated in the literature. In contrast, the advocates of rational planning mainly did not use the medium of plans, with Nelson Lewis a notable exception. Rather, they wrote.[330]

The present incarnation of US city planning emerged in the State of New York. In 1926, the State Legislature passed a bill enabling cities to engage in a planning process with the elements of that process spelled out in the legislation. Cities were to adopt official maps, create planning boards, improve and maintain records of plats, and undertake zoning.

The city planning movement affected transportation; the New York case is illustrative. *The Regional Plan of New York and its Environs* (produced by the Regional Plan Association [RPA], an elitist, nongovernmental group that had major roles in urban street and transit affairs) pushed the legislation. The argument for passage of the legislation was mainly transportation in content. The RPA argued that something had to be done to coordinate street development, manage the differences between public and private streets, and assure adequate street widths and street layouts. Issues of zoning, plats, and so on, had somewhat a secondary role, and they came in because of their transportation relevancy. Transportation questions were also highlighted as other states passed enabling acts for city planning, and most of the legislation referred to transportation and land use planning. So transportation was an important matter in the genesis of city planning organizations.

The rapid adoption of city planning enabling acts by the states was assisted by work at the US Department of Commerce. In the late 1920s, the US Department of Commerce published a model enabling act for planning, created a Bureau of Planning, and issued publications to assist the states and cities in undertaking planning.

There is a second way that transportation steered the development of city planning. Large cities began as a collection of neighborhoods, and in the early days these were mostly self-contained. There might be a mill or some other type of employment opportunity, with workers and managers housed nearby. Shopping and recreational opportunities were also close. Horse trams and then streetcars changed the scale of those neighborhoods, and they also enabled specialized downtown functions—shopping, banking, and so on. Next, the automobile brought a major change. It offered an order of magnitude more mobility than precursor technology and enabled the development of residential areas some distance from workplaces, the suburbanization of workplaces, and many other changes. The effect was to enable the development of areas with relatively homogeneous land uses.

Homogeneous land use was particularly valued for residences because individuals tend to select residential sites according to peer group criteria. Additionally, the externalities of noise, smoke, and so on associated with some land uses are particularly obnoxious in residential areas. Zoning took on the job of controlling externalities, and zoning and land use control began to be the central preoccupation of urban planning agencies.

In addition to supporting the emergence of relatively homogeneous land uses, there was direct consideration of transportation in planning agency efforts. First, a land use may have undesirable externalities because of traffic generation and the loads imposed on parking facilities. Planners and planning commissions give much attention to these matters. Second, there are transportation planning (or design) facets to subdivision developments. Some designs provide better sequestering of neighborhoods than others and fit the overall travel pattern in the city better than others. City planners have long attended to these matters, initially through the hierarchy of roads, now under the guise of traffic calming. Third, many city planning departments did engage in city-wide transportation planning and extended their interests to facilities such as parkways, expressways, and freeways.

The early days of city planning saw priority given to creation of the official map and zoning regulations. This was a time-consuming problem because it required the coordination of existing records (as well as a fair amount of legwork). Concerns about lack of coordination of infrastructure that lay behind the creation of planning activities tended to diminish as the Great Depression came along and growth ceased in most cities. Also, in undertaking transportation work, there was the potential of turf battles with existing public works agencies, and planning agencies were likely to be leery of these. Historically, when turf battles emerged, the engineering agency was usually the winner—the planning professional was overwhelmed by the technical arguments of the city engineers.[331] However, this has diminished in recent years as environmental agencies have established technical arguments that the transportation engineers can't counter. Ultimately, it becomes a question of values as much as of facts, and preferences of citizens and decision makers have changed over time as transportation has matured.

Even though transportation didn't enter much in early planning agency work, it was on the agenda. It was on the minds of early public works engineers who worked in city planning, such as Harland Bartholomew. With the resurgence of city growth after World War II, some city planning agencies began to make large plans for major transportation facilities. This aspect of their transportation initiatives, however, was co-opted by the federal government and metropolitan area planning—a matter that we will discuss.

16.5 City Planning versus Transportation Planning

As with most public works, the technical planning turf in transportation largely belongs to engineers. Planning applied elsewhere (for instance, health services, educational services, and land use) belongs to others. There's nothing surprising about that. Planning is organized in knowledge areas, and each domain holds special advantages embodied in notions of processes, available techniques, and professional traditions. Knowledge areas also have resources of institutional and financial arrangements. So, as expected, medical or paramedical professionals do health facilities planning, and macroeconomists play prominent roles in national economic planning.

We have observed that engineers played key roles in the development of urban planning and that transportation considerations loomed large in the creation of urban planning. For example, the first national conference on city planning had the title "National Conference on City Planning and the Problems of Congestion." Things have changed since then: engineers are only tangentially involved in city planning. Transportation planning, the engineers' domain, is a thing apart from city planning. So we have two questions: Why did civil engineers give up involvement in city planning, and why did urban transportation planning become an engineering-planning activity apart from city planning?

Perhaps engineers gave up interest in urban planning after they found that rational analysis failed when applied to social problems. Alternatively, perhaps the pendulum had swung too far from aesthetics, as embodied in the city beautiful movement, and engineers' talents missed the mark of the middle ground. Even so, engineers continued in the transportation planning field because of its technical content and the need for rational analysis.

Those explanations must have some truth in them, yet are insufficient. They beg the question of how the fields of urban planning and urban transportation planning got claimed and divided. Moreover, this division is not permanent; fields encroach upon one another.[332]

Our second question concerns the extent to which the power of concepts (memes) explains turf wars and their outcomes. Is there a Darwinian survival of the fittest at work? Do the most appropriate concepts and techniques devour less appropriate ones?

Our questions are large ones. We will deal with them by restricting their scope, discussing rational analysis applied to urban planning. We will see that the engineers gained and then gave up the turf. We think that was not so much the result of a Darwinian competition between concepts and techniques as it was the engineers' views of their social role.

Rational analysis refers to the application of science and its techniques to practical problems. This is analysis in the tradition of Galilei Galileo (1564–1642). Galileo gave attention to topics in structural engineering, but the development of the rational analysis tradition in civil engineering is usually dated from John Smeaton's work (1724–1794). Smeaton is regarded as the father of British civil engineering and (thus) the grandfather of US civil engineering. Careful problem definition, experimentation and testing, and calculations using physical laws marked Smeaton's work.

Above we referred to the conflict between Jefferson's rational analysis and L'Enfant's aesthetic design work, and also to the grand designs, city beautiful mode of architecture-based urban planning. The city beautiful movement in urban planning gave way to rational planning in the early part of the 1900s. One has to conclude that the debate was over the power of concepts, but perhaps the real issue was that of the image of what ought to be done. The early 1900s saw the birth of the progressive movement, and rational planning gained the turf because it was consistent with the progressive movement, the idea that "things ought to done in a progressive way!"

The debate is well documented, and some references to the debate were given earlier in the chapter. Ford states the rational case, and Olmsted states the design case.[333] The high ground, so to speak, was taken by Nelson P. Lewis,[334] who, more than any other person, argued that planning should focus on the physical facilities of the city, and the planning and deployment of those facilities should use sound engineering. His argument coupled nicely with the then growing interest in the application of science to business; human factors and industrial engineering began to permeate industry. In addition, it coupled with the progressive movement in city politics. Concerned publics were beginning to argue against the politicizing of urban governments and for the use of experts to manage and operate technical programs (and for civil service–like arrangements for managers of other programs). Because of these couplings, Lewis's influence on city planning was enormous.[335,336]

In spite of these developments, engineers did not retain urban planning. They withdrew to public works. The Lewis tradition passed to urban planners.[337] Was the engineers' withdrawal forced by the lack of efficacy of engineering approaches, or did engineers give up the turf because it did not fit their sense of social role?

An argument that's an alternative to the "lost interest in a social role" argument begins with the hypothesis that civil engineering concepts and techniques were inappropriate to a social role in the urban planning context. To what extent do concepts or paradigms from different fields clash, with the stronger driving out the weak, as in the world of Darwin?

Rational analysis is beloved by civil engineers, and one may say that it obviously doesn't fit artificial or socio-technical systems. That may seem true. But when one looks at what is called rational analysis, much of it is empirical. Empirical statements can be made about anything: urban land use just as well as the strength of concrete.

One might say that engineers are "nerds" or "geeks" and don't fit the kinds of social situations found when one is in social roles.[338] That's true overall, but professionals self-select their field of specialization: they want fields that fit their self-perceptions. So if civil engineering had a social role, then those who are politically effective would populate it.

We have a "perhaps civil engineers lost interest in a social role" conjecture. The alternatives that rational analysis doesn't fit social roles and that civils have the wrong personality seem weak. So we have the not very strong conclusion that the civils lost interest in urban planning because they backed away from social roles.

In applied fields, such as engineering and planning, outside criteria of success or applicability apply to a greater degree than they do in the sciences. For instance, irrigation engineering was established as a division in civil engineering in 1922. Outside criteria at the time said that irrigation was a great thing. Times have changed. Irrigation engineering still exists, but it has retracted and has been encroached on by others.

We will explore the temporal pattern of tasks as a system is created and deployed. Product engineering gets attention early on, for instance, designing bridges or ships. Process technology gets attention later, for instance, how to plan and construct systems. As a system moves to maturity, standardized products are tailored to their markets—market channeling.

Today, most transportation systems are well deployed and mature. As seen by outsiders (and too many insiders) the task is market channeling, for instance: get the capacities of the traffic signal system just right, balance the fine detail of airport capacity with demand, improve truck service schedules, and rebuild container yards. But in a mature system, these are not as important as the tasks of managing a static system, and economists, entrepreneurs and managers, and financiers (rather than engineers) are experts on the management of such systems. Given a static production set, certainly the applicability of economics is strengthened versus engineering.

Those who say that present problems are not so much engineering as they are social or economic are, in our judgment, not so much commenting on the changing outside environment as they are on the state of the system. The system is in a phase where the comparative advantage goes from engineers, who are good at designing physical infrastructure systems, to those who know about management, and how to handle social systems.

16.6 Other Varieties of Transportation Planning

It should be remarked now that what we have discussed to this point is not the whole story. First, we should remember that a role continued during that period for civic planning groups—the Metropolitan Housing and Planning Council in Chicago, the Regional Plan Association in the New York City area, and SPUR and the Commonwealth Club in San Francisco, for example.

Some varieties of transportation planning were not mentioned. There were antecedents for planning from the nineteenth century.

Small project planning know-how emerged as early ports, toll roads, canals, and railroads were developed. Engineering consultants did the planning; often they took part in construction. As the needs and opportunities for larger projects emerged (e.g., Suez Canal, Sault Ste. Marie, Union Pacific Railroad, ports to accommodate larger liners), projects were larger in scope. More complicated engineering designs were required, as were more complicated fiscal and management plans. Project analysis techniques were improved. Systematic work was begun on the interrelations of transportation and other activities, example, by researchers interested in location theory.

There were attempts at systematic construction of inter-city networks in the United States—Gallatin's and Clay's plans, for instance. In Ireland, the first flow map was developed.[339] Cities had surveys well before they had "plans" as we think of them now. New York's grid street system is one of the early systematic attempts that laid out a grid not only for present construction, but for future development,[340] though grids can be traced back to Roman, Greek, Chinese, and Indus Valley civilizations, among others, and in the United States, to early settlements like Philadelphia and Savannah, and the Northwest Ordinances.

The urban railroad problem was an important variety of planning. On the passenger side, there was an effort to coordinate service and promote "union" stations (see Section 8.4). With the growth of the city and the development of truck service, downtown teaming yards became obsolete. Also in the freight area, there were problems of coordinating movements among railroads and congestion caused by the operation of freight trains in crowded urban areas. Effort was given to the planning and deployment of railroads that would circumvent urban areas (belt lines) and/or provide better coordination of existing lines and facilities.

Urban airports started out as small facilities located on the fringe of the city. Cities grew and larger facilities were needed, and planning and development activities commenced during the 1930s after service became available from DC-3 type aircraft. The Hopkins Airport in Cleveland was a development that many cities attempted to emulate.[341]

Maritime port development was also on the agenda of a few cities. By the 1920s, inland waterway traffic was dwindling away and there was not much interest in inland ports, except for those on the Great Lakes. (Traffic later expanded in the 1940s and 1950s as larger dams and locks were constructed.) Although there was congestion around maritime ports, especially those with closely placed single or finger piers, technology development of maritime transportation stagnated and port development problems were not on many agendas. Where there was demand for development, much was accomplished.[342]

Finally, there is urban transit planning. During the first two-thirds of the twentieth century, not much new urban transit was deployed, although much was deprecated. As shown in Chapter 14, fleet conversion issues, financing issues, and ownership issues, which were as much policy as planning, dominated the conversation.

16.7 Discussion

> ... the operating expenses from bad railroad location come by a gentle but unceasing ooze from every pore which attracts no attention ... Errors ... are not likely to be discovered. ... the consciousness that there is danger of error becomes dulled.
>
> —A. M. Wellington, 1906 ed, p. 3

It is not too early to begin to forge templates for critiques. We use the word "templates" in the plural because planning is diverse. The word "template" suggests something with a shape or a figure in which things are to be fitted, and we have in mind criteria about "how planning puts things together and fits the situation."

Many endeavors link a perceived opportunity with a development. The clients (who used the plan and gained or lost from its implementation) and the sponsors (who wanted it) were the same. If we ask, "Did it lead to action?" the combination of client and sponsor into a single individual or organization greatly increases the probability that a plan will be implemented.

It should be noted that the words "client" and "sponsor" make useful metaphors. Who was the client for a transportation and traffic plan of the 1910s and 1920s? The immediate clients were the sponsor and the city council that would implement programs. But the urban population should be counted as a client also. Does a plan meet the needs of clients and sponsors? That would seem to be the question that precedes the "Did it lead to action?" question.

Implementation is one criterion. A plan is implemented; we then ask, "Is the result successful?" That question has a "relative to what?" content. Whether plans are financially successful depends on their context. Many projects were undertaken as systems were deployed, rapid growth was under way. Because of market growth, success would be expected. How could the planner go wrong?

However, the fact is that a lot of plans were accepted and projects implemented that were not financially successful, rail and port facilities in particular, and also some canal and toll road developments. In a networked system where there are cross-subsidies among links, lack of success may not be obvious.

What is "a lot that were not successful?" We have never seen numbers, and they would be difficult to obtain. But the literature is full of references to failures. In some canal and port instances, the engineering aspects of plans resulted in unexpected expenses and or shortfalls in performance. The lack of enough (inexpensive) water for canal operations is an example. These instances do not seem to have been numerous. We would not expect them to be, for there was a build-up of technical expertise as project followed project.

Most failures appear to have resulted from the project scope of planning and, thus, the lack of system considerations. Development of a network link (say, a rail route from here to there) or a trans-shipment node (say, a port) might seem favorable in a project-scoped context, but may prove not to be successful because of changes elsewhere in the system. That is clearly the case for ports, where projects were implemented although developments elsewhere did not bode well for the implemented projects. Sometimes system-scoped considerations boded success. Consideration of how the Suez Canal would interact with sailing routes around the Horn said it would certainly be successful.

There are dynamic system considerations. As the market becomes saturated and as most opportunities have been taken up, success comes harder and harder. The presence of agglomeration economies and route and node economies of scale works two ways in this situation. First, existing, competing facilities may have scale economies supporting their viability. Preexisting agglomeration economies and/or route economies of scale may so enhance competing facilities as to make the outlook for new facilities bleak. Second, projects late in the game often incorporate superior and expensive attributes in order to lower unit costs. To get those lower unit costs, high volumes of use are needed. That's the case today with high-speed rail schemes. This creates a risky situation.

Economies of agglomeration refer to the accumulation of advantages. The early development and activities of the ports in the Bay Area, for example, enhanced the growth of banking, brokerage, ship repair, freight forwarder, insurance, maritime management, and other activities. The advantages held by the port facilities were, in turn, enhanced by these activities.

The discussion has emphasized scale and agglomeration economies, but the point is that the life-cycle dynamic affects the likelihood of planning and project success.

The scorecard on system-scoping of planning shows mixed results. There are several ways that we can use the word "system." First, one might say that we have a system plan if we deal with the hard (facility) and soft (management, financing, control, etc.) aspects of what is to be done. Many plans achieve that system view, although what are called "plans" often deal only with specific subsystems.

We can also ask if a plan dealt with the mode as technological system: that is, did it treat guideway-fixed facilities, equipment, and operations aspects of the mode? The railroads did, but other plans mainly focused on guideways: canals, dock facilities, roads, and so on. Equipment and operations were (and are) taken as givens.

Plans were mainly link or node scoped. As rail service matured, questions of viable sub-networks owned and operated by individual firms weren't handled very well. The introduction of Interstate Commerce Commission–style control was an attempt to control behavior using common law precepts. The question, "How do we manipulate the network so as to manage problems?" wasn't raised until 1920.

A project here affects things there. More generally, the development of the railroads affected the fates of canals and toll roads. The latter were orphaned. Planning was and remains almost universally mode-specific; while it promotes development here, it fails to ensure graceful decline there.

In general, plans were made for projects and not for the evolution of systems: there was a project rather than a pathway stance. Some project decisions thus constrained development pathways.

Mr. Watson, come here. I want you

—ALEXANDER GRAHAM BELL (1876)

17 Telephone

17.1 Harmonious Bells

THE STORY OF the telephone's discovery is famous: a patent was granted to Bell on March 7, 1876. Elisha Gray, another inventor, had filed a caveat (an intent to file a patent) on a telephone, on March 7, 1876, as well, a few hours later. The first application of voice over wire, "Mr. Watson, come here. I want you," was spoken just three days after Bell had filed the patent. It is worth noting that this first spoken telephone conversation was a request for transportation, Bell asking Watson to travel to his location.

Bell and Watson had been working on a harmonic telegraph, and voice was not the first priority. The Bell Telephone Company was chartered in 1877. While Gray did not get the patent or fame, don't cry for him, he went on to work for Western Union, co-founded their subsidiary Western Electric, and invented an early fax machine. Patent suits raged between the Bell interests and the Western Union/Edison interests. By 1879, the parties settled, with Western Union selling to the Bell interests telephone exchanges in fifty five cities and a network serving 56,000 subscribers. In 1881, Western Electric (which acquired Western Union's telephone manufacturing arm) was sold to American Bell Telephone Company.

By 1880 there were 30,000 telephone subscribers in the United States; by 1881, 50,000; by 1888, 160,000.[343] In 1881, the telephone was only five years old, and there were tens, and soon hundreds of thousands of customers (see figure 17.1). In 1885, the American Telephone and Telegraph Company (AT&T) was founded to build and operate long-distance services connecting local telephone companies. Theodore Vail, a first cousin once removed of telegraph financier Alfred Vail, was the first president of this company, serving for four years. In 1899, AT&T assumed control of and became the parent company of the Bell System.

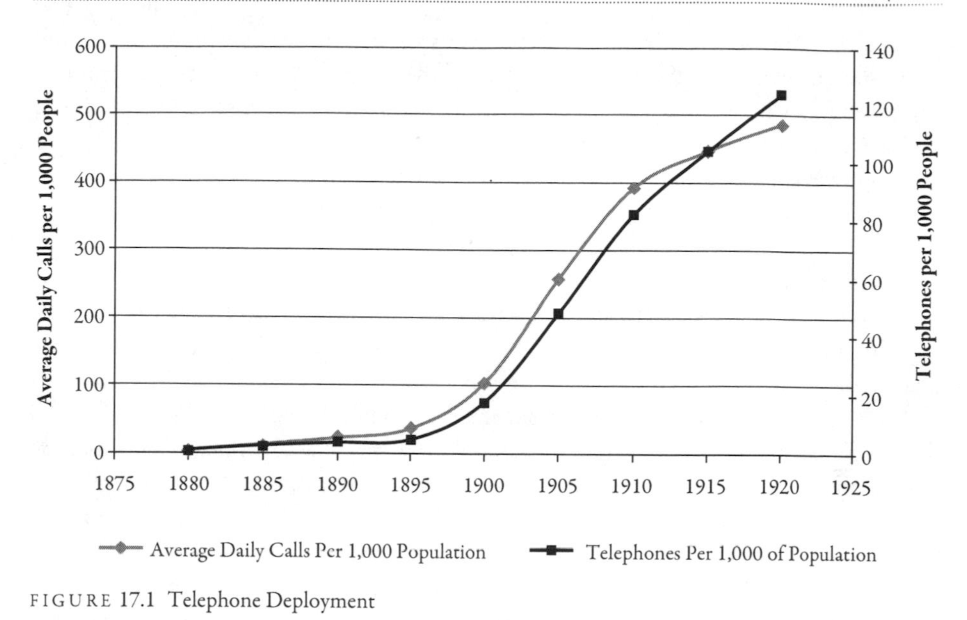

FIGURE 17.1 Telephone Deployment

The discussion in this chapter focuses on the telephone and some follow-on systems. The structural and behavioral parallels to transportation systems will be noted.

17.2 Regulation Ringing

In 1907, AT&T was in trouble and recalled an elderly Theodore Vail from South America (where he was running a transit company) to again head the company. Reviving their fortunes, in 1909 Western Union (and American District Telegraph [ADT] as well as some other telegraph companies) came under the control of AT&T, giving truth to the Telegraph in the AT&T name. However, by 1913, the Department of Justice required AT&T to divest its telegraph interests. With the Kingsbury Commitment, AT&T agreed to the divestiture, and agreed to provide long-distance service to non-affiliated local telephone companies. It could only purchase other companies if the Interstate Commerce Commission approved (which it did for 271 of 274 cases). In return, its telephone monopoly was ensured. Corporate stasis for some seventy years set in, as AT&T was able to maintain its monopoly through government protection.

The predominant technology of the telephone system hardened in the late nineteenth century and the government began to intervene. It is interesting that policy incorporated transportation learning; the roles of state public utility commissions and the Federal Communications Commission (FCC) mirror transportation roles. It is also interesting that the market for the telephone was difficult to imagine in the early days, and that's been true for transportation systems.

The product of the system was highly valued, and there was demand for deployment, locally and then at larger scales. While in the telephone industry, government assisted and enabled, almost no government financing was required in the early days (which contrasts with many transportation technologies and the telegraph). A few local governments developed facilities. With local use dominant, and long-distance expensive, the state Public Utility Commissions (PUCs) looked after policy problems.

As time went along, the long line (long-distance) system developed, and conflicts between firms emerged as a policy issue. AT&T used its long line monopoly power in attempts to acquire those local companies it had not already acquired. The Kingsbury Commitment logic has become known as the "essential facilities doctrine" in antitrust law. If one company's facilities are essential to the operations of another, use of the facilities must be allowed. AT&T was forced to allow non-Bell locals to connect to the long lines under the same terms as the Bell locals.

The story so far is quite transportation in character. A new service becomes available, and there is clamor for service. Perhaps we should remark that the essential facilities doctrine was a problem for the railroads, too. It was managed in a different fashion (creation of interchange rules by the American Association of Railroads [AAR] and Interstate Commerce Commission [ICC] referee of joint rates) but with little difference in substance.

At about this time, the corporate ownership character of the AT&T system had evolved to the form that held until the 1982 Consent Decree. AT&T owned the long lines, the Western Electric Company (equipment supplier), and Bell Labs; it controlled most of the local companies.

Is that ownership situation exceptional? We see it as a how-the-die-fell outcome. It's a matter of the point in time when concentration in industry was stopped. Who knows what the ownership situation in rail would have been if the ICC had not put a stop to change. The highway component of the auto-highway system did not see a stop put on it, and a layered system emerged somewhat like AT&T's. The national system and local systems developed.

An agency modeled on the ICC was created in 1934 to play referee over prices and services, the Federal Communications Commission (FCC). It operated in the "iron triangle" ICC style (see Section 13.7). (The FCC succeeded the Federal Radio Commission, but was given authority over both wireless and wired communications.)

Concentrated ownership got AT&T in trouble with the Department of Justice (DOJ) in the late 1940s, and the DOJ filed a suit to break up the company. The suit dragged on until Eisenhower became president, when the suit was dropped. To save face, the consent decree of 1956 forced AT&T to agree not to go into the computer business. At the time, that was taken to be a near-zero cost decree because of the limited and near saturated market for early mainframe computers.

Was the situation in transportation different? AT&T was the most ownership-concentrated of the facilities we are discussing, and there was the attempt to knock it off first. The railroads and truckers were concerned about the antitrust powers of the Department of Justice (DOJ), and they headed Justice off at the pass through special legislation limiting DOJ control over them. DOJ had from time to time attempted to limit the roles of railroad (and truck) rate bureaus. This was stopped in 1948 when Congress amended Title 49 of the US Code by incorporating the Reed-Bulwinkle Act, which permitted collective rate making under certain circumstances.

AT&T next got in trouble over its tight rules on equipment. Implemented to an extreme, only equipment manufactured by Western Electric was allowed to be connected to the system. The non-Bell locals (the largest of which was General Telephone and Electronics [GTE]) had their own company, a matter tolerated by AT&T. Those who wanted to be system suppliers complained to the FCC and to the courts. They got partial relief in the courts in the Carterfone decision of 1968, which allowed use of terminal equipment other than that supplied by AT&T.

The market was not opened up quickly. AT&T insisted that it supply interconnect devices to protect its system from foreign equipment, and charges were such that foreign equipment was not so attractive. The new suppliers, now part of the system, asked the FCC to play referee. That put the FCC in a messy situation, but it gradually tilted to support the foreign equipment interests. AT&T simply could not make the case that interconnect equipment was needed.

The story here is that of rules for system reliability, and such rules are common throughout transportation. The special Western Electric relationship was in part a matter of concentrated ownership, already discussed. AT&T truly perceived that it was in danger, and similarly, those in any transportation system would think they were in danger if there were the potential for rule violations.

17.3 The Bell Cracks

The next problem was competition in long line service. Some organizations have had long lines for many years. Railroads, in particular, had in-house long line service, and as microwave technology advanced, more and more organizations were internalizing service.

Microwave Communications Inc. (MCI) entered the private line business in 1968, and it developed a plan to provide service, using microwave towers, between St. Louis and Chicago. Its format would be that of an outsider providing private line service as an alternative to an organization building an in-house system. It went to the FCC for permission to provide service, and a 1971 decision enabled operations such as MCI's. Organizations providing service were termed specialized common carriers.

MCI soon found that it needed more. It needed many users to attain economies of scale. To manage many users, it needed to connect its between-city system to users in St. Louis and Chicago via the local Bell systems. Otherwise, it would have to construct local microwave (or hard line) collection and distribution systems. To solve that problem, it got a ruling from the FCC for FX services in 1973. The idea was that one office in, say, Chicago would use MCI facilities to reach St. Louis, and then connect to the local Bell system to reach multiple offices in St. Louis.

The AT&T system had for many years subsidized local service from its long line service for numerous reasons. Economies of route density on the long lines resulted in long distance becoming a "money machine" as traffic grew. Likely more important, though, was the Bell corporate culture and the value it placed on inexpensive, reliable, and nearly universal service. Cross-subsidy was the "right" thing to do. But if Bell charged MCI its local rates, then MCI would be skimming. It would be taking money from AT&T's grossly profitable long lines via a subsidy from AT&T to the collector-distribution part of Bell service. AT&T and MCI began discussions to see if they could work out a solution on rates before going to the FCC to get it blessed. In the meantime, MCI and other firms began to offer FX service. An example is a local number anyone could call from any phone in St. Louis and be connected with an airline reservations desk in Chicago. (AT&T offered this service too, i.e., its message toll service [MTS].)

MCI was one of several companies developing similar businesses; there was, example, SPRINT (which originally shared ownership with, and ran fiber optic lines along the right-of-way of, the Southern Pacific Railroad [the SPR in SPRINT]). We use MCI as an example because it was the most aggressive company and the first.

Nothing was exceptional here. Skimming is an old game in transportation—it is what the common carriers claim private truckers do. The Post Office says that UPS and Federal Express skim.

Cross-subsidy in the telephone system was and remains large. The charge for a long-distance call may have been by the late 1990s only a few pennies per minute, but charges to the long-distance carrier to pay for local access were (in 1996) three cents per minute on each end of the call (or six cents per minute per call).[344] This was in addition to a charge of $6.50 per subscriber line per year in access charges. In contrast, local calls, particularly rural services, are underpriced. The fact of cross-subsidy is not exceptional. Is the magnitude? (Later developments would effectively make the marginal cost of local, long-distance, and international calls free to consumers, and portend the beginning of the end of conventional wire-based telephony.)

Transportation cross-subsidies have varied. Long-stage air travel once cross-subsidized short-stage, and still does in some cases. Short rides on fixed-rate transit systems subsidize long rides. The largest magnitude of cross-subsidy is in the highway system. It costs a few tenths of a cent to provide for a vehicle km (mile) of travel on an urban freeway and, say, 20 cents on a lightly used rural road. Yet via the gas tax, users pay the same charge for facilities.

The 1973 decision by the FCC was soon given an interpretation by MCI (and, later, other specialized carriers) that shocked AT&T.[345] FX first was taken to permit one-to-many or many-to-one services. But suppliers decided they could put those together and provide many-to-many service. MCI called it Execunet. A subscriber could phone from any number in West Cupcake to any number in East Cupcake. He dialed a local number and accessed a private specialized carrier; the number given to that specialized carrier then accessed a city and a local phone. Costs were less than long line costs because there were no AT&T's long line tariffs, swollen by cross-subsidy to local phones.

AT&T was then shocked by a sweeping suit started in 1974, *U.S. v. AT&T*. Long embarrassed by the 1956 Consent Decree and with its hand strengthened by the flap over interconnect devices, the tension between specialized carriers and AT&T, and complaints about service, the Department of Justice sought to break up AT&T. The suit was started by the Ford administration, and it is said that the White House had no advance knowledge of it. Court actions got started, and it appeared that an unusually aggressive judge would be able to settle it in about four years, a remarkably short time for suits of this class.

With concentrated ownership, AT&T was somewhat riper for a suit than actors in transportation systems—a matter of degree. AT&T wasn't as well shielded from the Department of Justice as are transportation operators—again, a matter of degree. (The Department of Transportation reviews most transportation mergers.) To manage, AT&T brought on "star" lawyers and planned to make the case that its ownership accounted for the nation's wonderful service; breaking up AT&T would ruin the best of worlds.

To deal with this suit and the many-to-many competitive services that were expanding, AT&T sponsored the Consumer Communications Reform Act of 1976 (CCRA). That's a wonderful title, but the purpose of the bill was to legalize the AT&T monopoly—it was dubbed the "Bell bill." AT&T thought passage was certain. It brought managers and labor from every congressional district to lobby. It thought that the public held the in-house AT&T view that its activities were "right." Bell did not read very well the shift of public and Congressional views about monopolies. At the same time the Bell bill was at debate, there was well-supported legislation moving along that would increase government power over monopoly. The Bell bill did not pass.

The AT&T corporate culture held one worldview; the rest of the world held another. AT&T pointed to the excellence of the Bell Labs (see Section 17.5), the availability and low cost of domestic service, and leadership in the installation of advanced equipment. It compared its system favorably with those in other nations. Such internalization of information and values is a feature of transportation systems. For instance, the Federal Highway Administration (FHWA) sees its world exactly as AT&T saw theirs.

AT&T was in a difficult situation in the late 1970s. Failure of the CCRA hurt deeply, and Justice's aim was massive dismemberment. Justice had presented its case. Bell had begun its defense, a defense building from the image Bell had counted on for support of CCRA—what Bell does is right, and the proof is in the product. It would be unthinkable to ruin that product.

Two changes pulled the fat from the fire. First, there was a modest management turnover in Bell when Charlie Brown became chairman. We say modest, because Brown was an inside manager who had a similar background to his predecessor. Although committed to Bell ideals, Brown saw that the Bell strategy was not working, and he began inside discussions about giving up some local companies as the cost of a settlement. The other change was the return of the Republicans to office in Washington. During the campaign, the Justice case against Bell had been cited as a destructive anti-business extreme. New management at the Department of Defense said that destruction of AT&T would be damaging. Although the new administration judged it could not bear the political cost of dropping the Justice suit, there was a possibility of a settlement short of complete dismemberment.

Charlie Brown developed the inter-intra scheme—the spinning off of the local Bells—and sold it within AT&T. Next, that solution was sold to Justice and the White House. It was much more painful than the Kingsbury Commitment of 1919 and the Consent Decree of 1958, but was far short of complete dismemberment.

The 1982 Consent Decree created (as of January 1, 1984) an AT&T that included Western Electric, Bell Labs, and Long Lines, and seven independent regional Bell operating companies (RBOCs): NYNEX, Bell Atlantic, BellSouth, Southwestern Bell, Ameritech, US West, and Pacific Telesis. The RBOCs were prohibited from providing long-distance service, and AT&T could not enter local markets.

17.4 All the King's Horses

The RBOCs over time have done some merging and recombining among themselves: NYNEX, Bell Atlantic, AirTouch Cellular, and GTE formed what is now called Verizon.

BellSouth, Pacific Telesis, Ameritech, and Southwestern Bell formed what became SBC Communications. SBC Communications acquired what was left of AT&T and took that name. US West was acquired by a new long-distance company, Qwest (founded by SP Rail, the parent of the Southern Pacific Railway—the same people who founded SPRINT), which was later acquired by a rural telephone company CenturyLink. AT&T spun off Western Electric and Bell Labs as Lucent, which in 2006 merged with French equipment manufacturer Alcatel. Alcatel began life as La Compagnie Générale d'Electricité (CGE), and had acquired the telecommunications unit of ITT Corporation (which began as International Telephone and Telegraph, serving mainly the Caribbean and Latin America). CGE produced TGV train sets, among other things.

While there are lots of firms and undoubtedly too many names to remember, the story can be recalled due to its similarity of the merging behavior of railroads. The Communications Act of 1996 permitted the RBOCs to get into the long-distance business (and thus sell integrated services again) if they could demonstrate competition in local telephone markets to the satisfaction of the FCC. Some have been able to do so, though they clearly still, as of this writing, have an effective monopoly on local wireline telephone. AT&T tried to reenter the local market from two angles: acquiring a large cellular telephone company (McCaw Cellular) and acquiring large cable television companies (Tele-Communications, Inc., and Media One being the most important). Both have been spun off. (McCaw Cellular, renamed AT&T Wireless, was acquired by Cingular, a joint venture between SBC and Bell South. Those latter two companies merged, and later acquired AT&T, so it was, in a sense, spun back together). Comcast and others acquired AT&T's cable television assets.

17.5 Bell Labs

Bell Labs grew along with recognition of the importance of science and technology, and they were by no means unique. IBM (Watson), RCA (Sarnoff), DuPont, GE, and many other strong labs grew at the same time. Later, the automobile and drug companies developed strong labs. In these labs there has always been the tension between doing general, more basic research work versus product-oriented work for the corporation. Although well-known for its scientific work (counting 13 Nobel Prizes among its scientists), the Bell Labs did most of its work for the company, a fact not generally known.

The Labs were divided after divestiture. First was a Lab owned by the local companies (now Telcordia, a unit of SAIC, originally the Central Services Organization, and then Bell Communications Research, or Bellcore). Second was the Bell Labs owned by AT&T spin-off Lucent Technologies (the former Western Electric). A third lab, AT&T Laboratory, was created in 1996 after the Lucent spin-off. The cable TV industry has emulated the practice since 1988 with CableLabs.

The world of science and technology is so large that private labs cannot get the resources for broad command of relevant fields. Costs are too high for a private company, no matter how large and rich it is. The Ford Motor Company, for instance, has backed away from its commitment to a broad-based research program. The Ford research lab was downgraded decades ago; other automobile manufacturers more recently. There are now many reasons that firms more and more choose to buy when making make or buy decisions. In short, Bell Labs would have had problems regardless of the break-up of the company.

17.6 Discussion

From its point of view, what mistakes did AT&T make? First, it didn't protect itself from the Department of Justice. We suppose that's because it felt that what it was doing was "right," and no one would question it. When it made an effort via the Bell Act, it was too late. If AT&T had made that effort earlier, when telephone service was being deployed and services were rapidly improving, it might have been able to have favorable legislation passed. This strong feeling of "right" also led AT&T to play "hard ball" with foreign suppliers and MCI. Cooperation and co-optation might have been better tactics.

In the consent decree of 1956, AT&T agreed that it would not enter the computer business. Hindsight says this may have been a bad decision. But at that time, no one saw the subsequent growth of the computer business. At any rate, it isn't clear that AT&T would have been successful in the computer business during this period (a later acquisition and divestiture of computer maker National Cash Register [NCR] demonstrating the point). Computers were not a part of their corporate culture and mission.

We are struck by parallels between FHWA and AT&T. The FHWA has been in trouble from time to time. As the Bureau of Public Roads, it was in danger in 1920 after the break-up of roads during World War I. When Eisenhower was president, he brought in Commissioner DuPont to ease out Chief MacDonald and close down the Bureau. The nation was in a "get the Feds out of the way" mood. The Interstate program saved the Bureau. In a way, FHWA is in trouble now, and there is nothing like the Interstate waiting in the wings.

Recall MCI's specialized service. It invaded the money-making long line business and took advantage of the subsidy to local service. Isn't that exactly what a toll road does? It's no wonder that FHWA has, despite supporting a few experiments, an anti-toll road stance. FHWA has the feeling of self-evident "right" that so characterized AT&T. It plays "hardball" with those who disagree, and it has the Bell Labs "the truth is from our work" problem.

In 1989 there were 2,691,793 cell phone subscribers in the United States.[346] There are more now mobile phones than citizens in the United States (over 327 million mobile phones vs. 314 million residents, as of 2012). The one-way car radio, popularized by Motorola in the 1930s, has been standard equipment for many decades. Mobile two-way radio, untethering the communicator from physical network wires, has long been used in special applications (police radio, military, ship-to-shore, etc.). It has been seen as the future since at least Star Trek and the Communicator allowed Captain Kirk to speak to crew members on a remote planet (and ironically, much of the technology of 1960s Star Trek, set in the twenty-third century, already seems bulky and clumsy today).

The development of mobile phones, and in recent years Internet-enabled mobile phones, has changed how people communicate, how people interact with their environment, where and when people travel, and so on. The implications are numerous. The development has not, however, been one of inventors in the garage, but of systematic funding and deployment of networks by large, incumbent operators. While new players emerge (Silicon Valley companies like Apple and Google entering the handset market), the back-end itself is managed by large traditional telcos. In the United States, the two largest players are AT&T and Verizon (spun off from AT&T after the consent decree, and whose wireless division was for many years partially owned by Vodafone, descended from a British telecomm equipment maker, Racal). The next two largest players are Sprint and T-Mobile (a unit of Deutsche Telecomm). It takes large organizations to build, operate, maintain, and rebuild nationwide networks. There are multiple companies operating in most countries, providing vigorous competition in this fast-changing sector.

Standards are continually being upgraded, and there is some hope of convergence, so any phone can be used on any network (as has been possible in Europe, with GSM, for over a decade). Bandwidth is a potential bottleneck, as more and more bits are slung through the air, though many possible technological solutions present themselves, and will be employed as scarcity becomes costly.

The story of point-to-point communications continues in Chapter 29.

6 Phase 3 of the Life-cycle

The combination of highly permanent construction and lack of realism in its conception is one of the worse legacies our time is leaving to posterity.

—BJORN LINN, *Learning from Experience: The Use of Building and Planning History* (1978)

Mature systems suffocate nascent ones.

—Hughes (1989)

18 Aging in Place, Aging *sans* Grace

18.1 Managing

AS SYSTEMS TRANSITION from growth (see Chapter 10) to maturity, the administration of the system transitions from entrepreneurs and engineers to managers. Organizations become more and more risk averse: taking chances tends to be punished. Debate ensues about who captures the rents of this mature system (capital or labor) rather than how to grow the system. This chapter examines the management of mature systems and suggests strategies for coping. Readers are encouraged to observe the world and judge for themselves.

18.2 Increases in Productivity Diminish or Vanish as Systems Mature

Processes running prior to the maturing of a system run down, in a sense, as a system matures. Early on, the predominant technology is established (technological and institutional structure). Emphasis then is on the components of systems: first on the hard and then on the process technologies. The easy things get done first, and diminishing returns set in. So one characteristic of mature systems is the great difficulty of achieving productivity improvements.[347]

To obtain economies of scale, it's very desirable to have a highly standardized product, which is needed for production efficiency. It's risky to vary that product because economies of scale may be lost if unsuccessful.

18.3 Escape, Adapt, Accept are the Three Basic Strategies for Organizations at Maturity

There are three basic strategies for dealing with maturity, which map to Hirschman's (1970) *Exit, Voice, and Loyalty* (1970).

- *Strategy 1: Escape*: Reject the predominant system structure as given, and instead invest in research and development (R&D), explore systems options and markets so as to continually renew or break away from the tyranny of the Life-cycle. This should be a continuing effort so that new directions of development are available as the system passes its halfway point in deployment and the rate of growth begins to slow. Instead of having development and then growth to maturity, we seek development, growth, and renewal of development, growth, and so on.

  This strategy is becoming popular in the management literature due to Christensen's (1997) *The Innovator's Dilemma*, and the success of Apple Computer in the early 2000s. Christensen argues that leading firms can stay on top so long as innovations are not disruptive, but an innovation that changes the core technology, particularly one that significantly reduces costs for the task at hand and is technologically "good enough" even if not as capable as the mature product, upsets the mature firms in the industry.

  If a competitive product is invading the market (e.g., steam versus sailing ships; buses versus streetcars), then managers may invest in research and development to try to produce that product or buy a successful producer of the new product (e.g., railroads buying trucking companies).
- *Strategy 2: Adapt*: Suppose there is a situation where new producers invade the market with similar products offering higher quality or lower cost. That's the situation in numerous product lines as trade and competition have been internationalized. Managers seek to imitate lower cost producers and or dampen international competition via tariffs and quotas.

  An example is in airlines where the larger network carriers (United, Delta, American) are threatened by low-cost carriers using a different model (e.g., Southwest, Jet Blue). The larger carriers then try to restructure themselves in imitation.

  Managers in this case often have exhausted most sources for productivity improvements as a competitive edge; mature system managers increasingly fine tune services or products to markets, as getting the scale economies just right is essential. Airlines, for instance, try to price discriminate very carefully (people in adjacent seats may have paid thousands of dollars difference in fares) and aim to ensure a high load factor (as an empty seat is permanently lost revenue).
- *Strategy 3: Accept*: In general, the manager's tasks are to know the environment within which the organization is operating and the status of the organization. With that knowledge, the manager should take steps appropriate to the situation. That is a "do the best one can" strategy. It accepts the Life-cycle and maturity as inescapable.

  Suppose producers are in a monopolistic or nearly so situation and competitors are not so much a problem. The firm or other organization may be used as a "cash cow" to support taking up other endeavors that have more potential for growth. The cash cow may be spun off to raise money for other ventures.

That has happened in several rail cases. Through institutional reorganization, railroads have been stripped of their more promising holdings, example, land, pipeline, and communications and information systems companies. In a sense, the railroad cash cow was used by the company to create new cash cows that they then lost control of.

Often competing products are attacked as antisocial and management behavior appears complacent.

> The steam vessel was not a school of seamanship for officers or men. Lounging through the watches of a steamer, or acting as firemen or coal heavers will not produce in a seaman that combination of boldness, strength, and skill which characterized an American sailor of an earlier day; and the habitual exercise by an officer of a command, the execution of which is not under his own eye, is a poor substitute for the school of observation, promptness, and command found only on the deck of a sailing vessel.
>
> *Source*: U.S. Document 1411 of the 41st Congress, 2nd session: Morision, *Men Machine and Modern Times*. 98–99.

Ergo, steam ships are evil! (On the other hand, there is nothing like a sailboat to teach aspiring mariners the power of nature, and both the US Coast Guard and Naval Academy still maintain sailboats for training).

18.4 Obduracy Is Standard Practice in Mature Organizations

Firms may combine strategies. Different parts of firms may engage in different strategies. In general, the mature unit becomes the sacrificial "cash cow," which is milked for cash, and starved for new investment. That cash is reinvested elsewhere within the organization to grow new technologies and markets, or outside as the firm takes its profits and buys other organizations, buys research and development, or returns cash to investors as dividends so they may invest elsewhere.

We are sorry to say that the notion that managers should take early action in anticipation of eventual maturity has a "just words" character. There is much talk about strategic management, but little practice. For instance, it has often been said that the railroads should have thought of themselves as transportation companies much sooner. Redefining your mission may be a way out of an unprofitable industry. However, until deregulation, the railroads were prohibited from such a redefinition. Once deregulation took place, the redefinition soon followed. US railroads invested in moving companies, barge lines, and international shipping among other related businesses in the 1980s. They divested them a decade later.

We've said that managers are in general risk adverse, mainly because having the scale "just right" is so important. Risks exist if the scale doesn't work out correctly. Some managers, knowing their situation and the resources held or obtainable, engage in strategic planning and actions to reconfigure the organization for new developments.

We typically see that change in mature organizations is scarce, that is, *accept* is the most widely adopted strategy. The difficulty of changes has been called "obduracy" by Hommels (2005), who identifies three underlying causes:

- *Dominant frames*—"constrained ways of thinking and interacting" especially in design;
- *Embeddedness*—"close interconnectedness of social and technical elements";
- *Persistent traditions*—"long-term persistence of traditions."

All three factors have a role to play. Those who study the evolution of technology have developed several different theories to look at technological systems. These theories are in part contradictory, in part complementary. Some major ones are listed below.

Social construction of technology (SCOT) suggests that human action shapes technology (but not vice versa). It is associated with the work of Thomas Hughes and others.[348] Our Experiential Policy Model (Chapter 31) implies that human action shapes technology, but that technology and the environment also shape human action. There is physical evidence for transportation technology and the environment shaping human action, and that is the study of the brains of London's taxi drivers. London taxi drivers are required to have what is called "the Knowledge" before getting licensed, which allows them to deliver travelers safely to any location in London without consulting a map. The Knowledge has been required since 1865, and typical drivers practice for 34 months and take the test 12 times before passing. It has been shown that the hippocampus, the part of the brain associated with navigation, is larger for London cabbies than the average person.[349]

Actor-network theory (ANT) descriptively examines the relationships between actors, including both humans and the technologies.[350] The relationships (the network) are both material (between things) and semiotic (between ideas). It is associated with the work of Bruno Latour, who in one book describes the development of a personal rapid transit system proposed for Paris.[351]

Large technical systems (LTS) is a building block theory, which argues that systems are built upon systems in a cumulative hierarchical way. Thomas Hughes (2000) discusses large technical systems like the Big Dig. We discuss this in Section 18.7.

We need to mention another matter that holds for all managers, the effort to product differentiate. The hallmarks of maturity are product standardization and market saturation. Despite market saturation, the size of market may be growing (e.g., as population grows), steady, or falling (e.g., if a competitive product is invading the market.) To maintain market share in such situations, managers attempt to differentiate their product and mine out market niches. Confrontation results between the desire for standardization and the desire for differentiation. The result is that we get a standard product that may come in many colors, with or without fancy features, and so on. With maturity, then, comes product variety, even though the product is quite standardized. We see this clearly in the automobile market.

18.5 The Behavior of the System is Conditioned by Its Structure

Air, transit, and highway systems compare across a number of attributes.[352] The word "system" (or "mode") does not fit any of the systems very well. We noted the truck highway and auto highway systems. For air, there are commercial and private aviation systems and freight versus passenger systems. Transit has at least the varieties of suburban rail, bus, and elevated/subway. Often, diversity is ignored in policy debates. Diversity is suggestive of ways systems might change. It is also the reason for the distortions created by one-size-fits-all policies.

One way to recognize the problems of the air system is to see them as a result of force-fitting. That occurred partly when the system in the United States grasped the opportunity to use post–World War II former military airports. The commercial part of the system has been forced to work out of those airports. (Coerced might be a better word than forced.) The few major new airports that have been built have duplicated the World War II airports in important ways. The big force fitting was the introduction of the jet into the DC-3 system. The jet forced scale changes in non-equipment components of the system. In ways, the horse-powered highway systems have a parallel to the DC-3, and the motorized systems were force-fit as the jet was. Trucks and autos operate on streets built before motorization and on facilities built later, but which carry over premotorization protocols. Bus transit is in that situation also. The birthing of rail had more of a put-together-a-new-form-to-fit-a-function character, and rail mass transit grew from such beginnings.

We have stressed how once the structure of a system gets established, *the behavior of the system is conditioned by its structure*. Indeed, behavior reinforces structure. We have stressed how the exploration of new functional forms is thwarted, remarking mainly on limitations on the explorations of technology options and markets.

One way to state the urban transportation problem is to say that the systems available do not fit their environment very well. That's because the environment has changed and, importantly, perceptions of how environments ought to be have changed. Although there are several aspects to this, a simple statement to illustrate the point is to say that in the United States the urban market is not dense enough for transit and it is too dense for autos.

A desirable property of systems is the capability to adapt to changes in their environments. In ways, transit systems have adapted better to changes in the urban environment than the auto system. The adoption of the bus in the 1920s through the 1950s and the recent development of paratransit services are examples of changes responsive to the environment.

Following a point we made in Chapter 10, we have stressed the locked-in character of systems and their consequential inability to continue to improve what they do. The present point is that locking-in thwarts systems' abilities to adapt as their environments change.

Structure and behavior-conditioned rules are permissive of two classes of system-improving actions. These are actions seeking the economies of networking and actions seeking route economies of scale.

The two classes interact. The articulation of services over a network permits achieving economies of scale on routes on that network; the efficiencies to be achieved via route economies of scale may motivate networking.

Considered alone, the improvements from networking seem to be mainly service improvements. Trucking organizations that operate nationally have costs similar to those of regional operators, but they offer better service for long-distance movements. It's nice to have a networked international air system. One can travel the world with a single booking and ticket, airplanes can be repaired almost anywhere, and traffic rules are universal.

The low cost of networking results from the use of unitary technologies and the standards required by such technologies. Standards thwart change and the fitting of systems to local conditions.

Route economies of scale are often captured by upscaling equipment. If demand exists for a route, larger airplanes, longer passenger trains, or larger buses may be used. Commodities may be moved in unit trains, large tows, or large ships, such as very large crude carriers (VLCCs). In addition to capturing equipment-related cost reductions, the transportation-providing organization may also achieve cost reductions in passenger processing, managing, and so on. Movement cost may be lowered, but the queues may be longer. One waits longer, say, for a larger, lower (per-unit) cost aircraft.

Just as the freeway system has enabled sharp provision-of-facility route economies for highway departments, hub-and-spoke traffic patterns concentrate traffic so that route economies of scale may be achieved.

The point we wish to make is that these scale and network economies achieve service improvements or cost reductions, but also have costs associated with them.

18.6 Spandrels Create Hooks for Innovation

The "spandrel" is the approximately triangular space between the exterior curve of an arch and the rectangular framework around it. It is not intentional, but rather a necessary result. Architects have taken advantage of the opportunity of the spandrel to provide space for decoration. Gould and Lewontin (1979) have used the term as a metaphor to describe how certain features evolve—noting that not every feature exists for a purpose when it was created. In a similar way, transportation systems produce spandrels all of the time. In the most direct analogy, the space inside a highway cloverleaf is a spandrel most often used for landscaping, but doesn't serve a transportation purpose. These spaces can later be exploited (e.g., as a park and ride lot or a water detention pond). Innovation can take advantage of spandrels, hooks created by the existing design to attach new services. Another example is the use of runways at night by freight cargo services. Runways cannot just be rolled up when they aren't being used, taking advantage of this "temporal spandrel." Federal Express launched an overnight delivery service (see Section 20.4).

18.7 Consequential Developments Occur in Second Order Systems

With transportation systems largely in place (we have inquired, invented, and innovated), priority goes to questions about their reconfiguring, redesign, or re-energizing: How do we improve services and adjust to changing conditions? Here is some jargon bearing on structure that may help us understand what we are doing and what we might do.

Think of large technical systems (LTS) (and transportation systems in particular) as first order systems recognizable because of their common characteristics and behaviors.[353] Characteristics include their unitary technologies and products. Services are produced the same way everywhere and products are the same everywhere (one size fits all). The systems are used for many purposes. Extending, let's recognize second order systems. In contrast to LTS that have homogeneous technologies and services yet serve diverse purposes, second order systems are highly specialized and have heterogeneous technologies. They are formed by merging or combining aspects of LTS and other systems.

An example of a second order system is the US nitrogen fertilizer production and delivery system. It is built from fertilizer production technologies, as well as institutional and financing technologies. It combines or configures those with transportation and communication technologies in a very specialized and precise way. Communications and information systems play an important role. Fertilizer delivery in wet or dry form (depending largely on soil wetness, which varies market to market and year to year) by barge, rail, and truck service combinations arrives precisely when needed for the spring plowing season.

First order systems are made up of building blocks. With respect to transportation systems, we have referred to components (fixed facilities, equipment, and operations), and have emphasized that these involve both hard and soft ways of doing things. In the jargon proposed here, think of components as zeroth order systems. We would say that the highway delivery system made up of contractors, tax collectors, state and local agencies, and so on, is a zeroth order system. What does this notion of second order systems do for us? Is the jargon useful? Let's consider transportation improvements.

We strive to improve transportation for many reasons. Most people target zeroth order systems for improvement: operations (traffic flow), bridge structures, insurance arrangements, and so forth. Scoping to systems seen as first order systems is rare, and second order systems aren't considered at all. Does this observation help us understand improvements achieved and not achieved? In addressing that question, instrumental and consequential are operative words. With respect to instrumental improvements—those that can change things—we must be sensitive to development and growth dynamics. Relatively mature systems are not very responsive to improvements in zeroth order systems, and such improvements are thus not very consequential.

Even though we are surrounded by mature systems, there are some things at the first order level that might make a difference. The list includes achieving scale economies, reducing costs, and expanding territory using networking technologies. Today, such technologies (largely soft) are needed in response to NAFTA, the integration of Europe, emergence of developing and newly developed countries, and the increasing globalization of industry and commerce. Boundary conditions change, such as those supported by deregulation and imposed by energy and environmental considerations. We might seek to make systems more supple by specializing services and/or introducing new technologies. One is load leveling, such as that sought by airline fare schemes, congestion pricing, and flex time. These examples of ways to reconfigure apply to all LTS.

Of these, making systems more supple seems most interesting because suppleness might be instrumental in triggering consequential second order system development (where consequential implies improving matters by at least an order of two, for history seems to say magnitudes of that sort are needed to change things). The social uses of transportation are in second order system activities, of course. So if we are to find consequential developments, that's where we will find them.

Is the jargon useful? Does it help us understand the following dialogue?

Technologist seeking first order system improvement: "Increasing highway system capacity using Intelligent Transportation System techniques will increase consumer surplus."

Critic interested in making life better: "What's that going to do for us? Consumer surplus is the economist's fairy gold. Will the (second order system) things that are important to me and involve transportation work better?"

By giving attention to second order systems, one should be able to get a sense of desirable directions of development of LTS.

Clearly we like the jargon. We think:

- It provides interpretations of companion innovation and intermodal notions;
- It links communications and transportation in rich ways;
- It enriches understandings of the functions of LTS.

18.8 Government Policies for Maturity Are Negotiated Contracts with the Regulated

We think that as systems mature, "bloom off the rose" changes in public attitudes force systems to develop policies responding to the interest of diverse and not always friendly publics. As systems mature, more and more attention is given to problems that embedded policies do not handle well.

Although the division is not always neat, government policies can be divided into those that apply to activities generally (general policies) and to those that are activity or mode specific (modal policies). The former have increased in number greatly in recent decades, as has the balance between the activities of state and local governments versus the federal government. Sometimes, general policies are given a special twist when applied to the modes.

It seems useful to think of mode specific policies as contracts. The policies are adopted as parties to the contract give and take and agree on a course of action. Much of the policy action in the decades since the railroads began to mature, for example, can be thought of as:

- Efforts to modify policies (contracts) as things change;
- The reinterpretation of policies as the relative powers of the parties to contracts change; and
- The reinterpretation of contracts as social and economic values and views change.

Turning to the agencies that manage policy, agencies seem to go through a kind of Life-cycle: early they are full of vim and vigor, later they are dominated by bureaucratic and procedural matters, and they may be immobilized by inability to deal with conflicting demands. The reversal of policies in which much has been invested is difficult. That was the case for the ICC.

An asymmetric information problem evolved: Hayek's *Fatal Conceit* teaches that agencies cannot have as much information as is held by those for whom policies are made.[354]

This forces agencies to regulate in a broad-brush way (one policy fits all cases) and to miss or misread broad industry trends. This seems to hold in the rail case.

On the other hand, policies are responsive to broad changes in public attitudes and values (as expressed by legislators). In the rail case, the agency has been responsive to social cost, safety, labor, free enterprise, and monopoly issues as they have unfolded. However, the conflicts caused by agencies responding to many drummers are not well treated within agencies.

Government policies strived to handle problems that embedded (within-corporation) policies could not handle well. Government has been responsible for conflicts between and within other modes. Air traffic control is a government agency (or government-authorized company) in most countries. While it works in general (in that planes rarely collide mid-air) it has been criticized for being too conservative in policy, resulting in congestion, and too backward in technology, resulting in safety problems. In the United States, the Federal Aviation Administration has been regularly lambasted for its difficulty in upgrading computer systems. The criticism reached a peak with the 1981 Air Traffic Controller Strike. (For an overview of rail labor issues, see Section 13.5) Yet the problems were very slow to be solved. Newt Gingrich, an American politician, would routinely pull out a vacuum tube as a prop describing the state of FAA computers well into the 1990s, when private industry had long been using solid-state electronics.

Similarly, railroad-to-railroad problems were at the interfaces of the railroads and users. These seem to have been handled fairly well and, by the 1920s, the deeper problems of the "have and have not" railroads began to get attention. Government policies did not handle these latter problems well. Instead, changes in markets, opportunities for institutional improvements, and changes in technological tools were recognized by the railroads and yielded policies about car handling, traffic concentration, and network design and operations that government policy seemed to thwart. Mergers and acquisitions were essential to realizations of these embedded policies, and government policies were not strongly supportive of mergers and acquisitions, though they permitted them subject to conditions.

18.9 Large System Innovations Are the Work of Blind Giants

The term "blind giants," used by Paul David, an economic historian at Stanford,[355] has two aspects. We call them "giants" because it takes a giant to create something big and "blind" because the giant cannot possibly know the consequences of actions and how a system will grow and develop.

Large infrastructure networks tend be built and managed by these "blind giants." Certainly they may start small, but to deliver transnational transportation requires a transnational organization. Large organizations can arrange financing and supporting systems to deploy large networks quickly, by exploiting existing economies of scale.

Rational analysis can aid blind giants in only limited ways.

It goes without saying that we would like to improve innovation processes. Does rational analysis get in the way? How might it be helpful? Those are good questions, and we can only speculate about them.

As mentioned, Smeaton was very much a rational analyst. He took the steam engine as a given, and applied his rational, linear thinking to improve it. That contrasts with the more integrative approach taken by Stephenson and Pease and Fulton. Another case was the analysis Brunel applied to the *Great Eastern*. In the development of container liner services, both Malcolm McLean and the Matson Navigation Company used integrative approaches. But Matson's was backed up by considerable analysis. Did analysis provide risk information that dampened Matson's entrepreneurial efforts? Does too much analysis provide excuses for limited action? Should we conclude that rational analysis has great strengths if the task is to improve an existing system, and integrative analysis has its advantages elsewhere?

Pacey (1974) addresses the rational-linear analysis versus integrative-design thinking issue, and concludes that integrative thinking breeds innovations. However, we think this is not an either/or issue, for the ways of thinking might be complementary. Consider Brunel and the *Great Eastern* again. Brunel thought widely, and then took advantage of analysis.

As we see it, integrative thinking may run the risk of not using enough analysis to bound feasible spaces. It may generate ideas that won't work "because the numbers don't add." Rational analysis may conclude something won't work because of inability to consider possibilities. With a little innovation, steering became a non-problem.

The race goes to the first off the starting line and not necessarily to the swiftest.

We imagine a blind giant who is unable to imagine the future and understand the consequence of actions. That wouldn't be much of a problem if trial and error were disciplined by the "tooth and claw of the marketplace"; put another way, if pure evolutionary processes were at work. That is not the case, for blind giant decisions are subject to strong lock-in because of networking requirements.

Lock-in can occur in several ways. If one choice is as good as another in situations, tipping may occur. It just happens that when things went a certain way, others then saw that they were better off if they went that way, too. The choice of driving on the left or right side of the street may be an example.[356] Such a choice may be pushed by regulation, and it may force acceptance of second best. That's the case for TV standards in the United States and Japan (NTSC), which for decades yielded an inferior picture to the standards (PAL, SECAM) adopted in Europe. Why did the beta format give way to VHS in videocassette recorders? Video rental stores certainly disliked having to stock two formats, and were quick to jump on the bandwagon when VHS market shares marched ahead.

In the early days of systems, there are very rapid performance improvements—operating know-how, hard technologies, and so on. As the market begins to grow, economies of scale begin to be achieved. Pressures for standardization emerge in order to support production efficiencies (e.g., standardized employee uniforms and operating protocols) and also to support efficiencies to be gained by networking. The length of time the window is open for changes is small. Once the window is closed, it hard to reopen, even if something better comes along. After lock-in moots an option, it is often forgotten.

Perhaps the most costly result of lock-in is that it is unrecognized. It is assumed that pure evolutionary or competitive forces have been at play and what we have are optimal systems. So, people put priority on perfecting suboptimal designs (doing the thing right, instead of doing the right thing). A quite different cost is this: some assume that lock-in is so tight that change is unthinkable. History says that's nonsense. We can recall, for example, that it was the reason that jet aircraft would never have other than military uses.

Limits on cognition and personal energy translate into professionals viewing the set of possible transportation systems and the set of existing systems as the same. Professionals rarely think of new ways in which transportation services might be provided or ways that existing systems might be sharply altered. We find what-might-have-beens a stimulus to wider thinking and suggestive of alternatives.

Around the 1850s, there was much discussion of appropriate rail gauge, and in the late nineteenth century it was thought that dual systems of narrow gauge and wide gauge made sense.[357] This is one of many suggestive what-might-have-beens lost in the dust of history, so to speak.

What mechanisms are operating? First is pure chance, for example Bridgewater died. It just happened that road owners didn't like steam vehicles. Stephenson selected Roman cart gauge for his railroad. The list is very long. But pure chance doesn't take us very far unless there are some strong mechanisms that lock in the results of pure chance. One easy-to-see mechanism is returns to scale.

Other processes are at work to reinforce the increasing returns. Increasing returns from scale and scope economies are common in transportation, and as Marshall emphasized, things get locked in by the presence of monopolies. They also get locked in by the development of standards. So a new alternative might become an impossible choice because it doesn't meet the standards. For instance, today's standards call for multipurpose (trucks and autos) highways and lock in vehicle types. Other processes yield increasing returns with time. Technology and institutional improvements are focused on the choice made. The steam engine was developed rapidly for use on trains. It wasn't developed (very much) for use in on-road vehicles. Learning increases efficiency.

Limits on our cognition exist. The returns that matter are in what transportation does that is worth society doing. We would never learn about returns of an alternative (unpursued) technology; they are outside our cognition. Turning away from increasing returns, consider now changes in consumer expectations. An example may help. When the auto was first innovated, its early use was for social travel. So big cars were built. Standards were put in place for big cars and roads for them. We learned how to produce big cars efficiently. Today, cars are used for all kinds of specialized things. Often, as in the journey to work, a one-passenger car would do. That's an example of the processes that the idea of the change in expectations strives to catch. Changes in environments or situations need to be considered. Once a technology is honed and occupies the turf and limits our cognition, there is no room for an alternative.

18.10 Maturity Creates Imperatives

The word "imperative" often enters the discussion of transportation, energy, and the environment. In highway transportation, there are imperatives to keep vehicle producers healthy, find the money to fix the road system, decrease congestion, as well as to increase safety, clean the air, and achieve energy independence. All these imperatives are treating symptoms of the maturity dysfunction rather than maturity itself. What policies are needed to treat the maturity dysfunction?

Insights to the answers to that question are aided by the birthing experiences we will review: experiences for railroads, inland and coastal transport, steam-metal ship marine transport, and the auto-truck-highway system. Birthing is based on the old, yet often incorporates something new. System design is the process, and a new system design is the product. Market niches are involved. Sometimes a few actors are involved (e.g., Stephenson and Pease); in other cases, it is a many-actor situation (the auto-truck-highway experience.) The systems birthed successfully are consistent with economic and social trajectories of development; they do old things as well as new things, though often the new is discovered after a system is birthed.

Most experience is with incremental innovations applied to an existing system (variants of the strategies of "adapt" and "accept"). Conventional views of innovation, as well of policies to support innovative activities, are based on that experience. It is judged that the policies needed should speed up technology development and its diffusion. Policies should apply to advanced technology. There is little appreciation of the need for policies that support new departures or new designs.

The inability to obtain patents may be a problem. Modifications to existing designs may be thwarted by standards. The system birthing experiences each had to overcome barriers. They did so largely because their high payoffs pushed barriers aside.

The call for new system designs in market niches brings the response: We have tried that using demonstration projects. The experience with them has largely been unsuccessful. What needs to be different from present practice? For one thing, a large percentage of demonstration projects are just schemes for attracting funding. Congressmen arrange projects for their districts. Many projects have a technology transfer character. Where something new is involved, risk of failure may limit the richness of demonstrations.

Maturity creates imperatives. Managers must tactically manage the existing system, but they must also strategically seek new opportunities, innovating in new markets.

18.11 Maturity Creates Opportunities

We have used the words "successful" and "consequential," and such words are subject to interpretation. We meant them to be interpreted in very limited ways. An intervention was successful if there was a result that would not have otherwise occurred. It was consequential if the monetary and service results had nontrivial impacts. (Now, we have the word "nontrivial" to interpret, but we will not.)

Experience says that improvements are forged by taking what we have and "carrying out ... new combinations" (Schumpeter 1934). The designs (new combinations) that seed new development pathways are built from "old stuff" (in the beginning).

The auto-highway system, for example, built from existing highway facilities; wagon, bicycle, and buggy equipment and electric motors or steam, otto-cycle, or diesel engines; and know-how about the uses of transportation for passenger and freight purposes. The system also interfaced constructively with rail systems; it dealt with social, economic, and ecological problems of the day. Its deployment became an imperative—an imperative strong enough to break existing social contracts forming barriers to deployment.

There are many opportunities to produce "new stuff" which could be the "old stuff" in the forging of new combinations or departures. Going beyond that, it suggests that we should be alert to the opportunities.

While history underscores inevitability, it also tells us how to break the inevitability of the Life-cycle. We know that the instrument of choice for the intervention in the Life-cycle should be technology, for we have noted that technology is the major force for change. History certainly tells us that. Indeed, as we examine transit, the search for improved air quality, and energy conservation efforts, we will see that technology was one tool for planning. We will also see that knowing that technology is the tool to be used is not enough. One must know how to make technology work.

Setting aside issues such as market control by large suppliers, one cannot disagree in principle with deregulation, pricing, and, for that matter, other prescriptions for achieving economic efficiency in mature systems. Nonetheless, it must be recognized that the priority given these popular topics comes with opportunity costs. Other opportunities receive low priority. New ways of using technology, reducing the costs of labor, and reworking embedded policies may well have been more important to improved railroad efficiency than deregulation.

What's the overall bottom line for mode-specific policies? Reconsider the quotation opening the chapter. Perhaps it should be rewritten:

> The combination of highly permanent policies (contracts), lack of realism in their conception, and difficulty of revising them is one of the worst legacies our time is leaving to posterity.

As a mode matures, new policies are needed. We are probably better off with new policies than without them. We would probably be better off with less permanence, more realism, and an improved ability to revise obsolete, dysfunctional policies.

Maturity creates opportunities. One can identify at a general level things to try out and things to avoid when working with mature systems. One might work toward a set of rules.

- *Tune products to markets and specialize services.*
- *Shun standardized, one-size-fits-all design.*
- *Plan to achieve economies of scale. Many cost-reducing actions have already been mined out, and something new or something revised has to compete where turf is already occupied.*
- *Avoid high capital cost projects that will require high-volume use to cover capital costs.*
- *Recognize that managers are risk averse, emphasize tried-and-tested solutions to problems.*
- *Exploit the cash cows and fat cows that cover the landscape.*

Downsizing and changing money flows are options.

This is a sample from a possible larger set of rules. Note that what to avoid and what to do is sometimes left implicit.

18.12 Transportation Is Possessed with Zombies

In transportation, nothing ever truly dies as long as the line on the map is a memory in the mind of an advocate. A decision to not build a project is easily reversed, since "no" involves no investment in fixed costs, unless something is done in its stead. This is especially a problem if the right-of-way for the facility is being preserved, through either land purchases or prohibitions on development.

We might call these old ideas "Zombie Transportation," projects that are now "bad" (or at least no longer "good") ideas, effectively seemingly dead, yet still live in people's minds. Occasionally, like the California high-speed rail project, zombies get partial funding before their plug is eventually pulled (see Chapter 28).[358]

There are lots of other projects one can think of that were lines on maps for decades before being realized. Outer-beltway sections Maryland 200 (the Inter-County Connector) (see Section 22.4.2) and Minnesota 610 are two that come immediately to mind, on the map for sixty and at least forty years, respectively.

In many cases, the problem is simple, ruthless, benefit-cost analysis; the B/C ratio, which may have once been above 1.0, falls to a lower level due to changed circumstances, increased costs due to environmental or other concerns, or a change in demand associated with different finance (tax to toll) mechanism or price of energy. Yet because it is a "commitment" (political, or moral) to a community that their turn will come, they too will get their line built, the line on the map never comes off the map.

Once a project is completed and open, it is essentially irreversible. Few facilities are shuttered before they physically fail (a few exceptions to be noted: example, anticipating collapse, some urban freeways in San Francisco have been removed), or require replacement (streetcars in many US cities in the mid-twentieth century). And even then, many facilities which should be shuttered continue to be maintained and operated, and later reconstructed instead. The difficulty of implementing gravelization, which may be effective for low-volume rural roads, is an extreme example of this.

We need a better system for truly killing bad or obsolete ideas in transportation, for culling the losers or the no longer winners. Otherwise, agencies will look at decade-old maps, say to themselves: "what remains unfinished," and proceed along to build zombie facilities despite newer priorities rising to the fore and old ideas ceasing to be effective.

Some like aging systems; they provide a certain stability of career paths, social interrelation, and cash flow. Those are on-system attributes; off-system they have a stabilizing effect. Jim Mills, one-time president *pro tempore* of the California State Senate and promoter of rail in San Diego, said that rail was much better than bus service because it did not change easily compared to bus. That is, the great feature of bus, its flexibility, is also its weakness, lack of permanence. We counter by simply noting that while streetcars disappeared from urban streets more than half a century ago, almost all those routes have continuing transit service. So while the infrastructure may change, the service sticks.

The Life-cycle process is discussed in Chapter 25.

7 Wave Four: 1939–1991

Logistics comprises the means and arrangements which work out the plans of strategy and tactics. Strategy decides where to act; logistics brings the troops to this point.

—JOMINI,
Precis de l'Art de la Guerre (1838)[359]

19 Building Blocks: The Logistics Revolution

19.1 McLean's Insight: Inside the Box

MALCOM MCLEAN (1913–2001), an independent truck owner-operator out of Fayetteville, North Carolina, frustrated by the delays associated with waiting at the New Jersey waterfront to unload his cotton for shipment to Istanbul, and concerned about hauling empty truck trailers back to Texas, reinvented shipping. He identified the cargo as the trailer, rather than what the trailer was carrying, to reduce handling by the stevedores. In a sense, ships would be used as trans-oceanic ferries. The trailers could be stacked if the wheels were removed and the sides reinforced.

As with most innovators and system integrators, McLean assembled building blocks together in a new way, finessing constraints. In 1947, American Overseas Chartering Corp. (later Trailerships) offered overnight (roll on/roll off [ro/ro]) trailer-ship service on the Hudson River, but this service soon faded as the New York Thruway was faster. Another ship, the *Carib Queen*, offered a similar service between Key West and Havana in the late 1950s.

While ro/ro was the initial model, McLean thought beyond ro/ro. He obtained agreement from the Teamsters (who would benefit from additional trailer traffic). He also had to establish the transition between the Teamsters, who drove the trucks, and the Longshoremen, who would handle the cranes that took the truck from the shore to the ship.

McLean bought a tanker company and modified its ships to carry detachable trailers. On April 26, 1956, McLean's Pan-Atlantic Steamship Company's *Ideal-X* set "sail" from New York to Houston, inaugurating Sea-Land Service with 58 containers aboard. This differed from ro/ro intermodalism, which was if not well established, at least not unknown, in that the container was detached from its running gear on the pier and hoisted onto the ship using dockside cranes.

McLean's *Ideal-X* mounted containers on a deck of a T-2 tanker. Similarly, the first Matson ships placed containers on the deck of a Victory class ship (see Section 19.2). The first specially built container liners carried 200 TEUs.[360]

Financed by National City Bank of New York (later Citibank), McLean Trucking acquired Pan Atlantic Steamship in 1955, and then its parent, Waterman. To their mutual benefit, McLean's personal banker was Walter Wriston, who eventually ran Citibank. To avoid regulatory problems (inter-modal transportation cross-ownership was prohibited), McLean placed the shares of his own trucking company in a blind trust. Notably, McLean Trucking was prohibited from providing the on-road service from Sea-Land Services (newly renamed from Pan Atlantic).

The railroads petitioned the Interstate Commerce Commission (ICC) to stop container traffic, but the ICC ruled in McLean's favor. The ICC concluded that the trailer was the cargo (rather than the items in the trailer), using as precedent the ro/ro trailers that the railroads carried. While McLean's innovation was mostly raw capitalism, his firms applied for and received insurance from the Federal Maritime Commission, and tax advantages in terms of accelerated depreciation, as a result of using converted warships as approved by the US Office of Defense Mobilization.

As this converted T-2 was a new type of ship, it had to undergo months of testing to persuade the US Coast Guard and the American Bureau of Shipping of its seaworthiness.

The ships originally were to be called *Container Ship 1*, *Container Ship 2*, and so on, but maritime tradition demanded names, as captains were less likely to go down with a *Container Ship 1* than a properly named vessel. The first four converted T-2 ships were named *Ideal X*, *Almena*, *Coalinga Hills*, and *Maxton*. In addition to containers, these also carried oil in their tanks. From 1957, converted C-2 ships like the *Gateway City*, the first all-container ship, launched carrying 226 containers.

In addition to a new technology for the ships, new trailer-trucks with removable containers also had to be designed. These were built by Brown Trailer and Fruehauf Trailer, eventually in large numbers.

The spacing between containers on-board had to be carefully considered. Some space was required for the cranes to on- and off-load (before cranes moved shoreside), but too much space would be wasteful. A tolerance of about 1 inch (2.54 cm) between containers was established.

The economic efficiency associated with containerization was profound. Unloading and loading a typical break-bulk ship of the period would require 600 person days of labor (150 stevedores working 4 days each); the *Gateway City* required 14 person days (14 workers on 1 shift) (43 times less).

Ships, like all capital intensive transportation technologies, earn only when in motion. Time in port is lost, and so the profit-maximizing transportation firm aims to minimize it.

Another advantage of containerization is reduction in theft (both simple pilferage by individual workers and systematic pilferage by organized crime) and reduced damage on ships buffeted by winds.

McLean selected the 35 ft (10.6 m) trailer as a standard, as this was maximum length permitted by the State of Pennsylvania on highways. McLean forwent patents on his innovations to encourage industry standardization.

The American Standards Association established a committee to set standards, and any length divisible by 10′ (3.05 m) was permitted, and the industry converged on 20′ and 40′ as its practical standards (so that two connected 20′ TEU (trailer equivalent unit) would equal one 40′ unit. ASA also determined that any equipment such as refrigeration would be within the 20′ container.)

With standardization and profits, service expanded. Competitor Grace Line adopted 17′ (5.18 m) containers in its service to Latin America beginning in 1960. Union troubles in Venezuela delayed the unloading of the ship, and it took years for the issue to be fully resolved. Labor issues plagued the industry for years, as unions resisted labor-saving technologies. Sea-Land inaugurated transatlantic container service in 1966 from New York to Rotterdam. Domestic container service would soon fall by the wayside.

Sea-Land for many years preferred conversion of older vessels, while other companies purchased new. This basically took the stern and machinery of one existing ship and constructed a new bow and mid-body to create a new ship, while the bow and mid-body of the existing ship could be joined with the existing stern and machinery of another ship.

In 1969, RJ Reynolds took over Sea-Land, owning it until 1984; in 1986, CSX acquired it and held it until 1999. In 1984, newly deregulated and merged CSX acquired American Commercial Lines, a barge operator plying the US inland waterways. CSX was aiming for "seamless" intermodal transport.[361] CSX sold Sea-Land to Maersk, which become Maersk-Sealand until eventually dropping the Sealand suffix.

High oil prices constrained the market; all carriers ran into the problem that energy expenditure increased non-linearly with speed. After 1973, all new Sea-Land vessels were diesel rather than steam. McLean personally acquired the historic United States Lines in 1978, making capital intensive investments in Econships (large but slow to save energy costs). This did not pay off in the subsequent era of cheap energy, and the firm exited in 1986.

Taiwan-based Evergreen entered the market. Founded in 1968 by Yung-Fa Chang, it expanded from a single break-bulk ship. In 1984, it inaugurated, east-bound and west-bound "around-the-world" service with twenty four vessels.

To compare with the 200 TEUs of the first 1950s-era containerships, recently constructed vessels have been running 4,000 TEUs, with the most recent upscaling to 16,000 TEUs. Presently, about 90 percent of general cargo moves by containers stacked on ships. Over 200 million containers per year are now moved worldwide. Containerization was fast.[362] From a mere footnote in the 1956 book *Marine Cargo Operations*,[363] containerization was essentially complete in 1971, when all containerizable cargo on the transatlantic route was containerized (see, e.g., Figure 19.1).

Containerization takes an old idea (putting things in containers), which has been around since the first pots and barrels, and scaled it up using advanced technology. The scaling made many older, smaller ports obsolete and created a generation of new super-ports that acted as hubs in a packet-based freight transportation system. Container shipping also changed railroads, by bringing double-stack container trains to the railroads, with a low floor between the train's running gear.

In addition to scale-up and specialization, there have been increases in speeds, example, cargo ships from 20 to 66 km/hr (11 to 36 knots) and tankships from 20 to 26 km/hr (11 to 14 knots). However, fuel is a healthy part of operating costs, and to improve fuel efficiency,

FIGURE 19.1 A Containership Being Loaded in Vancouver, Canada

speeds have been reduced in some trades. The bulbous underwater bow section has been widely adopted, as have finer stern sections; welding has largely replaced riveting. Low speed diesels offer fuel efficiency advantages over turbines.

19.2 Matson's Innovations

The Matson Navigation's *Hawaiian Merchant* sailed on August 31, 1958, with 43 aluminum containers on deck inaugurating Matson's early entry into container shipping. The Matson Research Corporation had recommended converting ships for up to 75 on-deck container movements and the follow-on step of having ships specially constructed for container service. Almost two years after the *Hawaiian Merchant* sailed, the first specially designed Matson container ship, *Hawaiian Citizen*, launched in April 1960.

How could we have innovation and risk taking by an old liner company? Matson's experience was operating partly in domestic markets sheltered from foreign competition and in other markets where ship construction and operating subsidies applied. Tightly regulated by government and traditions, system-changing innovation was out of the question. But does regulation always stifle innovation?

For many years owned by Hawaii-based Alexander & Baldwin, Inc. (one of Hawaii's "big five" companies), Matson serves the California-Hawaii and US Pacific Possessions trades with specialized container and automobile ships. Starting over 100 years ago with sailing ships in the Hawaii sugar trade, Matson is perhaps best remembered for Hawaii, Australia-New Zealand, and Pacific islands passenger liners. Its rich and complex corporate history includes taking risks, betting on technology, and the spawning of associated companies for new ventures. For instance, betting on the future of air services, in 1936 it joined with Pan American Airways to offer air services in Hawaii and it established an airline subsidiary in 1943, only to have those endeavors thwarted by regulation.

Inflation in the 1940s and 1950s, along with labor strife at ports, forced Matson to seek and reseek increased tariffs, but these were strongly resisted. Increased liner size or velocity wasn't a route to increased productivity and cost control because of the time required in port for gangs of 16–20 laborers to load and unload ships—in the Hawaiian trade, a ship spent as much time in port as it did sailing. In the early 1960s, future Matson president Robert Pfeiffer led negotiations with International Longshoremen's and Warehousemen's Union president Harry Bridges to establish the Mechanization and Modernization Agreement, governing the transition to containers.[364]

Like any good story, there was tension, conflict, gambling, and mystery. Matson's Hawaii market could be and was challenged by competitors, so failure could wreck the company, which was already shaken by losses of other markets. Should Matson take the bold step of being the first to offer container services throughout the Pacific? It took steps by arranging port terminals in Japan and the Philippians. Tensions were high as its Board of Directors support waffled and bowed to the realities of current economic conditions, and there are Matson docks in places where Matson ships have never called.

Mysteries included how to manage opposition by labor, the pace of market development, and the development of land-side intermodal relations.

Small containers were already in use, serving mainly as lock boxes to thwart pilferage of valuable commodities. What size should general cargo containers be? Because many states had truck length limits, Matson elected 24′ × 8′ × 8.5′ (7.3 × 2.4 × 2.6 m) so that the containers could be moved as doubles—one on the truck and another on the trailer. Larger "high cube" containers came later.

At ports, Matson saw the opportunity for land-based gantry cranes replacing shipborne cranes. Container movement equipment and sorting and storage asked for large docks. Innovative managers at Matson Docks and the Port of Oakland took risks and made investments.

Those who stood aside, such as managers at the Port of San Francisco, were left behind as the predominant technology emerged in the format developed by Matson and others.

In 2012, Alexander & Baldwin divested its Matson unit, which remains in the West Coast to Hawaii market.

19.3 Alliances

There were many, many companies providing shipping services, and in one sense, the industry demanded consolidation to achieve economies of scale. On the other hand, they were international, and every country had local champions, and regulations, and labor arrangements, which worked against consolidation. As with domestic railroads in the United States, but unhindered by domestic laws, international conferences were established from the late 1800s to fix rates and activities. These still periodically bump up against anti-trust regulations. In the 1960s, some of the older firms established consortia to enter the container market.

Atlantic Container Line was a consortium of La Compagnie Generale Transatlantique, Cunard, Holland-America, the Swedish American Line, Rederi A/B Transatlantic, and Wallenius; it launched its first ship in 1967. Overseas Container Limited was a British consortium. Other container firms also launched in the late 1960s. Hapag-Lloyd was a merger of Hamburg American Line and North German Lloyd.

Like the more well-known airline alliances (see Section 20.5.1), the maritime alliance era began in the 1980s when multiple firms would coordinate and carry each other's containers.

Containership alliances in 1995:[365]

- Global: American President Lines,[366] Mitsui-OSK, Nedlloyd, OOCL (Orient Overseas Container Line);
- Grand: Hapag-Lloyd, Neptune-Orient Line, Nippon Yusen Kaisha Line, P&O Container Line;
- Tricon/Hanjin: Cho Yang, DSR-Senator, Hanjin Shipping;
- Maersk/Sealand: Maersk, Sealand.

By 2003, the alliance structure looked as follows,[367] with total capacity in TEUs given in parentheses.

- CHKY: Cosco, Hanjin, "K"-Line, Yang Ming (YML) (383,283);
- Grand: Hapag-Lloyd, NYK, OOCL (230,278);
- The New World Alliance: Hyundai, American President Lines, Mitsui OSK Lines, MSC (260,389);
- Gran Americana: A. P. Moller Group (Maersk/Sealand) (236,876);
- Evergreen: INCL, Lloyd Trestino, Hatsu (217,095);
- Zim: Cont Service, Norasia (100,430);
- CSCL: China Shipping Group (69,864);
- CMA/CGM: ANL, Andrew Weir, Caribbean General Maritime, FAS, Mac Andrews (50,279).

In 2011, the Grand and New World Alliances combined, as have the Mediterranean Shipping Company and France's CMA/CGM.[368]

The alliances are legal agreements, filed with agencies such as the US Federal Maritime Commission, allowing the parties to meet to discuss and rationalize in service in various ways, provide joint services, share revenue, distribute risk and liabilities, and act as a single organization when convenient. All of the agreements are now available online.

19.4 Container Ports

Containerization radically changed the scale and scope of maritime transportation. The growth of Asia as a manufacturing center led to the top four ports in 2000, and the top nine in 2010, being Asian (Figure 19.2). London, which is too far inland and served by too shallow a waterway, is no longer among the top ports, nor is New York. All of 2000's top ports remained in the top twenty one in 2010, but many of them slipped significantly.

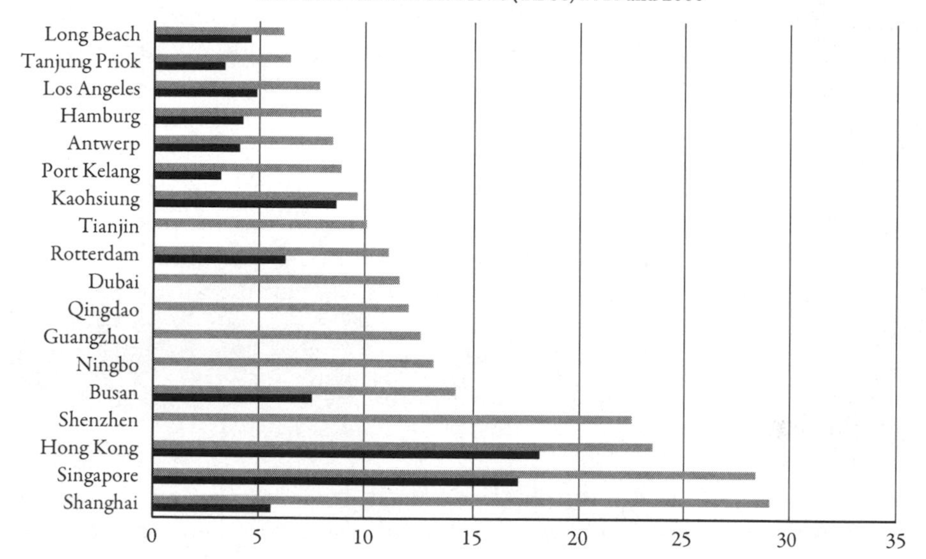

FIGURE 19.2 Top 10 International Ports and Containerized Cargo Throughput: 2000 and 2010. All cities in the top 10 in either year are shown, with 2010 TEU values for all ports, and 2000 TEU values for ports in the top 10 that year.

Source: World Shipping Council.

Port location affected port design. Port design figured into port evolution. Finger piers were popular in the United States, where ports were adjacent to cities. In Europe the ports were farther away from cities, and there was more land available for ships to side load against the land. The mechanization of European ports was thus easier, as they did not need to relocate to serve the larger ships with larger cranes. The United States saw many once capable ports (New York City vs. New Jersey, San Francisco vs. Oakland and Los Angeles) decline with containerization, as the finger piers were ineffective and the land scarce.

As far as policy goes, the American Association of Port Authorities[369] has called for:

- Expanding sources for port development financing and revenues;
- Balancing environmental regulation and economic development;
- Providing waterside port access through dredging and dredged material disposal;
- Securing resources for intermodal landside access to ports;
- Using transportation trust funds for infrastructure development, not deficit reduction;
- Enhancing free and fair trade worldwide.

Thus, for ports to grow, the industry association wants more sources for financing (they say that the markets are insufficient, and the ports would like subsidy), they want environmental regulations relaxed (but make no suggestion that they should pay for the externalities they generate), they want resources to expand land-side access, and they want a free trade policy. Only the last comports with free market competition.

Outside the United States, many ports, like airports, have been privatized. That may be a direction in which the United States heads in years to come.

A problem for the United States is that East Coast ports are limited to 77,000-ton tankships. Moreover, many important refinery facilities are located in ports that can only take 50,000 dwt ships. Japan and Western Europe have established VLCC facilities. San Francisco Bay can handle tankships of about 100,000 dwts. Prior to expansion, the Panama Canal was 32 meters wide (106 feet), and yielding about a 70,000 dwt tanker or 5,000 TEU. The standard for the maximum-sized ships that can traverse the canal is dubbed "Panamax," and as of 2014, "new Panamax." The new Panamax standard is 49 m width, allowing 13,000 TEU. Other bottlenecks in the system, including the straits of Malacca and the Suez, also have their own maximums.

19.5 Labor Arrangements

Steamship companies leased piers from New York City and contracted with stevedoring companies to provide labor, which was controlled by the International Longshoremen's Association (ILA).

Until just after World War II, longshoremen climbed onto cargo ships and using brute force, with the aid of nets and grappling hooks, moved small packages on and off ships. The process of loading and unloading might keep a ship in port for weeks. This "break-bulk" shipping was a major bottleneck in world commerce. It enabled the longshoremen to become a powerful union, and operations were prone to theft.

This was famously illustrated in the movie *On the Waterfront*, based on the reporting of Malcolm Johnson, beginning with murder of Thomas Collentine. Waterfront corruption was aided by a surplus of labor. Union bosses would allocate jobs to workers, who had to show up at the union hall daily without knowing whether there would be work. These bosses would allocate jobs to workers who were in debt to loan sharks or were otherwise bribed or given kickbacks. If you escaped debt, you did not get employed. The irregular schedule arises from the randomness of ship arrivals, requiring a surplus of labor. Ultimately, the reporting led to investigation by the Crime Commission, the expulsion of the ILA from the American Federation of Labor, and replacement of the union president Joseph P. Ryan.

With containerization, automation, and the rest of the logistics revolution, relations between management and labor remained raw. The amount of labor required per ton steadily declined, but labor remained unionized and a critical bottleneck in the transportation process. The question of who gets the rents (or profits) from the spatial monopoly is always contentious in transportation, and while strikes are becoming less and less common in industry, they remain relatively more visible in transportation.

On July 1, 2002, the contract between the Pacific Maritime Association (PMA), the industry association for the operators of twenty nine ports on the West Coast of the United States, and the International Longshore and Warehouse Union (ILWU) expired. This was of national importance because West Coast ports handle about $1 billion worth of freight daily; see Table 19.1 for details. It was in many senses a classic dispute of capital versus labor. The PMA wanted to modernize technology, such as introducing optical scanners and global positioning systems (GPS) to automate information flow at the ports. The ILWU was

TABLE 19.1

Selected Pacific Gateways: Percent Imports from Asia in 20-foot Container Equivalents

	Value of All Imports (Billion dollars)	Share Imports from Asia (percent)
Seattle	23	6
Tacoma	14	5
Portland	8	<4
Oakland	16	4
Los Angeles	119	35
Long Beach	49	30

Source: Journal of Commerces Port Import Export Reporting Service (Census Bureau, 2002).
Note: Total imports from Asia, 7.9 million 20-foot container equivalents (TEUs) for year ended July 2002.

concerned about employment of members (and especially the use of non-union employees to replace members). However, the situation was somewhat atypical because union salaries ranged from $80,000 to $158,000 for what are normally considered "blue collar" jobs. So the union had successfully captured many of the monopoly rents available from the ports during previous labor disputes.

The ILWU had been established in 1937, as a breakaway from the (East Coast–based) International Longshoreman's Association (ILA). The union was forged in 1934 by a bloody West Coast waterfront strike and general strike in San Francisco. After many years of control by communist figures such as the Australian-born Harry Bridges, the ILWU remained an aggressive and powerful union. To illustrate their control, in September 2002, the ILWU, which controlled day-to-day work assignments, required that they be randomized, and required strict adherence to safety regulations (a work-to-contract strategy). This policy, rather than allowing workers to conduct tasks in which they were skilled, resulted in productivity at the ports being reduced by 20–90 percent. The rationale for this was not simply to gain leverage over the PMA, but also to protest work conditions that had resulted in five deaths on the docks in the Los Angeles region in the previous five years.

In response to this work slowdown, on September 27 the West Coast port operators shut down cargo terminals for 38 hours, and then on September 30, the PMA closed ports until the ILWU agreed to either an extension of the existing contract or a new contract. This strategy "locked out" 10,500 workers from their jobs.

The irony of this management-labor dispute, which cost the economy hundreds of millions of dollars, was that the monetary difference between the two sides was estimated at only $20 million.

Intentionally evoking images of the movie *On the Waterfront*, wherein former boxer and longshoreman Terry Malloy (played by Marlon Brando) has to choose whether to betray the criminal union leaders on the New York/New Jersey ports or keep solidarity with his fellow workers, the PMA brought armed guards to the negotiations on October 1—offending the union (or, at a minimum, giving reason for the claim of offense).

United States Secretary of Labor Elaine Chao tried to intervene in the negotiations. However, she failed to negotiate a contract extension that would reopen the ports for 30 days. The ILWU supported an end to this lockout, but the PMA resisted, believing the union would continue to work to rules.

A Board of Inquiry established by the federal government to investigate the issues reported, "We have no confidence that the parties will resolve the West Coast ports dispute within a reasonable time." This finding set the stage for the beginning of the invocation of the Taft-Hartley Act. On October 9, President George W. Bush sought and received a federal district court order requiring the West Coast ports to reopen.

On November 1, a tentative agreement was reached. The management desire to cut clerk jobs (and introduce more computer technology) was agreed to, resulting in cuts of 400 marine clerk jobs. In exchange, however, it was agreed that the new jobs created would be union jobs. In addition, agreements had to be negotiated about pension security, health insurance benefits, higher wages, and safety provision. The details were worked out by November 24 for a six-year contract. Chief federal mediator Peter Hurtgen said negotiators "demonstrated statesmanlike leadership, which made this agreement possible." The ILWU membership voted after Thanksgiving in favor of the contract.

A short strike in Autumn 2012 by clerical workers represented by the ILWU (concerned about outsourcing) also shut down the West Coast ports, as longshoremen would not cross picket lines. This strike was quickly ended.

The 2002 West Coast Port Strike was the first application of the Taft-Hartley Act in ports since President Richard Nixon employed it in 1971 to stop another longshoremen's strike, the last major work stoppage at the West Coast ports, which lasted for 134 days.

A Republican Congress, overriding President Harry Truman's veto, passed Taft-Hartley in 1947. The law had a number of provisions: it outlawed the closed shop, in which only union members would be hired; but it permitted a union shop, which required non-union workers to join the union. More to the point of this discussion, Taft-Hartley provided for a 60-day cooling off period after a contract expires before a strike may be called. Moreover, the president may extend that period by 80 days. After that period, the National Labor Relations Board has 15 days to poll employees to see whether they will accept management's final proposal and an additional five days to count votes. Then, if the workers reject the proposal, they can strike.

However, this seldom-used procedure serves mostly as a threat. Prior to 2002, it was last used to calm a 1978 coal strike. Not surprisingly, unions oppose these provisions, referring to the Act as "plain ole slavery."

19.6 Discussion

Recent (and we expect continuing) trends in shipping include:

- More use of computers and communications, not so much to reduce manning, but to track cargo, including identity-preserved cargo, and for security checks.
- Better fuel efficiency through continued improvements in engine, ship, propeller, and rudder designs.
- Continued rise in ship sizes. This is limited by physical constraints (e.g., the newly enlarged Panama Canal or harbor depths). The increase in size should improve

efficiency, but other impacts are uncertain. East Coast US ports would need expensive dredging to accommodate larger ships. Historically, clipper ships were partly done in by opening of the Suez Canal, shortening routes for steam-powered ships.

- Further development of hubbing ports (such as Singapore and Rotterdam)—in part to gain economies of scale in shipping.
- Specialization and increasing sizes of containers, though again constraints (easy transfer to trucks or trains) play a role.
- Simultaneous loading of multiple containers (stacks) to speed up port turnaround.
- On-ship devices for stowing containers.
- Continued specialization of ships of all types.

Neobulk ships have emerged. These are bulk carriers specialized to new trades, and they range from scrap and lumber carriers to auto carriers. Such items were once carried as general cargo, and specialized chemical, wine, and so on, ships have been developed for other former general cargo commodities. Dry-bulk ships have been scaled-up to 270,000 dwts.

Passenger ships also scaled up to large sizes (today the Queen Mary 2 is about 150,000 tons), but with the advent of air competition, they are now largely deployed in the resort trades.

Since the logistics revolution, shipping has been reborn, grown, and matured again, and though it is still evolving, change is again coming slowly. There may be a new logistics revolution like containerization around the corner—speedy, giant ships from trans-Atlantic and trans-Pacific trade. If trends continue, the number of important ports will decline with the size of ships, but those few ports will continue to expand in size. The remaining constraints will continue to be the canals and the harbors.

> You define a good flight by the negatives: you didn't get hijacked, you didn't crash, you didn't throw up, you weren't late, you weren't nauseated by the food. So you're grateful.
>
> —PAUL THEROUX, *The Old Patagonian Express* (1979)

20 The Jet Age

20.1 Supersonic

PROVIDING SUPERSONIC FLIGHT at Mach 2, and travel at 18,000 meters (11 miles) above the weather, high enough to appreciate the curvature of the earth, the Concorde just oozed the future. Yet in June 2003, Air France and, in October 2003, British Airways retired their Concorde fleet (7 aircraft), a mere twenty seven years after they started regular service, after serving over 2.5 million passengers. The trend in transportation has been toward steadily higher and higher speeds. So is the retirement of the Concorde without a comparable replacement indicative of the general retrenchment in the aviation industry, or does it suggest cost and technological limits on speed increases?

The history of transportation is full of fits and starts; many US passenger rail routes achieved their highest routine speeds in the 1920s and have yet to match those records. Is the history of supersonic transport different?

The Concorde was perhaps the first major technology to be birthed jointly by two governments, the British and French, who in 1956 signed a treaty on the design, development, and manufacturing of a supersonic aircraft within six years. A bit ambitious, the first prototype was rolled out in 1967 in Toulouse, France, and took flight in 1969. In 1973, the Concorde made its first flight to the United States, landing at the new Dallas-Fort Worth Airport. By 1976, British Airways began supersonic service between London and Bahrain, and in 1977 from London to New York.[370] The first crash of an Air France Concorde occurred on July 25, 2000, and resulted in a grounding of the fleet until July 17, 2001.

The Concorde was the only supersonic passenger craft to get off the drawing boards and into continuing service, though there had been plans for a US supersonic aircraft at around the same time. In 1963, President Kennedy announced that the United States would develop a supersonic transport, or SST. Boeing won the contract over Lockheed and North American, though it turned out to be a Pyrrhic victory. Despite President Nixon's assertion on September 23, 1969, "The SST is going to be built," the SST was canceled in 1971 for both market and environmental reasons; there were and are concerns about the effects of sonic booms.[371] Later, in the 1990s, NASA, along with Boeing, spent $1 billion designing a "high speed civil transport," which was canceled due to high costs (expected to be $18 billion in development alone) and insufficient expected revenues.

The Soviets built the Tu-144, dubbed the "Konkordski" due to its striking similarities, which many assert are due to industrial espionage. The Tu-144 was the first commercial supersonic aircraft, flown in 1968. During the 1973 Paris Air Show, the plane crashed, probably because of an unusual maneuver to avoid hitting a French Mirage fighter that was secretly filming it.[372] The French and Soviets colluded to cover up the story, which only was revealed after the end of the Cold War. The Tu-144 did make semi-regular service in the Soviet Union between Moscow and Alma-Ata, Kazakhstan, between 1977 and 1978, but was retired soon after another crash. NASA later cooperated with the Russians to study the Tu-144, reactivating the plane in 1995. Other schemes have been proposed, but none have seen success.

The final decision to retire the Concorde is, of course, premised in market economics, which the Concorde, as a state-sponsored enterprise operated by state-sponsored airlines, had been fighting its entire life. The cost of operating the aircraft could not be recovered by fares; very few people had such a high value of time. The year of retirement, 2003, a period of fears about war (Iraq), terrorism (9/11), and plague (SARS virus), along with price volatility in oil, was an era of major troubles for the airlines. US Airways, United Airlines, and Air Canada all went through bankruptcy and American was tottering on the brink.[373] Rod Eddington, British Airways' chief executive, said: "This is the end of a fantastic era in world aviation but bringing forward Concorde's retirement is a prudent business decision at a time when we are having to make difficult decisions right across the airline."

In 2010, in the United States, personal expenditures for transportation ran $967 billion, $50 billion of which were for air transportation, about 5 percent.[374] The activity is relatively small but growing, total transportation expenditures rose about 20-fold in the last fifty years, but air transportation expenditures grew 75-fold (in current dollars). On expenditure grounds, it receives more than its share of attention. Perhaps that's because it is thought that its policy problems are more acute than elsewhere. However, we have difficulty seeing air transportation problems as much different and deeper than problems elsewhere. We think the reasons for relatively high problem visibility and debate include the status of users (the "jet set" are still image-makers, movers and shakers) and the feeling that air service is somehow more vital than other services. And recently the terrorism issue has made air transportation even more prominent.

20.2 Growth Pulses

> ... even considering the improvements possible ... the gas turbine could hardly be considered a feasible application to airplanes, mainly because of the difficulty in complying with the stringent weight requirements....
>
> —Gas Turbine Committee of the US National Academy of Sciences (1940)

Successful jet engine commercial aircraft came on the market around 1960. The B-707 was the first commercially successful jet aircraft, though the Comet and other aircraft had appeared in Europe earlier. Considerable improvement in service quality, faster, longer stage length, and a smoother ride resulted.

The longer runways required for jet aircraft created a problem for airports, as did the need for larger terminals as traffic swelled. To meet needs, in 1970 the federal government introduced a tax on tickets and developed an Airports and Airways Trust Fund, modeled on the Highway Trust Fund.

Figure 20.1 suggests two growth trajectories or pulses for the air transportation system—one with DC-3 piston type aircraft (the solid line), another with jet service (the line with triangle markers). We can think of other pulses—as the freight system shifted to diesel locomotives and as the nature of the automobile changed in the 1930s, for example.

We explore a logistic model for commercial aviation.[375] Our predicted jet era model (fit through 2010 data) has a very good fit (r-squared of 0.98) if we assume a final market saturation at 1,000,000,000 total domestic enplanements, and we reach within 10 percent of that value in 2024. This, of course, assumes no change in technology.

There were pulses in the equipment component of systems and pulses in regulation, which inspired a major reorganization to a hub-dominated system, but we can think of

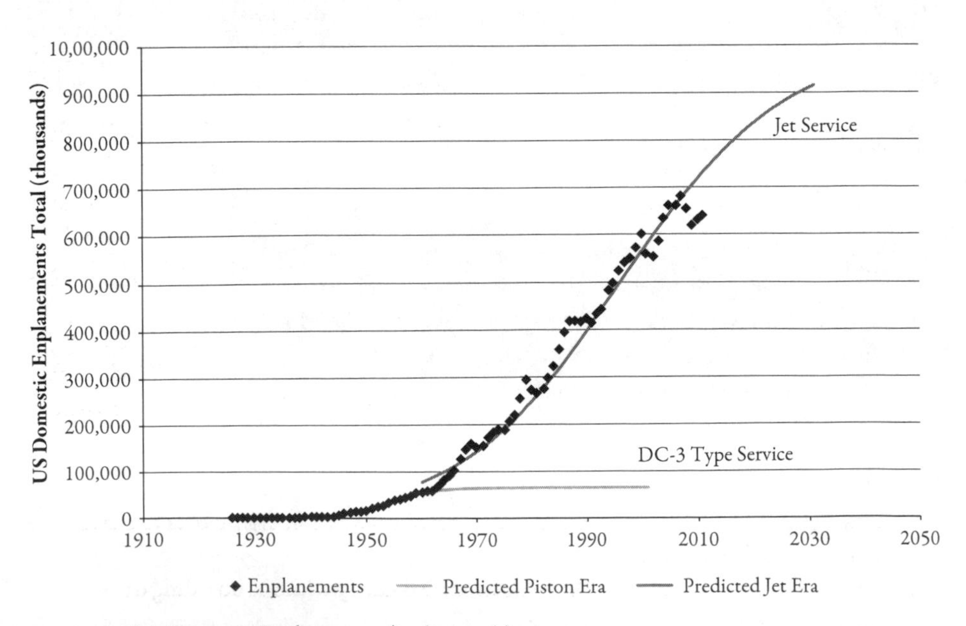

FIGURE 20.1 Domestic Enplanements (in thousands)

examples of pulses in other system components. A pulse in a component strains the incremental ways that components grow and develop in relation with each other. For instance, the diesel locomotive, the jet plane, and the interstate strained other components. Reactive public policy was undertaken—policy to ease strains.

Another policy question has to do with proactive efforts to achieve a pulse in air transportation. Recalling the strains set off by pulses, should public policy be addressed to more than one component of systems? Is seeking a pulse worthwhile when a system has almost run its growth trajectory?

20.3 Cabotage

Cabotage refers, in the maritime case, to coastal transport within a nation, and nations have long reserved that trade for ships flying their flags. Others may bring their ships to my ports, but they may not carry freight between my ports. That principle was extended to air transport in the early days. For instance, a foreign carrier may not offer service airport-to-airport within the United States. A carrier can, of course, say, fly from San Francisco to New York to Paris. But the carriage of passengers who just want San Francisco–New York service is prohibited, resulting in a waste of resources. The first six airline freedoms, described in the list below, were defined by the 1944 Chicago Convention of the Provisional International Civil Aviation Organization, which later became a UN unit. Later treaties defined specific relationships between countries.

The United States has established "Open Skies" agreements deregulating international air service, allowing US carriers to fly from any US airport to any EU airport (and vice versa), but still restricting domestic service to domestic carriers, with the European Union (2007), Australia (2008), Switzerland (2008), and Japan (2011). These international Open Skies agreements guarantee the First through Fourth Air Freedoms. The Fifth Freedom rights and beyond are not guaranteed by these arrangements, though some airlines have them through separate negotiations.

The EU has internal Open Skies arrangements, which allows up through Eighth Freedom (cabotage rights). Australia and New Zealand have a similar arrangement. A treaty between Brunei, Chile, New Zealand, and Singapore also allows cabotage. Generally, this is only permitted on a bilateral (or multilateral) basis (where two [or more] countries agree to allow each other's carriers to serve domestic markets). Airline code-sharing is another technique to avoid cabotage restrictions.

One policy issue is that of how the principle of cabotage will be resolved as the European Union evolves. Will the EU impose cabotage restrictions on non-EU airlines? If it does, that will hurt the US carriers. It might result in pressure to eliminate cabotage restrictions within the United States.

1. First Freedom (Transit Freedom): The right to fly and carry traffic over another country without landing.
2. Second Freedom: The right to land in another country without boarding or deplaning passengers (e.g., refueling).

3. Third Freedom: The right of an airline from one country to land in a different country and unload passengers from the airline's country.
4. Fourth Freedom: The right of an airline from one country to land in a different country and load passengers to the airline's country.
5. Fifth Freedom (or Beyond Rights): The right of an airline from one country to land in a second country, to then pick up passengers and fly on to a third country where the passengers alight.
    - Intermediate Fifth Freedom: The right to carry passengers from the home country to the second country (implied by third freedom).
    - Beyond Fifth Freedom: The right to carry passengers from the second country to the third country.
6. Sixth Freedom: The right to carry traffic from one state through the home country to a third country.
7. Seventh Freedom: The right to carry traffic from one state to another state without going through the home country.
8. Eighth Freedom (Cabotage): The carriage of air traffic that originates and terminates within the boundaries of a given country by an air carrier of another country.

20.4 Federal Express

> The concept is interesting and well-formed, but in order to earn better than a "C," the idea must be feasible.
>
> —A Yale University management professor
> in response to student Fred Smith's (1965)
> paper proposing reliable overnight delivery service.

This chapter has mostly explored the passenger side of aviation; freight has tended to stay on the surface. Aside from airmail, the aviation contribution to civilian freight transport until the 1970s was quite small. That was the environment faced, when, as noted in the opening quote of this section, Fred Smith, the founder of Federal Express, had trouble convincing his business school professor of the viability of his enterprise. After all, it was both technically difficult to coordinate overnight delivery at an affordable price (if it weren't, wouldn't it have already existed?), and there was uncertain demand, since we had had instantaneous communications for over a century. What needs to arrive overnight that can't be communicated by telephone?

The company began operations in Memphis, Tennessee, April 17, 1973, delivering 186 packages to 25 US cities using 14 Dessault Falcon aircraft. The company grew quickly, especially after the 1977 deregulation of the airfreight market.

The innovations of Federal Express were several. First was the use of a single national sorting facility (in Memphis) to assure economies of scale, guaranteeing that there would be at least one connection from every important local distribution center to every other. They used this hub and spoke architecture to minimize total flights (e.g., the overnight guaranteed shipments can be satisfied with one round trip from every city to Memphis per day, aside from a few of the larger markets. Smaller markets wouldn't necessarily require non-stop service). As a proposition, Federal Express would not succeed unless it began nationally. By its very nature as a last-minute shipping company, the shippers don't have too much time to figure out which company can serve which city overnight. Dropboxes, which were deployed in 1975, eliminated the need for carriers to pick up shipments from each individual (or requiring individuals to travel far to find a pickup site), lowering costs. The use of a Memphis hub led to apocryphal, though likely true, stories that to ship something between offices in the same building in New York was faster through Memphis than the company mailroom.

The company further took advantage of time zones to reuse planes. A plane going east at 4 or 5 AM, could return to Memphis, reload, and go west in time to do another delivery. This leads to the possibility of same-day delivery in these markets, as there is no reason to have a completely empty plane.

An innovation that came soon thereafter was the use of barcoding and scanning technology enabling the tracking of individual packages, minimizing losses because someone was responsible for a particular shipment. With the Internet, this reassured customers that their shipments were safely making progress.

In Chapter 10 we discussed the magic bullet, the positive feedback system that allows new technologies to take off. Perhaps the greatest magic bullet success was Federal Express.[376] Fred Smith, who ranks along with Juan Trippe and innovators described in Chapter 4, its developer, had a market niche in mind, and he fought the problem of getting scale to support quality service—a fight that lasted for several years. The technology tailored to scale and market was mostly soft technology—hubbing, efficient sorting, and so on. Privately owned Federal Express has had its own disaster (discussed below), suggesting that sector of the economy (private vs. public) is not the only factor in leading to the grail quest of the magic bullet.

A point we made in Chapter 10 is that if the wrong path is selected, a pathway disaster may result. An interesting thing about the Federal Express case is that there has been some action by managers that suggest they may sense they are on the wrong long-run path. In the 1980s, Federal Express invested in electronic document transfer (Zapmail) because of its concern about fax services, but their investment lost $294 million.[377] Federal Express, of course, remains a highly successful enterprise that simply misjudged or mistimed the market (overestimating the number of people wanting to send faxes to people without fax machines, or underestimating how quickly fax machines would be deployed).

While as a company, Federal Express has had some stumbles, like Zapmail, it has had in forty years become an important element in the global transportation/communications system, to the point that "to fedex" is a verb in common speech, if not yet the dictionary. The company has expanded internationally (though not with necessarily the same overnight guarantee), and into ground transportation.

Federal Express is given as an example; it has several worthy competitors. United Parcel Service began as a ground-based package company that moved into the overnight market following FedEx. Both have seen markets greatly expand with catalogs, and then Internet commerce, as individuals substitute shipping for shopping.

Internationally, DHL, now a division of the privatized German Post Office, Deutsche Post, dominates. DHL was originally founded in 1969 to deliver documents between

Hawaii and California. By 2009, the company abandoned domestic pickup and delivery in the US market.

20.5 Networked Organization

Airlines, even more than other transportation industries, have established highly networked organizations. Two major networks are alliances of the airlines themselves, and airline reservation systems.

20.5.1 ALLIANCES

Like other transportation industries, airlines have formed alliances to allow better international inter-networking, while still keeping individual airlines independent and nationally based. The alliances allow code sharing, in which a single airplane carries more than one flight number simultaneously, so you might be a passenger on a Delta booking, your seat mate might be ticketed through KLM. Loyalty (frequent flyer) programs are shared or linked, so frequent flyer miles accrued on one alliance partner can be spent on another. Alliance members will often have their gates staffed by partner airlines, and be worked on by partner ground crews, rather than maintaining such services themselves globally. Partial cross-ownership arrangements sometimes help glue the alliances internally and align interests.

That airlines don't just consolidate internationally, which might seem the simpler solution, is driven by policy, especially cabotage policy, and domestic regulation aimed at preserving jobs and flag carriers. Often these objectives are marketed under the guise of security, rather than improving customer service. Further, slot restrictions at major international hubs (e.g., London Heathrow) have been used to constrain competition, entrenching incumbents, thus driving up profits.

The earliest alliance began in the 1930s between Air New Zealand and its then parent Pan Am for Latin American flights. In 1989 Northwest and KLM began code sharing, while KLM acquired 25 percent of Northwest. This led to an *open skies* agreement, allowing freedom for airlines to provide services between the Netherlands and the United States American Airlines and British Airways, as well as United and Lufthansa, established similar partnerships.

There are now three major international alliances, which grew out of earlier bilateral (trans-Atlantic) relationships. Star Alliance (1997) is led by United Airlines, SAS, Singapore Airlines, and Lufthansa. Skyteam (2000) is led by Delta Air Lines, Aeromexico, Korean Air, and Air France/KLM. Oneworld (1999) is led by American Airlines, British Airways, and Cathay Pacific. Membership is not entirely stable, and as airlines merge or are acquired, they may change alliances.

20.5.2 RESERVATION SYSTEMS

Airline reservation systems, like urban transportation forecasting models (see Chapter 24) were an early application of information technology to the problems of transportation. The earliest system, *Reservisor* (1946), for American Airlines, used an electromechanical machine built by Teleregister. Later systems, under the same name, adopted steadily more advanced technology.

While originally owned by the airlines themselves, reservation systems were shared with hotel, car rental, and other airlines, and access was given to travel agents. There arose antitrust questions when they were used by travel agencies. Would a system owned by Airline A be fair to Airline B? The answer, of course, is "no."

As with airline mergers, airline reservation systems have formed and been combined into fewer and fewer entities in order to achieve various economies of scale. *Sabre* (Semi-Automated Business Research Environment) arose in 1964 from American Airlines and today provides the Travelocity service.

The *Apollo* system (formed 1971 by United Airlines) merged in 1992 with its European competitor *Galileo* (formed in 1987 by British Airways, KLM Royal Dutch Airlines, Alitalia, Swissair, Austrian Airlines, Olympic, Sabena, Air Portugal, and Aer Lingus) and is now owned by Travelport, which owns a large share of Orbitz. This organization merged in 2006 with *Worldspan* (formed in 1990 by Delta, Northwest, and TWA as a combination of *TWA PARS* [1976] and *DATAS*). As of this writing, the same organization still maintains multiple computer reservation systems (CRS).

Amadeus was a European competitor founded in 1987, with the aim of neutrality, by several airlines: Air France, Iberia, Lufthansa, and SAS. Asian airlines founded the *Abacus* system. There are others.[378] Notably, the mergers in the reservation system business do not align well with airline alliances.

20.6 Deregulation

Alfred Kahn (1917–2010), a Cornell economist, was chair of the Civil Aeronautics Board during the Carter administration. In his role, he pushed through deregulation of airfares and services, leading to a radical restructuring of what had been a comfortable industry that

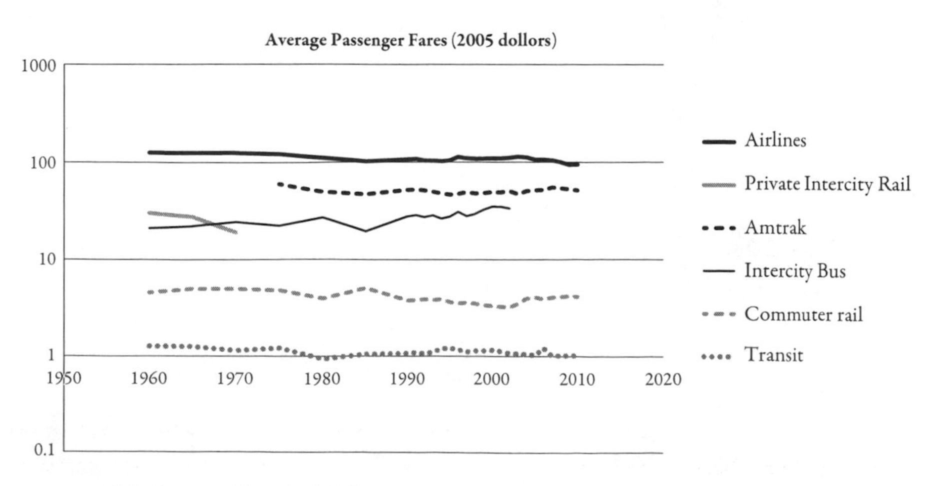

FIGURE 20.2 Passenger Fares by Modes

had captured its regulators to the detriment of consumers. The successful deregulation of airfares led to subsequent deregulation in energy, trucking, railroads, telecommunications, and other industries. While the airlines have been deregulated in terms of prices they can charge and services they can provide (within bounds), there are other components of

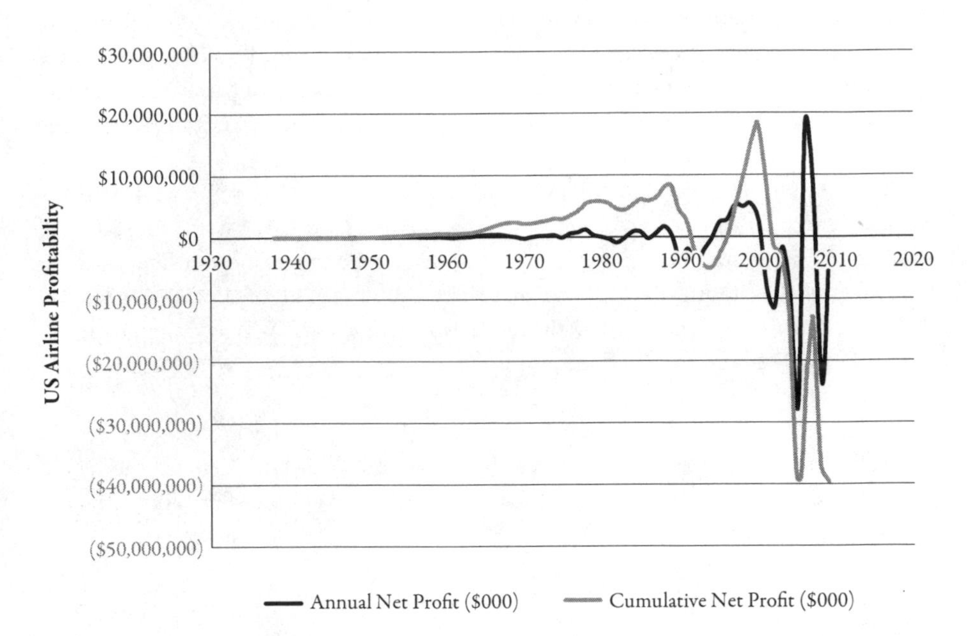

FIGURE 20.3 US Airline Profits (Losses)

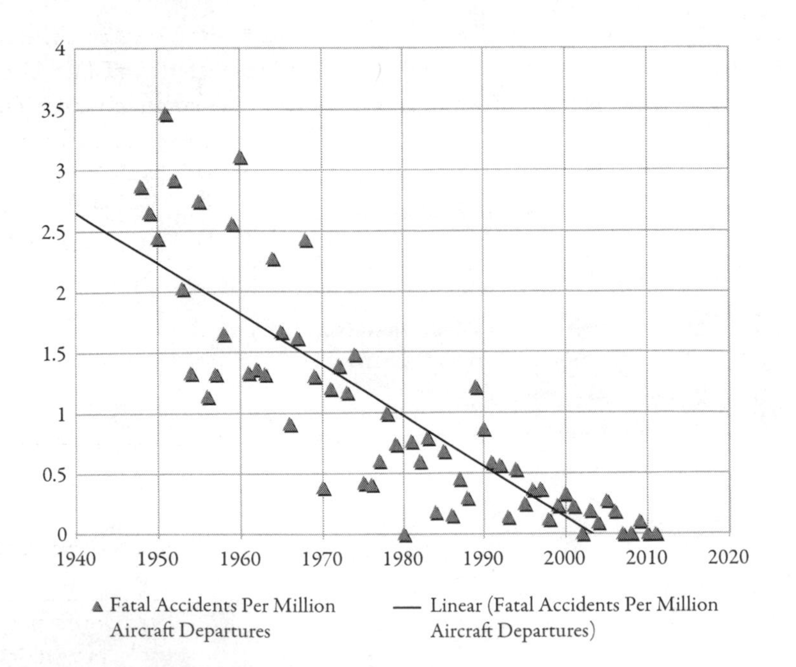

FIGURE 20.4 US Airline Fatalities and Fatality Rate

the system that have not been deregulated. Safety standards notably remain. So it's not proper to say that air transportation as a whole has been deregulated. Deregulation was to unleash competition and improve efficiency and service quality. To what extent can that be done through partial deregulation? Although many statements have been made about large consumer savings as a result of deregulation, consultation of data gives fuzzy results. Clearly, the increasing volatility in terms of firm profits (see Figure 20.3) is an indicator that the industry itself is not making continuously huge profits, and may indicate underlying instabilities in the market structure. Safety has continued to improve, to the point there are now years with no major commercial airliner crashes (see Figure 20.4). While number of deaths is not quite zero, that goal seems feasible.

20.7 Security Theater

> The point of terrorism is to cause terror, sometimes to further a political goal and sometimes out of sheer hatred. The people terrorists kill are not the targets; they are collateral damage. And blowing up planes, trains, markets or buses is not the goal; those are just tactics.
>
> The real targets of terrorism are the rest of us: the billions of us who are not killed but are terrorized because of the killing. The real point of terrorism is not the act itself, but our reaction to the act.
>
> And we're doing exactly what the terrorists want.
>
> —Bruce Schneier (2006)[379]

After 9/11, the US federal government established a Transportation Security Administration (TSA) to oversee airport security, which had been the responsibility of the airports, and generally contracted out to private firms. TSA was initially located in USDOT, but was subsequently moved to the newly created US Department of Homeland Security (DHS).

The politics of security are difficult. If you are in favor of security, you must be in favor of more spending on security, or on anything that will "keep us safe." If politicians or bureaucrats oppose a proposed security measure and something happens, they will be blamed, as summarized in Table 20.1.[380] Security ratchets up quickly. Ratcheting down can only really be by attrition.

TABLE 20.1

Security Politcs

		Terrorism	
		Yes	No
Security Supported	Yes	Blameless, we must go further	Measures have worked
	No	Blamed for any attack	Fades from memory, can never be proven right

Though the increase in security spending post–9/11 exceeds $1 trillion over ten years, DHS has systematically failed to study the costs and benefits of proposed security measures, most notably radiation scanners (Mueller and Stewart 2011). Other Western countries have had similar spending increases. The problem is not just the total, which is large, and might be spent in other sectors. The problem is the allocation within the security sphere. If you have $1 trillion to spend, what is the best way to do that to maximize security? The idea of opportunity cost rises again and again, and is never properly dealt with.

Rather than assessing both the probability of an outcome and its cost if it occurs, the agency has dealt with risk qualitatively, imagining worst-case scenarios and engaging in what Cass Sunstein calls "probability neglect."[381]

The artifice of travelers removing shoes and belts, unpacking suitcases, pouring out liquids, while agents frisk grandmas and children, and so on at security checkpoints has been called "security theater" because it aims to give travelers the impression that something is being done to improve security, when there is no evidence any of this has made a difference. The vector of attack from 9/11, hijacking planes and crashing them into buildings, has been unrepeated not because TSA has ensured safety. Secured cockpit doors and passengers who will not go calmly to their deaths has prevented this from occurring. Other attack vectors that remain (and with a little bit of imagination, you can think of them) are receiving less protection because so much money is spent in response to the previous attack.

Though the TSA monitors security, the lines leading up to security are managed by the airport or airlines themselves, and give priority to certain passengers (e.g., first class), though they pay exactly the same amount of security tax as everyone else. This has some analogy with HOT lanes (see Section 22.4.6); the key difference is that the car in the HOT lane pays more money to go faster, while in this case, first class passengers only paid more money for a better seat on the plane.

Security is not just an issue on airplanes; inter-city buses and high-speed rail sometimes have security measures, though rarely as stringent. Intra-city public transit generally does not, because it would quickly become unworkable. As a consequence, these other modes are much more vulnerable, as shown in the Madrid attacks (March 11, 2004), London attacks (July 7, 2005), or the Tokyo Sarin gas attack (March 20, 1995).

Security is the enemy of efficiency. Just as with safety, we want perfect security. That goal is unattainable, and security providers should rationally trade off between value of time and value of life. It's icky, because people like to think of life as priceless, yet no one acts that way at the margins. People take risks to save time, lower costs, and so on. Accepting that, rather than hiding from it, will move the transportation system toward a less irrational stage on which security theater is enacted.

20.8 Discussion

Aviation's first practical (civilian) use was airmail communications, followed by the substitution of air travel for ship and rail. Recreational flying and some business uses also evolved early. Service today provides much more than substitution, for many aspects of activities have been organized or reorganized and enabled by today's service. Aviation enables new activities that were previously impossible or impractical.

In the main, the air system of the early twenty first century shares properties of systems previously examined. Its market is not as fully saturated as other modes (international travel is still rising, even if domestic US travel may have plateaued) but the predominant institutional and technological structure seems well established. Less tightly tied to land routes than other systems, rationalization has somewhat different terms than is the case in other systems; it involves sometimes sweeping changes in the geographic structure of services along with changes in the firms. (Gate availability and international negotiation requirements do limit the range of geographic actions available.)

The commercial sector is as safe as any mode has ever been, and is improving in private aviation.

The airport problem has historic path dependence aspects. Once airports got going, a variety of forces "locked in" their behaviors, not the least of which was their association with particular cities (big, politically powerful), making location adjustments almost impossible. New airports are difficult to establish both because of opposition from incumbent airports and airlines not wanting the competition, and neighbors not wanting the externalities.

Deregulation of routes and fares, but not safety or air traffic control, came along during the Carter administration. Deregulation has had positive and negative effects, and there is discussion of re-regulation. As is the case in all systems, there is the issue of the amount of deregulation achieved. Many airport managers are implementing versions of deregulation, especially by seeking markets and efficiency through adjustments in landing fees, and airlines are seeking equipment and operations protocols tailored to niche markets. For instance, access to services through fractional ownership/rental of business aircraft (executive jets) is enlarging business aircraft services.

Ignoring that passengers feel like cattle being driven to slaughter, among the many other indignities suffered while at the airport, one must applaud the airlines; we have no objections to their search for efficiency. Fuel efficiency is a major consideration in aircraft buys. Many airlines are striving to reduce the costs of labor inputs, using means over and above specializing (standardizing) labor to tasks at efficient scale. At this time in the growth dynamic, we would expect most of the productivity gains from process and product technology to have been captured; emphasis would be on market channeling. That seems to be the case. Airports, equipment makers, and firms are competing in market channeling arenas.

Hubbing and careful fitting of aircraft to route segments, careful scheduling to maximize the hours per day an aircraft is in profitable use, and the sales of within-cabin seats to market segments (coach, business, and first classes) are examples of the ways in which air transportation firms attempt to squeeze profitability from a mature technology. The firms are grasping scale economy and market segmentation opportunities.

Emerging areas in aviation include:

- UAVs (unmanned aerial vehicles). While obvious applications, like spying or combat, have already been employed, there are many potential new markets enabled by this rapidly growing technology (delivery, information probes).
- Supersonic or hyper-sonic transport. Even if not supersonic, the development of a new aircraft requires a multi-billion-dollar front-end investment, a sum that extends

the reach of a single large company such as Boeing or a consortium like Airbus. A variety of institutional, financial, and technological issues appear.

- Commercialization of space. Private firms are undertaking rocket launches and placing satellites in orbit, and may eventually profitably offer humans the opportunity for space travel. These are still early, early days.
- Flying cars. The Piper Cub and some other aircraft developed in the 1930s suggested an opening path for a "plane for anyone" society, planes that cost about as much as an automobile and are as easy to operate. If not a plane, then a helicopter or autogyro. That dream held into the 1950s, but the path never opened. The technological advances of self-driving vehicles (SDV) and UAVs may converge to enable pilotless planes.
- Alternative fuels. Unlike surface transportation, electrification is not likely to be feasible for aircraft anytime soon. However, research is being undertaken in the use of biofuels to replace jet fuel, reducing environmental impact.

Might technological change in the form of the healthy interest in experimental aircraft combine with aircraft control technologies and new materials to yield the long dreamed of airplane in every garage? Like other technological inevitabilities (picture phones, domestic robots) surely we will obtain this vision, but it may take longer than hoped, and may not turn out as expected.

Our Chicago to Texas piggybacks run like the wind, covering 900 miles [1,450 km] in less than 36 hours.

—SCHORDOCK 1985

Shippers complain that a load of freight can make its way from Los Angeles to Chicago in 48 hours, then take 30 hours to travel across the city. A recent trainload of sulfur took some 27 hours to pass through Chicago—an average speed of 1.13 miles per hour [1.8 km/h], or about a quarter the pace of many electric wheelchairs.

—FREIGHT TRAIN LATE? BLAME CHICAGO[382]

21 Railroads Rationalized

21.1 Rationalization: Nationalization Style

BY 1917, US railroads, the country's largest industry, operated over 400,000 km (240,000 miles) of track and employed 1.8 million people. While the war was raging in Europe, war traffic was congesting America's railroads, especially at East Coast ports, even before the American Expeditionary Force went "over there." The American Railway Association tried to coordinate its member railroads to avoid car shortages. The Interstate Commerce Commission (ICC) was granted greater powers to ensure efficiency and that priority cargo received priority treatment. Yet the progress made by these organizations was insufficient.

On December 28, the federal government's director general of railroads, William McAdoo, took control of the largest private railroads and placed them under the control of the United States Railroad Administration (USRA) —a creature of Congress.[383]

Nationalization had more form than substance to it. Railroad employees were given military titles, and a few inter-railroad arrangements were forced on the properties. Some standardized locomotive designs were developed and engines were constructed. Labor arrangements were imposed that the railroads might not have agreed to under other circumstances. Railroads were granted compensation equal to their prewar profits. The shipping costs per ton-km rose, however, in part due to a rise in wages.

While the railroad regulatory regime clearly collapsed with the onset of World War I, the regulation initiatives of the late 1800s and the early 1900s worked for their times. They managed the gross inequities that worried the public. Government served as a referee for railroad rate-making disputes, and government had begun to assert its power over who gets the location-transportation rents (via maximum rate prescriptions). The list of problems that regulation did not manage remained long, of course.

One of these was the vast difference in the economic fortunes of the different properties. This was not just a matter of concern to the managers, stockholders, and bondholders of the railroads that were not doing well. Shippers had interest in strong railroads, as did regional groups. They were worried about good, stable service.

This chapter treats the notion of rationalization as attempted in several ways. The goal was efficiency; the means was often consolidation. But the shape suggested depended on who was suggesting.

21.2 Rationalization: Congressional Style

The Transportation Act of 1920 returned the railroads to private ownership. The debate in Congress went beyond the transfer of ownership to the problem of a viable system. It extended beyond the railroads to a general call for transportation improvements. It sought to improve the system rather than merely controlling its ills. The Act ordered waterway improvements and began what later became a remarkable renewal of inland waterway transport. The Highway Act of 1921 reflected the debate.

The Transportation Act of 1920 directed the ICC to prepare a plan for the consolidation of all common carrier line-haul railroads. It limited the number of systems to a few large systems of similar economic strength and overlapping turfs. Unlike the consolidation proposals that emerged in Britain and elsewhere at about this time, competition was to be preserved; in 1921 the British established four regional monopolies—Great Western Railway (GWR), London, Midland and Scottish Railway (LMS), London and North Eastern Railway (LNER), Southern Railway (SR)—which were subsequently nationalized in 1947.

The ICC engaged railroad expert and Harvard professor W. Z. Ripley to produce a plan. After some revision by the ICC, a 21-system plan emerged in August 1921, and the ICC began to hold hearings on it. The 1920 Act had given the ICC authority to regulate construction and abandonment, and the agency was supposed to adhere to the plan. That planning was in the "great men announcing results" style common at the time. However, Ripley's plan were not a win-win plan, in the sense that all railroads would be better off if the plan were implemented.

The hearings on the 1921 proposed plan did not go smoothly, and a final plan was not adopted until 1929—just in time for the Great Depression. It too, had 21 systems, but differed in that it excluded many small railroads and had different arrangements for the large ones. It never really had much effect, and was abolished by the Transportation Act of 1940.

The term "rationalization" is applied to the attempt of the 1920 Act. Rationalization has lots of different meanings. Perhaps, the sense of it is to change structure so as to get some desired behavior. The 1920 Act focused on spatial structure.

Another sense of rationalization emerged in the activities of the federal Coordinator of Transportation, Joseph Eastman. The Emergency Transportation Act of 1933 established his office. Eastman was to study rail transportation, identify economies, and recommend means for achieving them. Eastman's findings were to be submitted to regional committees for implementation (voluntary), and the committees were made of railroad representatives. Eastman quickly realized that there was little promise in grand plans of the Ripley-type, imposed from the top. So he put his energy into small improvements: in the main, consolidation of yards, terminals, or parallel routes here and there. He advocated pooling of cars, container-like service, and competition for the railroad-controlled Railway Express Agency (REA).

As expected from this voluntary structure, nothing happened. Even so, Eastman worked on the possibilities for consolidation of yards and terminals for which there were many options and the economies were quite clear. Nothing happened, yet Eastman felt so strongly about his proposals that he discussed issuing an order for implementation. (He had the power, and this would make a strong case if challenged in the courts.) Still nothing happened. It is said that he was blocked by rail labor, but railroads that would have been losers must have had a say in that.

He was interested in consolidation of parallel routes, but made no proposals. This option made more sense later, when new signal systems came along. Eastman proposed pooling of cars, another idea ahead of its time. Finally, Eastman had very interesting merchandise traffic and intermodal ideas. He proposed the use of containers, and, where the traffic warranted it, unit (one-car) trains. He wanted more competition in merchandise service and the addition of servers to compete with the REA.

When the Office of the Federal Coordinator was abolished in 1936, nothing had been accomplished. Perhaps this was because Eastman was an outsider. After all, by that time the ICC, the railroads, and interested congressional committees occupied the regulatory turf.

He accomplished nothing; yet many of his ideas have subsequently been adopted. The introduction of the diesel locomotive, long trains, centralized traffic control, and the large "hump" sorting yard yielded another round of operations planning, beginning in the late 1940s and continuing today. By and large, this work was done by the individual properties. Its results are clear from the ways in which operations have changed. It is also clear that some properties succeeded better than others.

21.3 Rationalization: Commission Style

The idea of Ripley-style railroad consolidation and the idea for Eastman's office came from Congress. We turn now to the Interstate Commerce Commission (ICC). What was their idea of rationalization? The answer to that question is hard to ferret.

The Interstate Commerce Act required the establishment of through routes as well as joint freight classifications and joint rates. The through route notion argues for moving freight in a spatially rational way. Why didn't the railroads do that? Lots of alternative routes between places were available. Let's move freight as far as we can on our own route; then, let's give it to a friend who gives us freight.

The ICC did three things. It used the shortest route to calculate cost and thus the tariff that would be charged. This is in the interest of efficiency. The tariff that can be charged encourages the railroads to move goods in a cost-effective way. Second, the ICC gave the shipper the right to specify the route over which freight would be moved. That gave the shipper control over the available service. The third thing was a bit more complicated; it has to do with the way friends are treated in the sense of the paragraph above. Early, it prohibited a railroad from discriminating among connecting lines (in those cases where the shipper does not specify routing). The question of how to identify discrimination was answered mainly through reference to historical patterns of handing off traffic.

Next is the special case of when railroads merge. A variety of things were tried. The conditions imposed on the Detroit, Toledo, and Ironton (DT&I) Railroad required that traffic would flow after the merger just like it did before. Those conditions were made standard. That is some sort of comment on what the ICC took rationalization to be. Efficiency must be an objective of a merger, and that surely has to with traffic management and concentration. Yet the ICC gave more value to stability—preserving historical relationships.

The Act of 1920 gave the ICC control over construction and abandonment, as mentioned. By that time, the network was pretty much in place and other modes had begun to claim traffic. Although the ICC looked for the convenience and necessity of new services, in the main its experience has been with abandonment.

Cost reductions result from abandoning routes (maintaining a route to be serviceable, even if unused, costs some money; nature takes its toll on track and equipment). Did the ICC operationalize rationalization as the capturing of those cost savings? The answer is clearly "yes." Although one hears horror stories about the difficulties in getting permission to abandon track, the empirical record says that the ICC allowed just about all that were applied for, some big ones and numerous small ones. Some question that record. First, it costs money and takes management time to abandon routes, so we do not know about routes not abandoned because they were too costly to pursue. Second, following the Transportation Act of 1940, the ICC was required to give "weight to the interests of carrier employees affected"; some conditions were worked out, and they are rather costly. These may have held down applications for mergers.

21.4 Rationalization: Corporate Style

Problems begging the rationalization of firms and routes remained and they began to receive attention after World War II. The first phase ran until about 1970, and its focus was on the consolidation of regional services, the Ripley thrust reborn. Initiated by the rail properties, railroads acquired or merged with competing properties in their service areas. The Norfolk and Western (N&W), for example, acquired the Virginian, essentially a parallel coal hauler in the N&W's service area. In this case, some redundant line was abandoned, and the N&W obtained a lower grade east-west line for heavy haul coal movements.

Ripley wrote as if he were a czar who could stand back and manipulate the entire net. That was not a real option, however. Merger control was an option. Mergers were allowed when they were proposed by two or more railroads. The conditions imposed by the ICC varied over time. By 1940, they had to with (1) the effect of the proposed merger on adequate transportation service, (2) the public interest effect of the inclusion or, more often, non-inclusion of railroads in the area served, (3) costs, and (4) the interest of railroad employees.

Item one was considered because mergers usually include abandonment items. Item two was almost always a question, and costs should not swamp benefits, item three. It is very difficult to judge the outcome of rationalization by merger policy. First, there was a long history of railroad mergers and consolidations.[384] Did policies and variations on them make any difference? The process was expensive and lengthy.

Often mergers had so many conditions on them that the economies proposed could never be achieved. Conditions had to do with the ability of the merger partners to alter traffic patterns. They may require that "landlocked" railroads be given track running-rights. The data we have seen suggest that merged railroads as a class have done a little better than they would have standing alone, but not as well as the merger proposal claimed.

A large number of system changes emerged, with the largest during the ICC-era being the merger of the New York Central and Pennsylvania systems, yielding the Penn Central. Even so, there was little payoff from the Penn Central and other mergers; the expected efficiencies were not realized. There were several reasons. First, the ICC often insisted that weak roads be included in mergers or acquisitions. In the Penn Central case, the New Haven and other weak railroads were included in the new system. Second, where there was redundant property and service as a result of mergers, the ICC was slow to allow abandonment. Third, the ICC insisted that previous traffic flow arrangements be honored. That is, the railroads had little control of the diversion of traffic to the more efficient routes available after mergers. Fourth, the merger took four years to complete.

This era ended with the financial failure of the Penn Central, resulting in government take over of the properties and the creation of Conrail.[385] The ills facing railroads included lack of capital (low profits, if any), excess physical plant (discussed in the case of low-density branch lines), and inability to achieve efficiency through consolidation of traffic flows. The solutions offered included more pricing freedom for railroads and a more hands-off role for the ICC in approving mergers and abandonment.

As a result of changes in the view of the problem and its solution, Congress pressed for reduced regulation (the 1980 Staggers Act) and there was another round of network rationalization. There have been many end-to-end mergers of properties. Inaction has resulted in the bankruptcy of firms (such as the Rock Island and Milwaukee Road) and the sale of some of their assets to other rail properties. The Illinois Central Gulf provides an example of the spin-off of low traffic routes (the Illinois Central was ultimately acquired by CN, the Canadian National Railway, in 1999). It has trimmed to a high-density north-south route by selling lines (serving collector distributor and/or east-west services) to small operators.

Mergers and acquisitions have redrawn the network map. Presently, a handful of major systems handle the vast majority of ton-km of freight moved. That's not to say that the number of railroads has decreased (though it has over the longer term); rather, the size distribution has changed. Class I railroads are the largest in the United States, having annual operating revenues of at least $250 million. As Figure 21.1 shows, there are fewer and fewer Class I railroads, though the total number has been relatively steady for the past few decades. Eleven North American railroads are Class I; eight operate in the United States. While mergers and acquisitions decreased the number of large roads, the spin-off of short lines has increased the number of small roads.[386]

21.5 Rationalization: Conrail/Amtrak Style

The railroads emerged from World War II in a strong cash position. This, plus the adoption of the diesel engine for line haul service and some labor productivity improvements, tilted them toward a "healthy" financial position for some years. But strains emerged.

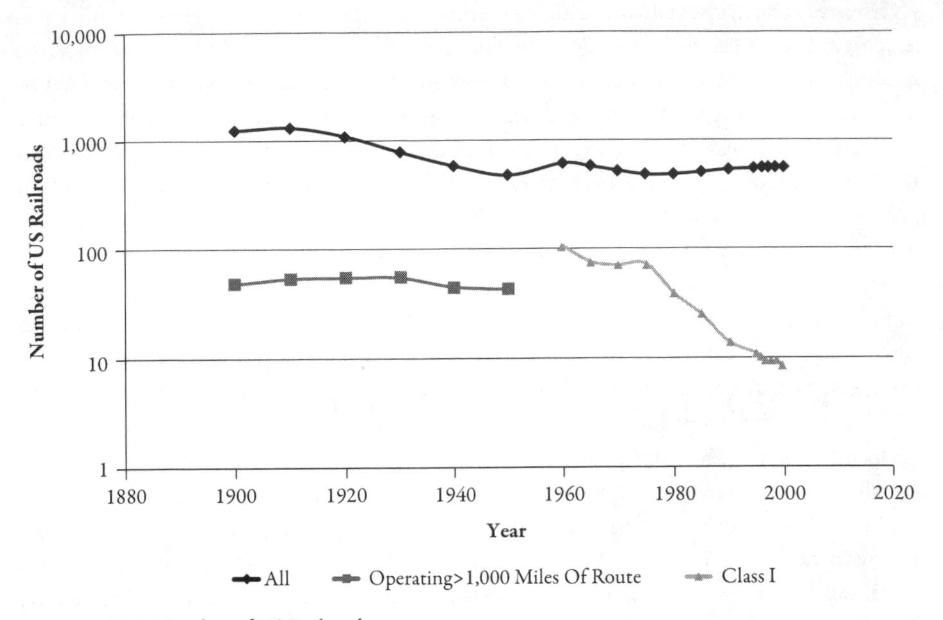

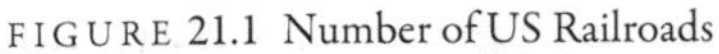
FIGURE 21.1 Number of US Railroads

Truck, waterway, and pipeline competition deepened. Passenger traffic eroded sharply with competition from air and inter-city bus and auto service. The beginnings of the Rust Belt to Sun Belt trend began to be felt. Regulation's slow price adjustments for inflation hurt. Finally, the ability of railroads to self-rationalize via mergers continued to be sharply constrained.

Crises and reactions to them first came bit-by-bit. The Rail Passenger Service Act of 1970 created Amtrak and relieved the railroads of their obligations for passenger service. (The railroads had already been gradually reducing service at the slow pace allowed by regulation.) Though of national scope, this Act was somewhat of a special law—one-half or more of the service abandoned was Pennsylvania Railroad service. The railroads were not all in agreement, and some railroads, such as the Santa Fe and the Southern, did not surrender service to Amtrak at the time. Next was the crisis of the bankrupt Northeastern railroad companies, and the Regional Rail Reorganization Act of 1973 (3R) established the Consolidated Rail Corporation (Conrail). It provided grants for interim operations and for loans for rehabilitation of some properties elsewhere.

The stopgap 3R Act was followed in 1976 by the Railroad Revitalization and Regulatory Reform (4R) Act. The 4R Act authorized $2.1 billion for Conrail. In addition, the Act had a Christmas tree character, something for everyone. Amtrak received money for its Northeast corridor, and the act assisted the railroads with abandonment of routes. Increased regulatory flexibility was called for and the ICC was instructed to give more consideration of the financial health of railroads in its rulings. A short timetable for merger decisions was prescribed. The 4R Act, in our view, was a response to an emergency, plus some side payments to gain the support of all players. The connection between the side payments and the problems was loose, though the connection was in the right direction.[387]

21.6 Rationalization: Iowa Style

The mainline mileage of US railroads peaked between 1920 and 1930 at about 400,000 km (240,000 miles); today there are about 160,000 km (100,000 miles). A main reason for the reduction is that a lot of redundant mileage was constructed. (Of course, trucking made tracks redundant as well, since it was no longer necessary to have tracks so close to the customer). Some long routes were constructed in areas where the market was already well served (e.g., the Milwaukee Road: Chicago to Seattle). At a smaller scale, many such lines were built in the Midwest. One force for over-construction was the money that construction contractors could make. Examples illustrate the scope of abandonments and other forces at work.

Iowa had about 16,000 km (10,000 miles) of route at the turn of the century and now has about 9,300 km (5,800 miles). The temporal pace of line abandonment is shown in Figure 21.2.

Much of Iowa is highly productive farmland, and much of the agricultural land was near a railroad. Wagons were used to move the farm products to the railhead, and the rail network then served as a collector system to move the products to yards and markets. Within state, movements must have been evenly distributed across links of the network. That is in sharp contrast to today. About 12 percent of the route mileage handles 50 percent of the gross ton miles (GTMs) moved in the state (1981 datum). That, however, overstates the within-state concentration case because about 62 percent of the GTMs are overhead traffic (traffic passing through the state) and overhead traffic was concentrated in the early days, though not as much as it is now.

The reduction in route mileage is not unique to the United States. By 1960–1970, British railways route mileage had been reduced by about 35 percent, Swedish state railways by about 20 percent, and the French national railways by about 8 percent.

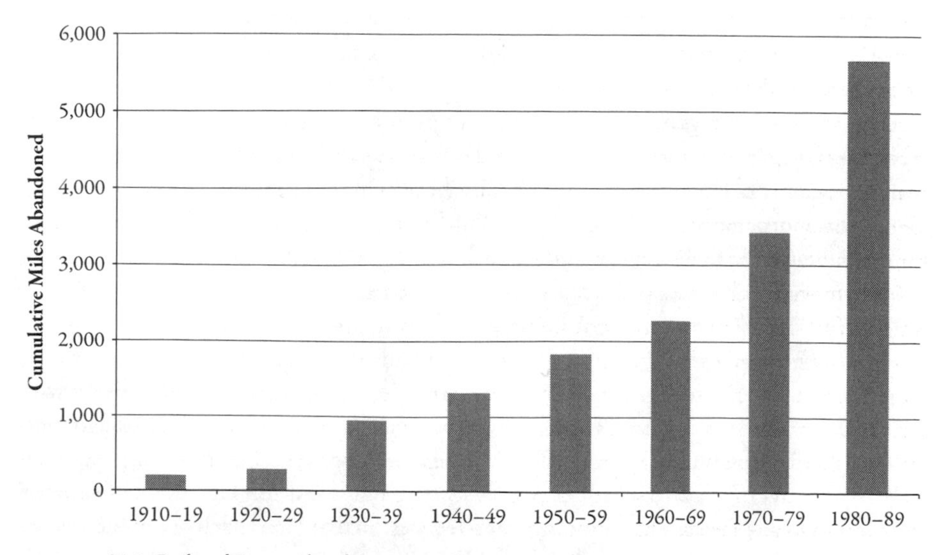

FIGURE 21.2 Railroad Route Abandonments in Iowa, 1910–1989

Returning to downsizing and the Iowa example, what is the cause? The data supplied and the map are not very helpful in answering that question because they fail to relate downsizing to anything other than time.

Things have changed a lot. The economy of Iowa has grown; demand has increased. Farm population has declined sharply. (More people lived on farms than in cities during the early decades of the period we are examining.) This has concentrated the demand for final products in medium size and large cities. The mechanization of agriculture and improvements in seeds and fertilizers have yielded this; more product with less people results.

The mechanization of the farm-to-market or farm-to-railroad interface was a major factor in changing the demand for low-traffic rail routes, of course. (That is a long way to say that trucks came along and substituted for horse-drawn wagons.) As roads and trucks were improved, truck service became competitive with rail over a range of about 500 miles (longer in the case of products such as grain), and collector/distributor rail lines became redundant.

The auto, bus, and airplane proved competitive to rail passenger service, and demand for passenger service no longer supported low-density routes.

That's the textbook explanation for the Iowa situation, and the same general explanation is used to explain most cases. However, passenger service has held on better in Europe than it has in the United States largely because of government policies and the closer spacing of large cities in Europe. Some exceptions to the textbook explanation exist, mainly cases where routes were redundant when they were built.

21.7 Rationalization: Community Style

Activities initiated in the 1970s included the accelerated abandonment of light density track and the elimination of failed railways, as well as end-to-end mergers and acquisitions. The railroads began to identify lines they would abandon and those they might abandon in the future. The states were very concerned about disruptions of service, so Congress had the Federal Railroad Administration (FRA) develop a procedure and fund state rail planning activities. The FRA presence was high because the feds committed to transition funding to assist the states during a transition period.

The states have now gone through several rounds of state rail planning, and it is fair to say that state transportation agencies have proved fast learners. They examine lines proposed for abandonment by railroads and consider the alternatives of abandonment or knitting devices for the continued operation of lines. Continued operation is contingent upon a benefit-cost study that scopes wider than a railroad's calculation of profitability. If there is potential for viability or some other reason to keep a line open, the states usually arrange for operation of the line by a short line operator. Subsidy to the operator may be involved.

State rail planning served its purpose. The railroads managed to spin off track of little value to them. The states figured out what was and wasn't worth saving and came up with action plans. Many of those plans did not lead to successful operations, and adjustments were required. Even so, operators of short lines gained experience, and the large railroads worked out ways to relate to short line operators. As a result and although the states are still involved, to a large degree state planning was so successful that it put itself out of business—deals can be cut to create viable short lines without the involvement of state planning.

The process for developing the state rail plans, as well as the content of the plans, was prescribed by the Federal Railroad Administration (FRA). That is another example of federal co-optation of what should be local planning. Even so, the contents of plans are influenced by what the states take to be their chief problems. Traditional concerns have been grade crossings and safety and the effects of railroads on traffic in towns. Some state plans were influenced by these concerns. Proposals for line abandonment affected some states very much, and concerns about the relationships between freight transportation, towns and industries, and major economic sectors influenced plans.

Let's look briefly at modern abandonment analysis. As stated, there was a prescribed procedure established by the Rail Service Planning Office (RSPO) of the ICC (and an alternative established by the US Railroad Association that was used in the Conrail area). Studies provided a description of the route. There was then treatment of the alternatives of continuation as is, continuation with subsidy, abandonment, acquisition of right of way for banking for future use, and so on. Rail revenues from the operation of alternatives were estimated and then compared with the avoidable cost of continued service. Avoidable costs include on-branch cost and off-branch cost. They generally include calculations of the costs that would be incurred to upgrade the line to a Class I facility (condition of track, ties, ballast, etc.) for acceptable service. On nearly all branch lines, the avoidable cost is greater than the revenue.

An additional calculation was made, termed avoidable community impacts. These impacts include the increased transportation cost if rail shippers were forced to use other modes, salary and wage shifts, and so on. This number was then compared with the difference between avoidable cost and revenue and a recommendation made about the future of the line.

The rail planning activities focused state DOTs on rail issues, and most have activities in the area. There was a healthy broadening impact on the agencies. However, state rail planning is now somewhat mooted because many of the urgently needed abandonments have been made. The big railroads continue to spin off thin traffic lines. Usually, these lines have potential for viable operations by a low-cost operator. Because the railroads want to keep the traffic generated on the lines, they assist a low-cost operator's entering the business.[388]

21.8 Rationalization: LA Style: The Alameda Corridor

Rationalization does not just require a reduction in route miles, it can also involve minor capacity additions (as other routes are abandoned). The Alameda Corridor is a system of rail routes that connect the California ports of Los Angeles, and Long Beach to downtown Los Angeles, 32 km (20 miles) north. In the 1990s, the ports of Los Angeles and Long Beach became the busiest ports in the United States due to the enormous expansion in Pacific Rim trade. Los Angeles and Long Beach ports are first and second in the United States for container shipments (see Figure 19.2). In 2000 they were third in the world for container shipments, behind Hong Kong and Singapore, though they have since slipped.[389]

Congestion on the rail routes spilled over onto highways because the railroads had track on the streets. Additional freeway congestion resulted from container traffic. The Alameda

Corridor project aimed to increase efficiency in movement of cargo throughout the United States and to overseas markets.

The Southern California Association of Governments (SCAG) formed the Ports Advisory Committee (PAC) in October 1981 to improve highway and rail access. PAC members included local elected officials, representatives of the ports of Los Angeles and Long Beach, the US Navy, the US Army Corps of Engineers, the railroads, the trucking industry, and the Los Angeles County Transportation Commission (LACTC). First, PAC dealt with the problems of highway access to the ports. The committee recommended numerous small-scale highway improvements, which were completed in 1982. The trains were consolidated on an up-graded Southern Pacific San Pedro Branch right-of-way in 1984.

The Alameda Corridor Transportation Authority (ACTA) was created in August 1989 by the cities of Long Beach and Los Angeles. A seven-member board representing the cities and ports of Long Beach and Los Angeles and the Los Angeles County Metropolitan Transportation Authority (MTA) governs ACTA. ACTA believed that the project was not going to be accomplished without federal intervention and invited Congressmen and other elected officials to the ports to see the seriousness of the situation. Congress in 1995 identified the Alameda Corridor as a "Project of National Significance," which secured federal funding for the project. Congress appropriated a loan of $57 million for the project in 1997. The USDOT authorized a $400 million 30-year loan for the project in 1998. The ports provided an additional $394 million. ACTA utilized the "design-build" process, wherein a single firm was responsible for both the engineering and construction of the project, and the engineering wouldn't be finished before construction began. To finance the project, ACTA sold $1.2 billion revenue bonds in January 1999, with additional funding from California state grants and sources administered by the Los Angeles Metropolitan Transportation Authority. The total financing package was approximately $2.43 billion.[390] The loans, grants, and bonds will be repaid by user fees from the railroads, ranging from $15 for a 20-foot and $30 for a 40-foot container. The constraints on user fees are competing modes; if the port raises fees for using the corridor, shippers will switch to trucks.

The Alameda Corridor Project began operation on April 15, 2002, with 33 trains using the corridor. As of 2012, it served about 45 trains with almost 12,000 containers daily.[391] It was estimated that 100 trains would use the corridor by the year 2020, but the current rate of growth does not look to meet that early forecast. As a consequence, there were problems with the financing; the $2.1 billion of debt has been downgraded. Markets are changing, extrapolations of growth from the 1990s were not met in the 2000s, perhaps there is a peak freight just as there is peak travel (see Section 24.6.2). The United Kingdom has already seen relative dematerialization of the economy, and the total freight shipped per person has declined since a 1990s peak. If the United States sees the same, the long-term growth in freight shipments will end, and the demand for new freight facilities will fall off. There will still be needs as markets move, but not the remorseless growth in tonnage the US saw for about two centuries.

The Ports of Los Angeles and Long Beach in particular face further competitive pressures, from US West Coast ports, from ports in Canada and Mexico not subject to US labor actions (see Section 19.5), and from a widened Panama Canal that may encourage more Pacific originating ships to sail directly to Southern and Eastern US ports.

The new enlarged Panama Canal may usher in increased use of larger ships that play the role of movable warehouses (in addition to its impacts on movement patterns in the US and elsewhere). That structural change is inducing behavioral change and enabling production and market change. New robotic technologies may encourage more localized production, as labor becomes a smaller share of production costs, and the advantages of cheap labor markets (e.g., China in the early twenty-first century) decline.

Nevertheless, construction is underway to extend improvements (especially grade separations) eastward from downtown Los Angeles, the Alameda Corridor East (ACE) project.

A similar project, the Brownsville West Rail Bypass International Bridge, a 13 km (8 mile) bridge that avoids existing bottlenecks and aims to increase NAFTA trade carried by rail from 6 percent to 35 percent, opened in 2012, the first new rail border crossing in a century.[392]

21.9 Rationalization: Tracks or Wires: Dakota, Minnesota and Eastern

The Dakota, Minnesota and Eastern (DM&E) Railroad was formed in 1986 as a spin-off of tracks from the Chicago and Northwestern Railroad. The C&NW was subsequently acquired by the Union Pacific Railroad in 1995. The DM&E has also acquired other connecting lines, reassembling much of the C&NW Railroad, which had been abandoned. It mainly serves the agriculture and ethanol industries.

In 1997, the DM&E first proposed to build a line into the coal-rich Powder River Basin of Wyoming. If built, it would be the first significant new railroad built in the United States in a century, since the Milwaukee Road of 1909.[393] It intended to lay 416 km (260 miles) of new track from Wyoming's Powder River Basin to connect to existing rail lines in South Dakota, and then rebuild track across South Dakota and Minnesota. Does this proposal suggest any rejuvenation in the freight railroad sector, or is it a mere tweak on a mature network?

The Powder River Basin Project was initiated in February 1998, when the DM&E submitted an application to the new Surface Transportation Board (STB) for the new track in southwestern South Dakota and northeastern Wyoming. This track would serve the low-sulfur coal reserves, and bring them quickly to market in the Midwest. (The advantage of low-sulfur coal is that it burns more cleanly than high sulfur coal.) About 100 million tons of coal per year could be moved by the line, more than one-third of the coal mined in the Powder River Basin.

The STB gave its approval in January 2002 (almost four years later). To supplement this goal, DM&E acquired the connected Iowa, Chicago and Eastern (IC&E) and IMRL (I&M Rail Link) railroads, and is now the largest of the Class II railroads in the United States. However, the hurdles to building a new railroad are large. There is the $1.5 billion price tag, which is especially large considering the size of the DM&E itself. There are additional challenges.

Getting the line approved has been no easy matter. In particular, getting approval for construction through the city of Rochester, Minnesota, home to the famous (and politically powerful) Mayo Clinic, has been contentious, as the city is worried about a constant stream of coal trains running back and forth across the city. The Mayo Clinic has been pushing for the railroad to build a bypass around Rochester. This is a far cry from the days when towns wanted train service, and is a different reaction from that of most towns along the line, which believe the railroad will provide a benefit. Nevertheless, a number of lawsuits have been filed by the city of Rochester.

Groups representing Native Americans in the basin have also filed lawsuits, arguing that historic sights and burial grounds will be crossed by the railroad. Farmers worry that coal will take precedence over grain, especially a problem in smaller communities where the line is a monopoly.

The Powder River Basin is already served by lines of two Class I railroads, the Burlington Northern and the Union Pacific. Is a third line necessary or in the public interest?

The DM&E was denied a multibillion dollar loan from the Federal Railroad Administration in 2007, which cited the risks to taxpayers. The DM&E was acquired by the Canadian Pacific Railway in 2008. Despite its ownership by a major railroad (the CP), it could not find financing to construct the line, indicating skeptical financial markets. The project was suspended in 2009. With recent shifts in US energy preferences from coal to less expensive and less polluting natural gas, as of 2012, CP has "defer[red] indefinitely plans to extend its rail network."

The proposed project also begs the question of why the coal can't be processed on site and the electricity sent back east. Certainly this would minimize transportation (and energy) costs, as electricity weighs less than coal and high voltage lines are cheaper than track. This would have less environmental impacts on most places, though of course more in the basin itself. Capital costs for new power plant construction would probably be less than the capital costs for a new railroad, though.

21.10 Rationalization: Coming to Peace with Modal Competition

The New York Air Brake Company (along with CSX) has developed an Iron Highway concept (Figure 21.3): a 366 m (1200 ft) series of flat cars with an engine at each end, linked by automatic couplers; hinged ramps in the middle of each train allow trucks to drive on and off. The market niche envisioned may be characterized as "thin market." The train has operating characteristics permitting stopping here and there for pickup or discharge of a few units. The Roadrailer is a competitor (sponsored by competing railroad Norfolk Southern) to the concept discussed here. The Iron Highway distributed power concept is a good one, and the train should have a good ratio of gross to tare weight. It might open a development path.

21.11 Rationalization: Laying a New Path

There are many paths for freight transportation that might be followed. The path discussed might evolve in different directions and it illustrates thinking about paths. The example to be used was considered seriously some years ago by a major railroad.

Imagine moving bulk freight in the vehicle illustrated in Figure 21.4. It is, say, 5 m (16 feet) in width (can fit on existing rail right-of-way) and has a near-square shape (which

FIGURE 21.3 Iron Highway Concept

FIGURE 21.4 A Bulk Commodity Vehicle and Guideway

maximizes content versus structure requirements). It is not trained in coupled cars behind a locomotive (avoiding the requirements for strength and, thus, car weight imposed by training). The single car arrangement permits rolling downhill and increased recapture of potential energy from velocity or hill climbing. The body of the car sits inside the wheels for a low center of gravity. It could use steel-wheel-on-steel-plates (low rolling resistance, no flanges to rub) or run on gravel or asphalt surfaces.

If the vehicle grossed at 100 tons, a 75 kW (100 hp) diesel engine would suffice (likely less). The engine drives an alternator, powering motors on each wheel. Clearly it could do much better than current practice.

Low velocity, say, 16 to 32 km/hr (10 to 20 miles/hr) avoids energy losses due to air resistance. Low velocity is acceptable. We have in mind the movement of low value bulk commodities, and their value of time is low. It is the time from A to B that is of interest and the precision of schedule-keeping that counts. Consider the opening quote by Schordock. The Union Pacific Railroad is noted for running fast trains, and it does (127 km/hr

(79 miles/hr) is common). But by our calculation 1,450 km/36 hours = 40 km/hr (25 miles/hr), and this is for priority rather than bulk freight. What's going on: stops in yards, crew changes, waiting on sidings, and so forth. We imagine cars moving at relatively slow speeds, but moving from point to point without stopping and without buffer time in yards.

The scheme is open to new technologies of many kinds: wire following and automation (no train operator needed), on-board (e.g. flywheel) or off-board energy storage (flywheel, pumped storage of water), alternative fuels, alternative propulsion devices, electrification of high-density lines, new materials, and so on. These technologies would be within a system design, where they could be very effective. The design (or variations on it) might be effective in many kinds of market niches and at different market scales. For example, if high throughput is required, cars could run on very short headways. Infrequent cars on simple guideways would do for low-density routes. Put another way, the design might develop on this pathway or that, depending on technology evolution, service requirements, and so on. To ensure safety, low speed avoids the need for grade separations. Velocity could be tailored to crossing situations.

The proposal uses existing guideways, is quiet, and opens a path toward reduced energy use. The shipment of merchandise (general freight) poses a more difficult problem than the shipment of bulk. Yet this bulk scheme may be a way to at least partially finesse the non-bulk shipment problem. Imagine a building design, say, in the Bay Area. Specifications are written for windows and they say when the windows are to be delivered. Rather than the designer looking up standard windows, the window designs are specific to the building (granting the designer more degrees of freedom). A local shop wins the bid. That shop (in tomorrow's world) has computer-controlled cutting and forming machinery. Rather than shipping windows from some distant factory, aluminum and glass are shipped in bulk and formed into windows on or near the construction site on an as-needed basis. For another example, wine is shipped in bulk and bottled locally on demand. We are suggesting that new forms of bulk shipment might combine with the computer and control revolution to produce more efficient and sustainable industry configurations.

Although striving toward sustainability may reduce coal movements, today's coal movements might provide an opening market niche. In Appalachia there are many medium-size mines that are difficult to serve because of their small output. Ideally, as is the case in the Western US, several unit trains are loaded at a mine daily. But many Appalachian mines do not produce enough coal for low cost unit train service at the mine. Cars have to be collected, often over some distance, to yards for train makeup. At first, using power cars pushing today's rail cars, a market niche might be found in Appalachia. That's blocked by high labor cost today, a minimum of two crew per train. But there are completely automated trains (e.g., in the Four Corners mining area), and automation might be feasible elsewhere. Short line railroads serve many market niches and are not so burdened by crew costs. They might provide a place to start.

The step between starting with conventional rail equipment and creating specially configured equipment and guideway (as in Figure 21.4) might be taken in a niche where vehicles are dedicated to service and not interchanges for haulage beyond the market niche. Such a niche might be at a gypsum mine, a cement plant, or where aggregate haulage is undertaken (there are already specialized aggregate haulage trains). Conventional tracks could be retained in the center of the guideway. The typical 5 m (16 ft) width of the vehicles would pose a problem in places where bridge structures and tunnels limit clearance. We see no particular institutional or fiscal problems, except for the current railroad low rate of return that limits the capital available to railroads and labor contracts. These would impose some transition friction.

21.12 Labor Rationalized

Responding to situations from time to time, policies were developed and adopted that preserved labor rights in mergers and acquisitions. The roads agreed to that in earlier days when excess labor following a merger could be spun off through attrition. Now, things have changed. For a variety of reasons, larger quantities of labor must be dealt with. There is the special problem of the short lines created when large roads abandon mileage. These short lines often can be profitable if paying local wages and not paying railroad wages and using railroad job classifications. It wasn't so difficult to modify policy in favor of the short lines.

But it has been difficult and it has taken many years to modify policy in the case of mergers and acquisitions, and revised policy has been tied up in the courts. In 1991 the Supreme Court let railroads ignore their union contracts when completing mergers approved by federal regulators.[394]

The contract modification responded to "things changing," and it was permitted by the weakening power of labor unions and changing public attitudes about the railroads and unions. It took a long time and it was costly.

21.13 Discussion

This chapter focused on a particular aspect of planning: rationalization to make the mode work better in current conditions. Toward the end of the 1800s, rationalization strived to control the behavior of the railroads and to bring behavior in line with common law norms. Subsequently, rationalization has meant many things: changes in the network and the spans of control of firms, bringing prices in line with costs, route abandonment, and the discard of old and the development of new service arrangements. The latest wisdom (from the last quarter of the twentieth century) is that regulation isn't rational, deregulation is.

Railroads would also point to several major post-regulation changes. New scheduling protocols focus on assembling trains more rationally based on the destination of cars and less on their weight. Railroads have introduced what they refer to as "compensatory pricing," and economists refer to as Ramsey or inverse elasticity pricing. Ramsey pricing allows the railroads to charge more not based on costs (which regulators demand), but based on willingness to pay of consumers (what the market will bear). This has been achieved with better information systems.

Where do the rationalization efforts of the twentieth century leave us? The good news is that there is a viable freight railroad system. The tonnage moved has increased and, although reduced in mileage, the physical infrastructure has improved in quality and capacity. Niche market railroads, example, mine to port, coal, iron ore, are doing very well. They are a good target for automation and marked cost reduction. The bad news is that the larger systems are barely viable. The return on investment of the large railroads, while positive, is not large enough to cover the cost of capital.

To that end, it requires federal subsidies for the United States' private railroads to implement new technologies such as Positive Train Control (PTC) technology, which uses sensors and computers to keep up with train locations and velocities, control velocity to allow passing, improving safety, and decreasing emissions. For a variety of reasons, this is not being deployed as quickly as was initially imagined. In some senses this is the rail analog to Intelligent Transportation Systems or NextGen Air Traffic Control, the application of information technology to help perfect an existing mode. Like ITS and NextGen, it is facing lots of practical difficulties.

As with other modes, new environmental rules are being put in place to reduce particulate and CO_2 emissions from locomotives

The story of the rail modes continues in Chapter 28.

Thanks to the Interstate Highway System, it is now possible to travel across the country from coast to coast without seeing anything.

—CHARLES KURALT

22 Interstate

22.1 Limited Access

TO BREAK THE monopoly of the dominant Pennsylvania Railroad, the Vanderbilts, who owned what became the New York Central Railroad, initiated the South Pennsylvania Railroad from Harrisburg to Pittsburgh, Pennsylvania. (The New York Central already faced competition from the Erie and related railroads.) Construction began on tunnels in 1884 and continued through 1885. However, while competition may be good for consumers, it is anathema to capitalists. Financier J. P. Morgan won a seat on the New York City and Hudson River Railroad and gained sufficient control to force the Vanderbilts to sell the unfinished line to the Pennsylvania Railroad. Although $10 million was spent and twenty six workers died, "Vanderbilt's Folly" remained dormant for fifty years.

In 1935, the Pennsylvania legislature authorized a feasibility study of building a new toll highway using the old, unutilized South Pennsylvania Railroad right-of-way and tunnels.[395] Construction started in 1937, and the road opened in 1940. This was the first significant new turnpike for over 40 years, as the old turnpikes had been reverted to free roads in the late 1800s and early 1900s. Moreover, it was the first of a new generation of limited access highways, generally called superhighways, that came to transform the American landscape.

The Pennsylvania turnpike was the forerunner of a major turnpike boom lasting from 1940 to 1956 (following earlier booms in the early nineteenth century, described in Section 3.8). During this period (and counting projects initiated in this period), some 6,400 km of freeway class turnpikes were built in the United States.[396] In 1956, the Interstate highway program was funded, giving states 90 percent of the money toward construction of freeways, provided they were "free," or untolled. While turnpikes are a great way to collect revenue from non-residents, who may not pay a state's gas tax, an even better way is to get federal cost sharing of this magnitude. Thus the era ended shortly after it started.

The now mature urban vehicle-highway system is large and intertwined in most facets of American life. Yet, the vehicle-highway system is almost never treated as a system. People talk about highway policy, automobile policy, safety policy, congestion policy, truck weight policy, and so forth. Historically, public policy has been directed to the provision of highways, although Interstate Commerce Commission regulation was extended to Interstate truck operations during the 1930s, at the same time the states began to regulate intrastate trucking. Beginning in the 1960s (and in response to Ralph Nader's [1972] *Unsafe At Any Speed*), US federal safety policy has been significantly extended.

We address the Interstate in this chapter. Though the formative experiences are similar, and result from the same laws, urban and inter-city systems serve very different purposes and have very different problems. We profile Robert Moses and Jane Jacobs, two icons of the highway and anti-highway movements, respectively. We discuss the formation of the Interstate system, and examine several cases.

However, by the end of the twentieth century, with the Interstate era ended, local governments again viewed toll roads as a reasonable way to finance roads. New toll road projects were begun in a number of cities, largely as suburb-suburb connectors, though some suburb-city routes as well.

22.2 Inventing the Interstate

Things have never been more like they are today in history.

—Dwight David Eisenhower[397]

22.2.1 SUCCESS HAS MANY FATHERS

The initiation of the Interstate highway system followed a long period of gestation. Some Interstate ideas can be traced to the 1920s: limited access designs, interchange designs, and regional highway proposals of that time. In particular, in response to the logistics problems of World War I, the *Pershing Map* of the 1920s proposed an extensive system of Federal roads.[398] Others, like conservationist Benton MacKaye (founder of the Appalachian Trail) advocated "Townless Highways," saying "The motor slum in the open country is today as massive a piece of defilement as the worst of the old-fashioned urban industrial slums."[399] Edward Bassett, appointed president of the National Conference on City Planning by Secretary of Commerce Herbert Hoover, coined the term "freeway" to denote a road exclusively for movement (and no land access),[400] in contrast to parkways, which were for recreation not movement, and highways, which were for movement but allowed land access.

The form of the Interstate began to be developed in the 1920s. In 1924, the First National Conference on Street and Highway Safety, chaired by Secretary of Commerce Herbert Hoover, advocated that communities construct "by-pass highways and belt highways which will permit through traffic, especially trucks, to avoid congested districts or even any built-up portions of the city or town."[401] Bypasses created new points of great accessibility, where

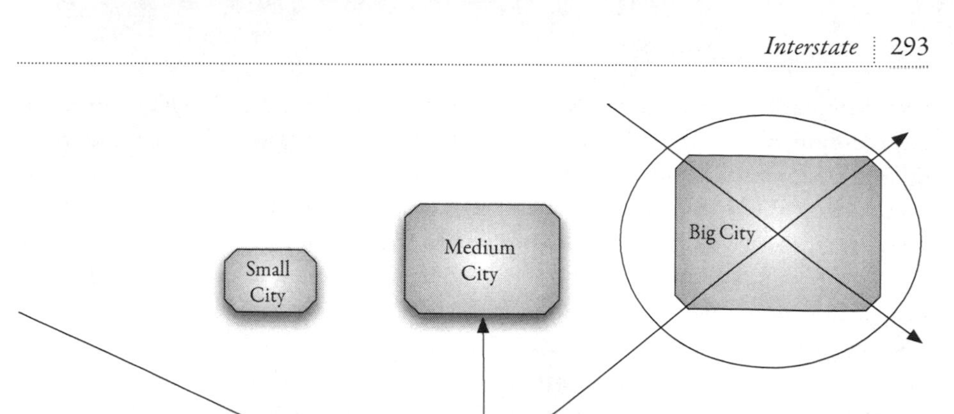

FIGURE 22.1 Three Approaches to Urban Links

businesses oriented to the automobile traveler could prosper. Inns were replaced by motels, sit-down restaurants by fast food, and gas stations just exploded across the landscape to serve the needs of the traveler.[402]

The Bureau of Public Roads's *General Location of the National System of Interstate Highways* was one result of the fifteen-year period of planning.[403] There were three approaches to urban links (illustrated in Figure 22.1):

1. Small urban places were bypassed;
2. Medium-sized urban places were served by a connecting link; and
3. Larger places (where links often connected) were served by circumferential routes, as well as routes that entered the city.

The Bureau went to considerable effort to review local plans, and it had in mind the coordination of the urban links of the Interstate with local plans.[404]

Bureau road programs faced problems during the Great Depression. There was the argument that Bureau work was not labor intensive; it didn't respond to the need for jobs. (The argument that the purchase of things such as cement and road building equipment created jobs was too complex for the political scene.) Road money began to flow to the Works Progress Administration (WPA), and that Administration began to supply funds for wages to urban public works departments. In response, rural road program interests began to press for Bureau-constructed toll roads. In turn, the Bureau issued its *Toll Roads and Free Roads*,[405] a feasibility study of a 22,500 km (14,000 mile) national toll road system (three north-south and two east-west routes). The study said that tolls would pay only 40 percent of costs, and the toll proposal died. As an alternative, the Bureau proposed a 43,500 km (27,000 mile) "free" system to be built to toll road standards.

Bureau Chief MacDonald disliked the toll road concept, for it was contrary to the institutional ethic of the Bureau: free roads for all. He was also concerned about the likely dysfunctions of toll road institutions: self-perpetuating monopolies, nepotism, graft, and so on. There is every indication that the Bureau did not want the toll road study to find feasibility. Examining that study, one finds that traffic growth projections were modest and tolls were projected to dampen traffic growth greatly. Also, the proposals were expensive and over-designed, certainly more than was needed at the time for much of the mileage. Interestingly, the proposed designs were one lasting impact of the toll road study. Aspects of those designs were eventually incorporated in the Interstate highway system.

Returning to Interstate proposals, the Bureau proposed that the system be built as a free road system, the White House asked for greater emphasis on rural roads (the proposal made for a rather thin net: two east-west routes and three north-south, as mentioned), and a revised proposal was prepared in 1936. In 1941, an Interregional Highway Committee was formed, and its 1941 report *Interregional Highways*, recommending a 71,000 km (44,000 mile) system built to toll road standards, was enacted into law that same year. The 1956 Act, fifteen years later, implemented the system. During that fifteen-year period, a fair amount of work was done on route locations (the famed *Yellow Book*), designs, and so on.

Advocacy on behalf of the Interstate in the 1951–1954 period was undertaken by the National Highway Users Conference (a coalition of industrialists from the trucking, oil, manufacturing, auto, and farm industries) under the banner Project Adequate Roads (PAR). Lack of quick success (in the form of a weak 1952 Federal-Aid Highway Act) led to the breakup of the coalition.[406]

Funding was an issue, perhaps the largest, as there was general consensus on the need for an Interstate Highway System on the part of most parties. To help move things forward politically, President Eisenhower asked the governors for their input. The governors wanted the federal government out of the road-building business, and out of the gas-tax collecting business (since half the gas-tax went to the federal Treasury). They needed to be persuaded otherwise. A new wave of turnpikes was being constructed: the Pennsylvania Turnpike before World War II and other turnpikes in the late 1940s and early 1950s, and this was viewed as a viable alternative to general revenue funding by many.

To establish his case, Eisenhower also established two committees. The Interagency Committee (within the federal government) was chaired by Commissioner of the Bureau of Public Roads, Francis du Pont. The President's Advisory Committee on a National Highway Program comprised industry leaders, and was led by the chairman of the Continental Can Company, General Lucius D. Clay, a well-connected friend of Eisenhower and well-regarded logistician from the Army Corps who most recently helped feed blockaded Berlin. Frank Turner of the BPR was executive secretary of the Clay Committee. The Clay Committee supported gas taxes to repay bonds issued by a Federal Highway Corporation. Congress had difficulties with this, and the Clay Commission proposal would fail in 1955,[407] after receiving major opposition from truckers, among others, against the gas tax. They had to be persuaded to come around as well.

An interesting proposal made off-and-on involved taking excess property for right-of-way and selling it to finance the system as the system was developed and benefits emerged. The final financing decision was to use the federal fuel tax and to match state funds on a 90/10 percent basis. In 1956, the main financing question was appropriate allocation of costs to vehicles of varying size and weight, although discussion of a benefit-based financing continued for several years.

There are many fathers of successful "inventions"; the Interstate is no exception. Senator Albert Gore, Sr., from Tennessee, was the Senate proponent of the 1956 Federal Highway Aid Act and the Highway Revenue Act. These acts authorized $31.5 billion in federal and state aid to build the Interstate highway system. There was much debate as to whether the roads should be "free" or "toll," dating from the 1930s. Gore fought against the plans to finance the highways with bonds to be repaid with tolls and was able to get enacted pay-as-you-go taxes on fuel, tires, and trucks. This "free" system clearly lowered transaction costs in collecting revenue, and increased use of the system compared with the "toll" alternative. However, it made management of the system more difficult as variable pricing (by time of day and by facility) is impossible with only gas taxes, resulting in the overconsumption of urban and suburban roads and congestion that confront commuters daily. The political trick, to ensure the gas tax revenue would be dedicated to roads, was the establishment of the Highway Trust Fund, which was solely dedicated to the Interstate until 1973, at which time some funds were allowed to be spent on transit projects in cities that canceled interstate projects. The amount was expanded in the 1980s, and less restricted. In the 1990s, some of the trust fund was used for deficit reduction. That money was paid back in the 2000s, when the trust fund moved from surplus to deficit, and highways again needed to borrow from general revenue because of lack of political will to either reduce spending or increase revenue.

Many attribute the Interstate system to President Dwight Eisenhower (for whom it has since been named). As an army colonel, Eisenhower made a famous cross-country journey in 1919 between Washington, D.C., and San Francisco on the Motor Transport Corps convoy, which took two months. The road trip in that era was very much an adventure. And while roads improved in the following thirty six years with the construction of the US Highway system, the German Autobahn was another inspiration for the US Interstate program. Eisenhower, among others, observed the relative efficiency with which Germans could move forces back and forth in a two-front war during World War II. The credit Eisenhower was granted surely exceeds the credit he deserved for the system, as it was designed before he was president, and his desire for a "self-liquidating" (by which he meant toll-financed) system was not to be and, some argued at the time, was not financially viable. The term "self-liquidating" was later interpreted to allow hypothecated gas taxes.

22.2.2 REGARD FOR THE BUREAU OF PUBLIC ROADS

Ellis Armstrong, commissioner of the US Bureau of Public Roads from 1958 to 1961, told a story to Garrison some years ago, one of many that Garrison has heard illustrating the independence and professionalism of Bureau leaders. After receiving letters and phone calls from (then) Senator John F. Kennedy, Armstrong agreed to meet with politicians and business leaders who argued for a shift in the location of the Interstate south of Boston.

Leaving the meeting and greeted by Boston reporters and TV cameras and asked for his comments, Armstrong remarked, roughly, "I've received lots of interesting information but I have to take it back to Washington for study. The Interstate is a carefully planned and balanced system. Moving a link 10 miles to the south here means we will have to consider moving the entire system 10 miles to the south."

Concerned that his attempt to be humorous might backfire, Armstrong was relieved that, when the story played on local television, no one questioned his remarks, nor did Senator Kennedy. That, it seems, is a comment on the confidence and authority that Bureau professionals had gained through years of service. Perhaps it also says that some citizens recognized that non-local concerns mattered.

22.2.3 THE BRAGDON COMMITTEE

A review of the work of the post–Interstate Act Bragdon Committee (actually, General Bragdon and staff) will assist in summarizing the late 1950s mood. The treatment of the work of that Committee is based on discussions with Ellis Armstrong at about the time of the events described, Lee Mertz, onetime head of planning at the FHWA, and Paul Sitton, in the Bureau of the Budget at the time.[408]

President Eisenhower depended on staff work very heavily, as many administrators must. He supported the Interstate Act of 1956 as a measure to counter unemployment, and his assistant for Public Works Planning, General J. S. Bragdon, began to give the Interstate his attention.

The 1956 Act had authorized the expenditure of $27.5 billion over thirteen years for the construction of the Interstate. Bragdon was shocked in 1958 when the Interstate cost estimate (ICE, the cost to complete) came in at $39.9 billion and the end date slipped into the 1980s. There was controversy when Congress raised the gas tax by a penny after the initial three cent gas tax to pay for the Interstate. The House of Representatives was concerned about the states wasting money, and Congressman Blatnik was named chair of the House Special Subcommittee on the Federal-Aid Highway Program.

Bragdon, working with the Bureau of the Budget, took it as his task to correct the situation, and worked for a couple of years to get the estimate under control and trimmed down.

Working with a staff of about twenty five, Bragdon made analyses and made suggestions to the BPR (through the Department of Commerce, where it was then housed). A dozen or so memos record Bragdon's recommendations and the BPR's reactions. Bragdon was very concerned about the money required for the program. He suggested expenditure reductions, and the BPR typically countered these by reference to the Act and Congressional intent.

Bragdon suggested that the states that wanted to accelerate programs build toll roads. They could use the money from those tolls to accelerate the provision of untolled facilities. The net effect would be to speed up construction.

Though he was interested in toll roads, Bragdon's main thrust was reduction of expenditures in urban areas. His argument was that Congress hadn't intended to manage urban problems. He made many proposals suggesting limiting the system, in particular: the number of lanes in urban areas, capacity for rush hour traffic, number of interchanges, and number of spurs into the cities (relying on outer loops only). All of this was to keep the cost down. He argued that the secretary of Commerce had the power to take routes off the Interstate and should take some urban routes off.

Bragdon didn't understand the progressive, cooperative traditions of the Bureau.[409] Many of his suggestions were unthinkable in that context.

A point of difficulty for Bragdon was the lack of information on where roads were needed in the urban areas. The BPR and AASHO had been working with the American Municipal Association and had plans for 149 of the 288 cities of over 25,000 in population, and 45 more were on the way. Bragdon did not think that was proper; the BPR was responding to what the cities wanted. He demanded that planning be required by the BPR. While it was to be comprehensive, it was to be in the frame of strict policy on state-local arrangements (i.e.,

Bragdon's ideas of how urban extensions should be allocated). Plans, as Bragdon imagined them, could be completed in two months.

As remarked, the Bureau was working with the AASHO and the American Municipal Association in a cooperative style. It had arranged the First National Conference on Highways and Urban Development at Sagamore, Syracuse University, New York, in 1958, which addressed design-planning issues.

Returning to Bragdon's demand that planning be mandated, the BPR countered that it did not have the power to require planning. That remark may seem strange from the perspective of today, for the federal government now requires significant planning efforts from local and state governments. The issue is a constitutional one. The constitution allows congress to tax to "... provide for the common defense and the general welfare of the United States." But there is another clause that the states have all the powers not specifically reserved for the federal government. The debate over the meaning of the welfare clause began with Hamilton and Madison, and there is a parallel long trail of court cases.

In the early days of the Bureau's programs, funding was allocated according to post road mileage, for providing for the mail was one of the powers of the federal government. Later, no questions were raised about matching money for the federal-aid system. An additional question to be settled was whether the federal government could withhold tax monies it has collected to get the states to do something. That was the leverage the Bureau would have to use to get planning started—either plan or no federal money will be available.

The power of the federal government to withhold funds was not settled by court cases until the 1970s, and it was that unsettled question that had made the Bureau go slow in requiring planning.

There was a "show-down" meeting between Bragdon, the BPR, and Eisenhower in 1960. It is recorded that Eisenhower said that running the Interstate through congested cities was not his concept and wish.[410] However, the BPR pointed out that the *Yellow Book* (formally: *General Location of National System of Interstate Highways Including All Additional Routes at Urban Areas*) was in the hands of Congress, and it was responsible for the 90–10 formula (90 percent federal funding, 10 percent state), rather than 60–40 as first proposed. Because Congress knew of the urban extensions, the program was committed to them.

We are interested in Bragdon's work because it represents some of the thinking of the times. There was the fear that urban expenditures formed a monetary black hole. Was urban congestion a federal government problem? The Interstate cost estimate kept going up, and increases in taxes or program stretch-out were not popular. In the early 1950s, there had been considerable questioning of whether a federal program was needed.

Our view is that the thinking of the times resulted in the urban Interstate being built "on the cheap," or at least contributed to it. The centerline mileage of links was limited, capacity was increased by adding lanes. A lower level of service was accepted in urban areas compared with rural areas.

Cheapness producing congestion created by channeling traffic onto a limited mileage facility is our main point. There was little or no debate about improved arterial, parkway, limited truckways, or other alternative designs. In spite of the multiple lanes constructed to serve estimated traffic for twenty years, Interstate links were soon congested.

How did we get into this mess? Schwartz (1976) says, "If there was a moving-party heavy for the urban Interstates, it was the cities themselves rather than the highway lobby."[411] He points out that city planners were generally in favor, and that politicians looked to free money from the federal government. Alf Johnson of the AASHO and some of the BPR managers that Garrison knew cooled on involvement in urban areas as the many problems became apparent.

22.3 Freeways Rising

22.3.1 FREEWAYS IN JAPAN

The growth pattern in the United States and Europe was replicated, and accelerated in subsequent national motorizations. After World War II, the use of the automobile in Japan took off, as seen in Figure 22.2. The number of vehicles rose from about 142,000 in 1945 to 922,000 in 1955. As late as 1955, the main road between Tokyo and Osaka was unpaved for one-third of its length (e.g. Figure 22.3). The first expressways weren't opened until the Meishin Expressway in 1963 (shown in Figure 22.4). Clearly, the vehicles drove the demand for expressways, which can be seen to be lagging vehicle growth significantly.

22.3.2 BUILDING URBAN HIGHWAYS: THE CASE OF I-94

Aside from the urban freeway revolts discussed in Section 22.3.3, the construction of the Interstate proceeded about as well as can be expected in most cities. Although over budget and late, it was eventually built. One such case is I-94 between Minneapolis and St. Paul, Minnesota.

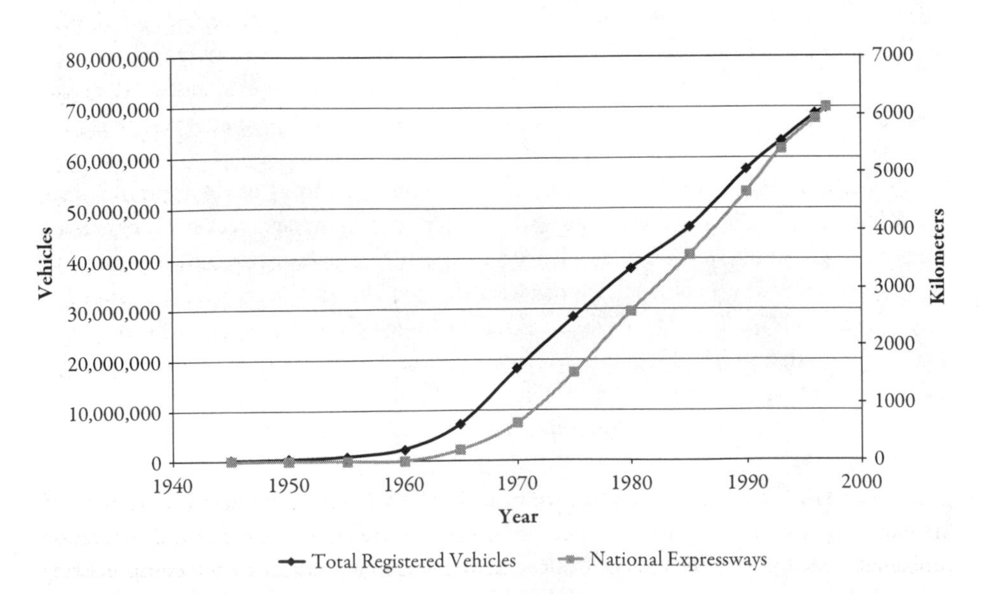

FIGURE 22.2 Japanese Vehicle Highway System, 1945–1997

FIGURE 22.3 Condition of Roads in Japan in the 1950s

FIGURE 22.4 The Meishin Expressway (1963)

Between 1947 and 1950, vehicle registrations in the Twin Cities increased 58 percent.[412] St. Paul officials realized that they needed to solve congestion and other transportation problems. Previously, city officials dealt with increased congestion by widening existing links. This option was becoming increasingly expensive as the city grew. Freeway plans were developed connecting downtown Minneapolis and downtown St. Paul, but it wasn't until the Federal Aid Highway Act of 1956 was implemented that it became certain a freeway would be constructed. The Federal Aid Highway Act ensured there would be funds available (90 percent federal to be matched by 10 percent state). With the construction of some freeway ensured, the next step was to determine which route it should follow.

St. Paul and state officials recommended that the route follow St. Anthony Avenue, a largely residential city street parallel to the busiest route between the two cities (University Avenue), which happened to run through a minority neighborhood (the Rondo).

George Herrold, St. Paul's chief planning engineer until 1952, argued against the construction of a freeway along this route. He proposed a plan dubbed the "Northern Route" about a mile to the north of the St. Anthony Route. The Northern Route, because of its use of existing railroad right of way and industrial land, would not displace many residents or sever neighborhoods. In St. Paul, the St. Anthony Route divided the state capitol and government buildings from the central business district. Despite Herrold's advice, St. Paul and state officials would not deviate from the proposed St. Anthony Route.

With the St. Anthony route all but built, concerned residents began to speak out. The St. Anthony Route would displace nearly one in seven of St. Paul's African American residents. African American community leaders quickly concluded that it would be nearly impossible to divert the freeway, so they devised a list of actions they requested government officials to comply with: help displaced residents find adequate housing; provide proper compensation; construct a depressed (below grade) freeway to enhance aesthetics. The displacement of the African American community members was especially significant because there were few options available to them. At the time (the 1950s, before fair housing laws were enacted), most white communities would neither sell nor rent homes to them. For this reason, officials feared that the African American community would become over-crowded. In the end, only the second and third actions were followed through.

The Prospect Park neighborhood in Minneapolis was also severed by the St. Anthony alignment, and residents were worried the freeway would turn this diverse upper middle class neighborhood into a low income one. Residents claimed that having a low-class neighborhood within close proximity to the University of Minnesota would make the University unappealing to students and faculty. The community had one request: that the freeway be placed over an existing railroad spur; however, limited funding disallowed this idea. The freeway did, however, skirt the Malcolm E. Willey House designed by Frank Lloyd Wright. Despite the freeway separating the neighborhood from the Mississippi River, the neighborhood did not deteriorate. The freeway was completed in the late 1960s.

Aesthetically, the I-94 freeway is a scar across the surface of the city; in addition to having displaced residents, it disconnected local streets, moved traffic from a relatively even distribution to a more hierarchical one, so that movement depends on fewer, now more critical links. While certainly more people are moving longer distances every day on the urban freeway, congestion has far from disappeared. One can ask in retrospect whether building the road was the right thing, or whether building it there was the right thing, but there is no real "control" for this experiment. Asking what-ifs are easy, answering them is harder. But no one seriously is calling for the removal of this element of the Interstate system, suggesting the collective intuition of those who think about the road daily suggest that "sunk costs are sunk," and while in retrospect not everything was done perfectly, leaving it in place is better than removing it.

22.3.3 FREEWAY REVOLTS

> People of Earth, your attention please. This is Prostenic Vogon Jeltz of the Galactic Hyperspace Planning Council. As you will no doubt be aware, the plans for development of the outlying regions of the Galaxy require the building of a hyperspatial express route through your star system, and regrettably, your planet is one of those scheduled for demolition. The process will take slightly less than two of your Earth minutes. Thank you. There's no point in acting all surprised about it. All the planning charts and demolition orders have been on display in your local planning department in Alpha Centauri for fifty of your Earth years, so you've had plenty of time to lodge any formal complaint and it's far too late to start making a fuss about it now.
>
> —Douglas Adams, *Hitchhiker's Guide to the Galaxy*

On November 2, 1956, the San Francisco Chronicle published a map of the proposed freeways through San Francisco while its editorial page was noting about the emerging revolt: "The remarkable aspect of these protests and claims of injury is their tardiness. They concern projects that have for years been set forth in master plans, surveys and expensive traffic studies. They have been ignored or overlooked by citizens and public officials alike—until the time was at hand for concrete pouring and when revision had become either impossible or extremely costly. The evidence indicates that the citizenry never did know or had forgotten what freeways the planners had in mind for them." The newspaper was referring to protests, including a petition signed by more than 30,000 residents of affected neighborhoods.

On January 23, 1959, the San Francisco Board of Supervisors voted to remove seven of the ten planned freeways from the City's master plan. The Embarcadero Freeway, which had already been constructed from Folsom Street to Broadway, was halted, mid-ramp, and left to sit there for another thirty years until the Loma Prieta earthquake prompted the city to pull down the remainder and reopen the waterfront.

The "Freeway Revolt" that sparked the Board of Supervisors move was a bitter fight between citizen-activists and the professionals. It was a revolt that was to be replayed in many other US cities over the next three decades, with different outcomes depending on the city.[413] Section 22.3.5 discusses the revolt in New York.

The revolt in San Francisco played in many newspapers and was emulated in many cities. It is often taken as an anti-urban Interstate revolt, but as the western terminus of Interstate 80, the Interstate mileage was minimal. The freeways were to be state and city planned and funded. The plan, first developed in 1947, was only partially completed.

22.3.4 PROFILE: ROBERT MOSES

Robert Moses (1888–1981) was born in New Haven, Connecticut, the son of a department store owner and grandson of a New York merchant. He studied at Yale and Oxford, and received a Ph.D. in political science from Columbia in 1914. His first job was with the Bureau of Municipal Research, a data crunching organization in New York that let him work on city budget issues. After World War I, he led New York's Committee on Retrenchment and Reorganization, and thus aligned himself with Alfred E. Smith and the good government movement that emerged during the 1920s. He rose through the political ranks, and in 1927 was appointed New York's Secretary of State.

Though he had tangles with Smith's successor as governor, Franklin Delano Roosevelt, Moses was able to leverage the power accumulated while under Smith to keep him in a dominant position in New York infrastructure for the next thirty five years. In 1922, Moses drew up the New York State Park Plan—which included the ability to issue bonds to rehabilitate parks, and to build parkways, and more important, highways leading to the parks. In 1924, Moses became chairman of the State Council of Parks, and also was president of the Long Island State Park Commission.

His parks, especially on Long Island, were very successful, and New Yorkers clamored to reach them. In 1930, Moses extended the parkways into the city. These parkways were 90 to 180 meters (300 to 600 feet) wide, gracefully curving, with no traffic lights and low overpasses (thereby prohibiting both trucks and buses from using them). Roadside commerce was also prohibited, aside from special rest stops.

His success with the parks and parkways led the new governor, Herbert Lehman, to make him chair of the Emergency Public Works Commission in 1933. This commission promoted new river crossings over the Hudson and East Rivers in New York, which would be toll bridges, which would require toll authorities, which naturally would be run by Moses. These authorities were self-sustaining and self-liquidating organizations, supported by user fees, which could issue bonds repaid by future revenue. They were new publicly authorized monopolies without elected leadership. In 1933, Mayor LaGuardia asked Moses to serve in his city cabinet. Moses agreed, provided he could retain his other positions. Unlike present-day politicians, Moses was able to hold state and city positions simultaneously, allowing his empire to grow. (This was permitted largely because he did not draw salary from most of his positions.)

Aside from an independent revenue stream and the control of multiple agencies at city and state level, Moses's tools included having his plans ready and fully designed before the money became available, thus allowing him to deliver projects quickly. Today, projects have to go through lengthy design and environmental approvals before they can be started. So even if cash were available, it would be years before the ribbon cutting. Somewhat more deviously, Moses tended to underestimate the cost of projects, so they would be started, and costs and quality desired would escalate: he offered the sponsoring politicians two choices, provide more money, or he would stop construction and the politician would have to explain why he wasted all that money building a half-finished useless project. Though this happened repeatedly, politicians were always willing to swallow the promise of the low cost because Moses had the reputation of delivering the ribbon cutting before the next election.

It would take far too long to list all of Moses's achievements here. Caro (1974) wrote a 1,344-page book, *The Power Broker*, describing and decrying what Moses did to the city. But whether he was loved or hated, he radically reshaped New York into a modern city using parks and road building as his tools.

22.3.5 PROFILE: JANE JACOBS

Jane Jacobs (1916–2006) was born in Scranton, Pennsylvania, the daughter of a doctor, who moved to New York at the age of twelve. Like Moses, she attended Columbia University. She then entered the publishing industry as an editor and writer for magazines such as *Architectural Forum* and *Fortune*.

A student of the city, she published several books, including the *Death and Life of Great American Cities* (1961). In that first book she celebrated the rich, textured chaos of urban streets, like her neighborhood in Greenwich Village, which had shopkeepers and other citizens monitoring the public spaces, "eyes on the street." She railed against urban renewal and freeway building (such as Moses's Cross-Bronx Expressway) that gutted organically arising places, vital centers of urban life, and replaced them with artificial inhumane spaces.

In 1962, she became chair of the Joint Committee to Stop the Lower Manhattan Expressway. Sponsored by the Downtown Lower Manhattan Association, led by David Rockefeller, the Lower Manhattan Expressway was proposed by the highway builders (including Moses) and incorporated into the Interstate highway plan to provide, as the name suggests, a new eight-lane limited access roadway from the west side of Manhattan (the Holland Tunnel) to the east (the Manhattan and Williamsburg Bridges). (It had sister projects, the Upper Manhattan Expressway and the Mid-Manhattan Elevated Expressway.) The road would have bisected neighborhoods such as Greenwich Village, Little Italy, Chinatown, and SoHo, displacing thousands of residents, merchants, artists, and artisans. In December 1962, the highway was canceled and Moses was defeated by a woman he had called a "busy housewife."[414] In April 1968, she was arrested in another protest against the highway. (In transportation infrastructure, "no" rarely means "no".) The collective protests ultimately led to the highway being canceled.[415]

22.4 The Interstate at Maturity

22.4.1 REBUILDING URBAN HIGHWAYS: THE CASE OF THE BIG DIG

> Wouldn't it be cheaper to raise the city than to depress the Artery
>
> —Barney Frank[416]

Perhaps the last major construction project on the US urban freeway system is the "Big Dig," Boston's relocation of its urban freeways from an elevated highway to underground.

The story of the Big Dig began with the pre-Interstate construction of the Central Artery, an elevated highway through downtown Boston that was funded by the state. Like other highways constructed in the United States in the 1950s, the highway reduced traffic congestion in Boston for a while, but by the mid-1960s, Boston's highways and local roads were again heavily congested. Highway planners proposed a ring road around the downtown central business district as a solution to the congestion, and a third tunnel was proposed in 1968 for construction between downtown Boston to near Logan Airport.

The question of whether additional roads should be built raged in Boston during the late 1960s and early 1970s. Two projects survived: a restricted use tunnel from downtown

Boston to Logan Airport, and reconstruction of the Central Artery that would relocate the highway underground, eliminating the division between downtown and the North End and Waterfront neighborhoods, which had been determined to be a detriment to the city by this time.

Over the next twenty years, the Central Artery/Tunnel project evolved through complex negotiations into what is being constructed today: the most expensive urban highway project ever undertaken in the United States. The project was only made possible because in the early 1980s the Speaker of the House of Representatives, Thomas P. "Tip" O'Neill, served the Boston area, and on retiring was able to get the highway as a "present" from Congress and then President Ronald Reagan. Nearly every component of this project underwent numerous modifications to address concerns of special interest groups—modification that often resulted in raising the concerns of other special interest groups. Mitigation costs mounted. Ultimately, it is expected that most of the project's more than $14 billion cost is due to mitigation. The engineering challenges of constructing a tunnel underneath a highway, while keeping that highway operating, should not be underestimated.[417]

The Big Dig is an exception to highway planning in the last quarter of the twentieth century. The Clean Air Act Amendments of 1990 (CAAA-90) restricted highway development in air pollution non-attainment areas and promoted actions to restrict vehicle uses. The Intermodal Surface Transportation Efficiency Act of 1991 (ISTEA-91) reformulated funding to allow for a "level playing field" between transit and highways, allowed for direct relations between the feds and councils of governments in large urban areas, authorized increased funding for transit, and also authorized a National Highway System (NHS). Legislation in 1995 implemented the NHS. The Transportation Equity Act for the 21st Century (TEA-21) basically followed the formula of ISTEA.

According to conversations with project champion Fred Salvucci, no consideration was given to just tearing down the Artery and not replacing it, as has been done with many urban freeways since.

22.4.2 BUILDING SUBURBAN HIGHWAYS: INTER-COUNTY CONNECTOR

In 1950, the National Capital Planning Commission proposed an outer circumferential freeway (Outer Beltway) for the Washington, D.C., area. This beltway was to fall beyond the radius of what later became the Capital Beltway (I495/I-95). The Capital Beltway has been called Washington's Main Street.[418] Would the Outer Beltway be similarly profound? Importantly, the Outer Beltway was placed on land use maps, so that residents would know where the right-of-way of this route was to be. Local government began to acquire the land in the right-of-way of the road so that it would be preserved.

However, some parts of the Outer Beltway appeared incompatible with local land use plans, in particular, building new roads through an area designated as rural. Further, Virginia residents in the wealthy district of Great Falls opposed the Interstate highway in their neighborhood, where the bridge would land. Thus, in 1968, the Maryland National Capital Park and Planning Commission (a bi-county agency serving the Maryland suburbs, and distinct from the National Capital Planning Commission) took sections of the

Outer Beltway connecting what become I-270 with Virginia (which would better connect to Dulles Airport (Section 22.4.4) controlled by the Metropolitan Washington Airports Authority, an agency controlled by Virginia and Washington, D.C.) off the map. Still, the sections connecting I-270 in Montgomery County, Maryland, with I-95 (and nearby US 1) in Prince George's County (reducing travel times to Baltimore-Washington International Airport, controlled by Maryland DOT) remained on the plan, and were renamed the Inter-County Connector (or ICC—not to be confused with the former federal agency of the same letters). This is about a 29 km (18 mile) highway, depending on the final right-of-way chosen.

The Capital Beltway was opened in 1964, and by 1979 transportation officials were beginning to seriously consider the ICC. The Maryland State Highway Administration (SHA) initiated the first of two project planning studies for the ICC. By 1983, the first Draft Environmental Impact Statement (EIS) was issued and public hearings were held. A small section of the road, I-370, connecting eastward from I-270 to the Shady Grove Metro Station, was constructed; however, budget and environmental issues kept most of the road as proposed.

The original alignment, and later southwestern spur of the ICC (connecting to I-270 south of the City of Rockville) dubbed the "Rockville Facility," would cross Rock Creek Park and the Northwest Branch. The planning agency and state government had acquired much of the land in this Master Plan alignment to preserve the right-of-way. Government ownership of land does have its drawbacks, and in 1987, Maryland state senator Idamae Garrot was able to pass a state law creating Matthew Henson State Park in part of the road right-of-way, and prohibiting the state from building the highway. The county has constructed part of the western segment of the road (now the Montrose Parkway) in a suburban area without significant environmental impacts.

Another study took place in the 1990s, and in 1997, the SHA issued a second Draft EIS. But the road remained controversial. Residents were mixed; some sought the road to relieve congestion, others opposed the road for its environmental impacts, in particular, its crossing near the headwaters of otherwise protected stream valleys (the road runs east-west, most streams in Montgomery County run north-south). Proponents claimed, "The ICC is not about promoting tomorrow's growth—it's about providing for yesterday's growth."

The Army Corps of Engineers, which regulates wetlands, was a major player in the EIS, and proposed a number of new alignments for consideration not in the master plan right-of-way. Their objective was to protect the streams, and the spawning grounds of German brown trout, which were introduced at the beginning of the twentieth century and now reproduce naturally.

The additional rights-of-way greatly increased opposition, as now more people would be affected by a possible road than previously; moreover, people bought land and homes believing they would not be affected by the road because of Montgomery County's long-standing practice of building roads at the place they were on the plan.

In 1999, Maryland governor Parris Glendening halted the ICC planning study and canceled the middle section of the road, while keeping the eastern and western portions on the map. "I will not build the Inter-County Connector. As far as I'm concerned there is no ICC."[419] His "Solomonic" decision did not have the effect hoped for (killing the road). Democratic Lt. Governor Kathleen Kennedy Townsend, running for governor when Glendening's term expired, was more politic: "If I had an answer, I'd give it to you.... But with the legal impediments to the ICC, you have to cut your losses and build all of these other roads."

In 2002, Townsend lost to Republican Robert Ehrlich, a road proponent. A third Draft EIS was conducted by the SHA, which, under Ehrlich, made this Maryland's highest priority new project. The federal government fast-tracked the environmental review process. Lawsuits were won. To pay for the road, and restrict traffic, tolls were imposed. Construction contracts were let. In November 2011, the road was completely opened (as Maryland 200) between I-270 and I-95.

Maryland 200 is a very nice ride, perhaps one of the more pleasant roads to be on. Compared with most long-distance roads in Montgomery County, it does not (yet) suffer congestion (and with tolls, it may almost never). So long as the original opponents of road remain alive, there will be grumblings about the road. But, like I-94 above and all infrastructure, it will eventually become part of the landscape. As of 2012, it carried 34,000 vehicles a day in its peak section, lower than the traffic flow before tolls were started (44,000 vehicles a day).[420]

The fears about induced development (development that would not take place but for the road) and induced demand (trips taken or lengthened that would not take place but for the road) will eventually be realized. Transportation creates accessibility; new projects change the accessibility landscape. Developers and travelers will take advantage.

The road, a dashed line on the map for over fifty years, in serious consideration to move forward for over twenty five years, is finally a solid line on the map, and on a map more complicated than previous. It took so long to build not because of a lack of information, or a lack of public participation, but because of a lack of political will and paralysis by analysis. If you can't win on the facts, you propose another study. As researchers, we think studies and information are good things, but studies should not be used to justify, rather to inform. It is the decision makers who must decide.

Transportation and land use interact. Foremost, transportation creates access, and by creating access, transportation embodies that land with value. Second, transportation consumes land; it does this for the linear facilities themselves (roads and rail networks) as well as the nodes (ports, airports and parking). In urban areas, the land devoted to parking is surprising (see Section 15.5). This is a cost of transportation facilities, but is paid for during construction. Third, transportation causes negative externalities for those who are adjacent to facilities, including noise, pollution, the potential for fire (from steam powered trains), and severing communities (both human and ecological) along their routes.

The debate about the Inter-County Connector touched on all three issues, and basically asked the question: Does the improved access offset the negative externalities? Proponents would argue that the negative externalities from the road are offset by reduced negative externalities on other routes. Opponents would argue that the gain in relative accessibility in Montgomery County is offset by the loss in relative accessibility to places not on the route. Further, the road would encourage workers in Prince George's County to work in Montgomery County (thereby making the rich richer, since Montgomery would get jobs without the cost of supporting residents). How these qualitative arguments balance quantitatively depends on where you sit.

22.4.3 THE LAST INTERSTATE

I-69 runs from the Canadian border at Port Huron Michigan to the I-465 beltway around Indianapolis. Advocates for freeways, led by David Graham (descendent of the founders of automaker Graham-Paige) in southeastern Indiana, proposed extending the route in that direction.[421] Recognizing there is no national interest in a small highway segment, they built a coalition for a 2,700 km (1,700 mile) highway through Tennessee, Kentucky, Mississippi, Arkansas, Louisiana, and Texas. The Southwestern Indiana Highway Coalition was expanded to a Mid-Continent Highway Coalition. This local road was now of national interest, perhaps as a NAFTA highway (connecting Mexico with Canada, though one imagines few trucks would make such a long journey), but garnering the interest of Congressional delegations from eight states.

Federal "High Priority Corridors" 18 and 20 designated in 1991's ISTEA became part of the proposed extension of I-69. Much of this involves upgrading of existing routes, but there is new construction required as well, which has attracted opposition from environmental groups and local preservationists. The entire route has been divided into "sections of independent utility" (or SIUs), which are supposed to be constructed independently, and as the name suggests, be justifiable independently as well.

Whether the route as a whole is economically justified is also debated. Opponents note that existing Interstates can get you from Laredo to Port Huron in 2,746 km (1,706 miles), while the proposed I-69 is 2,881 km (1,790 miles).[422] Laredo (one of the possible termini on the Mexican border) has Interstate service, McAllen and Brownsville in Texas (other possible termini) do not. There are obviously local time savings for individual communities that would be on I-69 and are relatively far from other Interstates. Whether that justifies the cost, or the federalization of the highway, is not clear.

22.4.4 PRIVATE ROADS? THE CASE OF THE DULLES GREENWAY, VIRGINIA

The first private toll road in Virginia since 1816, the Dulles Greenway is a 22.5 km (14 mile) western extension of the Dulles Toll Road, connecting Washington Dulles International Airport (opened in 1963) with US Route 15 in Leesburg. The Dulles Greenway originated in 1988 with the Virginia General Assembly authorizing the private development of toll roads. Construction began in September 1993 and the road opened for service on September 29, 1995, which the owners (Toll Road Investors Partnership II) note was six months ahead of schedule and under budget. On opening, it comprised seven interchanges, thirty six bridges, a toll plaza, twelve ramp toll barriers, an administration building, and four operational lanes. It further allowed for construction of two additional lanes, two additional interchanges, and for a rail system in the median.

Dulles Greenway is one of the first toll highways in the United States that was designed, built, and financed in the private sector since the end of the nineteenth century turnpike era. Dulles Greenway is the fourth highway segment comprising the Dulles Transportation Corridor, Virginia.

- The Dulles Airport Access Road (DAAR) was built by the Federal Aviation Agency (FAA) and opened in 1962 along with the Washington Dulles International Airport. It is a no-toll, 12-mile long, four-lane expressway serving only airport traffic.
- The Dulles Toll Road (DTR), which opened in 1984, was built by the Virginia Department of Transportation (and now owned by the Metropolitan Washington Airports Authority). It consists of four and six lanes paralleling both sides of the DAAR. It has full interchanges and no restrictions on vehicle size or type.
- The Dulles Access Road Extension (DARE), a no toll, four-lane, 2-mile long expressway extending eastward from the DAAR and DTR to I-66 by Falls Church with both truck and occupancy restrictions was completed in 1985.[423]

The initial toll was $1.75 each way and did not vary with the length traveled along the highway or the time of day. In comparison, the toll on the Dulles Toll Road was only $0.85. The volume of traffic averaged 11,000 vehicles daily, much less than estimates of 25,000 vehicles per day (*Washington Post*, Jan. 14, 1996). Suggestions for increasing patrons included lowering the toll for non-rush hours and weekends, pricing comparable for length traveled along the highway, better marketing strategies, increasing the speed limit, and establishing electronic tolling. The investors believed that the highway might have been five years ahead of itself, since development was just beginning along the project. Michael Crane, one of the proprietors, when asked what he would do differently, next time, stated, "I would never do it as a totally private toll road. I would do it as a public-private venture."[424]

The Virginia Senate, in February 1996, passed bills allowing the Dulles Greenway to have a speed limit of 65 mph and to obtain Federal Highway loans. The toll was reduced from $1.75 to $1.00 in 1996, and volume doubled, but the owners missed a $7 million interest payment to its creditors and a $3.6 million payment to the State of Virginia in July 1996.[425] The owners were given time to refinance the $350 million debt and succeeded in April 1999. A higher speed limit of 65 mph, electronic toll collection, and a frequent user program were credited with the highway having 40,000 vehicles per day in 1999.[426]

Development expanded significantly in 1999 with then fast-growing Internet companies MCI WorldCom and America Online (AOL) having offices in the corridor. Tolls were increased for the Dulles Greenway to $1.40 for cash and $1.15 for electronic payment. Loudoun County began analyzing large developments for the corridor.[427] Future population projections resulted in the owners expanding the Dulles Greenway. Construction for an eastbound lane began in June 2000 and was completed in December 2000. The westbound lane began construction in the spring of 2001 and was completed in August 2001. Construction costs for the additional five miles of two lanes were estimated at $10.4 million.[428] Usage of the Dulles Greenway increased to over 60,000 vehicles per day in June 2002, a 7 percent increase from a year earlier.[429] Loudoun County has experienced rapid growth in the last few years, which significantly increased highway usage.

The Dulles Greenway is one of the few experiments with private building and financing of a project that had in the past been accomplished with public resources. The risks that the private partners incurred were an extremely large leveraged debt, a long time frame before profitability, a project subject to economic downturns, and competition from untolled roads. In addition, Dulles Greenway could not raise its toll above $2.00 unless the Virginia State Corporation Commission (SCC) gave the owners permission. The main advantages that the Dulles Greenway highway realized were that the lenders were willing to negotiate and wait for payments and the highway was built in an area that was expanding. Though a bit premature, the road may yet turn out to be profitable.

The Dulles Corridor itself continues to grow fast, as government contractors and high-tech companies choose to locate with easy access to downtown Washington, suburban Tysons Corner, and the airport.[430]

Tysons Corner, Virginia, exemplifies the pattern of suburban cross-roads emerging as centers are widespread. Tysons Corner is the largest job center in metropolitan Washington, D.C., and the twelfth largest business district in the United States.[431] Tysons Corner was at the intersection of Route 7—Leesburg Pike (running from Alexandria, Virginia, to Leesburg, county seat of Loudoun County) and Route 123 (Chain Bridge Road/Dolley Madison Boulevard), running from the Chain Bridge (in Washington, D.C., just north of Georgetown) to the City of Fairfax, Virginia (county seat of Fairfax County). The value of this once rural crossroads was enhanced with the construction of the Capital Beltway (I-495) and the Dulles Airport road in the 1960s near the crossroads. First a shopping mall and later offices were located at Tysons. In this case, the nucleus occurs because Alexandria and Georgetown, both part of Washington, D.C., when it was created (and before Alexandria was retro-ceded to Virginia) were small centers in their own rights as ports on either side of the Potomac River. Since cities are not points, routes which travel from a hinterland to one point in a city may cross another route connecting a different hinterland point to a different point in the city.

Proposals to tie the Washington Metro (discussed in Section 23.3) to the Dulles Airport via Tysons Corner have been around since the earliest days of the airport planning, when the access road preserved in the median a right-of-way for potential rail transit. Plans were proposed in the 1990s by private firms to build and operate the line. Ultimately, most of the profits of the Dulles Toll Road were dedicated to funding the new Metrorail Silver Line. (One of the major debates around the line was whether the section through Tysons Corner should be elevated, at less cost, or tunneled, which would provide a better urban form). In the event, cost won out. Project-related construction began in 2008, with expected completion date of Phase I in 2013, and a hoped for completion of Phase II, to the airport in 2016.

22.4.5 HIGH-OCCUPANCY VEHICLES

On December 30, 1940 (just prior to the Tournament of Roses parade), the Arroyo Seco Parkway connecting Los Angeles and Pasadena, California's first freeway, opened to traffic. The road was perceived to be so successful that a full-fledged freeway network followed. The Santa Monica Freeway (I-10) in Los Angeles appeared on planners' maps as early as 1956 (as the Olympic Freeway), saw its first segment opened to traffic in 1961, and was completed by 1966.

While freeways were still widely (though not universally) lauded, transportation professionals recognized that different vehicles had different values of time, and there might be some gains to give priority to those with a higher value. To that end, High-Occupancy Vehicle (HOV or Diamond) lanes were first deployed on the Shirley Highway (I-95) in Virginia outside Washington, D.C., in 1969 as an exclusive bus lane. Today, there are over 3,200 km (2,000 miles) of HOV lanes in the United States. By restricting traffic and ensuring freeflow speeds, HOV lanes provide an incentive for travelers to form carpools, and thus reduce the number of vehicles on both the freeway with the HOV lane, as well as the roads leading to the freeway. They also reward existing carpools, as well as users of buses and vanpools, who take advantage of the faster routes. By privileging these high value vehicles, total social benefits are supposed to increase.

Following on the success of HOV lanes such as the Shirley Highway, and with little advance warning to motorists, in 1976 the California Department of Transportation (Caltrans) converted a lane of the Santa Monica Freeway to HOV 3+, requiring three or more persons per vehicle to use the lane. This provided an incentive for carpoolers. By converting a lane, Caltrans made the time difference between the HOV and other lanes greater, but thereby "punished" those who did not use HOV by reducing the available capacity and increasing their travel time. Evidence showed rising HOV use and transit ridership soon after. However, this "take-away" strategy led to vociferous protests from motorists, who felt a real loss, and within a few weeks, Caltrans relented and reverted that lane back to general-purpose traffic. The lesson that was learned (and learned perhaps too well) is that "take-away" strategies are unacceptable, and HOV must be the result of additional capacity. While HOV users can benefit, other users must not be harmed.

The consequence is that when an HOV lane opens, bus passengers and carpoolers of course benefit, but those who drive alone also benefit because there are fewer carpools congesting their lanes. This greatly diminishes the potential time savings that HOV provides. HOVs only work well (better than general purpose lanes) within a small window.[432] If the mainline is uncongested, or will be so after the HOV lane is added, there is no benefit. If the HOV lane itself becomes congested, there also will be diminished benefit.

An example of this is an experiment that took place on the El Monte Busway demonstration project, along the San Bernardino Freeway in Los Angeles. The Busway was opened in 1973, and HOV 3+ were permitted to use the facility in 1975. Still, the lane had the appearance of being under its vehicle-carrying capacity. In 1999, the California legislature instructed Caltrans to lower the vehicle occupancy required to use the facility. A study of this experiment found that the addition of HOV 2+ carpoolers caused congestion in the carpool lanes, worsening reliability and increasing travel times for both cars and buses, yet freeway travel times in the general purpose lanes did not see significant improvements. In July 2000, the road became HOV 3+ again.

The HOV lane enabled a new transportation phenomenon, dubbed the "slug line" in metropolitan Washington (or instant or casual carpoolers in somewhat more jargon-laden transportation talk). Slugs, or members of the slug line, are commuters who wait at known locations (typically at or near bus stops) to be picked up by single occupant vehicles (SOVs) hoping to use the HOV lane, but lacking a passenger.[433] The phenomenon emerged as SOVs cruised bus stops looking for passengers, but soon individuals with no intention of riding the bus would queue up. The phenomenon emerged spontaneously, without government organization or posting of signs. Whether this is a good or bad idea we will leave to the reader's judgment. However, the slugs on average double the value of the cargo carried by passenger vehicles. Whether this process eliminates another vehicle on the road, or a passenger on the bus, depends on the slug. Slugging briefly emerged in New York City after Hurricane Sandy and the resulting closure of many subways and the conversion of streets in Manhattan to HOV3+.

22.4.6 HOT LANES AND HOT NETWORKS

In the late 1990s, High-Occupancy/Toll (HOT) lanes that would allow single occupant vehicles to use HOV lanes (after paying a toll that depends on the level of demand) began to be considered in many US metropolitan areas. HOT lanes were conceived of by Ward Elliott at Claremont McKenna College in the 1970s, and reinvented in the 1990s by Gordon Fielding and Dan Klein as part of a Reason Foundation study.[434] HOT Lanes were introduced on I-15 in San Diego, for instance, using the same facility that had been the testbed for the Automated Highway Systems experiment described in Chapter 30, and have generally been viewed as successful. The revenue generated after paying for operating the facility helps subsidize transit in the corridor.

Seeing the success of I-15, as well as a handful of other HOT lanes, Poole et al. (1999) call for networks of HOT lanes (or HOT networks) in major US cities. They conclude that the benefits of such a system (including congestion reduction for those not using the system) outweigh the costs.

While pricing every congested road using marginal cost pricing may increase system efficiency, providing differentiated services further enhances efficiency. We recognize that different types of freight have different priorities (overnight, two-day, and ground are choices for shipping); that same kind of differentiation applies to drivers. Different drivers have different values of time at different times of the day. Thus the ability to pay a premium and travel at a better level of service during peak times provides a service not currently available, a service enabled by bundling several ideas. The old idea is tolls themselves, whose original intent was simply to raise revenue. Electronic toll collection complements that; its original intent was simply to automate the collection of tolls at traditional tollbooths, reducing both traveler delay and agency operating costs. HOV lanes aimed at giving priority to vehicles carrying more passengers (vehicles which had a higher value of time, since two people are more than one).

The HOT networks also provide the ability to provide bus rapid transit (BRT) services in metropolitan areas, by providing the high speed limited access routes that give transit a travel time advantage over the automobiles not paying for the HOT lanes. In many situations, buses are more cost-effective than fixed rail alternatives, but the lack of a fast right-of-way leads people to perceive rail as inherently faster than the bus. With BRT, that perception can be made to disappear. Several cities are now building-out HOT networks through a combination of new construction and conversion of HOV lanes.

22.5 Ideas: Privatization versus a Public Utility Model

A 2012 proposal by Prime Minister David Cameron to "privatize" UK roads, by contracting out management of the roads in exchange for a stipend of taxes (but notably not tolling existing roads, only new construction)[435] promises a short-term revenue fix (and possibly better managed roads) in exchange for less funds downstream. In Great Britain, after World War II, public corporations managed most utilities (electricity, gas, water, and rail) while others remained within the public sector (post and telecommunications, roads). The Thatcher administration successfully privatized British Telecom in 1984 and other public utilities in subsequent years, including bus transit and some rail transit, but not roads. The government retained the power to regulate these natural monopoly industries.

In many countries, freeways are operated by private sector firms under a franchise or concession agreement with the government, which usually retains underlying ownership of the road.[436] As of 2004, more than 37 percent of motorway length in the EU25 plus Norway and Switzerland was under concession, and 75 percent of that was privately operated.[437]

There is even limited experience in the United States with contracting operation of existing roads, which has not been without controversy; the most notable examples are the long-term leases of the Indiana Turnpike and Chicago Skyway.[438] New toll roads built and operated by private firms are much more widespread, and include the Dulles Greenway and Pocahantas Parkway in Virginia, and the Adams Avenue Turnpike in Utah. This experience applies well to toll roads, and variants such as High Occupancy/Toll (HOT) lanes, as discussed above, and Truck-only Tollways.[439] California's SR-91 median toll lines were privately built on public right-of-way, and later bought out by a public toll agency. Presently, the MnPass HOT lanes in Minnesota manage toll collection under a concession to private organizations. A large share of the few new limited-access roads built in the United States have adopted the toll model, and more could follow suit.[440]

Yet, most roads, and even most freeways, in the United States are not toll roads. Strategies such as mileage-based user fees or vehicle mileage taxes,which replace and improve upon existing motor fuel taxes have been vetted, and may ultimately be implemented. But allocating funds to particular roads, while technologically straightforward, may face resistance from privacy concerns.

There are technical solutions to privacy issues, but implementing these, in the face of the desire of security agencies to be able to track individuals, will be difficult. It may turn out with cameras, cell phones, and other devices, we lose privacy about our whereabouts well before road pricing is implemented. A solution that David Brin (1998) suggests is a *Transparent Society*, where everyone can watch everyone, so the state does not have a monopoly on monitoring. Based on historical experience,[441] implementing tolls on existing untolled roads is likely to be politically difficult and unpopular. A 2007 petition in the UK to then Prime Minister Tony Blair beseeched:

> The idea of tracking every vehicle at all times is sinister and wrong. Road pricing is already here with the high level of taxation on fuel. The more you travel—the more tax you pay.
>
> It will be an unfair tax on those who live apart from families and poorer people who will not be able to afford the high monthly costs.
>
> Please Mr Blair—forget about road pricing and concentrate on improving our roads to reduce congestion.
>
> —The petition, now closed, can be found at: `http://petitions.number10.gov.uk/traveltax`

This petition to scrap "the planned vehicle tracking and road pricing policy" was signed by more than 1.8 million UK residents by 2007, more than any other petition in history. It clearly has informed Cameron's proposed policy.

Further, the problem of rates differing by route (such as marginal cost prices, the theoretical ideal from a micro-economics perspective), would undoubtedly increase system complexity and distrust, with likely only small gains from system efficiency. Our best estimate from computer models is that moving from a user equilibrium solution, where each driver selfishly chooses his or her own route, to a system optimal solution where each driver chooses a route that is best for society is less than 5 percent reduction in total vehicle hours traveled in the Twin Cities. This suggests the "price of anarchy" (the ratio of user equilibrium to system optimal travel times) is not large on real road networks, despite externalities such as congestion, and imperfect competition among roads. Much larger gains are to be had if travelers shifted to different times of day, but that need not be route-specific.

If the rates were set by private firms in an unregulated manner, monopoly links would have higher prices and be rightly perceived as exploiting their position. In a robust network, monopoly routes are scarce, often there are many viable paths between given origins and destinations, but local monopolies remain, especially on poorly designed, or geographically constrained networks. While there are innovative economic solutions, it is likely that a disjoint system of too many road operators, in addition to being complex and unpopular, may be inefficient as economies of scale and network externalities are not fully realized.

Recent toll road privatizations indicate a change in government intervention which sees "transitions from internal control on processes and inputs to external control on performance outputs."[442] Toll privatization results in an increase in price regulation. In Europe, privatization entails transfer of management and operation (through concessions) for a time period, while underlying asset ownership is retained by the government. It is widely observed in the public management literature, which found that more agency autonomy is accompanied by an increase in external controls. Still, focusing on the outputs (the performance measures) rather than on how those measures are achieved should, by decentralizing decision making, produce a more efficient outcome.

Economic solutions to the monopoly problem include auctions for the privilege for operating routes, which would allow the public to recover these monopoly profits, or reverse auctions where firms would bid to charge the lowest rate to operate the route. Future franchising such as Present-Value of Revenue (PVR) auctions may entice government agencies to reconsider the toll finance mechanism. In PVR auctions, private firms compete through bidding for the present value of toll revenue they want to obtain from the project.[443]

A model that has been insufficiently explored in the United States is that of public utilities. Many utilities share with transportation systems the characteristic of having a networked structure. Most, if not all, of these utilities are operated on the basis of a payment-for-use system. Utility pricing varies regionally, some locales vary prices by time of day, and users often have the option of choosing different rate plans. These models are never strict marginal cost pricing, but they may improve upon average cost pricing. There are strong parallels between public utilities and transportation services, though some differences exist in the nature of the services consumed, the role of technology, and the structure of institutions and decision making.[444]

Water faces similar difficulties to transportation in the ambiguity of appropriate property rights. Institutional reforms began in the twentieth century to better allocate water resources and to improve the efficiency of water use. The perspective of water changed from being perceived as a free good to a scarce economic good—a change that took place around the world.[445] Institutional reforms differ by political setting and social environment;[446] decentralization (from central to state and municipal governments) took place in Mexico and Brazil, while corporatization and privatization occurred in Chile, Brazil, France, the United Kingdom, Australia, and New Zealand, among others.

Hillsman (1995) suggests four categories in which utilities have developed to manage demand:

- Altering infrastructure,
- Packaging services,
- Substituting technologies, and
- Changing the price of service.

Transportation agencies have considered all of these, but implemented them weakly. In reverse order: Prices are largely invariant, technological (modal substitutions) are not viable for most passenger or freight users, bundling and packaging of services is not considered when looking at pricing, and infrastructure is hidebound to engineering standards, and difficult to modify. One could easily imagine more creativity on the part of road providers in all of these aspects. The constraints on the application of creativity are due to the engineering culture in a public agency, where risk-taking is discouraged if not punished, and certainly never rewarded.

With some modification, it seems possible to transfer the utility model of governance to road transportation. This model separates the organization delivering the service from the client, is subject to rate regulation, and implements a more direct, user-pays system of financing. This model could depoliticize management of the existing transportation system. Whether rate regulation is in fact *economically necessary* is the subject of debate; for instance,[447] arguably there is no difference in prices in the electrical sector due to regulation, because electricity is competitive with other energy sources in the long run. One expects from experience with other utilities, toll roads, and road concessions in other countries that it would be *politically necessary* to have some public guarantee of an upper bound on the rates a road utility could charge, as provided by a regulatory agency. The risk is that an upper bound on revenue would be too tight, resulting in financial losses (and one of the causes of municipal takeover), as occurred in the then private mass transit sector throughout the United States in the early to mid-twentieth century.

Such a system would transform but not replace public highway or transportation authorities as the party responsible for providing and maintaining roads. One example of a transportation system that has transitioned to more of a utility-based model is the road authority in New Zealand.[448] This system was designed to be self-financing, with what was originally called the National Roads Board allocating charges among users on the basis of costs incurred. Three types of costs were identified: load-related costs, capacity-related costs, and driver-related costs (covering signing and other costs not related directly to road use).

There are other elements of costs not included, such as *access costs* (the cost of accessing the network from land) and the cost of a connected network, which can be separated from capacity costs (related to the width of the roadway), and *load costs* (related to the thickness of the roadway), and *environmental costs* (both how the system deteriorates due to weathering independent of use, and how the environment is degraded due to use).

Vehicles are split into two classes on the basis of weight, with vehicles less than 3.5 tons paying a charge in the form of a fuel tax. In the United States, Oregon has had for many decades a weight-mile tax for heavy trucks. Heavier vehicles pay a distance license fee, which is essentially a form of weight-distance tax. Such a system is relatively straightforward and requires minimal new technology, leading to low collection costs compared with most proposed road pricing systems. Researchers have also estimated the costs and relevant charges for a similar, though hypothetical, system of user charges for the United Kingdom.[449] Despite the long experience with weight-distance tax in Oregon and outside the United States, it has yet to be emulated.

These types of road user charging schemes contrast with user charges based on a mileage tax concept utilizing GPS systems.[450] There are a variety of potential technologies for assessing mileage taxes; most use GPS (or an equivalent such as cell-phone triangulation) to identify location, since one of the advantages of these types of systems is the ability to charge different rates for different locations (city vs. country, freeway vs. local street, congested vs. uncongested road). GPS receivers do not normally transmit information. GPS-equipped vehicles can log the vehicle location internal to the vehicle. Some additional communication technology, which might report a reduced form of information (e.g., total amount owed) would be used to complete the transaction. For instance, a pilot study in Oregon[451] had a chip in the vehicle log distance traveled by zone (an aggregated version of location) and time of day, without storing the precise location. The chip only reported to the external source the total charge owed, calculated by an onboard algorithm. So no detailed tracking information was shared. Simpler technologies, such as a mileage-based user fee, would simply record the odometer reading, but this would not allow differentiation by time of day or location.

While the road user charging concept remains an attractive prospect, its application may still be many years away due to a combination of privacy concerns, implementation and transaction cost issues,[452] and technological development issues. Some of these concerns might be obviated under a different governance structure, where it was neither the legislative nor executive branch of government making these decisions. Public utilities have a "mean level of trust" of 42 percent,[453] which is much higher than the trust in the federal government, which hovers in the 20 percent range.[454] Dynamic pricing, as suggested for toll roads, significantly reduces consumer's trust in an organization,[455] as prices are no longer predictable and feelings of price gouging take place.

The discussions of road pricing for financing and congestion management in the United States are still largely under the guise of existing institutions doing the pricing. To date, this has essentially been a non-starter. Perhaps with institutional reforms, reconfiguring state and local DOTs as public utilities rather than departments of state and local government, the logic the public applies to roads will change, from one of a public service paid by the pot of general revenue to a fee-for-service proposition paid for by direct user charges.

22.6 Freight

22.6.1 FEDERAL TRUCKING REGULATION

> There was Anthony Stracci, who controlled the New Jersey area and the shipping on the West Side docks of Manhattan. He ran the gambling in Jersey and was very strong with the Democratic political machine. He had a fleet of freight hauling trucks that made him a fortune primarily because his trucks could travel with a heavy overload and not be stopped and fined by highway weight inspectors. These trucks helped ruin the highways, and then his road building firm, with lucrative state contracts, repaired the damage wrought. It was the kind of operation that would warm any man's heart, business of itself creating more business.
>
> —Mario Puzo, *The Godfather* (1969), pp. 283–284

The extension of federal regulation to trucking appears to have posed no very difficult problems. By 1935, the industry was sizable and growing in spite of the Great Depression. Unregulated, there was fierce competition among trucking firms—competition claimed to be threatening the stability of the industry. At least a few railroads were beginning to feel the threat of truck competition. An important matter was the availability of a model and an existing organization—railroads and the Interstate Commerce Commission (ICC). The Depression grew and the truckers and the railroads sensed a crisis of competition. With little debate and the agreement of all concerned, regulation was enacted. Protocols previously developed for the railroads were applied. The ICC granted antitrust exemption, rate bureaus were created. The freight classes used by the railroads were adopted. Shippers paid about 110 percent of the rail carload rate for truckload shipments and about 115 percent in less-than-carload/less-than-truckload situations. The situation stabilized.

Unlike the railroads, which were no longer growing, the truckers wished to be protected from new entries. The Commission applied a good part of its resources to actions relative to operating rights. A firm could only expand through the purchase of valuable operating rights. The creation of such monopoly values hardly seems to have been in the national interest.

It is one thing to create a federal activity; it's another to change an activity. Deregulation of trucking appears to have been an extreme case on the simple side. It was difficult to develop a case for a national role for regulation of trucking, for the truck business is hardly a natural monopoly; nor do the fortunes of a few firms have national impacts. In addition, there was the recent model of rail deregulation, which occurred in 1980.

Efficiency gains from trucking deregulation are said to save shippers about $25 billion per year (about 10 percent of the truck freight bill). Several old line, unionized LTL carriers had difficulty adjusting to the deregulated environment and went out of business. The adjustment period seems to have passed quickly.[456]

Yet, the desire of the trucking industry for a trucking agency in the USDOT is an example of a desire to have an advocate with influence in power, even when regulation is slight. This modest change in structure, promoting an existing group, resulted when the Federal Motor Carrier Safety Administration was created in 1999 (taking responsibility out of the Federal

Highway Administration). Here the dysfunction was that other modes were represented in the Department, but trucking wasn't. The American Trucking Association heralded the agency as "a major victory for highway safety."[457]

22.6.2 IDEA: BIFURCATION

> ... nothing seems more certain than that many special highways will be constructed for motor trucking.
>
> —*Roads and Streets*, December 1928, p. 569

Per distance traveled, large trucks are three times more likely to be involved in fatalities than autos. Trucks take twice as long (or more) to stop as autos. They congest urban traffic. Truck stability on freeway off-ramps at posted speeds is a well-documented safety problem. Operators strive to achieve the benefits from operation of longer combination vehicles (LCVs). We are unable to make the investment in road facilities needed by trucks.

The auto-truck-highway system was birthed and grew as a multipurpose network. The truck versus auto problem is the result. If the system had the adaptive capability of bifurcation (splitting into specialized networks), those (and others) would be non-problems.

Conventional actions focus mainly on improved truck brakes, and stability when braking, and are achieving modest improvements. If we really wanted to manage that problem, the action to take would be to change system structure: separate autos and trucks.

Major actions have multiple effects. This same action would offer opportunities for reduced costs: some roads tailored to larger, heavier vehicles, other roads limited to cars and light trucks only. Results might be more efficient road provision and efficiencies in equipment and operations areas.

Trucks for rural use began to become available in the 1910s and 1920s, ranging upward in size from the Model-T Ford with a simple truck bed in back. At first, these were substituted for wagons for existing farm to market and urban pickup and delivery services.

Although trucks moved materials by road from the Midwest to East Coast ports during World War I, as late as the mid-1920s the wisdom was that only small amounts of inter-city freight traffic would ever move by truck. Rail had a firm hold on that market, with air transport having long-term potential for high-priority shipments.

However, by the mid- to late 1930s, the development of inter-city freight movement by trucks was well under-way, well enough under-way that Interstate Commerce Commission regulation began. The development of the state primary highway system played a major role in the growth of intercity truck services, as did improvements in truck equipment and the skills developed by trucking firms.

The inter-city business grew at first in market niches—movement of household goods, agricultural livestock, and automobiles, in particular. Other services and the emergence of the common carrier, private, and owner-operator segments of trucking activities were in place prior to World War II, and they grew very rapidly afterward.

Questions of the interrelations of equipment and roads emerged as truck axle weights began to be controlled to protect pavements. By the mid-1910s, the cities had begun to control weights—their problem, at first, being solid-tire coal hauling and construction vehicles.

By that time, the rural road problem was regarded as tamed by the Federal Office of Public Roads. However, movement of trucks during World War I broke up roads badly, and a search was started for appropriate rural road designs. The Bates (Illinois) and Pittsburg (California) road tests provided information for pavement designs; long-standing bridge structure knowledge was available.

This knowledge, together with information on truck dynamics, was used in the design of state primary and secondary systems and, later, the Interstate. Interstate design also drew on the findings of the AASHO road test undertaken in the late 1950s. Compared to the primary systems, the Interstate was designed with wider lanes (minimum of 3.6 m [12 feet]), lower grades (3 percent maximum rather than 5), and higher overpass heights, all affecting equipment size. A design velocity of 112 km/hr (70 mph) was incorporated in the Interstate. (There are design exceptions on parts of the Interstate, e.g., some grades are greater than 3 percent, and many primary and secondary routes, have been upgraded to Interstate or near Interstate standards.)

Although the federal government working with AASHO developed construction standards for the Interstate and also standards for maximum truck size and weight, the latter remained a state matter in many cases, for existing state regulations were grandfathered when Interstate regulations were set. Another way to put the matter is this: federal weight and size regulations said that the states must allow trucks meeting those standards on the Interstate, but the states could allow larger and heavier trucks. (The 1956 Interstate Act limited width to 2.48 m [98 inches], single axle loads to 8,200 kg [18,000 lbs], tandems to 14,500 kg [32,000 lbs], and gross vehicle, weight [GVW] to 33,300 kg [73,280 lbs.]) In general, the states west of the Mississippi River allowed higher gross and axle weights than those east of the Mississippi, and a North-South band of states (Illinois and southward) had lower axle weight standards than others.)

Responding to the trucking industry's desires, the Federal State Transportation Assistance Act of 1982 (STAA) co-opted state control of size and weight on the Interstate and on "reasonable access" routes. It raised the maximum axle weight on the Interstate, although where states had higher weight limits, these continued. It increased width of equipment from 2.48 m to 2.59 m (98 to 102 inches). It permitted single trailer lengths of 14.6 m (48 feet) and double 8.7 m (28.5 foot) trailers. However, there continues to be pressure for increases in size and weights by the truck community, citing productivity and energy efficiency gains from uniform standards.

The trucking industry also points to the contribution of truck taxes to financing the road system. The DOT and the FHWA find the pressure for increased size and weight a political hot potato.

The trucking industry reports productivity increases of up to 30 percent from increased size and weight. Actions needed to obtain those productivity increases involve providing appropriate access routes to the Interstate and sites for terminals. So far as we know, there has been no planning supporting systematic approaches to these actions, and to the coordination of increased truck weights, stronger pavements, and user charges. There should be because of great efficiency gains.[458] There are also many market niche opportunities.[459]

The 1982 STAA asked for a study of the costs and benefits of a nationwide truck route system for longer combination vehicles (LCVs). Double 48s or triple 28s were the vehicles in mind. These LCVs would run about 33 m (110 feet) in length. Because of the number of axles, they might run 55,000 kg (120,000 lb), well above the 36,000 kg (80,000 lb) usual gross weight limit. There was much interest in the trucking community because a 50 percent gain in productivity was suggested, compared to the pre-1982 STAA situation.

Maio (1986) provided a rather full but inconclusive analysis of the LCVs question. It laid out what could be said about safety, fuel savings, productivity, geographic availability of services, and costs. The study examined the curving and tracking behaviors of vehicles using examples of LCVs in some Western states. The cost data address mainly the geometries of interchanges and staging facilities. Costs ranged from about $0.3 to $0.6 billion dollars as a one-time investment (about 1/80th of annual highway expenditures.)

While the FHWA study was aimed at concluding "no way," it turned out inconclusive because so many signals said "good idea." We think that there is a major opportunity here, and that well-conceived planning would define it. There is nothing new about the idea. Henry Ford was quoted in the *Washington Post* in the 1930s as supporting a regional system of truck highways, and the topic was debated in Congress at that time.

While some circa 1900 designs for roads separated autos, trucks and wagons, and bicycles, and discussion continued in the 1920s and 1930s of the needs for truck-only highways, and auto-only parkways, nothing has materialized. Might those old ideas be interesting ideas today? Samuel et al. (2002) call for such a network. Frank Turner, former head of BPR, has also endorsed the idea.[460] We think the opportunity will remain vague and ill-defined until some sort of tactical planning provides concrete ideas.

The STAA standardized and increased the size and weights of truck vehicles allowed on the Interstate system, and deregulation legislation in 1980 changed the situation in ICC regulated trucking.

The 1991 Intermodal Surface Transportation Efficiency Act (ISTEA) recognized and went beyond the Interstate by authorizing a more comprehensive National Highway System (NHS). The stated intent was to focus federal funds on roads that are most important for Interstate travel and the national defense. These roads also serve as intermodal connectors and are regarded as essential to Interstate commerce.

There is no question but that the NHS advantages trucks. The NHS consists of the Interstate and many primary highways, of which trucks already make intensive use. However, the opportunity to concentrate resources, by creating stronger pavements and bridges and, in turn, allowing heavier or larger truck vehicles, seems not to have been seized.

22.6.3 IDEA: TRUCK COLLECTOR SERVICE

Fawaz and Garrison (1994) explored designs and market niches building from today's roads and trucks. The market niche examined was the haulage of grain from farms to local (country) elevators and from local to large elevators. Optimizing equipment size and weight, road characteristics, and operating protocols, it was found that large trucks operating at low speeds on unsurfaced roads offered considerable efficiencies. (Figure 22.5 shows the truck design.)

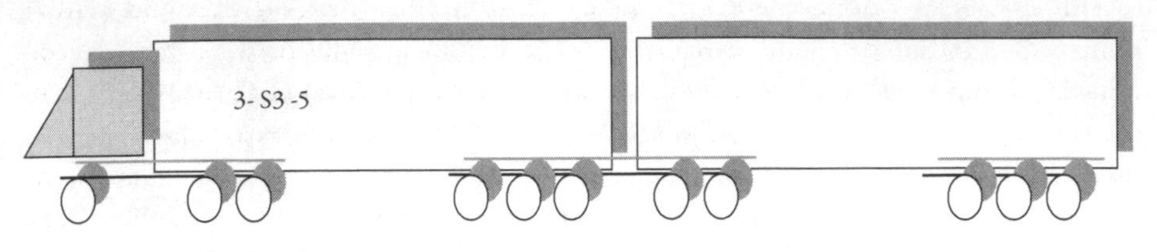

FIGURE 22.5 Truck Collector Services

This was a very conservative approach, in the sense that it built from existing ways of doing things—equipment technology, the road system and its protocols, and the ways that grain is moved. The decisions to be made if such haulage were to emerge and the distribution of costs and benefits were considered and other types of designs and market niches were imagined by the researchers.

22.7 Discussion

Friedlaender (1965) found the Interstate a good investment because it saved time overall. She noted that urban efficiencies offset the high cost of time saving by rural travelers. Yet we suggest that ignores all of the externalities and effects of the urban freeway. Simply considering time forgets the human cost of those displaced and replaced as the urban freeway restructured the urban activity system.

Still, a time calculus is very handy for transportation analysts because often faster is better and the time value of capital and labor is considered when designing facilities, buying equipment, and such. Prior to the Interstate, road improvement calculations ran on the savings in equipment cost, usually just wear and tear. (The UN and pavement engineers still use such calculations, ignoring the effects of their designs on users' time.) Facility maintenance cost was also considered. But even without traffic, pavements experience wear due to weather.

The consideration of time for users is critical in understanding the reason we have transportation. As the expression goes, "he who wastes my time, wastes my life." To argue against transportation improvements because they "induce" more demand is to argue against transportation projects because they do what transportation projects are supposed to do, connect people with their destinations. It would be instructive to compare the lives cost by the additional pollution generated when a new project is constructed and compare it with the lifetimes lost stuck in traffic, or the opportunities lost stuck at home.

Let's review the overall situation and inquire about prospects.

The rapid growth of cities beginning in the late 1800s accelerated the provision of urban roads, and cities built institutions and took actions appropriate to their needs. They created public works agencies, imposts on property owners paid for the construction and maintenance of local streets, and with experience, ways were found to fund and build parkways, viaducts, and arterial streets and to broaden funding bases using property taxes on vehicles and tolls on costly facilities. Design and planning concepts emerged; every Grand City needed a Robert Moses-like Grand Plan. Planning was strengthened as the states passed legislation enabling city planning and zoning for the control of land uses.

Although there was much progress, there were nagging problems—street and railroad conflicts, state highways that dumped traffic on city streets, congestion, and graft and corruption in street programs. Suburbs enabled by streetcar services were quickly becoming automobile suburbs and there were hints of the suburbanization of employment. Running faster just to keep up with the increasing use of automobiles was the response of street construction and traffic control programs. But their progress stumbled when the Great Depression of the 1930s reduced funding from property taxes.

Urban growth resurged in the late 1940s and 1950s and by and large the cities had ambitions, institutions, and plans, but little money. But there was hope because state and national political stages were changing as the balance of political power gradually shifted from rural to urban areas. (There was lag; the population of urban areas exceeded the rural population by the 1920s.) By 1950, political support for federal and state expenditures in cities was in sight and almost in hand for highways and many other programs.

The picture just painted was the context for snapshots presented in this chapter—the Bragdon Committee reports, the freeway revolt, the building of limited mileage high capacity state/federal facilities in cities, the actions of Robert Moses and his imitators, and almost one-half century later, the status of the urban road system.

The heavy hand of the experience is there. Have we learned from it?

One might say that we have learned to balance local, state, and federal interests and to pay more attention to the social and environmental impacts of large programs. That's a matter of style and equity. It is important. But how well balance and compromise are achieved is at debate. One thing is certain: project and program approvals require resources of time and money. Obtaining action requires a trail of studies and plans, a posture of local, state, and federal cooperation and funding, and providing funding to compensate for negative impacts. Burdened by "what's mine is mine and what is yours is open to debate" negotiations, obtaining approvals is costly, and one wonders if high transactions costs constrain the search for innovations.

Imperatives for congestion relief, improved public health, the city beautiful, and large rural-like parks affected city building in the late nineteenth and the early twentieth centuries, and imperatives continued to drive programs. Defense, congestion, and employment in public works construction and equipment manufacturing energized creation of the Interstate and its urban extensions. Today, emphases are on constraining urban sprawl and the automobile. The cynic might say that we have learned that programs require claiming imperatives for action.

Coupled with high transactions costs, imperatives have given us "megaprojects" such as Boston's Big Dig and the Alameda Corridor. The "to tame the automobile imperative" constrains funding, leading to an underinvestment in roads. Of smaller size are projects and investments in non-road-based transportation projects (transit) in response to congestion and environmental imperatives. Congestion and funding imperatives drive interest in tolls, and congestion drives interest in advanced traffic control systems.

The paragraphs above strived to say where we are, using the languages of programs and events. Speaking in a more general fashion, the authors see the spinning out of the maturing of the auto and truck highway system in urban settings. The system's institutional and technological aspects are rather fixed, and strident claims of imperatives are needed to nudge its evolutionary path. With maturity, productivity growth comes hard and there is market channeling as efforts are made to fit the system to market niches.

Will market channeling uncover new formats that renew urban transportation services? Might the building blocks of abandoned railroad routes and yards, congested highways, heavily subsidized transit services, pre-automobile walking streets, and other relics from urban transportation history merge with modern sensing and communication technologies and create new futures?

The story of roads continues in Chapter 30.

A man who, beyond the age of 26, finds himself on a bus can count himself as a failure.

—MISATTRIBUTED TO MARGARET THATCHER.[461]

23 Recapitalization

23.1 Federalization

IN THE UNITED States, the appropriate federal role in transit, even in interstate passenger service, remains a long-standing question. As noted in Chapter 14, transit was beginning its long post–World War II decline from 1946. Ridership was down and costs were up. In the late 1950s and early 1960s, private commuter railroads faced the beginning of what seems to be a perpetual funding "crisis." While the population was more urban (and suburban) than rural, there was still resistance to urban programs. Even the Brookings Institution argued against federal support for commuter trains.[462]

President-elect Kennedy supported a $100 million mass transportation loan program proposed by New Jersey senator Harrison Williams. This eventually passed as a $50 million loan, and $25 million demonstration-grants (not for long-term capital improvements), as well as a small planning program, but "While loans might be more palatable to Congress, Budget officials pointed out that they would offer little incentive to communities which had reached their debt limit or which could float municipal tax-exempt bonds at a lower interest rate than federal loans. They also emphasized that the only criteria in the Williams bill for setting the magnitude of federal financial participation had been how much Congress might be persuaded to accept."[463] Loans were nevertheless endorsed by the urban coalition of metropolitan areas and the private railroads. Authorization was subsequently cut from $75 million to $42.5 million by the House, which was less sensitive to urban issues at the time than the Senate.

Mass transit was then housed in the Housing and Home Finance Agency (later the Department of Housing and Urban Development [HUD]). This positioning of transit and housing together is echoed by the Obama administration's Livability Initiatives cutting across USDOT and HUD. The concern of locating mass transit within the Department of Commerce was the influence of the Bureau of Public Roads.

In 1964, a $375 million urban transportation program was passed and signed by President Johnson. However, this program took the form of grants rather than loans, which were by their very nature more favored by states and metropolitan areas.

In the post-Interstate era, there has been less new construction, highway expansion has slowed. More federal funds go to transit (about 25 percent of federal surface transportation capital funds go to transit), and the match requirement for states has increased.

The spread of slums in the inner cities of metropolitan areas, the relative decline of central business districts (CBDs), and the relative or absolute loss of central city population, together with the many social problems of the older low-income parts of central cities, focused national attention on "urban problems." The federal government was asked to come to the rescue of the cities. It did so with a variety of crime prevention, health, welfare, physical redevelopment, and other programs. Included was action to aid transit—the creation of the Urban Mass Transit Administration (UMTA) in 1964.

The creation of the Urban Mass Transit Administration (now the Federal Transit Administration) and the working out of its relations involved a diverse group of outside-government interested parties. Large rail operators in old cities, commuter bus and rail, the downtown businesses interest groups, and auto opponents all had their own interests. At the time, there was no single voice for the operators, they were split between rail and bus. The National Conference of Mayors served the urban development downtown business interests. A crisis was sensed, but the crisis had many interpretations. Finally, there was no clear model and set of existing suitable institutions. The closest model was the Department of Housing and Urban Development's (HUD) programs. HUD knew how to work with the cities; it had subsidy programs. It was chosen as the first home for UMTA, with a study of the advantages of remaining in HUD versus moving to the new USDOT to be made. The first home seemed to satisfy the urban development interests that had worked with HUD. However, at the end of the first year, UMTA moved.[464] Absent a close model, there was the problem of figuring out a government role satisfactory to diverse parties. Charles Haar, a member of the Harvard Law School Faculty and the first UMTA administrator, worked hard to establish an urban development role. Even so, the mood that held was that better management and technology was needed, and that helped pull UMTA into the DOT sphere. (At the time, DOT was striving to play a major role in research and technology development, fallout from the "If we can put a man on the moon, why can't we..." thinking.) UMTA learned to work with the states and the cities, the American Public Transit Association (APTA) has emerged as a force in the industry, and it has learned how to work with other groups, such as the American Bus Association.

A national need for UMTA was arguable, but it was seldom challenged. Changed programs in HUD seem to have provided models for change in UMTA. For instance, change, such as that driven by emulation of the private sector in program activities, seems to be in the HUD model. The Americans Disabled for Accessible Public Transportation (ADAPT) fought for wheelchair lifts on all buses. (See also Section 23.7 for some consideration of the problem.) UMTA accepted that in Section 405 of its regulations. APTA was forced into the role of protecting the capital-short operators who cannot afford lifts. APTA's interesting defense was that this is a local matter—the properties ought to figure out the mix of bus and special van services. That's interesting, because it departs from the "all issues are national" stance of APTA.

A good bit of what UMTA was to do was established in its organic legislation; bureaucrats worked out the rest. Steps taken early on had lasting impact.[465] Most of those steps were mandated by one or more of the interest groups that supported the creation of UMTA. We now list some decisions taken early on.

1. Students of urban affairs and government administrators had the notion that transit management was "up through the ranks" and unskilled—trained managers with business school backgrounds were needed. As a result, UMTA and local agencies didn't take advantage of experienced employees who knew what did and didn't work. "Political hacks" were also rampant in the industry.[466] This perceived need to turn over management resulted in lots of "know nothing" managers, especially at UMTA.[467]
2. The renewal of capital was urgent. Some thought that a one-shot renewal might be enough. Others had the notion that technological improvements were needed. Interest groups composed of traditional suppliers, aerospace, and engineering and contracting firms pressed this view, garnering public support. The Disneyland monorail was new, and that was the model in many minds.

   The capital grants program resulted. Another result was a round of new systems studies and the encouragement of defense contractors. This emphasis, together with the "hoops" one has to jump through in government procurement, put the traditional suppliers out of business—suppliers such as St. Louis Car and Pullman. Later, the aerospace industry found it couldn't make money. Lacking technology management capability, UMTA engaged in some disasters—its bus development program was an example.
3. People who wanted to manage the urban problem in a holistic way and those who wanted independent programs for UMTA were engaged in a behind-the-scenes tug-of-war. Do we combine transit with health, housing, education, and other urban social programs? UMTA kept a distance from those programs, and the void created was filled mainly by the then Department of Health, Education, and Welfare. School bus programs operated by local governments continued, of course. This debate continues today, pressed by those who want to coordinate public agency decision making about land use and transportation services.
4. Next was the question of the geographical organization of UMTA programs. Political realities said these could not just be big city programs.[468] When UMTA was created there were many central city transit agencies, both public and private service providers in the suburbs, and also a mixture of service providers in small towns and rural areas. How could UMTA interrelate with so many entities? How would it work with state and metropolitan area organizations? How would transit facility investments fit into transportation planning activities? By and large, UMTA was careful to fit in with existing arrangements, support both large city and other constituents, and go along with a large amount of congressional earmarking of expenditures and pressure for the expenditure of so-called discretionary grant funds.

UMTA did adopt the public administration dogma that single public agency organizations were best for urban areas (avoid duplication, obtain economy of scale), but that's very questionable.

5. Another question to be settled was that of the size of UMTA and whether it should be restricted to equipment-facility programs. Giving way to political pressure, subsidies have been made available for operations. The size issue remains at debate, although transit has access to highway trust funds, partially relieving the funding constraint on program size.

23.2 San Francisco's BART

The Bay Area Rapid Transit District (BART) subway-elevated/commuter train system began providing service on September 11, 1972, and it is a Bay Area icon. Although there is no place to take such a picture, one imagines sleek trains with the Golden Gate Bridge and palm trees in the background. Sponsored by central city politicians, property owners, and newspapers, it was the first new transit system in the United States after World War II. BART, and to a lesser extent the Washington Metro and Atlanta MARTA systems, served as a model for the wave of rail transit improvement projects promoted in subsequent decades, and that is why we emphasize it. Also, it illustrates that the gap between hype and reality can be great but invisible to publics and policy makers.

Centered on Oakland, BART service fans out westward under the Bay through San Francisco to its airport, north and south along the East Bay, and to suburbs to the northeast and southeast of Oakland. It is rather a skeletal mainline system, more so as extensions have been added over the years. It is essentially a commuter railroad serving San Francisco, Oakland, and other of the older centers of employment in the region. In a metropolitan area of about 7.4 million people, it serves about 342,000 one-way riders per day.[469] Since riders are one-way, and most people make round trips, the number of people served is about half that number. Thus, in a region with about 22,000,000 total trips,[470] about 1.5 percent of weekday trips are by BART. BART serves about a quarter of the 1.2 million transit trips in the Bay Area.[471]

Initial investment capital was backed by imposts on property in three Bay Area counties, and funds for the completion of the system and for extensions have come mainly from the federal government. The 2010 net operating expenses are about $647 million per year;[472] about 57 percent is recovered by fares, which is very good by US standards. Capital costs are not recovered.

In today's contentious world, anyone questioning transit costs is taken to be ignorant and pro-automobile. Even so, let's do a bit of math on costs anyway. It will give a feel for magnitudes.

The BART extension to Pittsburg/Bay Point to the northeast of Oakland cost $505 million,[473] and it serves about 3,000 round trip riders daily. Certainly many of these patrons are regular commuters to centers of employment.

Dividing capital cost by the number of riders says that about $170,000 has been invested per rider; considering interest in order to calculate annual cost, there is a capital subsidy of

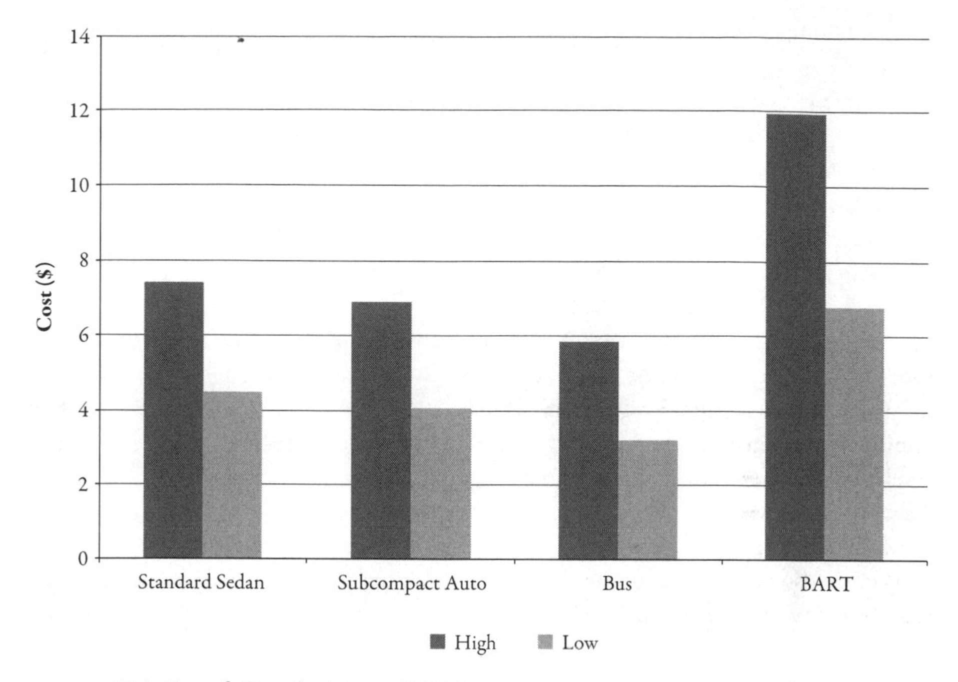

FIGURE 23.1 Cost of a Transbay Trip on BART

about $17,000 per rider per year. It is left as an exercise for the reader to compute the daily subsidy and per one-way trip per rider. What's more, there is an operating subsidy applying to each rider, so the total subsidy is somewhat higher. Numbers in this range apply to many transit investments, especially rail investments.

The $505 million divided by the 6.8 million people in the Bay Area works to about $74 per person in the Bay Area, but that is ignored because much of the money went to Washington and came back as free money. An irony is that the Yamamoto (2003) article highlighted BART Board director Roy Nakadegawa, who is a transit supporter but questions the wisdom of BART expansion. If no one pays attention to him, what's the use of our math?

The best-known early work on comparative costs was done by T. E. Keeler in the early 1970s.[474] The figure (redrawn as Figure 23.1) from that publication compares automobiles, bus, and BART. It needs to be remarked that this service niche is favorable to the bus. It's the comparison that is important—rail is expensive compared to the bus.

There is never a critical word of the iconic BART in the newspapers, citizens assume that the service is essential, and although they are more than illusive to analysts, BART is assumed to have had great positive impacts on community and regional development. Ignoring the probability that, if desired, private sector investors and consumers would have long ago energized developments at BART stations, concentrated development near BART stations is high on Bay Area planners' wish lists. Moreover, there is clamor for extensions to swing BART through San Jose and complete connections around the South Bay.

23.3 Washington's Metro

The Great Society Subway, as Schrag calls Washington, D.C.'s Metrorail,[475] is the East Coast sister to BART. The technologies are similar, though not identical (and one wonders why the opportunity to achieve economies of scale in car purchases, procurement, and design was not sought), and they opened a few years apart. Metro is the more successful sibling, with 762,653 weekday riders as of March 2012.[476]

Washington had long been served by commuter rail service and streetcars (from 1862 as horsecars, through 1962). Subway plans for Washington first emerged in 1959, when lines were proposed for downtown. A 1962 plan proposed an 89-mile (143 km) rail network. The 1968 Adopted Regional System Metrorail plan (97 miles [extend to 103 miles in 1984]) was constructed essentially unchanged, completed by 2001. In one sense, a plan is a contract, so abiding by the plan gives certainty to those who rely on it (e.g., prospective landowners and developers, or other transportation planning agencies). In another sense, this lock-in reduces the flexibility of Metro to adapt to changes. The planned highway network of 1968 did not materialize, nor did the land use pattern, so why is one rail network, optimized for one set of conditions, also optimal for another? Unlike the San Francisco Bay region, which is tightly bound by geography (water and mountains) channelizing development and laying obvious where lines would go, Washington is much less constricted.

Since the original system was fully built, one major new line has been started (the Silver Line to Dulles Airport, see Section 22.4.4), there have been some extensions and infill stations, and some additional services have been added.

Like BART, Metrorail is costly, and it is debatable whether it is worth the cost. Increased development in the City of Washington would be difficult without Metrorail or other public transit services. The District of Columbia employment ranges from 687,000 in 1990 to 660,000 in 2002 to 733,000 in 2012, an increase that would have been very difficult in the absence of public transit.[477] In 1990, the Washington region led the United States in carpooling with 16 percent, due in large part to its system of HOV lanes,[478] while its regional transit mode share was 13 percent. In the absence of Metrorail, one could imagine buses (probably in dedicated lanes) providing the additional service, but not private vehicles. To serve that growth in the central city, transit is necessary. In the absence of transit, that growth would not have occurred there. But that job growth undoubtedly would have occurred somewhere. The land use pattern and the transportation technology to serve it are joint processes.

23.4 Other People's Money: Rational Behavior in Irrational America

Working out the matters just discussed partly describes the situation today; most of those matters are not completely resolved, of course. Conservative members of the national government question the federal role in transit, but we do not see that as much of a threat to the program, partly because conservative downtown property owners are supportive of transit.

As is well known, today, there are subsidies covering fixed and much of the operating cost of transit. They run about $70 per capita per year and are increasing. Although many

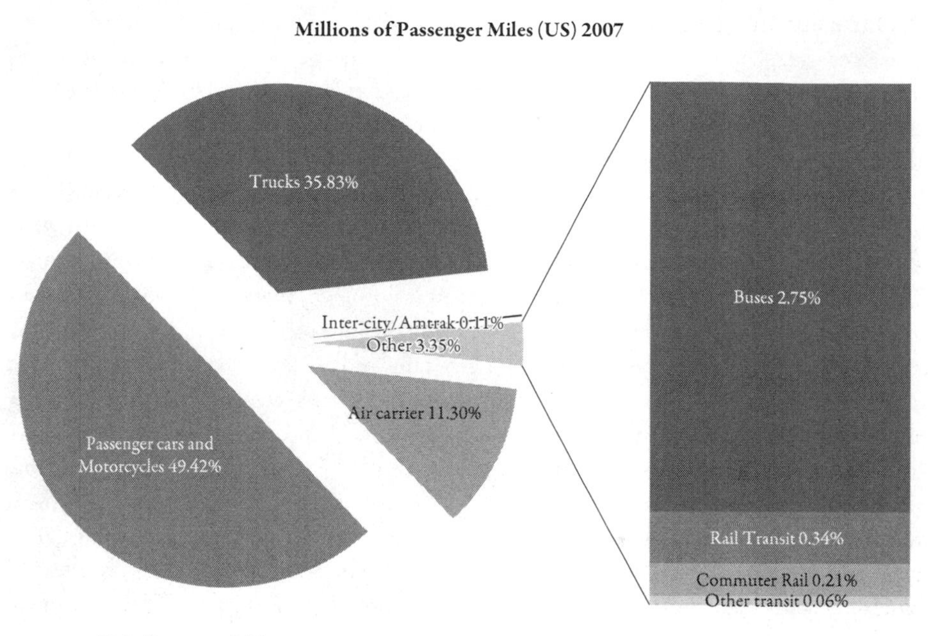

FIGURE 23.2 Passenger Miles

who do not have automobiles available and/or lack the means to purchase and use private transportation are transit users, overall there is an income transfer to the wealthy—especially commuters.

Ridership has declined to the point where about 2 percent of all trips are by bus and streetcar and about 0.3 percent by other rail.

In 2007, passenger miles break down as shown in Figure 23.2. Buses (including schools buses, inter city buses, and transit buses) and rail transit comprise fewer than 3.5 percent of all trip distance. Buses are the vast majority.

Thirty-five percent of all US transit trips are in New York City and over one-half are in New York, Philadelphia, Boston, and Chicago. Cities are planning new systems or expansions of old ones, and vast funds will be required for these.

It seems clear that we need to preserve and improve services in urban environments such as those found in San Francisco and other old, densely populated cities. Citizens who do not have access to private automobiles need services. But nowhere is it carved in granite that the federal government is necessary to achieve these goals and that its programs should be continued. One might say that federal transit programs are a success because transit service has been preserved, but might there have been other, better ways to do that? Might downsizing and market specialization as illustrated by the US freight railroad industry served better?

How did we get into this situation? We are there because of lack of *cognition*.

A major part of the problem is "other people's money." In the US funding environment of the past four decades, transit capital expenditures are in large part paid for from Washington, but operating costs are local. Rational local governments, acting in an irrational system, have every incentive to make projects capital intensive and minimize operating costs.[479]

One meets with policy makers, and discusses modal costs and services.[480] It is stressed that fixed rail, people movers, and monorail are off-the-chart expensive when cost effectiveness is considered. Alternatives such as improved bus services, or even buying used vehicles or paying for taxis for those who can't drive now are seldom considered. Real improvements, those that make a factor of two or better changes, should be tried.

But policy makers pay no attention to such suggestions. They have already decided that autos and buses are non-feasible solutions to problems. Rail is the only answer. Exactly what the problem is isn't always clear. In cases the authors are familiar with (Denver, Minneapolis, Chicago), CBD access was the main issue, but this is a problem that has not worsened in several decades as CBD employment is flat (while, in contrast, other transportation problems have worsened). In other cases, congestion generally and air pollution have been at issue. There also seems to be the feeling that modern cities have streetcars; what else is there to discuss, especially since the feds will pay much of the bill. Expressions like "Minneapolis—St. Paul is the largest city in North America without rail transit" were bandied about, illustrating the competitiveness question—to be a "world class" city one must have a new rail system, a new major league baseball stadium, a new professional football stadium, a new college football stadium, a new basketball arena, a new hockey arena, a new minor league baseball stadium, a beltway with three lanes (Minneapolis—St. Paul was at one time apparently also the largest city in North America with a two-lane beltway), a convention center, a festival marketplace, and so on. (and in Minneapolis—St. Paul, if Minneapolis has it, then St. Paul needs it as well). Transit is just one more instance of *feature-itis*.

We judge that policy makers, newspapers, and others see transit as the only solution, and rail transit as the superior choice. We think that is because they can't imagine seeking new solutions.[481]

23.5 Docklands Light Railway and the Jubilee

The experience outside the United States differs greatly. The privately operated (under franchise agreement) Docklands Light Railway (DLR), opened in 1987, after three years of construction and just a few years after conception (in 1982 according to TfL, though antecedents can be seen as early as the 1973 Docklands Study, according to the London Dockland Development Corporation), to serve the emerging Docklands regeneration project centered at Canary Wharf. The existing system and proposed extensions are shown in Figure 23.3. All of the built extensions since 1987 were foreseen by Jolly and Bayman (1986), but not the subsequently proposed ones.

Containerization of shipping changed the nature of that industry, which migrated in the 1960s and 1970s from London to Felixstowe. The Docklands development replaced the newly abandoned shipping docks in East London with an emerging financial center, an American-like downtown for London. Almost immediately upon opening, construction started on extensions. Success begat success, and proposals and funding for extensions to this new, Docklands-centered automated public-transport technology continued to flow in. By 2010, the DLR system was serving 80 million passengers per year (300,000 per weekday). As can be seen from the map, extensions continue to be proposed, notably to Charing Cross. The early DLR segments took advantage of abandoned railway lines, and while not a green-field, it was a relatively open brown-field canvas on which to work.

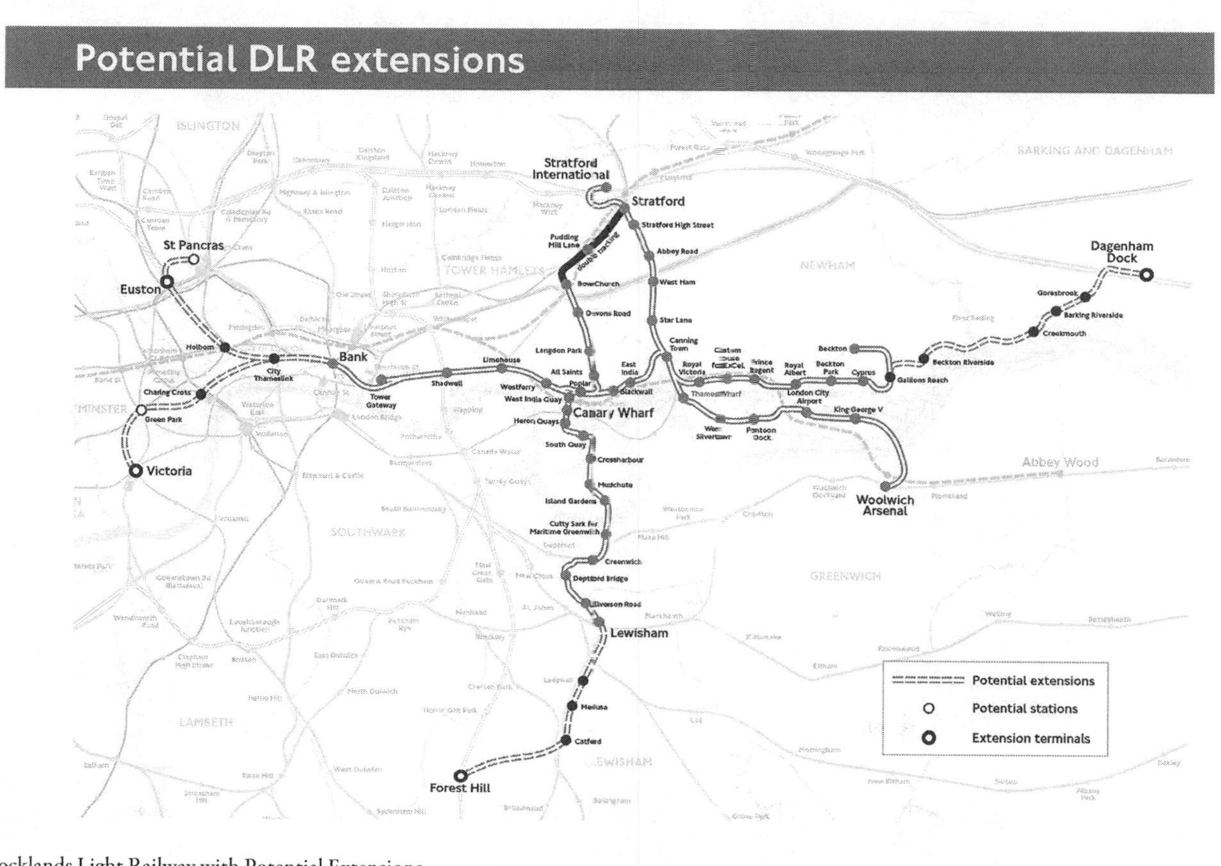

FIGURE 23.3 Docklands Light Railway with Potential Extensions.

Source: Transport for London unofficial map posted on London Reconnections Blog, "Extending the DLR." May 2, 2011 (used with permission of London Reconnections)

The planning for the Jubilee Line, then the "Fleet Line" (named for Fleet Street and the River Fleet), apparently began around 1965. The first section opened in 1979, two years after the Silver Jubilee of Queen Elizabeth for which it was named. The line temporarily terminated at Charing Cross station. It was to continue eastward, along Fleet Street.

In 1999, the Jubilee Line was extended to Canary Wharf as well, but this eastward extension resulted in the abandonment of the then twenty-year-old extension to Charing Cross, as a new routing split from the old in Green Park. The decision to do this can be seen as early as the 1990 Jubilee Line Extension options map. At this point, the expensive Charing Cross station was just over ten years old. Stations are often more expensive (and require more tunneling volume) than the lines which they connect, so this is a major rethink, not some temporary train halt that was subsequently bypassed. Jubilee Line Extension is said to have cost £3.5 billion, which delayed other significant potential construction projects, such as Crossrail (currently under construction) and Crossrail 2 (Chelsea-Hackney).

There are several points to be made about this:

- No one in the 1970s anticipated abandoning the Jubilee line portion of Charing Cross station, certainly not so soon after construction.
- No one in the 1970s (or 1980s or 1990s) anticipated that the DLR might one day reoccupy Charing Cross station, or even construct such a large cross-London network centered on the Docklands. (Speculation about this emerged in the first decade of the 2000s.) While the Underground system is centered on the Square Mile of the City of London, with many branches and a Circle Line, DLR has a new hub to the east at Canary Wharf, with a new driverless technology, that inter-connects with the old at several key stations, and if the extensions are built, at several more.
- In order to build anything, you must have a vision. Rarely does construction start without a fixed end point in mind (this is not SimCity). That construction once made is largely irreversible, though it may be abandoned, and it provides opportunities for reuse; it creates "facts on the ground" that are hard to undo. The City of London street network is a perfect example. Even after the Great Fire of 1666 or the bombings of World War II, it greatly resembles the facts on the ground at the time of William the Conqueror.
- In order to move forward, you must be willing to abandon old visions. The DLR and Jubilee Line Extensions were new visions, built on top of (and beneath) old constructions, not old visions. They were even willing to abandon one major fact on the ground (Charing Cross station) so as not to be wed to a vision that no longer worked.
- Facts on the ground create new constraints, new opportunities, new ways of looking at the world. There are many possible lines in a network, but only a few can actually be built. Transportation networks are far from optimal, always will be, but the planner needs to consider what is the best decision given the world as it exists, not the world as it might be according to some plan.

We ought not be too locked into our plans; we forgo many opportunities by clinging to the zombie maps of long-dead officials.

23.6 Challenge: Serving the Disadvantaged

The urban highway system substituted only in part for transit, and many of transit's problems result from a failure of the auto highway system. For instance, there is an inability to provide high-quality personal transportation services to the poor, physically (and mentally) disabled, the elderly, or children. (Which is not to say that fixed route transit provides high-quality services for these groups either.) Moreover, this is a growing problem; in the developed world, the populations of the elderly and non-native speakers are increasing rapidly (Figure 23.4). It is important to emphasize that these groups, while indicative of the disadvantaged, are not generally incapacitated, and (aside from the under 5 population) contain individuals fully able to use any and every mode of transportation.

Transportation service providers serve multiple groups; each group has separate problems.[482]

The elderly find giving up the flexibility of the automobile very traumatic. Rather than seek assistance or use transit (which may not go to the desired destination anyway and is perceived as dangerous), many will forgo trips and activities.

The challenge facing the disabled is not the loss of flexibility, it is its absence. The transportation system is designed for the physically able; the disabled are considered an afterthought. The trend toward community-based living compounds the problem, as many communities are poorly served by transit.

Many developmentally disabled individuals require supervision when traveling, greatly restricting options, and thus flexibility. Transfers become especially burdensome, as a single transit driver no longer can be counted on to keep an eye out for the traveler.

Poor families, especially those working poor who have multiple jobs but no vehicle, may need to work shifts during which no transit is provided. Day care services may be available,

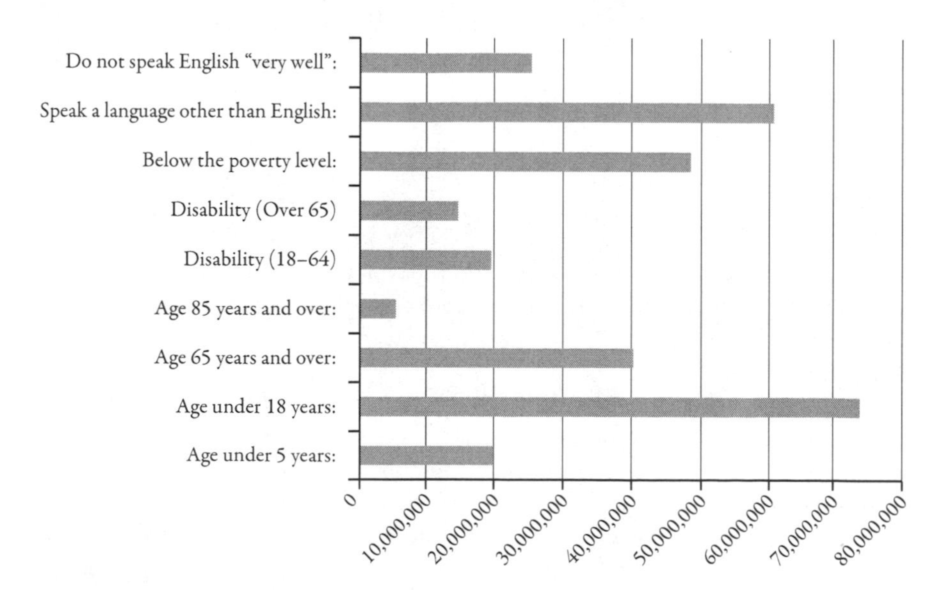

FIGURE 23.4 Size of Transportation Disadvantaged Populations

Source: US Census 2010 and American Community Survey 2011

and provide a safe environment for poor children, but a safe means to get between the service and home is also required.

The automobile still requires both driving skills and capital. In previous eras, the transit network was sufficient to serve almost all users. However, the reduction in transit demand and the transformation of the urban landscape have worsened transit service for those who still depend on it. A negative effect of the auto is worse transit. Fixed route public transit systems cannot provide the kind of point-to-point services that the transportation disadvantaged require.

If we relax a policy constraint: only fixed route public transit systems can provide transit services, we imagine many more services for specific populations of the transportation disadvantaged. A number of suggestions along these lines have been made in numerous research articles, yet the transit monopoly is reluctant (for obvious reasons) to let go.[483] There is a reluctance to allow taxis to be the service provider for the transportation disadvantaged. At best, there are paratransit (on-demand, dial-a-ride) services. While often considered a backwater when compared to the fixed route services used by advantaged (or less disadvantaged) travelers (itself considered a backwater by others in the transportation field), providing this service efficiently is one of the greatest technical challenges in transportation, but also promises an alternative technological path that might become mainstream in the future. The field has seen its share of innovations, but not just by vehicle makers. For instance, Everett Rogers and collaborators Rice and Rogers (1980)[484] and Rogers et al. (1979)[485] (best known for the *Diffusion of Innovations*) identified 77 innovations in Dial-a-Ride developed by users, rather than manufacturers or system integrators. Notably, most US cities (and many outside the US) prohibit private jitney services, preventing the innovations in dial-a-ride from being deployed to service a wider market, and from the scaling up of markets to generate further innovations.

Over the longer term, we imagine that autonomous vehicles (that can operate both on and off special facilities) might be of use. If the physical and mental abilities of the driver were no longer factors in ensuring a safe trip, travelers (or their guardians) could simply instruct the vehicle where to go.

23.7 Universal Design

> Universal design is the design of products and environments to be usable by all people, to the greatest extent possible, without the need for adaptation or specialized design.
>
> —Ron Mace

The problems that transportation disadvantaged individuals face are similar to, but more severe than the problems that the rest of the population sees on a day-to-day basis. Take, for instance, the simple question of which bus goes from the bus stop nearest my home to downtown. When going to the bus stop, is information provided other than the simple sign saying "Bus Stop"? Many stops don't even identify routes, much less schedules. Without a guide (either person or documentation), using the system entails taking great risks, among them the risk of winding up across town from your desired destination and being hours late. The relatively simple, but seemingly revolutionary idea of providing information with the service may help. For instance, operators should make the signage clear so the bus stop can be found, make clear the route number, route end points, and the direction toward which the user needs to travel at the bus stop; make clear when the bus is coming, especially when service is infrequent; and clearly convey to the user, when he or she should get off the bus (which may require more than drivers announcing the bus stops as the bus travels the route). While this may be critical for those who are unfamiliar with the location (tourists) or the language (immigrants), it is also important for those who are cognitively challenged, and would probably provide a much better travel experience for those who lack special difficulties. Some transit systems do this, especially in downtown areas. Others would rather spend money on new rail construction (seeking new, wealthier customers) than make the existing bus system (serving existing, poorer customers) work well.

To provide another example, when the Americans with Disabilities Act was passed, a major concern was retrofitting buses with elevators to accommodate wheelchairs. Buses traditionally had steps leading from the ground to the level where passengers sit. A more universal design would lower the floor of the bus (and gradually raise the level of the ground at the bus stop) so that wheelchairs could roll onto the vehicle, the way that occurs on many subway systems. Such a system would benefit many others with poor knees who can walk but find steps difficult. The lowering of the floor of buses is becoming more common, the raising of bus stops less so.

Having a universal design assists those who need the assistance while benefiting others. The principles of universal design[486] are a set of values, but they are hard to disagree with:

1. Equitable Use: The design is useful and marketable to people with diverse abilities.
2. Flexibility in Use: The design accommodates a wide range of individual preferences and abilities.
3. Simple and Intuitive Use: Use of the design is easy to understand, regardless of the user's experience, knowledge, language skills, or current concentration level.
4. Perceptible Information: The design communicates necessary information effectively to the user, regardless of ambient conditions or the user's sensory abilities.
5. Tolerance for Error: The design minimizes hazards and the adverse consequences of accidental or unintended actions.
6. Low Physical Effort: The design can be used efficiently and comfortably and with a minimum of fatigue.
7. Size and Space for Approach and use: Appropriate size and space are provided for approach, reach, manipulation, and use, regardless of user's body size, posture, and mobility.

23.8 Personal Rapid Transit: Imagination in Search of a Market

Personal Rapid Transit (PRT) systems are a hybrid between transit and the private vehicle. In a sense, they can be thought of as horizontal elevators; you enter a small car, push your destination, and the vehicle takes you there. The vehicle carries one to four individuals, typically along an elevated guideway. Unlike traditional elevators, the guideway is such that vehicles can pass each other, so that passengers can reach their destination without stops along the way. The uses for such systems vary, from within-airport transportation to urban circulators.

Promoters claim that PRT can become an important part of the urban transportation scene. Yet despite some demonstrations of technical feasibility, they are not considered outside of narrow niches.

Sydney has constructed a lightweight urban monorail, which serves similar market niches, as do people movers (e.g., Detroit and Miami). However, those systems have fixed stops, and have not been widely lauded or imitated outside airports and amusement parks.

Serious PRT development can be traced to the 1950s, when Edward Haltom developed the "Monocab," which was suspended from an overhead guideway that needed to move for switching. Its disadvantages were its height, resulting in cantilever posts that increased the cost and visual impact of the system. Under the Urban Mass Transit Administration, a different PRT system was deployed at Morgantown, West Virginia, in the 1970s. It was very simple with stops at two ends, serving a college market, and though technologically successful, not the kind of large market promoters sought. US federal government funding for research in the area was withdrawn in the 1980s, as efforts focused on more conventional rail transit systems.

In France, a great deal of effort went into developing the technology called ARAMIS, a PRT that allowed vehicles to couple and decouple to gain some of the advantages of trains. The system was sufficiently complex, without enough political champions, and suffered from scope-creep as the design specifications of the project were a moving target. Despite, or perhaps because of, its visionary nature, it failed before deployment.[487]

Over the course of time, a number of other attempts at deploying these types of systems have been made.[488]

PRT applies imagination to envision a new mode, comprising new networks and new vehicles, and for that reason is to be praised. But maybe it reaches too far. A truly new combination of networks and vehicles has seldom been deployed in the history of transportation. One might suppose the railroad to be the best example of such a deployment. As we saw, even it drew on antecedents in mining operations, where animal-drawn carts on tracks were used to move coal. Prior to that, there of course were already animal-drawn carts on roads. Elevators are another similar technology, where automation replaced labor (and in general, elevators were not feasible when relying on animal or human labor, and so were rare prior to automation).

Other technologies either used natural networks (shipping, aviation), or existing roads (automobiles). While the modes using natural networks still required new nodes, a point-based facility (port or airport) is much easier to deploy than a line. Eventually, as autos began to be widespread, old roads were upgraded and new roads constructed to accommodate them, but it can be seen as a process of co-evolution: vehicles use existing network, the network is upgraded to accommodate new vehicles, the vehicles are upgraded to best fit the new network.

Among the PRT promoters are mechanical engineering professor Edward Anderson of the University of Minnesota, who holds patents on a number of related technologies and formed the company Taxi2000 (which used to sound more futuristic than it does presently). Taxi2000 is pushing its Skyweb Express system and has constructed a test facility.

Proponents such as Anderson claim that government policies are preventing their technology. While government may not be helping much, we think the problem is more fundamental. PRT, like automated highway systems, requires developing vehicles that can only use the PRT network. Because of network effects, the value of building a PRT system increases with the size of the system, and when fully blown may actually meet some of the promoters' goals. The problem is the lack of value in a small system, and any system will initially be small. Yet it will require a high fixed cost just to get going. Technologies are first proven in small market niches and then expanded, not deployed widely on promise. Thus there is no deployment path to PRT.

To date, operational PRT systems have been deployed at London Heathrow Airport (connecting to some parking lots), and Masdar City, Abu Dhabi, United Arab Emirates, underneath this planned community. These are both subsidized systems, and more experimental than practical. Other deployments are planned or in process (e.g., Amritsar, India, to serve the Golden Temple). The market niche where a PRT is better than a people mover, the automobile, or conventional fixed route transit is limited. The most likely outcome is the convergence of autonomous vehicles and PRT systems, though these will differ from the original tracked, grade-separated PRT vision.

23.9 Reinventing Fixed Route Transit

23.9.1 RETRENCHMENT

Mass transit systems in the United States are collectively losing money. Yet some individual routes (including bus routes) earn enough to pay their own operating (and even capital costs). While we see new fixed route rail systems being constructed in some markets, with federal support, agencies are also financially strapped and cutting services. In general, this is not being done strategically.

We can divide individual systems into three sets of routes:

1. Those routes that break even or profit financially (at a given fare, i.e., there is a fare at which this is true). This is the *core*.
2. Those lines which are necessary for the core routes to break even, and collectively help the set of routes break even. These are the *feeders*.
3. Those lines which lose money, and whose absence would not reduce profits on other routes by more than their costs. These *equity* services are there as a social support system, rather than economic efficiency, serving social and especially spatial (political) equity objectives.

Since public takeover of transit systems, their role has changed. Originally, the private services were meant to make money for their owners, and they did this by providing services that were sufficiently valuable to consumers that they generated more revenue than they cost to operate.

The role of transit as a transportation agency or as a welfare agency continues to be debated, though usually not out loud. It is pretended the agency can serve two masters. But the alternative goals—providing spatial coverage to serve everyone equally, or providing high frequency to serve some markets well—are in conflict. The goal of covering costs is at odds with providing subsidized services. We think to be successful, transit agencies should clarify their objective. If they are to be considered primarily transportation services, they should be considered (and reorganized as) public utilities rather than departments of government. Utilities by definition provide a useful service for a price to their users. If they are instead to be welfare services, that should be made explicit.

As transportationists, we have a clear opinion on the matter. The agencies should identify and propose to retrench to the financially sustainable system, and present local politicians with a choice.

If local politicians want additional equity services (e.g., covering low ridership suburban routes), they should be given a list with a cost of subsidy per line, and then can collectively choose which lines to finance out of general revenue, as this is primarily a welfare rather than a transportation function. In other words, public transit organizations would present the public with a bill for these services (the subsidy required in order to at least break even on operating them (i.e., the difference between their revenue and their cost), and not be expected to pay for them out of operating revenue.

If the cost of those lines is deemed too expensive (i.e., the politicians are unwilling to pay for them with general revenue tax dollars), they should be canceled. Transit agencies would no longer be losing money, they would now be break-even or slightly profitable. They might even pay a dividend to their owners (the general public). This is roughly how transit is organized in many other countries, for example, England (discussed below).

General revenue (the Treasury) would of course now be losing money: we didn't pull money from thin air, but since this is a social welfare/redistribution function, that is perfectly appropriate. This would entirely change public and political perception of transit services. It might also result in fewer bad routes being funded, since it would be crystal clear where the subsidies lay.

The "which routes to fund" decision should be revisited regularly.

23.9.2 SUBSIDIZING THE TRAVELER NOT THE SYSTEM

Transit fares presently cover only a small fraction of the operating costs, even in the best US transit markets. To provide an efficient set of choices, all modes should cover all of their costs, including both direct and negative externalities, unless a coherent reason is provided for subsidy (such as evidence of large economic spillovers). Transit is probably furthest from covering its full costs of all the land modes covered in the book. Capital costs are already sunk, and in a market with decreasing average costs (i.e., economies of scale), recovering capital costs might not be warranted. On the other hand, operating costs are roughly proportional to the amount of service provided (while the marginal traveler imposes no additional costs, assuming an available seat, the marginal fifty travelers almost certainly require an extra bus). To break even on the system, transit fares would need to cover average costs of operating services.

If these resulting fares are too high, politicians should give the users they are concerned with money directly. It makes more sense to do that than to subsidize people who are perfectly capable of paying so that some group doesn't face a full cost that they cannot pay. If they don't want to give poor people money (e.g., because they don't trust poor people with cash, or fear they would use the money for something other than transit [say, food, or housing, or heat]), they should top-up their transit smartcard (e.g., by adding funds weekly). These funds would come from a separate government agency (let's call it the "Transportation Opportunities Office") which is completely separate from the transit organization. The recipients of government subsidy would use the same smartcard as everyone else, so no public stigma is attached to using the card. Moving toward smartcard systems is efficient all around, saving boarding times and reducing transit run times.

Currently, transit is sometimes subsidized by employers and university campuses, particularly with the use of seasonal passes. Income-based vouchers are also used for people who might be riding paratransit services in rural areas.[489] The same kind of subsidies can be extended to other groups.

23.9.3 CONTRACTING OUT

US public transit services are currently provided by the agencies directly; they hold a monopoly on providing services, were staffed by a public sector union, and had little incentive to innovate. Before the 1980s, the United Kingdom was facing the same set of issues. In response, a new model was instituted, in which transit services would be bid out, and private firms could compete on how much they would pay to provided service on given routes, or what subsidy they required in order to provide the service. In exchange, they would keep the fares they collected, motivating them to increase demand by providing high-quality services. This model has been widely credited with improving the quality and reliability of bus services in London, as reflected by the long run increasing demand, and reduction in costs, though has not been without detractors.[490]

Figure 23.5 shows the trends in ridership throughout the United Kingdom. Clearly, London is doing something different and successful, having doubled ridership since its early 1980s nadir, while ridership elsewhere has tended to drop. This model could be adopted elsewhere.

23.9.4 THE PRODUCT

If you're not paying for it, you're the product.[492]

Transit has several sources of revenue. One, of course, is the farebox, but in most markets this fails to cover even operating costs. Another is advertising. While this is relatively small (less than 5 percent of agency revenue), it could grow. A Transportation Research Board study[493] makes a number of recommendations to increase this share of revenue. Just as much content on the Internet, in newspapers, and in magazines is subsidized by advertising, some transportation service is subsidized by advertising as well, and more could be. This phenomenon of advertising supported services is encapsulated in the famous recent quote at the top of this section.

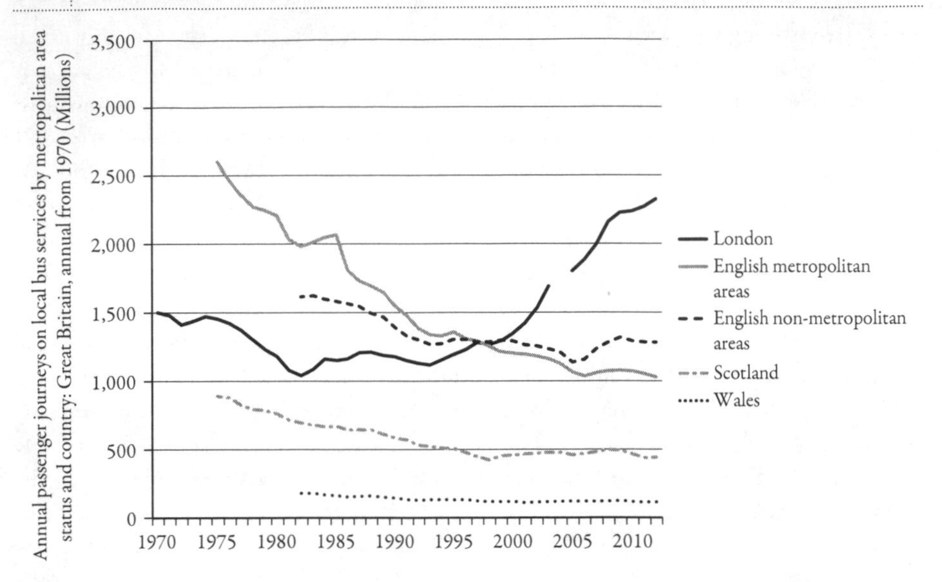

FIGURE 23.5 Ridership Trends on buses in United Kingdom (millions of riders)
Source: UK Department for Transport[491]

Transit advertising comprises several major components, interior and exterior. Interior ads inside the vehicle or station are aimed at customers; the exterior ads on the vehicle or outside at bus stops, bus benches, and so on, are aimed mostly at non-customers.

Historically, interior advertising has been static. With smart-payment cards, we now know who is on the vehicle or in the station, and with RFID tags we could even know where they are sitting. Department store magnate John Wanamaker is reputed to have said, “Half the money I spend on advertising is wasted; the trouble is I don’t know which half.” But now, just as with the Internet, advertising can now be customized. Electronic screens instead of cardboard can provide dynamic advertising customized to both the transit user and the location of the vehicle. It should be more effective. It might be more profitable (if the ability to sell advertising outweighs the extra costs). Both “captive” and “choice” riders are the classic “captive” audience; they can turn their heads, but can’t really escape.

Users should appreciate this because it can (1) be more useful (i.e., advertising is not inherently bad, it might inform you of a product you actually would want but were unaware of, or of a discount on the product), and (2) the extra revenue from the advertising can keep fares down. We should expect to see more, and more customized, advertising in the next generation of transit vehicles.

There are other, less innovative, but still possibly lucrative sources of funds for transit agencies. Some agencies have sold naming rights to stations. While this is generally a bad idea from a wayfinding perspective, it does provide revenue for agencies seeking funds. Another strategy, station sponsorships (such as Apple in Chicago) may help, but likely won’t generate as much money.

23.10 Discussion

Simplifying the broad sweep of history, we imagine for passenger transportation a walking (and draft animal) society, a transit (and electric inter-urban and inter-city passenger train society), and the auto society. Walking gave way to transit, and transit gave way to the auto. In each case the transition was not graceful. It stranded older fixed investments and was costly to those for whom lifestyle changes were neither easy nor desirable. "Gave way" was destructive of desirable aspects of life, such as the activities best served by walking and transit.

With this preamble, we may state the transit and walking problems as a failure of automobilization. The automobile system innovation failed to the extent that it could not fully replace transit or walking.

One response to the transit (and walking) problem presents itself: tweak the automobile highway system so that it can more effectively substitute for transit and walking.

Niche-fitting is another response. Suppose we focused investments in the niches they best work—highways in low-density environments, fixed-route transit in high-density areas. Interfaces and boundaries will always remain an issue. How does one travel from a low-density to a high-density environment, or vice versa? What happens when environments (land uses, relative prices, technologies, policies) change over time? What constitutes low and high density?

Cars need not fail for transit to succeed. Each mode has its use; the problem comes in deploying it where it doesn't fit (e.g., urban freeways, cars on campus, low-volume suburban fixed route transit). If we don't acknowledge the misfit, we will waste scarce resources (time and money) that could be better spent elsewhere.

There is not necessarily one answer that fits the entire country. A policy option is the devolution of federal activity in the transit (or even surface transportation) area. This is an idea periodically discussed in Washington, seriously in 1982, most recently in 2012. In Senator Lamar Alexander's proposal, the states would pick up transportation (and education) costs in exchange for the federal government funding Medicaid.[494] While the linkage seems strange, and we are accustomed to dealing with transportation policy within the transportation field, there are broader society issues at work. The reasoning behind one objection to devolution runs this way. Healthy cities are in the national interest. Therefore, a federal role for transit is in the federal interest. That claims too much, and it ignores the question of whether the federal government has or could create an effective role. Another objection is on equity grounds. Some cities aren't as wealthy as others, and it is only fair to send money from here to there. However, in conflicts between the central city and its suburbs, state governments can and do deal with such issues (to no one's satisfaction).

Consider, for instance, the long-planned, many times started, and now under construction Second Avenue Subway in New York. This is an important piece of infrastructure for New York, with its crowded transit system. It is expected to cost $17 billion; more than a quarter of the first phase (and presumably subsequent phases) will be federally funded.[495] All of its users will be in New York, almost all will be New York residents. Why is it an important national priority? We must remember, most transportation is local, and most transportation funding is local, and most federal dollars are returned to the states that generate them. Is the Washington middleman necessary? (Yes, the same argument applies to roads as to transit.)

Overall, we see transit in the United States as a niche market business, serving the few remaining dense areas of US cities, mostly those built before 1920, and otherwise one that should be more focused on serving those without choices (the transportation disadvantaged such as the poor, elderly, disabled, and children) rather than providing additional and largely worse choices to those who already have options (the car). Policy ought to treat it that way. It should recognize that in spite of enormous expenditures, the transit market is not expanding (transit trips are up, of course, which is not surprising given the massive investments and growing population, but trips per capita are essentially flat for almost four decades). Niche markets offer appropriate environments for innovation and system improvements where lessons from other modes and other technologies may apply. Transit can certainly do better than it has, but even if it were to double or quadruple the share of US travel it serves, it would remain a niche.

All forecasts are wrong, some forecasts are more wrong than others.

—ANONYMOUS

Prediction is very difficult, especially if it involves the future.

—NIELS BOHR

It has been said that he who lives by the crystal ball soon learns to eat glass.

—ZOLTAN MERSZEI, PRESIDENT OF OCCIDENTAL PETROLEUM

24 Lord Kelvin's Curse

24.1 The Limits of Knowledge

> I OFTEN SAY that when you can measure what you are speaking about, and express it in numbers, you know something about it; but when you cannot measure it, when you cannot express it in numbers, your knowledge is of a meager and unsatisfactory kind; it may be the beginning of knowledge, but you have scarcely, in your thoughts, advanced to the stage of science, whatever the matter may be.
>
> —Lord Kelvin, *Addresses*, 1883

Previous chapters have discussed a number of attributes of systems, example, presence of slow and fast variables, Life-cycle, level of contingency, and within-system decision making. Taken alone or together, these may assist in forming perspectives on planning and the system to which planning is applied. This subsection will not restate those system attributes and ask about the insights they offer. Rather, we attempt to respond to the often-heard comment that these systems are social ones and analysis won't work.

Suppose these were physical systems. How well could they be studied; how well could forecasts be made? Garrison was once a meteorologist, and he still follows that literature in a casual way. What does the meteorology experience tell us? Meteorological or weather processes are completely understood in broad outline. Processes are those applicable to systems, and, given initial conditions, laws describe changes in states due to changing energy flux. Given that situation, there would seem to be two tasks. First, fill in gaps in the science and data. Recent work has clarified the relations between sunspots and weather. Second, improve the calculations required when large systems are analyzed. Larger computers allow us to solve those equations in a more precise way and in a shorter period of time, so there should be more precision in weather forecasting than previously.

There is a parallel to work in the Urban Transportation Planning System (UTPS), the standard forecasting methods used in transportation. UTPS analysts take the processes to be well-known, example, traffic assignment. A lot of effort is going into numerical methods for solving equations and forecasting, and there is work to polish up empirical details.

Returning to meteorology, large computers began to come along in the 1970s, and for a time there was a flurry of effort in numerical weather forecasting. But with experience, it was found that this forecasting was no better than forecasting using previous techniques. What seems to be the problem? After all, the meteorological system is a physical system. It obeys laws, so one ought to be able to predict states.

To date, we have been unable to make much progress on long-term weather forecasting. We know winter will on average be colder than summer. And we know it will rain in 10 minutes. But will it rain tomorrow evening at 6 PM? Should we cancel the game? Should we irrigate the crop? Those questions remain unanswered. Advances in models and remote sensing may help, but we are not there yet.

There is a parallel between meteorological forecasting and the UTPS. We have some exact techniques for (parts of) the UTPS, but they are applied to systems where the initial state is far from known.

There is another aspect of the weather forecasting problem. There has long been the assumption that the system could be described with continuous, well-behaved functions. Now researchers think that chaos and its mathematics are an inherent part of system behavior.

There is some parallel to that in the UTPS.

We find the weather forecasting perspective useful because it suggests a good bit. There are problems that would exist in the UTPS even without the complications introduced by the social elements in urban and transportation systems. Do we know initial states; do well-behaved functions describe the dynamics?

We have observed that the steps toward the end of the UTPS process involve rather exact task statements and methods for doing calculations. We know exactly how to assign traffic to networks, make mode choice studies, do economic analysis, and undertake zone-to-zone traffic distribution analysis. One might claim that exact calculations are also made in earlier steps, but on comparison, problem statements and calculations are rather fuzzy. One could suggest several reasons for this. The one we usually state is that the near-the-end-of-the-process steps are more exposed to clients and general publics and thus beg precision. One could also suppose that the content of these steps is closer to the peer group interests of workers and/or that the earlier steps treat inherently more difficult topics.

All of this of course implies that we have precision without accuracy, which is not very useful at all.

We refer to the precision called for by Lord Kelvin as "Lord Kelvin's curse," for it yields a perspective that too often generates work with much attention to numbers and little attention to thinking. Even with its dangers, a small dose of analytic thinking and number crunching may offset planning steered by advocates and assertion.

24.2 Policy Wants to Control

Policy is a rule or set of rules for control. The term "cybernetics" (which is the study of feedback and automated control) comes from the Greek word *cubernetes*, meaning pilot or steering or control. The word "government" comes from that same word's Latin translation *gubernetes*, hence the word "gubernatorial." We have run across two physical models of how to control systems (or parts of systems) that shape policy thinking.

24.2.1 LAW OF INERTIA

The system to be controlled is seen as moving along a development trajectory, and policy aims to accelerate/decelerate development or steer the trajectory if there has been a perturbation pushing the system off track. That's a control problem very similar to the "man on the moon" problem, and the Newtonian model applies (from circa 1686 and Newton's *Principia*). We agree with the interpretation that policy makers and analysts tend to be Newtonian,[496] but we have never tried to explain that to a policy analyst, and would not dare to try.

An object has attributes of velocity (v), position (r), and acceleration (a). We can move it along its trajectory, $r(t)$, in any direction we want to. A point on the trajectory is a state, and the objective of policy is to move from one state to another. That thinking underlies the way policy is discussed. The debates concern desired states and ways to move things. Time's arrow has no meaning; a dynamic object is controlled by initial conditions. So we accelerate transportation development by pumping money into it.

The reader may react that ours is an outrageous and unneeded abstraction. We do not think it is. A large proportion of the policy work we have experienced involves thinking that the problem is to tweak the direction things are going (steer within a narrow lane) or to accelerate by subsidies or decelerate by taxes. The change of structure and behavior in fundamental ways is not considered.

24.2.2 SECOND LAW OF THERMODYNAMICS

We have also observed another way of thinking. It comes up in debates about transportation and energy and the inevitability of congestion. Carbon fuels are limited in supply, and they are being exhausted. A new facility is provided, and there is more travel and we haven't eliminated congestion. In these cases, the problem to be controlled is that of managing a bleak future. The future is bleak because technologies are inherently self-limiting. It's inevitable that things will get worse, and we must adjust expectations downward.

Although we may be pushing the abstraction too far, perhaps the implicit model is an entropy one, dating from Clausius in 1850. In particular, the second law of thermodynamics influences thinking, roughly: any physical system left to itself distributes its energy in a manner so that entropy increases; the available energy of the system diminishes. Things run to homogeneity and death.

The theory of Thomas Malthus (1766–1834) on the inevitability of overpopulation and the work of William Stanley Jevons (1835–1882) on the effect of sustained growth in English coal consumption represent the independent development of ideas that yield

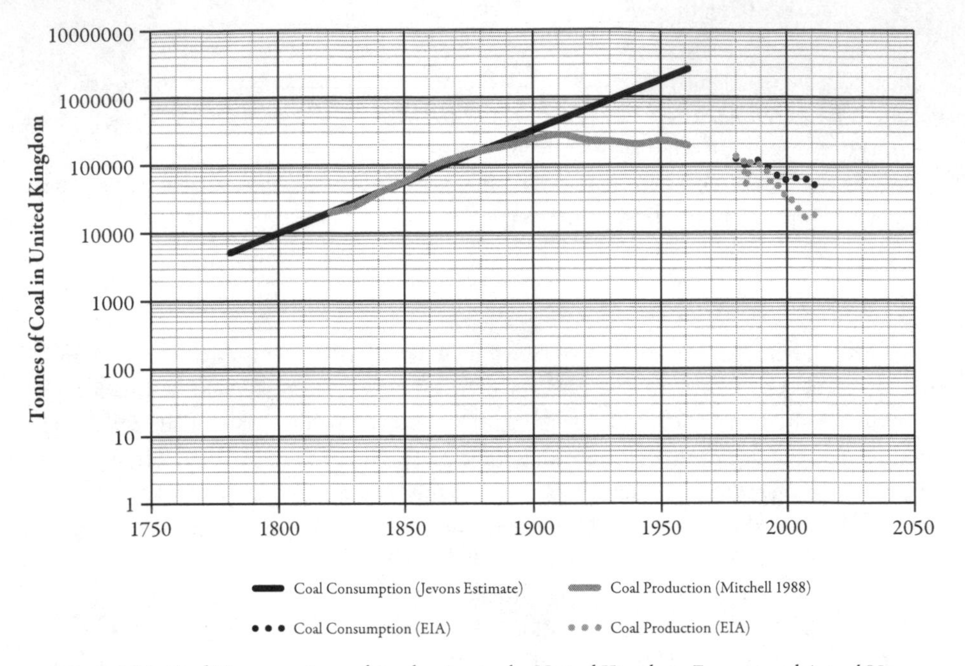

FIGURE 24.1 Coal Consumption and Production in the United Kingdom: Forecast and Actual Use

end-states similar to those that entropy-think yields. Jevons (1866) book, *The Coal Question*, had wide influence. Jevons was highly regarded in academic circles, and government policy makers paid much attention to his point that the growth of coal production driven by industrialization would soon exhaust English supplies. His broader point was that technologies are inherently self-limiting. His was a theory of technology that continues to hold coin for many. He posed the policy issue: "... we have to make the choice between brief greatness and longer continued mediocrity."[497]

Figure 24.1 is adapted is from Jevons's book and more recent data. It's presented here because of its similarity to population, energy (e.g. Figure 24.2), land use, and other many other projections we have seen. The magnitude of the forecasting error is huge (observe the log scale on the Y-axis). Similarly, as can be seen in oil price forecasts, even when conducted by experts in the field, they tend to be extrapolations of recent trends. Every forecast (from Delphi I to Delphi IX) expected prices to rise in the long term, despite twenty years of largely falling prices. This is not to say prices won't rise, just that the forecasts have been consistently pessimistic. If we knew with certainty the price of gasoline a year from now, we would be rich.

How is this pattern of thinking reflected in policy? We hear debates about how we are running out of petroleum (energy available for the system is diminishing); it is assumed the highway system is fully deployed (things run to homogeneity and run out of steam). The only thing that can be done is to control boundary conditions (for that is the way thermodynamic objects can be controlled) and accept mediocrity. So we control energy use in automobiles by seeking improved use of propulsion energy and accept the mediocre service of congested transportation facilities.

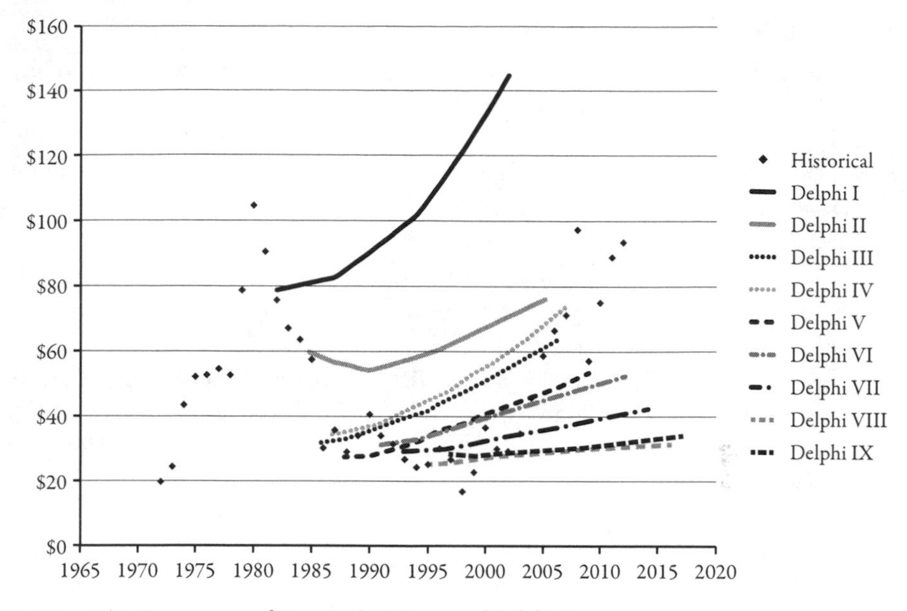

FIGURE 24.2 Comparison of Historical Oil Prices and Delphi Forecasts

24.2.3 WE CAN THINK BETTER

We do not dispute scarcity and limits (e.g., we talk about peak travel in Section 24.6.2), but we don't want to forget about innovation as a mechanism for addressing it.

This discussion is not saying that policy analysts should ignore resource matters, for certainly resource questions drive important parts of policy agendas. We are discussing self-limitations and inevitability. Again, the reader may think that our abstraction is outrageous and unneeded. The authors would turn that thought around and say: It is outrageous that we think about policy and forecast processes in physical system terms. We do indeed do that, even though transportation systems are socio-technological systems.

The broad point of this discussion of control is that people accept things as they are and limit the consideration of policy instruments to those that can shift states and/or change boundary conditions. Movement is along an equilibrium path. Those ways of thinking are not wrong; they are just too limited. We can think much better!

24.3 Forecasting Travel

Traffic forecasting has tended to apply Newtonian thinking to the problems of traffic. While methods of extrapolation have become more sophisticated: moving from forecasting traffic on roads directly (a method still used in rural areas and by pavement analysts), to forecasting traffic between origins and destinations, to forecasting the behavior of individual travelers; in the end, we are dealing with a paradigm of extrapolation.

24.3.1 UTPS EMERGES

It is useful to describe the modern Urban Transportation Planning System (or Model) (UTPS) as a clean break from precursor urban transportation planning experiences, although some exceptions will be noted. As a clean break, the emergence of modern UTPS ignored the well-honed planning techniques for arterial roads and local streets lodged in urban public works offices, the transit experiences, and the beginnings of transportation and land use planning in the style of Nelson Lewis, lodged in emerging planning agencies.[498] The UTPS also ignored some urban traffic analysis experiences, to be discussed shortly. In this "ignoring," more than one kit bag of techniques replacing another was involved, for there were institutional and conceptual breaks.

There was a break, and we can think about that in two ways. Modern UTPS can be viewed as revolutionary compared to precursor planning, following the body of literature on how revolutions in analysis (science) occur.[499] People who have discarded old dogmas develop new, more powerful ideas. The new drives out the old over a generation, as the bearers of old ways retire and the new move up the ladder. In the UTPS case, the rearrangements during World War II may have aided the displacement of the old by the new. The establishment doesn't change its collective mind; rather, young researchers who create and join "new schools" form a new establishment. There is evidence for this process in common experiences, as well as in the sciences.

On the other hand, there is evidence, especially in the transportation experience, that the new is built from the old. The new is built by putting existing building blocks into new designs that work about two times better than the old. Adoption is pushed by this more effective construct. That's the way, for example, modern container shipping was developed. It put hard and soft building blocks together in a new way.

Actually, a close examination of change reveals that both behaviors are usually found. For example, contemporaneously with the emergence of UTPS, a truck company manager, Malcom McLean, who put liner company dogma aside, as well as assembling building blocks from the transportation experience, developed container shipping (see Chapter 19).

The actors involved in UTPS development had not been involved in pre–World War II urban planning endeavors, and they did not have the burden of existing dogma defined by those endeavors. They weren't empty-headed, of course. They brought ideas to planning and they borrowed ways of thinking about the UTPS problem.

The realization of modern UTPS is often explained by saying that an analytic process pushed a weaker, less efficacious process aside. We do not find that explanation convincing. The analytic revolution was underway in social science and engineering, and we conjecture that no matter the origins of modern UTPS, an analytic process would have been developed. Indeed, as we will see when we examine work by urban transportation planners of the "old school," they strived to work in an analytic fashion and sometimes to be more analytic than the UTPS innovators. We also may point out that the break was not just analytic versus non-analytic, for breaks in concepts and institutions were involved.

The technical, institutional, and conceptual frames for what is called urban transportation planning, but is really simply urban transportation forecasting, emerged in the Chicago Area Transportation Study (CATS), which was initiated in the 1950s. The CATS approach, which was adopted nationwide and, later, worldwide, has evolved over time; techniques have been refined, and often they have been renamed. Indeed, techniques have been so modified that many practitioners are unaware of their origins. The CATS approach has been modified to accommodate scales and circumstances that differ situation to situation, mode to mode, and time to time. Creighton (1970) is an example of a textbook using the "pure CATS" approach.[500]

24.3.2 CATS AS FORECASTING PARADIGM

With the upswing in urban growth following World War II, a number of cities began to engage in a new round of facilities planning. Even prior to World War II, the State Department of Highways in California had begun to construct freeway-like facilities in Los Angeles, and planning for a more extensive system was begun in the late 1940s and early 1950s. The Detroit Metropolitan Area Transportation Study (DMATS) developed an extensive freeway plan. Minneapolis–St. Paul did some preliminary work, as was the case in the greater New York City area.

Many of the actors who worked on the DMATS study moved to Chicago when the Chicago Area Transportation Study (CATS) work was initiated by the Illinois Department of Transportation in cooperation with the Bureau of Public Roads, including the director, Douglas Carroll. CATS undertaking had the advantages that actors had already gone along the learning curve, resources were sized to the tasks, and it was the right style of study at the right time when the Interstate program was initiated.

With respect to facilities, the CATS planning effort was a broad one. Attention was paid to the entire street network, as well as to transit, and over the years there was attention to freight and air transport topics. This breadth was in spite of the state-BPR sponsorship and their limited responsibility for urban facilities. CATS investigated a wide breadth of research topics, as a review of CATS reports will reveal.

In 1956, when the Interstate Act was passed, CATS was in place and engaged in a widely scoped study in the BPR style. Although broad, it did not extend to institutional and financing matters. Also, it did not interact with existing urban institutions: public works agencies, city planning departments, the metropolitan planning department, and public interest organizations such as the Metropolitan Planning and Housing Council of Chicago.

Federal legislation and the 1955 *General Location* document gave the centerline mileage of the urban Interstates. The technical questions about the urban extensions were their final locations, including interchanges and how many lanes to provide.

To answer these questions, seven analysis steps were adopted by CATS:

- Economic and demographic forecasting;
- Land use forecasting;
- Trip generation forecasting;
- Trip distribution forecasting;
- Mode choice forecasting;
- Route assignment.
- Economic evaluation.

To answer questions about concepts adopted, one must refer to the CATS planning scheme (Figure 24.3).

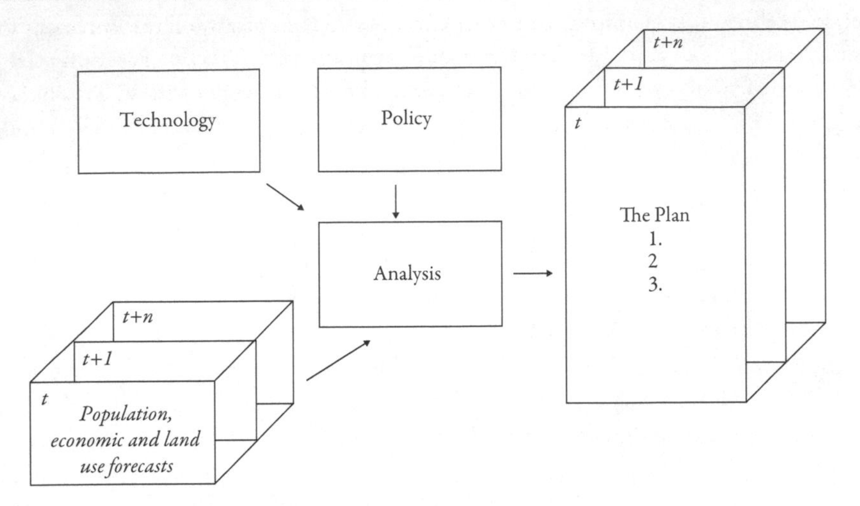

FIGURE 24.3 CATS Planning Scheme

Careful attention was given to the unfolding of the physical future of the city and of travel demand: do *X* in year 1, *Y* in year 2, and so on. The analysis involved putting the traffic on the network and determining needs for lanes. The plan was to be a one-time thing based on one-time analysis. A twenty-year time horizon was taken. The technology was a given (freeways, vehicles, etc.) as was policy (agency responsibilities, funding source, revenue source [new toll roads were a non-option], fixed lane mileage, etc.) and the determinants of behavior (trip rates, willingness to travel, preferences).

It is important to see this planning scheme in a larger frame. It suggests how the planner/forecaster apparently imagined that the worlds of policy, technology, and institutions worked: there is no feedback from the plan to policy or technology. It also suggests how planner/forecasters were thinking of urban growth and development and its relation to transportation: there is no feedback from the implementation of the plan to forecasts. It was imagined that the plan (a map based on the study of "facts") was a one-time project rather than an ongoing process. It was accepted that an outside institution would draft the plan. Finance and control of facility development resided with the state and federal governments, largely external to the city. The plan aimed to satisfy demand rather than manage it, or balance supply and demand at an optimal level. It was a given that the freeway was the technological solution, and the problem was the journey to work (other travel purposes and freight were ignored).

Across town in Chicago, a completely different analytic framework was being developed by Beckmann, McGuire, and Winsten (1956); however, these two approaches to modeling developed independently, and the Beckmann path remained academic.[501] In part this is because CATS was uninterested in congestion pricing, which was a main point of the Beckmann et al. research. The divergence was also due to the practical matter that, at the time, equilibrium flows could not be handled within the planning models. The not-invented-here syndrome and differing conceptual bases also delayed interest on the part of agencies in the use of linear and dynamic programming.

Rather than building on the urban experience, CATS built mainly on the Bureau of Public Roads (BPR) and state highway department experience. CATS was sponsored by the Illinois Department of Highways; and it is easy to see why many of the UTPS building blocks were from the state-BPR experience (see Section 15.8) (and similarly, why building blocks from the urban experience were not used).

24.3.3 EVALUATION PARADIGM

The theoretical roots of transportation evaluation are in microeconomic theory and in engineering economics. The consumer surplus idea (dating from the 1850s, and developed by civil engineer Jules Dupuit[502]) is on center stage. There are applied roots in railroad work (e.g., Wellington's work[503]) and in the analyses undertaken by the Bureau of Reclamation (1930s and 1940s, especially).

Martin Beckmann and later William Vickery and many others applied Alfred Marshall's comfortable and well-known notion of consumer surplus (following Jules Dupuit). Figure 24.4 provides a vehicle for the discussion of types of surpluses.

Consumer surplus is the difference between what a consumer is willing to pay for some quantity of a product (as defined by the demand curve) and what he has to pay (as defined by the price that must be paid for that amount). Producer surplus is the excess of total revenue over total avoidable cost that accrues to the seller as economic profit. At market price P_0 and producer cost C_0, resulting in level of consumption Q_0, areas A and B represent consumer and producer surplus for an "as is" situation.

Areas G, H, J, and K show increases in surpluses with lowered costs (the supply curve moves outward to S_1, costs drop to C_1 and prices to P_1, leading to an increase in consumption at level Q_1. Areas G and H show costs avoided (or resources saved) over the "as is" situation. J is consumer surplus on new traffic, and K is producer surplus on new traffic.

Railroad and water project analyses developed and honed benefit-cost and similar techniques. With this technique, a transportation system or facility can and should be thought

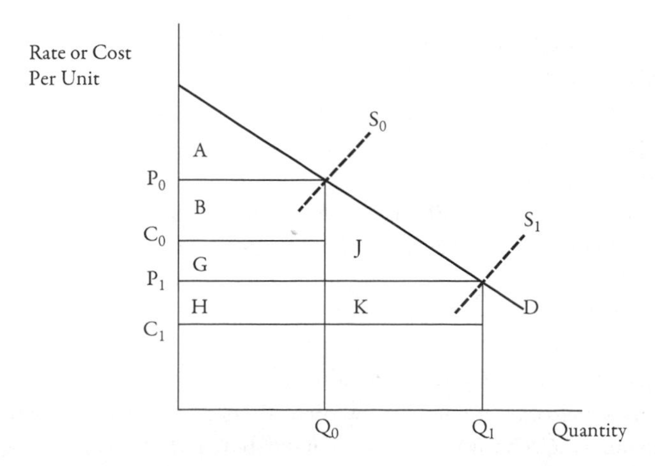

FIGURE 24.4 Consumers' Surplus

about in the same way one treats widget production. It's one activity in a world full of atomistic activities. There is nothing special about it. The efficient manager takes stock of the costs of inputs, examines marginal costs, and selects the optimal output. If the values of that output are to be measured, they are in the calculi of benefit-cost and consumers' surplus.

The 1956 interstate legislation called for the analysis of cost allocation, and it referred to user and non-user benefits. The legislation set off a good bit of empirical work and debate about benefits. In response, the "widget production, micro-economic view" was revisited and written in terms explicit to transportation. The products of at least ten or so authors might be noted, including Mohring, Harwitz, Meyer, Kain, Wohl, Kuhn, and others.

But incidence usually is ignored by the broad generalization that society is better off. So we have a conceptual scheme that lets us answer benefit questions, say, "What will be the result of Intelligent Transportation Systems (ITS) improvements in traffic flow?" ITS will create surpluses of various kinds. This is how the benefits of investment are most widely thought about—decreases in user costs are entered as benefits in benefit-cost calculations.

In addition to stress on benefits and benefit-cost analysis, attention extends to optimal pricing policy. Vickery applied these notions to urban highway congestion, and others addressed other modes. One result of this work was the deregulation of several of the modes to encourage their bringing prices charged in line with costs. There has also been an expansion of interest in congestion pricing on urban freeways and at congested airports. Time of day variation in transit fares has also resulted.

Although this conventional view can claim considerable adoption, we continue to see debate that ignores it. Not too long ago, for example, an issue of *Metro* questioned why voters in Houston turned down taxes for the construction of a rail transit system, even though the system would have 87,000 direct and indirect employees. Such employment should be counted on the cost side of the benefit-cost equation, of course, from a micro-economic perspective. It should be noted that many macro-economists consider additional employment a benefit, so long as the economy is not at full employment.

We also see some petty things one might quarrel about, things having to do with the assumptions and limitations of microeconomic theory: distribution of costs and benefits ignored, increasing costs assumed, complete market assumed, and so on. Such challenges are easily deflected. For instance, side payments can be developed to "even out" equity, the distribution of costs and benefits.

At a larger scope, one may question how the paradigm catches the large impacts of transportation: bringing new resources into the economy, increasing the size of labor sheds, offering greater choices for jobs, shopping, recreation, and socialization, increasing market areas, and so on. The "conventional wisdom" answer to those questions is that these impacts are measured in the flow of consumers' surplus. They are the basis of the elasticity of demand.

24.3.4 DIFFUSION OF CATS

In 1956, the states were ready to build many rural Interstate links, and were thinking about the urban extensions. CATS was ready to be emulated, and it was. Emulators took from CATS what they needed. The institution was copied, as was the size of the effort. But all this took time. Studies were initiated here and there as the states could get organized. Some studies involved cooperation with local agencies, especially metropolitan agencies where they were available. Even so, the local agencies were hardly full partners in the activity, for the Bureau and the states had control of funding and personnel.

Many leaders in CATS moved to positions of responsibility in the newly initiated studies. Some of the studies relied heavily on consultants, and previous CATS actors turned up that context.

The Bureau pushed urban planning in the same way state planning had been pushed, and the Federal Aid Highway Act of 1962 required long-range comprehensive planning in cities of 50,000 or over in population as a condition for the receipt of federal funds. The Bureau began to offer technical assistance, and these steps greatly increased the transfer of CATS concepts and techniques. By the 1970s, freeway-type facilities were being planned in cities worldwide, and there was more and more emulation. By this time, of course, CATS experience had been modified by experiences in other cities.

It needs to be noted that institutional and "size freeways" aspects of the planning endeavors were modified as time passed. The Federal Aid Act of 1970 required that planning integrate highways into other transportation plans and the evaluation of the social, economic, and environmental impacts of highways. The Act of 1973 opened highway trust funds to use by mass transit. It emphasized highway safety and completion of the Interstate.

24.4 UTPS and Its Discontents

> [T]here are known knowns; there are things we know that we know. There are known unknowns; that is to say there are things that, we now know we don't know. But there are also unknown unknowns, there are things we do not know, we don't know.
>
> —US Secretary of Defense, Donald Rumsfeld, February 12, 2002

The UTPS, including its land use and transportation components, have been heavily criticized. The critiques of the land use models have been sufficient to significantly diminish application in that realm.[504] However, despite criticism, the "four-step" transportation planning model remains. In Churchman's writings on systems analysis, defining and analyzing systems absent considerations of their interactions with and changes in their environments yields an environmental fallacy.[505] With this point in mind, we begin by positioning UTPS within its broad environment.

A project sponsored by the American Public Works Association recorded the history of the Interstate highway system as part of the *History of Public Works in the United States, 1776–1976*.[506] One strong feature of the study was the recording and interpreting of the oral histories of many actors active in the building of the system. At the descriptive level, the study does a pretty good job of reporting how institutions, financing, and knowledge were used to deliver the system. On size, dollar, and other scales, that's a remarkable story.

It's not a unique story, for there are many examples of social agreement and organization for the delivery of products and services for the common good. Perhaps the Interstate pales as a relative effort in comparison to the churches, defense walls, and markets produced in medieval cities and canals and walls in China, where commitments were made over decades for the investments and large fractions of social surpluses were invested.

Considering those instances and others, as well as the Interstate, the notion that the experiences are unique obscures the search for systematic ideas. Ideally, we would like to formally know how society can find agreement about and construct and manage infrastructure systems in acceptable and efficient ways. Such systematic knowledge is elusive; it remains hard to see how medieval churches and the Interstate are variations on a theme.

To give a specific highway example, one motive of the sponsors of the Interstate history (AASHTO) was that it would provide systematic understandings useful to the debate about post-Interstate highway programs. It is not at all clear that such understandings were developed. Actually, the writers of the history were skeptical about achieving such findings.

The problem here was that the Interstate was only a snapshot of the larger processes of the growth and development of auto/truck/highway/land use systems. We see it as the penultimate phase of deployment, more growth than development, and oriented to economies of scale. The UTPS follows the period when most emphasis is given to product (deploying highways), and it falls in the period when greatest attention is given to the process of delivering the product (planning highways). Considered in that light, the UTPS is essentially what would be expected. It's a process for delivering a standardized product. If something is said to be wrong, it's not with the UTPS, but with the overall product cycle.

Our second consideration will position the UTPS process within the environment of legislative actions, policy debates, and economic "mood" of the times. We have already reviewed the evolution of the idea of some sort of Interstate system and the toll roads versus free roads debate. There was the decision to build facilities at service level C, rather than B as in rural areas. That was to keep the cost down. Finally, there was the decision to stick with limited mileage. That, of course, was also to keep costs down. So the UTPS was oriented to supplying routes "on the cheap," and that's what it did. The result was the construction of a limited mileage freeway system.

24.4.1 THE UTPS IS BIASED IN FAVOR OF THE INTERSTATE AND SCALE ECONOMIES

Considering the positioning of UTPS within a particular environment at a given point in the Life-cycle, we expect that as the market grows suppliers standardize product and process technologies to achieve economies of scale, as discussed in Chapter 10. The on-the-cheap, high-capacity, limited mileage facilities are consistent with that expectation.

An interesting, but not unexpected, feature of the UTPS was the use of scale economies to decrease supplier (highway agency) costs. It may have been neutral with respect to equipment supplier costs, and its effect on user cost was mixed.

Scale in the highway enabled improved service, so users got something from scale. However, that scale was achieved in a fashion so that as traffic grew, diseconomies set in. Unlike the other transportation systems, highway designs are such that adjustments are difficult. We are observing long run average cost increases for new construction.

The highway system differs from other systems in this important economy of scale way, but this is not due to the UTPS.

There is another point about scale, a simple one. The scale at which the highway suppliers' product is efficient doesn't seem to fit the scale of the city very well. Scale creates conflicts with the environment.

24.4.2 THE UTPS DOESN'T MODEL MUCH OF HUMAN ACTIVITY

The UTPS didn't fit the state of the city when the UTPS was birthed. The gap is now greater. That's an impression we have stemming from two thoughts.

Our first thought requires that we distinguish between growth and development. If we look at a city at time t and then at $t+n$, to what extent is the difference due to development and to what extent is it due to growth? Suppose a city is small at time t, its shopping is mainly downtown. It's larger at $t+n$, and there are non-CBD shopping centers. Our impression is that the decentralization of shopping is simply because cities are larger, it's a growth matter. We suspect that's true of many things that are purported to be fundamental changes in the ways cities are developing.

But there are surely some development-like changes. There's the moving of "back office" transactions from the CBD, and the development of transactions generally.[507] We need a modeling process for a transaction economy.

Our next thought asks the question, what's the city for? The UTPS answers that question when we examine its product—the UTPS focuses on work travel; a city is a place to work. There are 8,760 hours in the year. In the United States, those who work full-time do so for about 2,000 hours (40 hours a week for about 50 weeks a year, discounting vacation and holidays). Fewer than half the people work (the rest are children, retired, stay-at-home parents, unemployed, or independently wealthy) so the average person works about 800 hours. So about 9 percent of urban time is used for work.

Average work travel runs 0.66 hours per day (maybe a bit more, but in most cities, well less than an hour round trip), and about 240 hours per year. Say a third of that is due to congestion delay, 80 hours. That's about 0.9 percent of urban time.

What all this rough arithmetic says is that UTPS is concerned with a small percent of urban life. To be sure, there is leverage in that focus because the ease of the work trip bears on choices of housing and jobs. Also, the peaking of work trips is the time of the worst congestion—the point when people see the transportation system "failing." Even so, we would like a UTPS that is at least more explicit with respect to the leverage and that scopes to things like housing and social and recreational travel.

We would like a UTPS that focuses on the city as a place for living. Perhaps the work-trip orientation was the appropriate focus at the beginning of the UTPS, but that seems less so now.

24.4.3 THE FLAWS OF UTPS ARE MYRIAD

That positioning complete, we strive for a level of criticism where UTPS bashing is in order only to the extent that we wish to illustrate undesirable outcomes. Examples of UTPS bashing include:

- unrealistic assumptions,
  - behaviors are unchanging
  - technologies are unchanging

  - inputs (e.g., travel times used for trip distribution) are inconsistent with outputs (e.g., travel times from route assignment), which leads to a failure to deal with induced demand (see Section 24.4.4)
  - ...
- human behavior can't be reduced to numbers,
- failure to deal with fine detail problems, and
- conflicts with community goals.

These examples are ad hoc, and are surely drawn from some bounded but not closed universe. The point we wish to make is that we have no systematic way to deal with bashing messages considering them on their own.

But we can begin to make some sense of bashing on three dimensions: the UTPS within its immediate environment, the UTPS in relation to analysis know-how, and the UTPS within the larger system logic.

Chapter 16 has prepared us for this criticism, for we have discussed the way urban problems were viewed just before the development of the UTPS, and we discussed transportation developments for the same period (BPR surveys, city traffic studies).

24.4.4 UTPS INADEQUATELY ADDRESSES INDUCED DEMAND AND INDUCED DEVELOPMENT

Transportation forecasts often implicitly assume that demand does not respond to capacity increases (i.e., there is no feedback from supply conditions to demand). Yet when capacity is added, drivers switch routes, times, and modes to take advantage of the more desirable capacity,[508] as shown in Figure 24.5. When capacity is taken away, or congestion rises, people divert from those routes, the peak spreads, and use of alternate modes (which are not caught in the same congestion) increases. After some time, people move or change jobs to

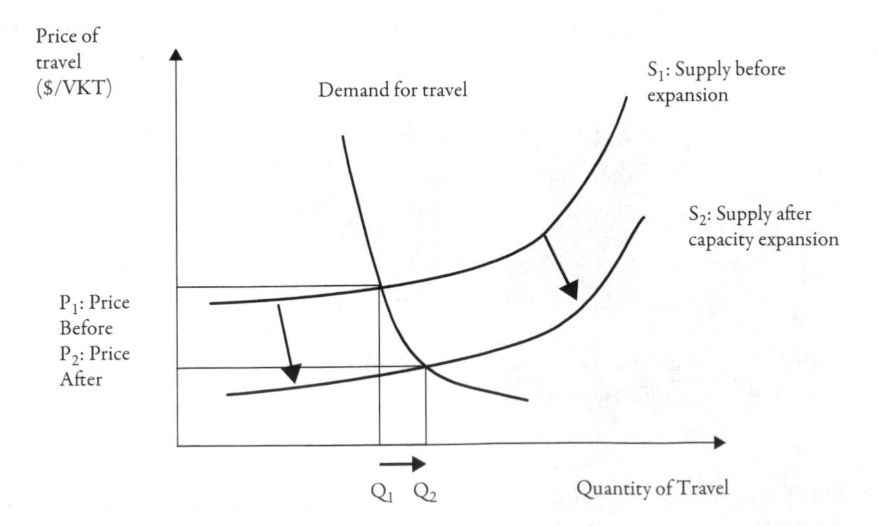

FIGURE 24.5 Induced Demand Curve

take advantage of new infrastructure or to avoid rising congestion. As transportation professionals, these shifts are what we want and expect. Otherwise, why would we build new capacity, or have any hope for policy?

In the long run, there is induced development, as new construction takes advantage of new accessibility (and people abandon newly useless places).

Capacity (highway or transit) may be added for multiple reasons. It is often sold on "congestion relief," that is, this piece of infrastructure will save existing commuters time. It may further be promised as opening up areas for development or redevelopment (the latter being especially important for transit promoters). However, the issue of transportation doing what it is supposed to (that is, move people quickly between two points) creates problems for those who value the negative externalities of travel (pollution, land consumption) more highly than the benefits of transportation.

It is not always clear, especially in mature networks, whether capacity expansions provide net costs or benefits to society, as additional travel generates negative externalities.

24.4.5 WHERE YOU STAND DEPENDS ON WHERE YOU SIT

A "worldview" is a perspective on how the world works. As has been often noted, *where you stand depends on where you sit.* Ellis (1996) discusses the differences between professions, which we discussed in Chapter 16.

There is an interesting, but difficult to find, essay by Hillman (1979) ("Psychological Fantasies in Transportation Problems") that points out that transportation problems are in the minds of the beholders. Hillman notes that the typical transportation planner is trained in rational analysis, works about 250 days a year in a white collar job, and typically drives to work at some place in or near the CBD. It is no surprise, says Hillman, that the archetypal urban transportation problem is the journey from a suburban home to the downtown workplace. Attributes of the problem to be managed are things such as facilities that aren't big enough or have potholes and making sure that other travelers are more "rational."

Hillman also points out that the transportation planner gets "marching orders" from politicians who share the planner's lifestyle.

Hillman's remarks lead to two conjectures. First, the problem to be managed is the state of travel when commuting. That problem and its context aligns more closely with the latter steps in the UTPS process than with early steps, so one might expect that latter steps are less ad hoc in concept and technique than early steps. Also, the journey to work gets priority. Second, there is the rational view of processes. One would expect that formulations of processes similar in style to what's taken to be rational analysis would predominate.

Finally, one would expect planners to think that transportation plays a key role in processes. For instance, most transportation planners think of access to work as the key consideration in household location decisions.

24.4.6 UTPS IS NOT DESIGNED TO ADDRESS SYSTEM MANAGEMENT

The UTPS analysis was state-of-the-art early on; it's now much less so. It didn't fit the city very well at the beginning, and the situation is worse now.

Understand that trajectories are not independent. Just as the state of the city affects forecasts, UTPS affected the state of the city by justifying particular investments (at the margins, some investments must have been affected by the forecasting procedures used).

As the UTPS was developed, the urban transportation system was drifting into the mature phase where attention is given to tailoring service to markets, to the extent that unitary products can be tailored to markets. In the urban environment, that was realized under the Transportation System Management (TSM) rubric. No discussion is needed to conclude that the UTPS was disconnected from TSM.

There is an interesting bit of history. TSM was triggered by UMTA and the lack of money to fund the flood of transit capital proposals. The FHWA was a reluctant partner in its development. We would say that UMTA's action was a triggering event and not the reason for TSM. TSM would have come along without the event; it was the next step in the logic of system development.

Our summary statement is that the UTPS was very state-of-the-art when first developed, and especially in its CATS incarnation, but as the world has advanced, UTPS has not kept pace. It dealt with the behavior of aggregates, was a logical or rational approach to problem solving, used numbers and statistics, modeled formulas on the physical sciences, and used serial steps. It used normative ideas. It was state-of-the-art for social analysis and normative planning, and it was state-of-the-art for transportation technical analysis, too. The latter is because it built so closely on BPR work.

24.4.7 CRITIQUES OF MODELS VEIL CRITIQUES OF POLICY

Is the transportation planning model objective or subjective? Is it used to inform decisions or to justify them? Do forecasts predict the future or create it in a process of self-fulfilling prophecy?

In 1991, the Montgomery County, Maryland, transportation planning model (*Travel*), a then state-of-the-art descendent of UTPS, was being used to assist the county Planning Department in the development of a Master Plan for the suburban area of North Bethesda.[509] Policy called for much of that area, which incorporated three Metro subway stations on the Red Line to downtown Washington, to receive higher density zoning, and thus more development. Policy in Montgomery County suggested the additional congestion resulting from the development could be tolerated in areas with better transit service. The alternative of providing additional highway capacity was too costly, would undermine support for transit, and faced numerous environmental and other hurdles. Additional transportation demand management approaches (TDM) were also put in place, but were thought to be inadequate.

Citizens groups in North Bethesda, led by activist and transportation safety engineer John Viner, were unhappy with the direction the plan was going. While the area would become more developed, existing residents would face many of the negative externalities (more congestion) and receive few of the benefits (since their own property was already built out and not eligible for rezoning). Informed by previous studies showing significant congestion with even less intensive land use, the community came to the belief that the model forecasts they were being shown would underestimate the amount of traffic produced by the proposed development. The consequence was that more development would be approved than the roads could handle at a given level of service (even a worse level of service). If the model forecasts were low, the land use approved based on those forecasts would be high.

There are several possible reasons for an underestimate. First, the model could be flawed. Second, the modelers could be misusing the model, either accidentally or intentionally. Intentionally misleading the public and officials can be achieved by making unreasonable hidden assumptions. The reasons for this have been documented in a number of sources,[510] and are discussed later in this chapter.

After testimony by citizens in hearings before the County Council about the inaccuracy of the model, in what was dubbed locally as "The Trial of Galileo," County Council member (and future candidate for county executive) Bruce Adams called a special meeting of the Council's Planning, Housing, and Economic Development Committee on the model. The community argued that the model was flawed, rather than arguing the policy was flawed (which they believed as well), or that the modelers were intentionally misusing the model under pressure from superiors. The strategy behind this was to go after the weaker target, the model. The Council, which established policy, would be more open to hearing a critique of the model than its own directions or criticism of planners who were doing their bidding.

The "Trial" involved numerous maps, charts, tables, and memos being prepared for Council consumption by both sides. The citizens compared forecasts against observed traffic and suggested the model was *prima facie* wrong. Citizens argued that the model underestimated traffic (the forecast year counts were lower than today's on some links, and surely traffic won't decline if population increases). The citizens also argued that officials buried earlier studies that produced different results (and higher traffic estimates). Planning staff said otherwise, and pointed out that they were assuming numerous travel demand measures that would be imposed on the new development, and also that more congested level of service standards would be in place.

In the end, not much was decided; the community still opposed the plan, and still believed the model was flawed. The Council took the results and the plans with perhaps a more skeptical eye, but in the end did agree to a significant upzoning of North Bethesda, along with additional travel demand management policies. Few new roads have opened in the area, and some very large developments are now getting started, twenty years later.

24.4.8 STRATEGIC MISREPRESENTATION AND OPTIMISM BIAS PERVADE FORECASTING

The experience in rail transit projects (and toll roads) says that projects will come in over budget and will not meet initial ridership projections. The game has become sophisticated, so that initially there are unofficial high forecasts for demand, which are lowered for the time of approval, and then the lowered "official" forecasts beaten on opening, allowing politicians and planners to claim "success." The literature on transit strategically overestimating demand and underestimating cost is vast.[511] Notably, forecasts of demand have been improved since planners were first called out for what might be nicely termed "optimism bias."

Optimism bias is the notion that people who want a project to succeed will make favorable assumptions, drawn from the range of possible assumptions, when forecasting. Compounding a slew of favorable assumptions will tilt the outcome in one direction. The problem is compounded by the need for a single forecast, rather than a range of potential outcomes associated with probabilities, which is somewhat more rational, given our knowledge that there are known unknowns, unknown knowns, and unknown unknowns.

Observing that subsidies are highly concentrated, ridership is limited, and new facilities very expensive, one pundit commenting on the Los Angeles subway observed, "Never will so many have paid so much for so few."

Road forecasts have not been terribly accurate either for different reasons. Flows on Interstates have been underestimated (and on non-Interstates overestimated), according to one of the few *ex post* evaluations.[512] This is largely due to technical flaws in the UTPS process, in particular the assumptions of unchanging behavior (e.g., female labor force participation), as well as poor specification of networks and bad input forecasts of demographics.

The problems of strategic misrepresentation and optimism bias are generic, and not specific to UTPS.

24.5 Traveler Behaviors

UTPS attempted to forecast traveler behavior, but it missed several major aspects that structure travel. First, it was unstructured concerning how time is used for activities by travelers, and missed the notion of diminishing returns to travel (if not outright travel time budgets) that constrain how bad congestion can get. Second, it did not consider systematically how those activities occur across the day.

24.5.1 TIME USE

Figure 24.6 summarizes some long-term trends in time use in the United States. Despite the differences in methods, some clear trends emerge. In 1990, adult Americans are working more on weekdays, and less on Saturday than in 1954. The weekday rise is principally due to the larger number of women working outside the home. Although Schor (1993) has argued (controversially) that time at work has risen for men as well, this may not show up in a travel or activity survey, but rather in wage data. The Saturday drop reflects the widespread adoption of the five-day workweek since 1954. The amount of time spent shopping has held remarkably steady, although even small time differences in this category represent larger percentage differences. Americans would appear to be shopping more on weekends. This in part is due to Sunday shopping, which was rare in 1954 due to Blue Laws, but this also seems to be true on Saturdays.

The two most curious categories are home and other. Given the increase in female labor force participation, time spent at home from 1954 to 1990 should be expected to decrease on weekdays. This is supported by the data. However, several interacting factors make the issue more complicated. Saturday work has decreased, which makes more time available on Saturdays (for home and shop), while the opening of stores (and other activity locations) on Sunday enables people to get out on Sunday.

Figure 24.6 suggests that the amount of time in travel is almost identical between 1954 and 1990. It raises the intriguing possibility that there is some form of "travel time budget." There are 24 hours, or 1,440 minutes, in a day. Thus, per day, there is a fixed amount of time

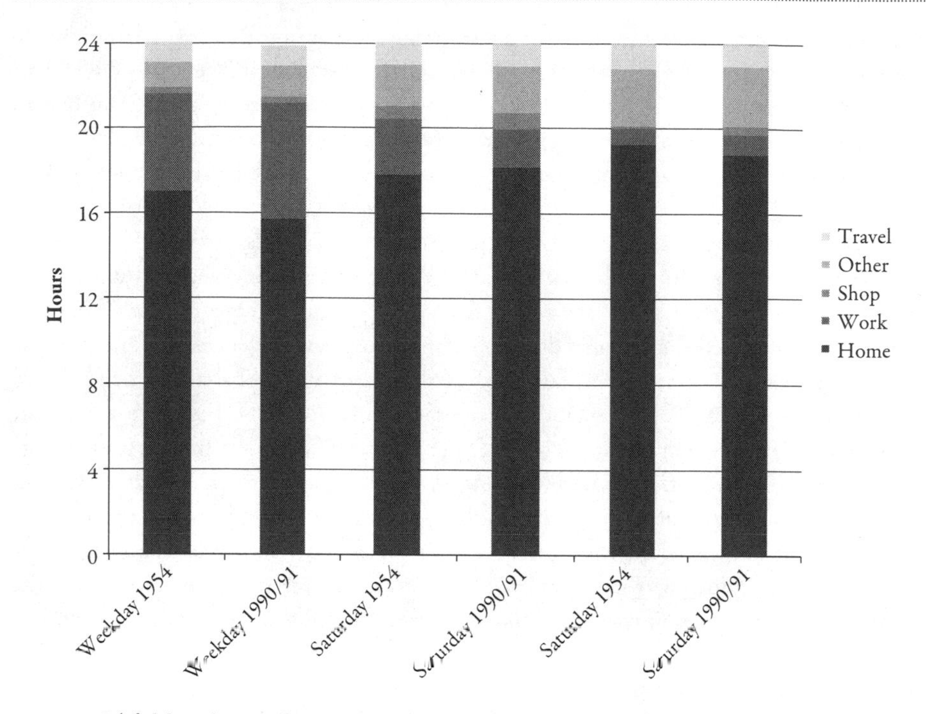

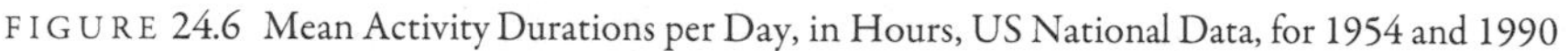

FIGURE 24.6 Mean Activity Durations per Day, in Hours, US National Data, for 1954 and 1990

to do things. Consider a typical childless adult. Taking out 8 hours for work (production) and another 8 hours for sleep (Marx's reproduction) reduces flexibility. Another portion of that time must be spent on consumption (shopping and eating). There are further household maintenance activities that must take place (cleaning, hygiene, medical care, etc.). Thus there are maybe 4 discretionary hours per day. If that is the case, there is a practical upper limit on the amount of daily travel one can undertake (4 hours) that is split between travel to work, travel to non-work activities, and other discretionary activities.

A lot of evidence supports that individuals spend a fixed amount of time per day (just over one hour) in transportation (see Section 24.5.2).[513] Yet aside from the fixed bounds of the problem, the theoretical causal mechanism for a fixed budget is weak.

Marchetti (1994) suggests that humans have a nesting and a roaming instinct. At night we return home; by day we stake out a territory for a fixed time. When people walked, the distance was short. But as technology has allowed us to move faster (horses, streetcars, automobiles, airplanes), we stake out a larger area.

Other researchers[514] disagree with the notion of a travel time budget, echoing Becker (1965), who argued that trade-offs are made between travel time, other time, and expenditures for the full gamut of activities, depending on relative price and income changes and the valuation of time.

Levinson and Kumar (1994, 1995) note the stability in journey to work times over a long period in metropolitan Washington, D.C. (the 1968 to 1988 data shown here, as well as similar numbers from 1958 by William Whyte[515]). Rational locators act to shift jobs or homes to maintain commute times. Thus, for instance, suburbanization of jobs helps solve the problem of long duration commutes by bringing work nearer to labor. Further, while congestion is rising almost everywhere, roads in suburban areas are still generally faster than roads in urban areas. As a greater share of travelers uses the faster roads, the average overall average speed rises. That hypothesis provides a mechanism for stability in commute times, despite both rising congestion and longer distance commutes. The stability in commute times might be consistent with Marchetti's notions; since it is the longest trip that stakes out territory, the shorter trips typically operate within those bounds.

However, some data from other cities (e.g., the Twin Cities) show that commute times are increasing over time. Can that be reconciled with a commute budget notion? We recognize that larger cities do have longer duration commutes than smaller cities, if only because there are more opportunities (jobs) farther away that have some probability of being taken. Further, the equilibration process noted in the rational locator hypothesis, the movement of jobs and homes, also requires time. If there is a budget, which is on average at least as large as the commute times in the city with the longest commute times, then residents of smaller cities may not have consumed their travel budget yet, and travel could be expected to rise.

Mokhtarian and Chen (2004) suggest that the aggregate data mask individual variations. And though individuals may have personal budgets, those budgets are not necessarily the same. As different mixes of people comprise the population, the aggregate may shift, but if similar people occupy similar shares of the population, on average, the aggregate number won't move. The most important information for travel behavior analysis, the amount of time spent traveling, is ironically the least clear. There are, needless to say, many ways to cut the data.

24.5.2 PROFILE: YACOV ZAHAVI

Yacov Zahavi (1926–1983) developed UMOT (Unified Mechanism of Travel) in an effort to deal with supply and causal model issues.[516] His first work was with aggregate data; before his death he began work on micro underpinnings.

With respect to causal argument, Zahavi notes that the travel money budget runs about 12.5 percent of household expenditures for households that own cars, and it runs about 4 percent for households without cars. He also observes time budgets. His analysis brings these "demand" measures to the supply of travel facilities to yield travel information.

The data Zahavi musters have a "things are the same everywhere" character. Time and money expenditures are similar across nations; implying participation levels are market clearing in the same way everywhere. Household aspirations and capabilities tend to be similar, as do opportunities.

Zahavi observes that the travel time budget ranges from 0.8 to 1.1 hours, and the TT/HH (travel time [TT] per household [HH]) is a function of household size and cars per household. As mentioned, the travel money budget (TM) varies between car-owning and non-car-owning households. Zahavi extensively uses data from cities in Europe, Asia, North and South America, and the Soviet Union.

What did Zahavi accomplish? Zahavi starts with household income and size and the notion of a travel time budget. He examines the attributes of the supply of facilities. From these considerations, he calculates travel quantity measured in time, distance per traveler, and velocity dimensions. Time varies little. It is the daily travel distance that is the main variable explained. His argument is very convincing from an empirical, among cities and households view. It gives satisfaction also because it is based on a utility maximization scheme. (As stated earlier, Zahavi had begun work on within-city relations, and first within-city empirical results are convincing.)

Put another way, households may be thought of as being in one state or another: with auto or without. The decision to shift from without to with is based on distance and not time considerations. More money is spent if the household has an auto, so money is traded for distance. What does the household gain from distance? That's the question, and its answer must have to do with access to more options for housing, work, shopping, and so on, and/or opportunities to specialize.

24.5.3 TEMPORAL AND SPATIAL AGGLOMERATION

The UTPS says nothing about time-of-day. Models have subsequently added simple models to allocate travelers to time of day, or complex models to predict individual activity patterns. This matters, because facilities are seldom above capacity for an entire day, only for peak times.

We typically observe peaks in travel demand on weekday mornings and evenings (as well as other times on select facilities: Friday late afternoons leaving town, Saturday afternoon on shopping streets). Figure 24.7 illustrates this information. From 1968 to 1988 there was a spreading of the peak in Washington; as roads became more crowded, people were willing to start work earlier or later.

But why does travel peak at all? A commuter could save travel time by leaving for work at 10 AM (or 4 AM), for instance. The first answer is that workers show up when their firms want them to. So why do firms want all of their employees at work at the same time? The reason that firms have for setting a regular schedule both internal for their employees and

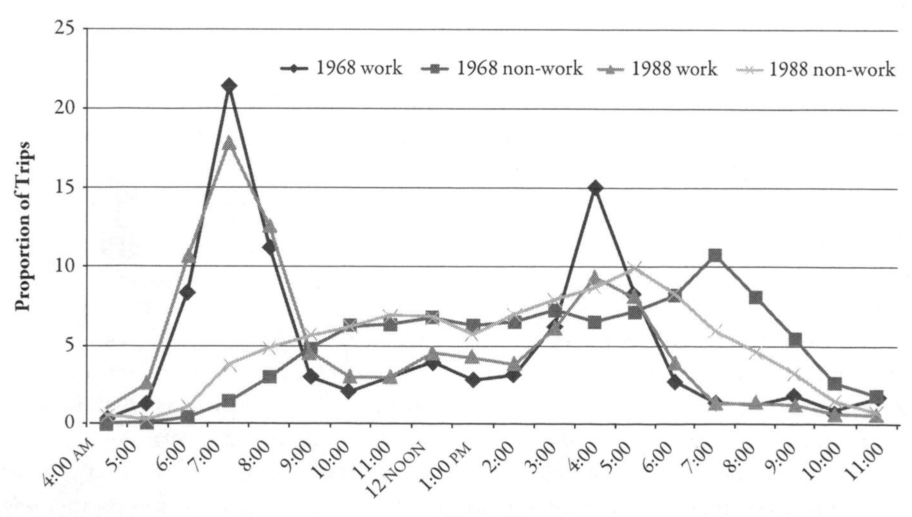

FIGURE 24.7 Diurnal Curves for Work and Non-work Trips in Metropolitan Washington, D.C.

external in common with other firms is an "economy of temporal agglomeration." A firm needs to coordinate activities internally and to deal with customers and suppliers at the time when both are open; this is all made easier if everyone is at work during approximately the same time frame. This doesn't apply to every organization: bars (which coordinate schedules to serve customers after work) may have later hours; police, who need to arrest rowdy bar patrons, among others, must be on the job 24 hours a day.

This economy of temporal agglomeration has benefits that outweigh the congestion costs they impose. Programs to introduce flextime and variable work schedules have had some effect, but not enough to flatten out the curve at a regional level.

Yet the congestion effects are not small, and, in the absence of road pricing, are not fully considered by the travelers who create them. The congestion externality drives capacity investment. Without peaking, major roadways wouldn't need to be as wide; capacity would be more evenly utilized throughout the day. Pricing roads higher during the peak hours (and lower during the off-peak) would provide incentives for travelers to consider whether peak-period travel was necessary.

Another point to note from Figure 24.7 is that non-work travel also peaks during the afternoon, only slightly after work travel. Many of the travelers at a given time are non-work oriented. These tripmakers presumably have, in general, somewhat less time pressures than those making work trips. Pricing may provide sufficient incentive for trips with a high "elasticity of demand," or sensitivity to price, to move to an alternate time, smoothing out the peak, and reducing the need for new construction.

In a sense, peaks in time (rush hour) are like peaks in space (downtown). Both exist to obtain economies of agglomeration—the ability to do business with others because they are nearby and open. Both produce costs, and make it more expensive for others to do business, but hopefully the gains from specialization and interaction outweigh the losses. Peak spreading is the temporal equivalent of suburbanization, without putting the same demands on creating new infrastructure.

24.6 Futures

> It is difficult to get a man to understand something, when his salary depends upon his not understanding it!
>
> —Upton Sinclair (1935)[517]

24.6.1 FORECASTING THE FUTURE OF FORECASTING

After decades of stagnation, urban transportation models are changing in the early twenty first century. There is a move in practice from a "trip-based" paradigm to an "activity-based paradigm" (moving toward a fully agent-based model, including dynamic traffic assignment). From a modeling perspective this is better, as it allows better intra-personal and intra-household substitutions, and better accounts for underlying behaviors.

The needs for models have moved from determining the number of lanes to determining resultant pollution levels for air quality conformity analysis. To do that, models need to track which vehicle is used on which road by whom for how long.

It is computationally more intensive, and significantly more complex to program, but if done correctly should be no worse at forecasting, and probably better than conventional aggregate techniques.

Models are not going away, but they are not getting better at capturing behavioral changes. They are still constructed assuming human behaviors are fixed. A 1960s person dropped into a 2010 model will behave the same as a 2010 person, given the same set of demographic and socioeconomic inputs. The underlying preferences are fixed over time. The underlying technologies are fixed.

Moving from forecasting to scenario testing (i.e., from "what will be" to "what if") would be a useful change in perspective. Nevertheless, legal requirements dictate the use of these models as long-term forecasting tools despite inherent inadequacies.

24.6.2 FORECASTING THE FUTURE OF BEHAVIOR: PEAK TRAVEL

One of the predictions of travel time budgets and the rational locator theory is that individual travel will not inexorably increase, but instead will only increase so long as the benefits outweigh the costs. One anticipates this process has diminishing returns, ever increasing travel (cost) will result in ever decreasing additional benefits from travel. Passing a third supermarket is only marginally better than the second, and the fourth even less better than the first three. There are diminishing returns to opportunities. People will, given their location, trade off between those benefits and costs, and over the long term, relocate to a place that gives them the options they prefer.

- Travel is correlated with the monetary cost of travel, so if the price of fuel goes up, the amount of travel goes down.
- Travel is correlated with system utilization, so if the amount of congestion rises, willingness to travel diminishes, keeping supply and demand in an equilibrium.
- Travel is correlated with available network capacity. If the network is not growing, the rate of travel increase will be limited.
- Travel is correlated with population growth; if that slows, the growth in travel should slow.
- Travel is associated with wealth; if people feel poorer, they will consume less.
- Travel is correlated with Life-cycle, so as fewer workers and parents with children are part of the system, we expect slightly fewer trips.
- Travel is associated with need for travel. If some activities (working, shopping, playing) can be done online instead of in person, we expect less travel.

In the United States and other advanced countries, travel per person, and perhaps overall, as shown in Figure 24.8, appears to have peaked in the late 1990s or early 2000s.[518] Other resource consumption has also peaked, people have discussed peak materialization (the tons of material consumed has declined), peak oil (oil production may have peaked), and peak coal (the US is burning less now than a decade ago and CO_2 emissions are falling. Per capita CO_2 emissions in the United States returned to 1966 levels as of 2008,[519] and total emissions are also down since 2008). The US Energy Information Administration reports total

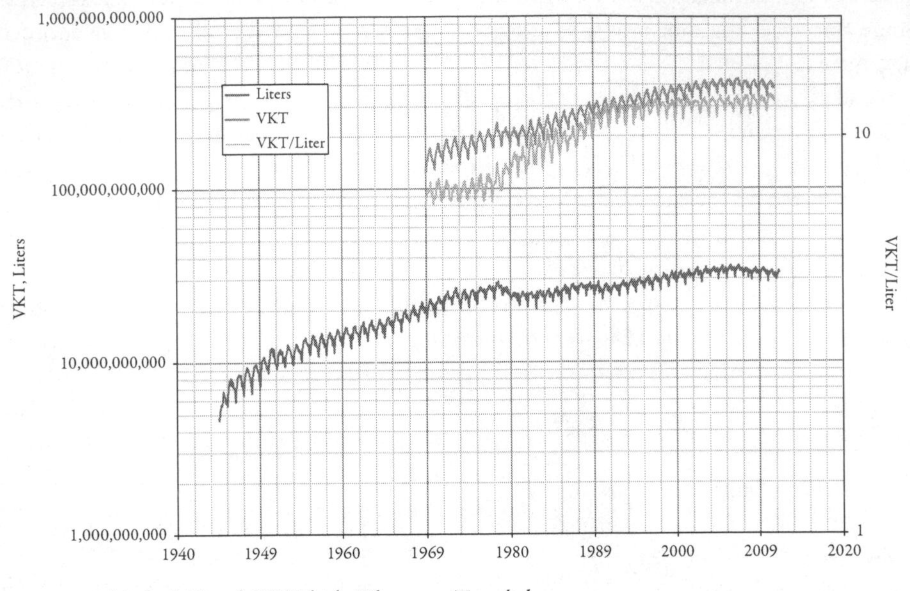

FIGURE 24.8 Peak Travel: US Vehicle Kilometers Traveled

US carbon emissions have fallen from a 2007 peak of 6 billion metric tons to 2012 estimate of 5.2 billion (near 1990 levels as petroleum and coal have declined and natural gas has risen),[520] and so on. The logic of this follows the life-cycle model we have used throughout the book. As a system matures, we expect its curve to flatten, and we expect it to decline as substitutes for the service the system provides are found. Oil production may peak either because we are in fact close to running out, or because alternatives (such as newly inexpensive natural gas) acts as a substitute, or both. The search for substitutes is not unrelated to anticipation of either decreasing oil or increasing costs of oil extraction.

The difficulty in forecasting is anticipating the peak before it occurs. While most people get the existence of the peak, not all do—as Upton Sinclair put it: "It is difficult to get a man to understand something, when his salary depends upon his not understanding it!"[521]

The harder part is establishing the timing or magnitude of the peak. Is the current downtrend just a temporary change or permanent? The difficulty in politics is telling people that something that has been regularly growing for seventy years (like automobile travel) has peaked.

24.6.3 TELECOMMUNICATIONS AND TRANSPORTATION: COMPLEMENTS OR SUBSTITUTES

Mokyr (2004) describes the industrial revolution as first involving innovations in the factory system in the late eighteenth century, followed by transportation innovations in the first half nineteenth century, with telecommunications (telegraph, and then telephone) in the second half of the nineteenth century. The peak of the "factory" system in a sense is when factories and transportation enabled and required agglomeration, and before telecommunications enabled more remote real-time action. The gains in transportation implied that the relative costs of moving people initially declined faster than the costs of moving information. Expertise was more likely to be supplied in-house rather than using hired consultants when production was very specific and local. Knowledge-pooling (for instance, a hospital with many specialist doctors rather than individual general practitioners) arose in many professions (law, engineering, medicine, architecture, teaching). This was another economy of scale of centralization enabled by transportation.[522]

Couclelis (2004) argues that new technology increases time fragmentation, so activities are increasingly interspersed, what some call multi-tasking. The argument about transportation and communication as substitutes or complements has yet to be resolved; however, more evidence is suggesting substitutability, as younger generations travel less and delay the purchase of a vehicle and spend more time online.

Work at home can be compared with the factory (work out of home) systems, basically as trade-offs in scale economies and transport costs. To the extent that scale economies are less achieved with people commuting out of the house—because manufacturing requires less and less labor, shopping can be more effectively conducted virtually rather than in store, and telecommunications become better and better substitutes for in-person contact—we should expect the substitution to become the more dominant feature of the relationship between transportation and communication.

Mobile phones, though, are obvious complements to transportation, and mobility is becoming an increasing significant component of the information technology landscape. As of this writing, Apple Computer, the most valuable company in the world, earns most of its profits from mobile devices, having surpassed transportation energy supplier ExxonMobil in market value in 2011.

24.7 Discussion: From UTPS to Inquiring Design

We have stressed the notion that present-day planning concepts and techniques result from interacting precursor experiences with the problems at hand. An interesting aspect of modern UTPS is the way it drew on the rural road, Bureau of Public Roads, and state experiences rather than urban experiences.

We have also stressed the role of CATS. It was there to be emulated when the need for urban Interstate planning emerged. That emulation was selective. CATS was a broad-based undertaking, emulators took what they needed.

It is important to understand the task posed by urban Interstate planning. Interstate designs were rural designs; the Interstate involved space-consuming, limited-access designs. These designs were imposed in urban areas; centerline mileage was limited. So the central question became "How many lanes?"

Why limited access? Land development and accompanying turning traffic had unwanted impacts on state highways, and the Bureau wanted to avoid that. Also, abutting structures and curb cuts had made it difficult to add capacity when needed.

The Bureau's rural experience had been to make plans and implement. Twenty years was the expected life of pavements, so they made twenty year plans. At the end of twenty years, more planning and implementation would be undertaken: renew pavements, add more lanes, straighten curves, new bridges, and so on. That sort of thinking was taken into the urban area where it was inappropriate. Essentially, the Interstate built out so far as the capacity of areas allowed.

Previous remarks have said that the UTPS was a product of its positioning within the environment: the general environment said "do it on the cheap" and the transportation environment said "do it this way at this stage in the Life-cycle." In a sense, UTPS was "what had to be," and we can understand it completely. If we were to do it again, or if another situation comes along in which something similar is called for, it would seem improvements would have to be addressed to the working of the system logic rather than to UTPS logic.

So we leave the UTPS for the moment and jump to the model we have in mind for large system planning or planning working on the system logic. Rather than tackle the impossibility of forecasting and running into Lord Kelvin's curse, in our ideal, one operative word for planning would be "inquiring"; another would be "design." We would like analysis that scoped from all building blocks (knowledge, available technologies, physical facilities, institutions, etc.) and that produced designs.

We imagine such designs having physical content, and extending to institutional, policy, and other "soft" matters bearing on the working of a design. The designs should be consistent with deep running social and economic trends and constraints of all sorts. They should open options (development pathways) rather than mine out options. They should offer choices to the public. It's by offering choices about pathways that planning is inquiring.

Let's be clear about what we are saying. In previous discussions we presented the system logic. It has a disjoint character, and the micro processes at work yield a macro realization of an S-shaped sort. There is no reason to think of that realization as having optimal properties, but there is reason to think that it's very stable.

We are saying UTPS-like activities are an outcome of system properties, and the way to improve UTPS-like activities is to develop a scope and style of planning that changes system properties, rather than simply forecasting within the constraints of existing system properties, subject to the limits of cognition and self-serving assertion.

(a)

London Population Density and National Rail In 1850

London National Rail Stations
London National Rail Lines
Population Density 1850
Person/Sq Km
0 - 1000
1001 - 2000
2001 - 3000
3001 - 4000
4001 - 5000
5001 - 7500
7501 - 10000
10001 - 12500
12501 - 15000
15001 - 20000
20001 - 25000
25001 - 45000

Generated By: **NEXUS**

(b)

London Population Density, London Underground, and National Rail In 1900

London National Rail Stations
London National Rail Lines
London Underground
Bakerloo
Central
Circle
District
East London
Hammersmith & City
Jubilee
Metropolitan
Northern
Piccadilly
Victoria
Waterloo & City
Population Density 1900
Person/Sq Km
0 - 1000
1001 - 2000
2001 - 3000
3001 - 4000
4001 - 5000
5001 - 7500
7501 - 10000
10001 - 12500
12501 - 15000
15001 - 20000
20001 - 25000
25001 - 45000

Generated By: **NEXUS**

FIGURE 8.3 The Growth of London

(c)

London Population Density, London Underground, and National Rail In 1950

London National Rail Stations
London National Rail Lines
London Underground
Bakerloo
Central
Circle
District
East London
Hammersmith & City
Jubilee
Metropolitan
Northern
Piccadilly
Victoria
Waterloo & City
Population Density 1950
Person/Sq Km
0 - 1000
1001 - 2000
2001 - 3000
3001 - 4000
4001 - 5000
5001 - 7500
7501 - 10000
10001 - 12500
12501 - 15000
15001 - 20000
20001 - 25000
25001 - 45000

Generated By: NEXUS

(d)

London Population Density, London Underground, National Rail, and Docklands Light Railway In 2000

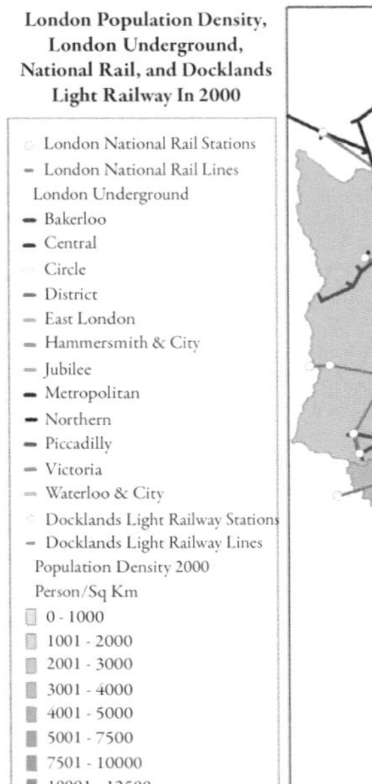

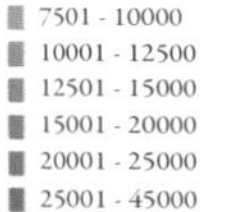

Generated By: NEXUS

FIGURE 8.3 (*Cont.*)

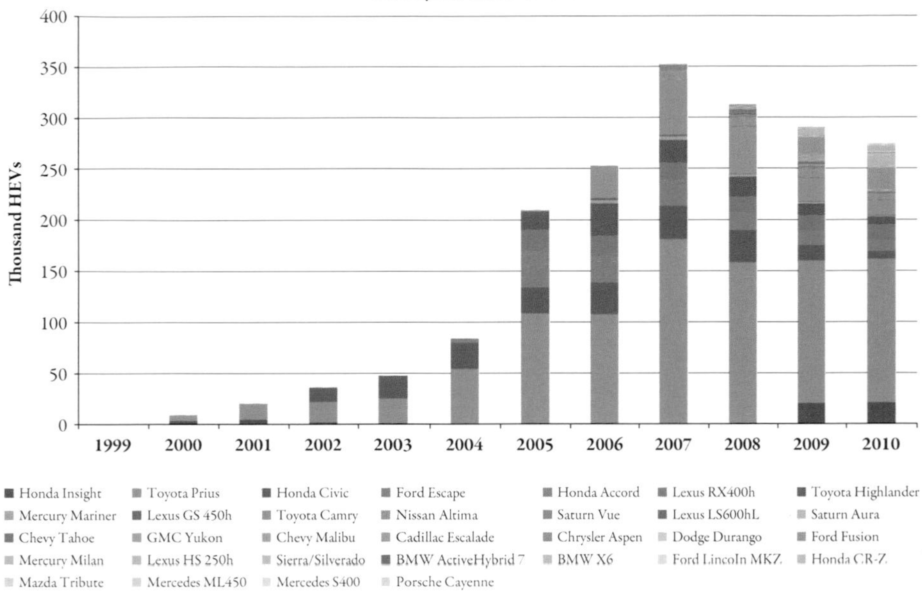

FIGURE 27.13 US Hybrid Electric Vehicle Sales

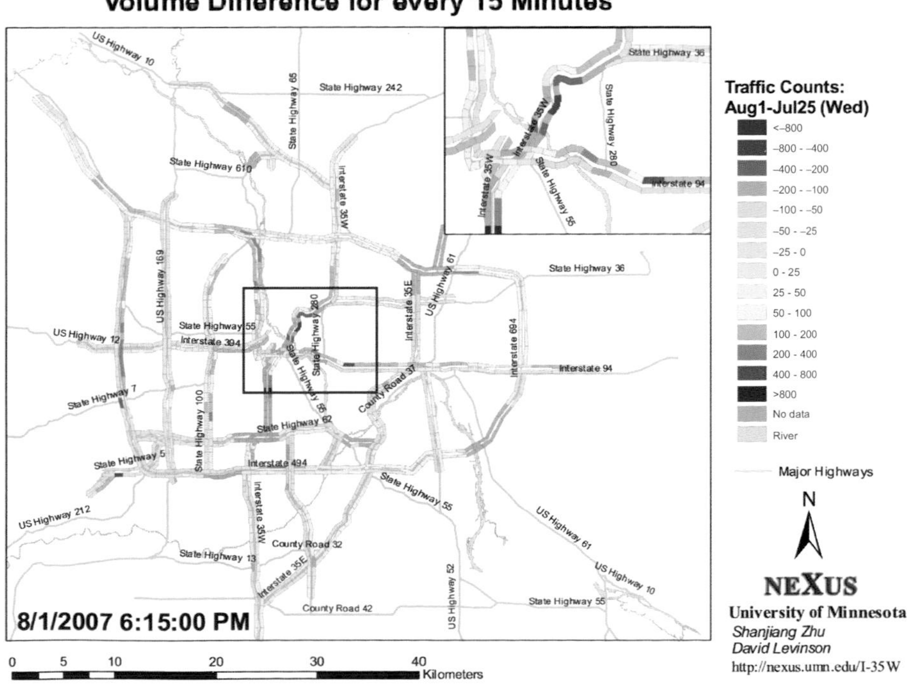

FIGURE 33.1 News of the Bridge Collapse Traveled Fast. Parallel (substitute) routes saw an increase in traffic. Upstream and downstream links saw a decrease in traffic.

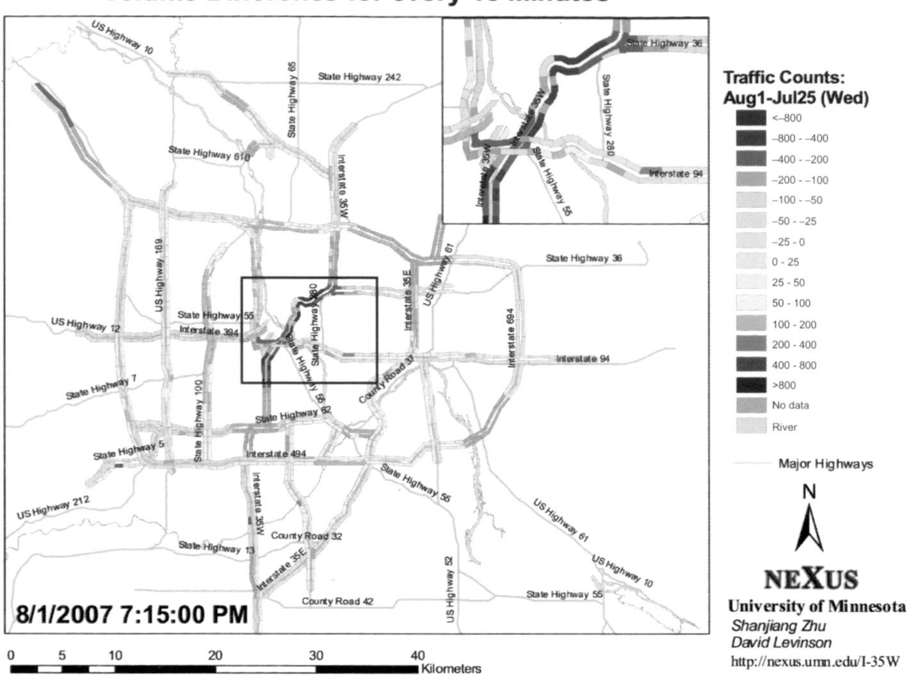

FIGURE 33.1 (*Cont.*)

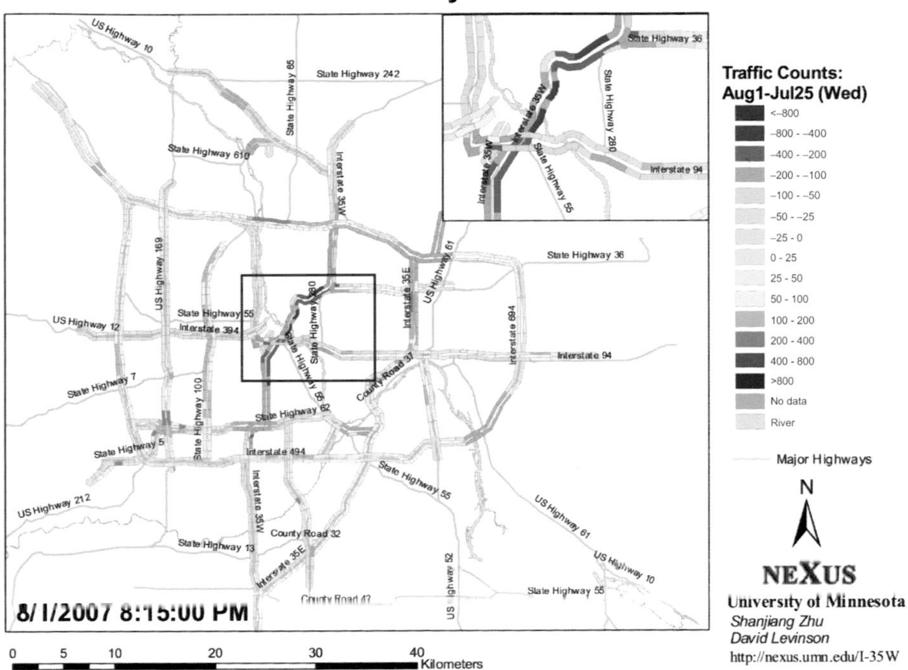

FIGURE 33.1 (*Cont.*)

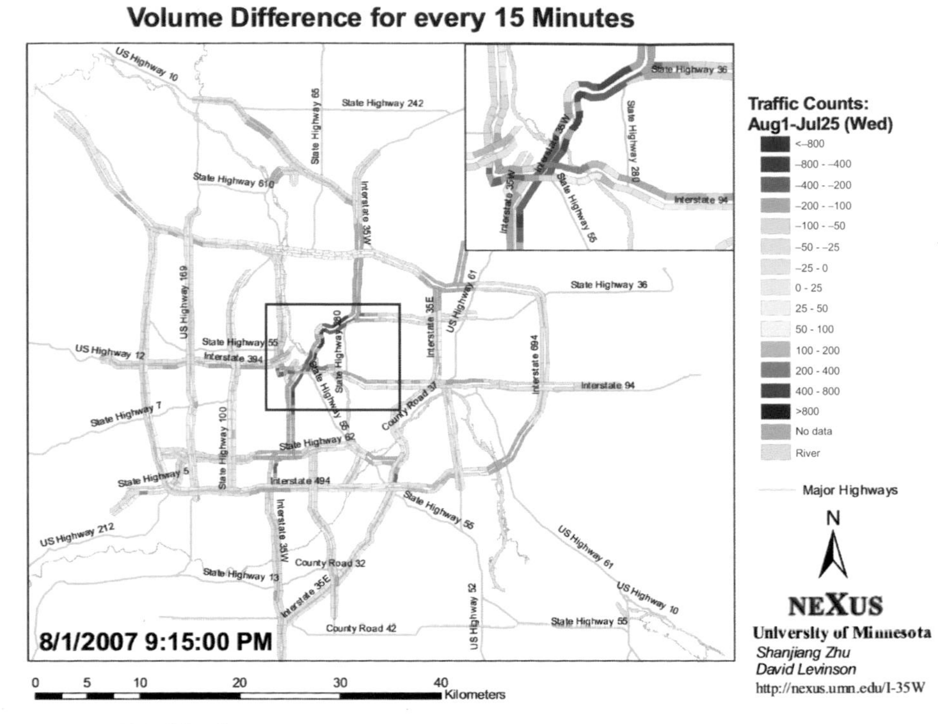

FIGURE 33.1 (*Cont.*)

8 Life-Cycle Dynamics

That he not busy being born is busy dying

—BOB DYLAN,

It's Alright, Ma (I'm Only Bleeding)

25 The Life-Cycle of Transportation Systems

25.1 A Metaphor

TO AID IN trying to understand the behavior of a system over time, we have found the life-cycle metaphor useful.

- Innovation: There is a period of development, the early days. It's a period of discovery: What should the system's predominant technologies and institutions be like; what will be its markets?
- Growth: Innovation is followed by a period of growth. Markets are captured, the technology is honed, and technology honing and the magic bullet (economies of scale) provide for increased efficiency.
- Maturity: Finally, the system fills its market niche. Spatial monopolies are established, and profits are generated and not reinvested. Technologically, stasis tending to senescence (politely called "maturity") describes the situation. Marginal improvements may yield a "polished" present that looks healthy but is failing to advance very much.

Recalling the disjoint pattern of decision making, we can say quite a bit about system change and system properties as a system moves along its life cycle. One property of systems, for example, is diminishing returns from investment and technology as a system moves through the latter half of its life cycle. Things move against technological limits; cost-effective things get done first. For instance, the fan jet engines used by jet aircraft are approaching the limits on the thrust that can be obtained from them, the Otto cycle engine is just about as fuel efficient as it can be made to be.

On the other hand, there would appear to be increasing returns in the first half of the life-cycle, perhaps as the fixed costs of investment (including both "real" fixed costs such as plants and research and development, and more amorphous fixed costs such as changing the mind-set of producers and consumers) decrease.

The life-cycle framework sometimes appears in different guises. Christensen (2012)'s model suggests that there are three types of innovations that capitalists may invest in:

- *Empowering*, which take elite products and transform them to the mass market, and tracks with what we call innovation in Chapter 4.
- *Efficiency* innovations, which reduce costs. This drives what we call the magic bullet, discussed in Chapter 10.
- *Sustaining*, which replace old products with new, and can be seen as the weak type of innovation we see in mature systems, discussed in Chapter 18.

25.2 Temporal Dynamics

Throughout *The Transportation Experience* we trace the life-cycles of various modes. S-shaped curves are a key feature of this work. These are diffusion curves showing the realization of a system over time.[523] The earliest known S-curve to track the diffusion of ideas was developed by De Tarde (1890), and Rogers (1995) looks at the diffusion of the S-curve itself. One would expect curves of this shape, such as in Figure 25.1, for transportation is a product like others; it enters and floods a market.

Some questions that arise from observing the process have to do with the positioning of the curve in time, the time it takes for the process to run, and saturation levels.

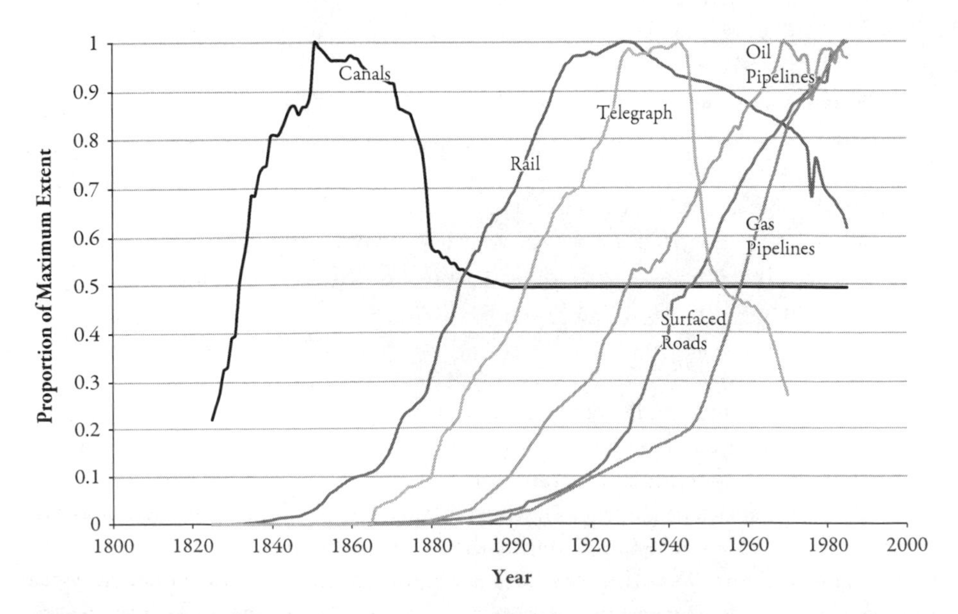

FIGURE 25.1 Networks in the US as a Proportion of Maximum Extent

Rule-of-thumb, we like to say that it takes sixty to seventy years for a transportation system to run its growth cycle. That time period corresponds to the "waves" defining parts of this book. But for instance for automobile populations, late adopting nations compress the time required. The growth of China in the late twentieth and early twenty first centuries is remarkable in regard both to the rate of motorization[524] and its rate of infrastructure deployment (at cost to both quality control and environmental and social consequences).[525]

Having observed empirical regularities and having in mind a market flooding logic for them, forecasting comes to mind. Interestingly, however, S-shaped curves seem not to have been used much in transportation. Forecasters mainly extrapolate data linearly, their thinking is that things just grow and grow forever (see Chapter 24). That kind of thinking reveals itself in present debates about future highway needs in which it is assumed that vehicle travel will continue to grow explosively. We challenge that thinking. There are underlying causes to growth in travel (for instance, the rise in female labor force participation) that ultimately must saturate. Thus we see a tapering off of vehicle travel (at least until speeds significantly increase). This has been referred to in the literature as "peak travel," for which there is increasing evidence in the developed world,[526] and complements discussion of Section 24.5.1 on travel time budgets (although the phenomena are different).

Rather than use interpretations of the behavior of systems available in the transportation literature, let's intermix comments on transportation with comments on other activities.

S curves of the type shown in Figure 25.1 fit the temporal realization of transportation systems very well, as well as the deployment of many other technologies. Three remarkable things about S-Curves have been observed:[527]

- Stability: the curves fit the data very well through good times and bad, shifts in energy costs, and technological evolution.
- Symmetry: the curves are symmetric around a central inflection point.
- Stimulation: deployment time decreases with time. For a place and system, the later in time deployment begins compared to other places, the less time it takes for deployment to be completed.

We would like to know why the realizations of systems have those characteristics. Naturally, our interest goes beyond drawing curves to ferreting out the causes of their realization. What is there about structure and performance that yields "perfect" S-curves? We see that service quality improved and costs dropped because:

- Rapid improvement occurs in the hard technology (e.g., the railroads began to use more steel and less iron, fuel efficiency increased, etc.).
- Lagging somewhat, rapid improvement appears in the soft technology (control, organizational arrangements, governmental regulation, etc.).
- With networking and the growth of the market, economies of scale and scope are found.

Investment in product development is high during early days of the life-cycle, but as time passes more and more attention goes to processes of production.[528]

For example, we saw (see Chapter 15.8) that, during the first decades of modern system growth, product development was an issue for highways: "What should highway designs be like?" That was answered by the end of the 1930s when freeway designs had emerged, although they still receive attention. "How to build" was the question somewhat later; it was pretty much answered in the 1950s when the interstate and related planning, institutional, and fiscal arrangements were put in place. The urban transportation planning process is an example of market channeling, and tailoring the product and process to its markets is very much at issue in today's mature system.

A related issue is how a region adopting a product might switch from an importing to an exporting posture as the life-cycle progresses. Japan's early import and now export of automobiles provides an example. For a within-nation example, the Western US imported rail cars from the Eastern US in the early days. Now, manufacturing is more spatially distributed.

Bejan and Lorente (2011) argue that the S-curve can be decomposed for many processes into two components. The first is "invasion," where a flow invades a territory. This is analogous to construction of backbone systems. This is the left-hand side of the "S-shaped" curve associated with deployment. The second is "consolidation," which is diffusion perpendicular to the invasion lines (which we can think of as local or feeder routes). This is associated with the right-hand side of the S-curve. This two-stage deployment process is physical in nature, but may describe examples such as the construction of the trans-continental railways in North America. On the other hand, some processes are locally emergent, out of which the backbones emerge, so we would be hard pressed to give this two-stage process the status of "law."

Similarly, Perez (2002) decomposes the S-curve into a two-stage process denoted "installation" on the left side of the inflection point (turning point) and "deployment," on the right side. We call the whole thing after birthing "deployment," so the terminology is confusing. The installation period is subdivided into "irruption" and "frenzy." Irruption is the period of disruption of the old, along with period of high unemployment. The frenzy period gives us financial bubbles. The deployment process is divided into "synergy" and "maturity" (which is also confusing, as we use maturity for the period a bit later in the cycle). Synergy is described as a golden age when gains can be consolidated, while maturity is the end of the cycle, leading to disappointment or complacency.

Neither of these two posited processes require, or imply, the symmetry that is observed in the deployment phase.

25.3 S-Curve Math

The S-curves in the book are largely displays of data. On occasion we fit smooth curves to the somewhat noisier observations. We seek a curve that best fits the data, assuming the data take on a logistic shape. The life-cycle model can be represented by the following equation:

$$\frac{S_t}{S_{max} - S_t} = e^{bt+c}$$

or

$$ln \frac{S_t}{S_{max} - S_t} = bt + c$$

We estimate the latter using an ordinary least squares (OLS) regression model and solve for b (the slope) and c (the intercept).

Where:

S_t = system status at time t (e.g., network length at time t)
S_{max} = final market size (e.g., ultimate network length)
t = time (e.g., year)
c, b = model parameters.

The objective is to solve for c and b to best explain the relationship.

One concern when using this for forecasting is to identify the final market size (S_{max}). While we may know the current system size (S), to use an S-curve requires knowing how large the system will be. The models can be fit for alternative final system sizes, one will best fit the data, but this is an inherent limitation in this kind of forecasting. In back-casting, explaining the deployment of already built systems, S_{max} is apparent.

To apply the model, it is helpful to estimate the midpoint or the inflection year (t_i). It turns out that:

$$t_i = \frac{c}{-b}$$

We can then predict the system size (S_t) in any given year t using the following equation:

$$\widehat{S_t} = \frac{S_{max}}{1 + e^{(-b(t-t_i))}}$$

25.4 Spatial Dynamics

This book largely explores the macroscopic factors that affect network growth, from birthing to maturity. However, these macroscopic explanations may not give us the specific information we desire, like where is the next link on the network going to be built or expanded, or what is the next link that should be constructed. For that we need more explicit data and different kinds of models.[529]

The planning or engineering question of where all of this activity should take place is addressed by the Network Design Problem (NDP), which has received a great deal of attention in the operations research literature. It is hoped that if clear objectives can be stated for networks, they can be planned and constructed in an optimal fashion. A cursory glance at most large metropolitan areas suggests that road and transit networks are far from optimal, and that this is both an investment and a pricing problem, which need to be solved jointly. Early applications of network design in Chicago are described in Boyce (2007), which were a trade-off of user life-cycle travel cost (which decreases with road length) and up-front construction cost (which increases with road length).

The analysis of network growth has received recent interest again with the rise of the Internet, and it is thought that all transactional networks (transportation and communication) have similar structures and processes governing them.

Broadly we can think of several problems about predicting network growth:

- node formation: where will the next node form;
- node expansion: how do nodes grow;

- link formation: what two nodes are next most likely to be connected with links; and
- link expansion: of the existing links, which will be widened.

As networks decline, the same questions can be asked in reverse, leaving us with the link contraction, link abandonment, node contraction, and node abandonment problems.

Node formation and node expansion are in many ways geographic questions, as many nodes depend on natural resources (e.g., the location of free energy) and nature's networks (the location of harbors, easy places to ford rivers, and river junctions). Other nodes are formed by the intersection of manmade network elements, the crossing of two roads, for instance. The geographer's theory of central places is an important element here.[530]

Link formation describes how and which nodes are connected. Within more recently built towns cities, many networks are in the form of a grid system. But over larger areas, example between cities, the shape of the network is not so predetermined. Garrison and Marble (1965) observed that connections to the nearest large neighbor explained the sequence of rail network growth in Ireland. Yamins et al. (2003) developed a simulation that grows urban roads using simple connectivity rules proportional to the activity at locations.

One illustrative network morphology dynamic is worth noting. In the case of toll roads, there is a corridor, say, between two major commercial cities (as shown in Figure 25.2). Early toll roads targeted end-to-end links along the corridor. For instance, the need for improvements might be greatest in link D–E, and a toll link is introduced there first. As other links are filled in and as collector-distributor roads are improved, say, H–D projects work toward marginal returns.

In the case of canals, river basins form the context, and link-by-link improvements are made. For two basins, at some point the question of an upland link enters (link between two river basins). Such links are generally expensive, for example, water supply is a problem in upland areas, and there are the uncertainties of traffic between basins.

Link expansion describes which links will be widened. As many points are already connected, it becomes the sizing of the connections that matter. The analog in scheduled services is the increasing frequency of service along a route. Yerra and Levinson (2005) simulate link expansion, showing that a network becomes a hierarchical network even if it begins as a uniform network with uniform land use over a finite area, with all links identical except for location. The network, like observed networks, exhibits power-rule type of behavior, a few very fast links, some moderate speed links, and many slower links. In brief, a hierarchy of roads would exist even if no planners or engineers intended it, and even if there were no "central places." The question has also been attacked empirically by

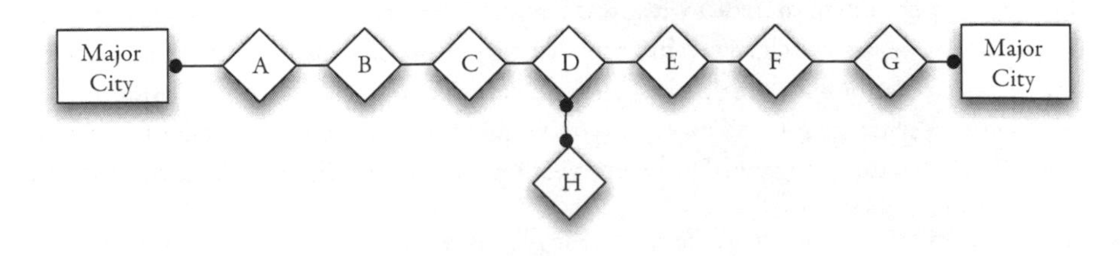

FIGURE 25.2 Step-by-Step Toll Road Improvements

Levinson and Karamalaputi (2003a, b). They found that existing capacity deters expansion (there are diseconomies associated with expanding wider links). Similarly, cost deters expansion, while total budget favors it. Importantly, it was found that increasing congestion on a link, and on upstream and downstream links, leads to expansion, suggesting that decision makers respond to demands placed on the network.

25.5 Infrastructure and Economic Development: Macro-economics versus Life-cycle Economics

The decades since World War II have seen the growth of a considerable economic development literature. Transportation enters that literature in several ways; in particular, it is conceived of as part of social overhead capital (SOC), which is a basic investment that is necessary to development—it's one of the universal input industries supporting all activities. Other such industries include production of clean water and provision of sewage treatment and electricity. Infrastructure might be another name for SOC. Additional characteristics of SOC include its tendencies to be associated with large capital expenditures that have technical indivisibilities, are non-importable, and are provided (or at least regulated) by public authorities.[531]

A related idea is that of social savings, used in Fogel (1964)'s study of the US railways as an *ex post* analogue to *ex ante* notions used in the analysis of projects: Identify costs with or without a project; compare the costs with the savings. Fogel used the idea of a counterfactual hypothesis. However, this research method has its problems (we don't have an identical United States that has evolved from 1830 without rail). While we may be able to determine the marginal contribution of additional infrastructure today (which may be small for mature systems), determining the marginal contribution 100 years ago is fraught with challenges. It is especially difficult to identify all of the indirect or external effects of the technology. It is also difficult to value qualitative differences in transportation. The questions of "How large was the social surplus from railroad development and to whom did it accrue?" remain unanswered.

Discussion of transportation in the economic development literature often introduces the concept of linkages. (The often-used word "spillover" captures the same concept.) Backward linkages lead to investment in input-supplying activities. For instance, investment in highway construction may require investment in cement production. Much "benefit" analysis focuses on backward linkages. However, a critic would point out that any investment, even if it only involves digging and filling holes, would have backward linkages.

Backward linkages should not be ignored, for some investments may have more socially desirable backward linkages than others. With respect to economic development, for example, investment in truck-freight activities many provide an incubator for entrepreneurship compared, say, to a pipeline.

However, the forward linkages matter more. How can we get a handle on forward linkages? One route is through the economist's production function. The neoclassical production function takes the form:

$$Q = f(K, L)$$

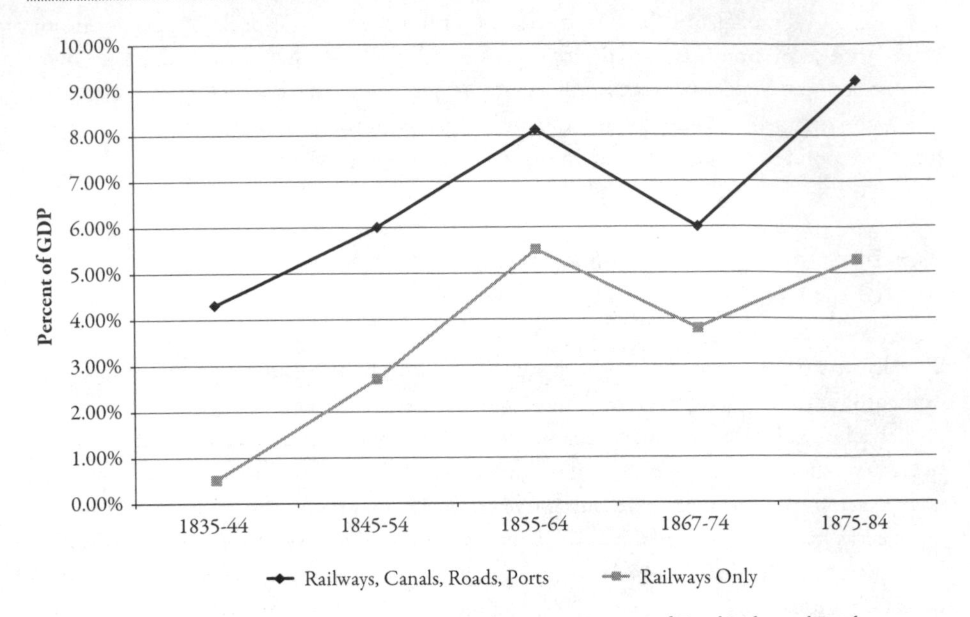

FIGURE 25.3 France—Gross Public Works Expenditures as a Percent of Total Industrial Product, 1835–1884

where: Q = quantity of output K = capital inputs, and L = labor inputs. The partial derivatives of Q with respect to K and L are the demand schedules for input factors. We may introduce transportation (T) as an input, along with capital and labor, so $Q = f(K, L, T)$.

We know that transportation investment accompanied the growth of output. In France, for example, Figure 25.3 shows transportation expenditures increasing as industrialization ran its course. Observations such as this observation about France say that there is some optimal investment level. In the United States, for example, Figure 25.4 illustrates that investment in highways peaked in the 1960s. So this is viewed by some as a problem; investment is not at the level where it ought to be.

To counter the reasoning that infrastructure investment should be fixed over time, the point was made that emphasizes change as times change. In particular, in recent years there has been private investment (forced by government requirements for cleaner cars, factories, etc.) in response to environmental protection requirements. However, during the 1970s and 1980s it was reasoned that previous investment in infrastructure was endangered, and underfunding didn't make economic sense because investment in maintenance and repair is more economical than rebuilding. That was a strong enough argument to drive some increased funding. Along with the potential for job creation, that argument has become politically salient.

Thanks to David A. Aschauer, an "it deeply matters" view of the investment shortfall crisis began to emerge in the 1990s.[532] In a series of papers, the first appearing in 1988, Aschauer pointed out that public capital expenditures on infrastructure ought to be included in economic calculations of the inputs to production. He specified an aggregate production function that used labor, public capital expenditures, and private capital

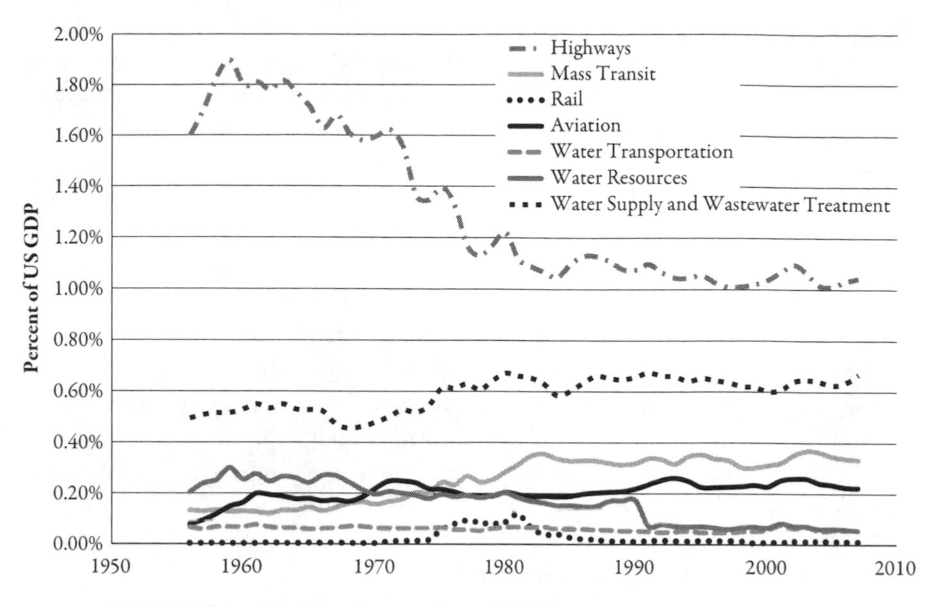

FIGURE 25.4 US Infrastructure Spending as a Share of GDP

expenditures as inputs and examined how these inputs related to outputs during post–World War II years. Previously, economists had stressed private investment. Including public investment enables asking, "Does public capital matter?" Aschauer deserves credit for asking the question. He points out that the ratio of nonmilitary fixed public capital stock to fixed private capital stock reached a peak of about 32 percent in the late 1960s; it has subsequently fallen to about 23.5 percent. He argues that the decline has adverse affects on private investment, profits, and productivity. He supports the argument with a series of time series regressions made using aggregate, national data. In addition to that finding for aggregate public capital, Aschauer identifies core infrastructure with a stress on transportation, especially highways, where public capital investment matters very much.

Swarms of naysayers and Aschauer supporters have had their say also. Aschauer's publications and findings, the findings of others, and debates about those findings are reviewed in a set of conference papers.[533] Nadiri's research claims that "the average cost elasticity with respect to total highway capital for the U.S. economy during the period 1950- to 1991 is about −0.08." That is, increasing highway investment by 1 percent will reduce costs by −0.08 percent. The average net rate of return from highway capital fell from 54 percent in the 1960 to 27 percent in the 1970s to 16 percent in the 1980s. The last number is close to the private rate of return, indicating a near optimal level of highway investment. The question continued to be debated through the 1990s. In the 2000s, it has faded, and those advocating major new infrastructure investments have largely been successful.

Productivity advances are the central issue in this literature, and productivity is improving when, year after year, output increases faster than inputs of capital and labor. The issue facing the United States and many other developed nations through the 1980s was the tapering of productivity advances. As Aschauer and others have pointed out, the Group of Seven industrialized nations had average productivity growth of about 4.0 percent per year during

1960 to 1968 and growth declined to 1.4 percent per year during 1973–1986. We add that in 1987–1995, US productivity fell even further to about 1.2 percent. But in 1995–2000 it rose to 2.3 percent. There was a sharp turning point in the late 1960s and another in the mid-1990s. Decreased productivity growth was the root of the failure of real incomes to increase, and is seen as a culprit in lack of competitiveness of the United States compared to its trading partners. Information technology may be behind the rise in the late 1990s, as new telecommunications technologies (the Internet, mobile communications) are deployed.

To get from raw comparisons and conjectures to processes, Aschauer and others have estimated the coefficients of aggregate production functions. Data and procedures differ. Some studies made international comparisons, some used aggregate US data, and others were on a state basis. Results range from Aschauer's claim that a dollar of public capital investment increases output as much as does $3.30 of private capital investment to the finding of no relationship. That is, some adopt the position that stepped-up investment in infrastructure, especially transportation infrastructure, will reenergize productivity growth; others say that's wrong.

The previously cited Federal Highway Administration paper and Conference Report summarize and compare studies; they seem to point to a weak positive effect. The comments made by researchers are interesting. Most suggest that there must be a relationship, but that large effects are not plausible. Henry Aaron, for example, remarks that few would question that the road building of the 1950s and 1960s contributed to economic growth, but the claim that economic growth slowed because of the wind-down of road construction goes too far.[534]

Perhaps there is a simpler explanation. Let's return to the S-curve. As the Interstate highway system grew and matured, the marginal benefit decreased from each additional roadway (most places were already well connected), and the marginal costs increased (the cheap roads were already built). It is natural that the economic benefits from new construction in a mature system would be smaller than the same construction in a nascent system. As the benefits of new construction decline, investment declines with it. The cause of the decline of growth is not the lack of investment; the lack of investment is the symptom of a mature system. We can explore this idea graphically in Figure 25.5.

On the left, there are few roads, and travel has a significant cost in backtracking. In the middle, the number of roads is doubled, but the travel cost is less than halved. On the right,

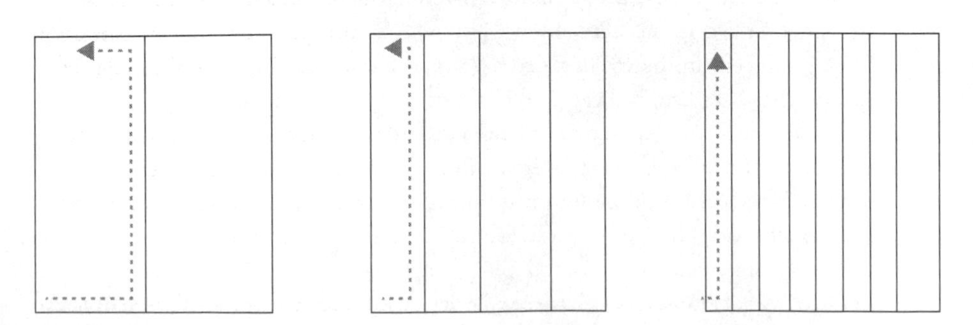

FIGURE 25.5 Diminishing Returns to Network Deployment

the number of roads is again doubled, but the travel cost diminishes only marginally. A similar idea happens with transit schedules. If we have a bus running once an hour, the average schedule delay is about a half hour (the difference between actual arrival time and desired arrival time). If we double the number of buses, the schedule delay is halved, but that means it drops by 15 minutes. If we double the number of buses again, the schedule delay is again halved, but this is only a drop to 7.5 minutes. We have a clear case of diminishing returns (in terms of rate of return) to additional transportation in both space and time as a technology is deployed. Of course, there is little incentive to double the number of buses (or roads) again.

It seems to us that the economist's production function, while useful to economists, is too simple a representation of reality to capture the causal relations involved in transportation impact questions. Even economists have questioned its use in productivity analysis. Regressions specified as neoclassical production functions and applied to time series data have low correlation coefficients. There is about an 80 percent unexplained residual, and the thought is that technical change, not captured in production function analysis, is driving increased output.

When interpreting results such as those shown by the figures above, Aschauer refers to core infrastructure, and especially to transportation. "Is a shortfall of capital expenditures on transportation facilities hurting the US in international competition?"

The following possibilities might be considered:

1. Aschauer's interpretation is correct. Congestion and other things are hurting productivity growth. That is surely the case for an organization such as Federal Express, but is the impact of congestion and the like on productivity direct, widespread, and large?
2. Underlying common causes yield the downturns in investment and productivity. A long list of causes comes to mind, ranging from federal budget problems, the high level of investment of engineering and scientific talent in defense matters, and high interest rates to Americans' attitudes about "not in my back yard," taxes, and the environment.
3. There is an underlying relationship that is common among the compared nations. The nations differ because the relationships are driven by clocks with different settings. The United States pushed infrastructure development hard, and it has already achieved most of the productivity gains that can be obtained from present technological formats for infrastructure, example, the technological format of the highway system. Western European nations are a bit behind in timing, and they are still capturing productivity gains. Japan is yet farther behind, and now China is capturing healthy gains.
4. The premise is wrong; the United States is not suffering in competition with other developed countries. Although the transportation system is far from perfect, productivity growth rates of the late 1990s suggest that it is investment in the next wave of technology, not transportation, which is most important.

A good case can be made to establish the "clocks with different settings" assertion. All one has to do is look, say, at the pace of automobilization in the comparison nations. Similar timing patterns appear in railroads, domestic containerization, air traffic, and elsewhere. If (3) is true, the need for increased spending on infrastructure implied by (1) makes little sense unless the increased expenditures are focused on new ventures.

25.6 The Idea Queue

Queueing models describe how much delay there is in a system, by looking at when the system is entered and when it is exited. For instance, if a vehicle enters a queue at 6:00 and exits at 6:15, it experiences 15 minutes of delay. One could sum up that time for all vehicles, and estimate total delay. Rogers (1995) has implicitly extended that concept to ideas or technologies themselves; they enter the queue when the knowledge is gained, they exit when the technology is deployed. Since knowledge itself faces a deployment process, this would give us an indication of the speed of take-up of a technology (see Figure 25.6).

The risk remains of over-adoption—the case when technologies are taken up too soon. Plank roads for instance, discussed in Section 3.9, were adopted before their full life-cycle was understood, and they failed prematurely. Some patience with deployment would have been warranted there. In other cases, deployment of what turn out to be successful technologies are choked off and slowed because of a lack of capital. Delay is not of itself to be minimized, but rather traded off against the costs of rushing deployments, costs that arise because not all cost-saving innovations have yet to be wrung from the system and because of the extra expenses incurred in speeding deployment (for instance, more construction workers are used for a shorter period of time, rather than training and employing fewer workers over a longer period).

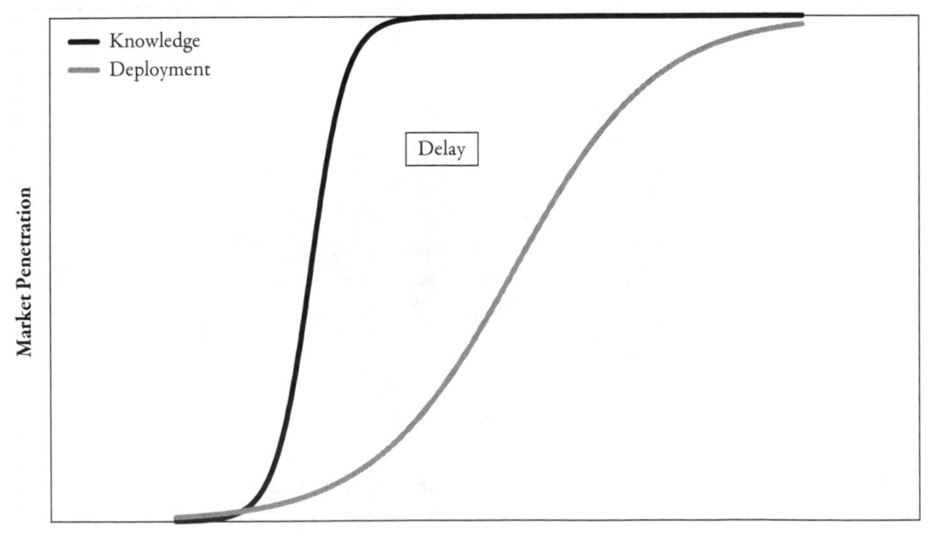

FIGURE 25.6 The Idea Queue

Another perspective on the idea queue is what technology consultancy Gartner calls the "Hype Cycle." Gartner identifies five stages: Technology Trigger, Peak of Inflated Expectations, Trough of Disillusionment, Slope of Enlightenment, and Plateau of Productivity.[535] Autonomous vehicles and mobile robots are still in Gartner's first stage as of 2012. So if Gartner's thesis holds, we should expect disappointment before deployment. This point of view is useful for disentangling short-term and long-term changes.

25.7 Social Impacts

> We tend to overestimate the effect of a technology in the short run and underestimate the effect in the long run.
>
> —Amara's Law

We observe that the social gains from improvements in transportation occurred fairly rapidly, that is, hard on the heels of system deployment, as illustrated in Figure 25.7. There is a big impact early on, and impact then tapers. Those impacts reach beyond reduced costs to consider bringing of new resources into the economy, creating new options for social and economic activities, enabling enlarged labor sheds, and so on.

The "winding down" of impacts as systems are deployed focused our attention on how new systems are created, what happens to them once they are birthed, and how transportation technology "works" generally. The reenergizing of social impacts seems important. Also, attempts to improve existing systems late in their life-cycle seem difficult to useless.[536]

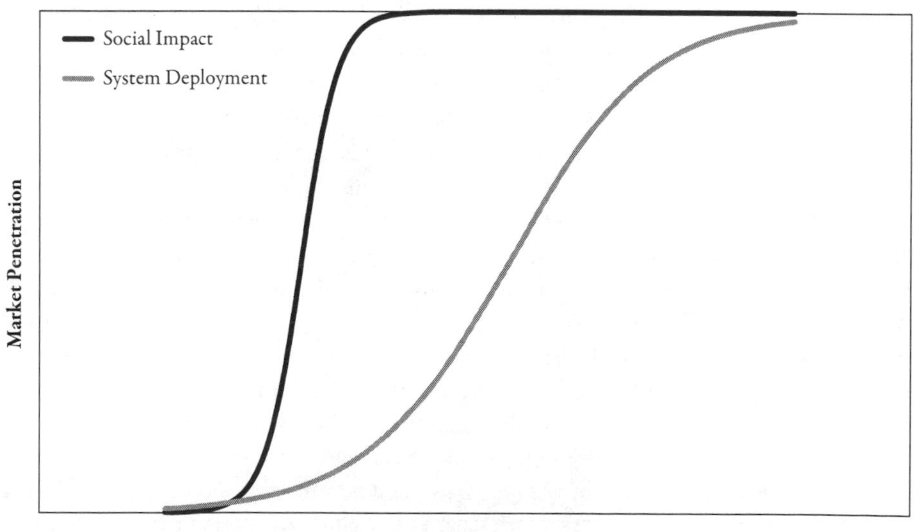

FIGURE 25.7 Social Impact vs. System Deployment

25.8 The Value of Being at the Center of the World

The 1990s was a decade of explosion in information technologies; led by the Internet and World Wide Web, a slew of related technologies were invented, innovated, and began to be deployed. Where they were first deployed was often the centers of high technology themselves, example Silicon Valley in the San Francisco Bay Area. These areas often considered themselves at the "Center of the world," and in terms of leading-edge technology, at that time they may have been. Residents there would have advantages such as broadband connections with cable modems, or in the early 2000s, wireless connections for computers to the Internet at selected "hot spots," before people elsewhere had much of a notion of what was going on. This leads us to a hypothesis: If the rate of creation of (information) technology is accelerating, the leading places will pull further and further ahead. If it is decelerating, the trailing places can catch up, and the relative value of being in the center steadily diminishes.

As we have noted throughout *The Transportation Experience*, most technologies follow an S-curve. During early growth, technology accelerates. It seems reasonable to suppose that some place gains the new technology first, and that place will usually be near where the technology was developed. We can call this the center. Once this may have been Detroit with the automobile. Today, in the most recent wave of technology, this would appear to be places such as the Bay Area with the. To date, with previous technologies we have observed that during late growth and maturity phases, growth decelerates and the technology is spread around. There are a number of reasons for this, including diminishing marginal returns to innovation (at least in the short run), and the desire of capital to reap returns on investment by spending money on deployment into new markets rather than development of new technology.

Will the advantage of being at the "center" remain? Are there advantages that accumulate at those centers that keep them dominant? In the United States, Detroit is still the center of action for development of the automobile, but aside from a fairly good highway network, offers little real advantage to those who wish to use one. While there are still spatial elements to the Internet, one of its great advantages is as a "destroyer of distance"; anyone connected to the network (which still may require being in the developed world) can obtain the latest software, information, or entertainment virtually instantaneously, regardless of location. But the physical infrastructure is still physical, cables must be laid, wireless hubs or towers must still be constructed, and the system cannot be assembled everywhere simultaneously.

Michael Porter attempted to explain the *Competitive Advantage of Nations* with a diamond model, in which factor (supply) conditions, demand conditions, firm structure and rivalry, and related and supporting industries were arrayed at the vertices of the diamond, and interconnected.[537] He made the point that you are more likely to get an advantage if there is a strong home market. So one expects snowmobiles to be manufactured in Minnesota (they are) rather than Jamaica. This might be expected to spill over to other small motor industries, like lawn-mowers (it does). These factors are mutually supporting, so complementary industries help lower supply costs and may increase local demands. Having competitive firms, which may at first blush seem a disadvantage, may result in thick markets, and attract consumers. Being first certainly aids in mutual causation that supports a locational advantage during the life-cycle of a technology. Whether it sustains over the evolution of technologies is not at all clear. Some seem to (Silicon Valley), others (mainframes in Minnesota, minicomputers in Massachusetts) peter out.

25.9 Specialization

Chapter III of Adam Smith's *The Wealth of Nations* (1776) shows rather clearly what (improved) transportation does for society and its production and consumption activities. Transportation improvements are constructive because they support specialization of production and consumption. Efficiencies result from specialization.

Transportation improvements also have a destructive side because they tear down existing arrangements, shift the holding of resources and wealth, etc.

There are gainers and losers, and that puts strain on the social contract. In theory, we might expect government action to ease that strain—side payments would be made to those who lose, so that no one is worse off. That is an equity, not an efficiency, matter, of course.

(Side payments are discussed when there are undesirable externalities. That's an efficiency matter, for actors should pay the full costs of their actions. The only side payments we have seen in our discussions so far are payments to those forced to move by urban highway and railroad construction. Interestingly, the need for payments was argued by social activists rather than by efficiency activists.)

The real world pattern seems to be this: As systems begin to be deployed, the voices of those who are gainers and those who expect to be gainers swamp those who lose or who will lose (but may not know it yet). It is only when most gainers have already gained and their voices are muted that the voices of losers can be heard. For example, the expansion of the railroads to the west destroyed most crop agriculture in New England. But the voices of New England farmers were swamped by the voices of gainers.

When examining the literature of the nineteenth century, we have seen specialization mentioned, but not much, and it is not mentioned much at all today. Our speculation is that it doesn't fit the prevailing paradigms of economics. Indeed, in 1951, George Stigler looked back at Chapter III in an article sharing its title "The Division of Labor is Limited by the Extent of the Market" with Smith's Chapter III.[538] He took the title as a theorem and took the tack: if it is true, industries are characteristically monopolized; if false, Smith is wrong. Of course, he found Smith incorrect. He had to; otherwise economics would have to deal with industries having spatial monopolies.

As will be discussed later (Chapter 26), we think the opportunities created when transportation (and communication) systems are innovated and deployed trigger waves of innovations and development cycles broader and deeper than those recognized for the model building cycle.[539]

The movement from one life-cycle to another is discussed in Chapter 26.

26 Meta-cycles

26.1 The Logic of Capital

AS WE HAVE seen throughout *The Transportation Experience*, each technology faces a life-cycle of birth, growth, and maturity. Birth requires innovations, growth requires learning and deployment, maturity engenders management. What moves us from one technological life-cycle to the next? Is there any order in this? We have broken this text into "waves," overlapping chronological periods of a little more than fifty years.

The general argument is that yes, there is a logic to this process. Capitalists (those providing capital, typically money to do things) can spend more or less resources on "research" and innovation, or capitalists can spend those same resources on "deployment," building out the system. When you have a "hit" technology, you spend resources to build out and deploy widely. Once diminishing returns to expansion set in (as they eventually do, according to the life-cycle logic), scarce capital is spent more on speculative ventures to find the next "hit" to generate future profits.

That said, everything is contingent on events, making the world a much noisier place than idealized models might suggest. A new hit can come anytime, whether or not the previous technology is fully deployed. Just because more resources are spent on innovation is no guarantee new hits will be found. It is just that the likelihood of finding a hit increases with the resources spent looking for one.

Further, we talk about the technologies as if they are a single thing here, but we have already identified the building block nature, so innovation can come in small, medium, or large sized packages (e.g., the railroad was a large package, cruise control is a small package). Both directed exploration as well as unexpected serendipity lead to innovations, so it is difficult to demand regular innovations on time and on budget, Moore's Law notwithstanding.

We see in transportation steadily increasing maximum and average speeds over the long term, though not so much we want to attribute a "Law" status to it. Speeds increase both within and between technologies (modern trains are much faster than those of 1830, as were those in 1920), as shown earlier in Figure 10.2.

26.2 Waves Short and Long

We are not the first to observe cyclicality. There has been a long, non-mainstream history in economics looking at both business cycles and long waves.

- 3–5 years: Kitchin waves
- 7–11 years: Juglar waves
- 15–25 years: Kuznets waves
- 50–60 years: Kondratieff waves

Given the variability in the size of the waves (5 years is 67 percent longer than 3 years, 25 years is 67 percent longer than 15 years), it is hard to say much about regularity from many of these cycles. Many have, with good reason, asked the question "Are cycles real?"

Kondratieff's posited cycles track those of long-term investment. As markets reached saturation, new investment in existing technologies would decline. The evidence for these is difficult to read in the macro-economic data, but some researchers have found evidence for all kinds of cyclic activity in the economy.[540]

Perez (2002) describes the cascade of big bangs. The process begins with an innovative big bang, and ends with the next, which sets off the next surge. These big bangs map to the long cycles of Kondratieff and others, and point where the techno-economic paradigm shifts include a "constellation" of new technologies and infrastructures. Perez, for example, dates the second-wave's big bang (1829) to Stephenson's Rocket; the first is Arkwright's mill (1771), the third is the Bessemer steel process (1875), the fourth is Ford's Mass Production of the Model T (1908), and fifth is the Intel Integrated Circuit (1971). Since we are interested in the birthing process as well as the surge of "installation" and "deployment" following the big bang, and we don't really believe in singular transformative transitions, our calendar is roughly skewed to the left and centers on Perez's big bangs and other key events, (we ignore Perez's first big bang as it is not transport related). In *The Transportation Experience*, Wave One runs from about 1790 to 1851, Wave Two from 1844 to 1896, Wave Three from 1890 to 1950, Wave Four from 1939 to 1991, and Wave Five from 1984 onward.

Mensch (1979) extends the Kondratieff (1935) model and suggests that the economy evolves through a series of intermittent innovative impulses that take the form of successive S-shaped cycles, what he calls "the metamorphosing model." The model further suggests, "surges of basic innovations will come during the periods when stagnation is most pressing, that is, in times of depression."[541]

The upper portion of Figure 26.1 displays long waves in the economy, Kondratieff cycles. When the economy's is on the upswing there is prosperity. That's followed by recession and depression. Finally, recovery sets in, followed by prosperity.

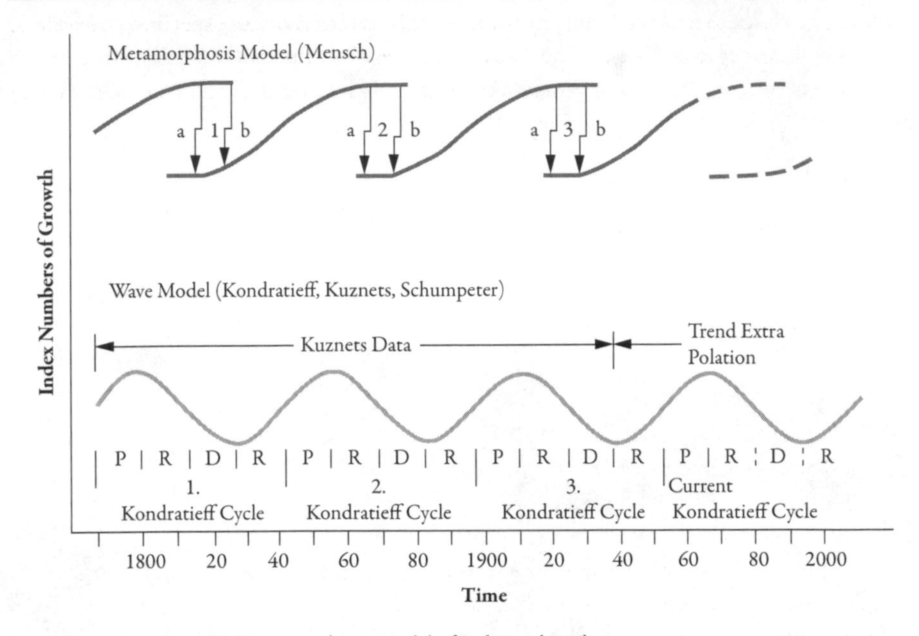

FIGURE 26.1 Mensch's Metamorphosis Model of Industrial Evolution

What is the mechanism? Mensch refers to waves of innovations that trigger investment, jobs, and so on. As those technical systems begin to age, there is recession and then depression. Recovery begins as another wave of innovations begins.

Mensch's reasoning leaves open the question, "What causes waves of innovation?" We argued above that it is the search for profits. The idealized model of Mensch is interesting, but much tidier than reality.

Mokyr discusses another possibility: explain the obverse, why are so many places not innovative? His thesis, which he refers to as "Cardwell's Law," is that individual societies have brief bursts of technologically creativity before conservative forces drag them back. In particular, prospective losers from a change, an innovation, will resist and attempt to suppress it. Few innovations are Pareto-superior, benefiting all players. A classic example (turned into a movie) is Preston Tucker, whose post–World War II automobile is said to have threatened the established ways of doing business with prospective technological changes.[542]

26.3 Emergence of a New Paradigm

The difficulty lies, not in the new ideas, but in escaping from the old ones.

—John Maynard Keynes (1936)

Since the 1980s, a new worldview has begun to emerge. The new paradigm partly rides on the coattails of the increasing interest in the importance of innovation and technology in economic development. Nelson and Winter (1982) provide an example of increasing interest. They refer to progress resulting from the introduction of novelties into the system. Now there are many working with these concepts.

Nelson and Winter remark on the many neoclassical growth models, "... (they have) enhanced our understanding of economic growth." But they go on to say:

> However there is a peculiarity about the success story, which we noted earlier. By the later 1950s, it had become apparent that it was impossible to explain very much of the increase in output per worker that had been experienced over the years in developed countries by movements along a production function resulting from increases in capital and other inputs per worker, if constant returns to scale and the other assumptions employed in traditional microeconomic theory were accepted. The "residual" was as large as that portion of total output growth explained by growth factors of production. For the growth of output per worker, the "residual" was almost the whole story. The researchers working within the theory found a way to resolve this problem. Earlier, Schumpeter (1934) and Hicks (1932) had proposed that innovation (technical change) could be viewed as a shift in the production function. In the late 1950s Solow's work (1957) made this notion an intellectually respectable part of neoclassical thinking about economic growth. In the empirical work, the "residual" was simply relabled "technical advance." Instead of reporting to the profession and the public that the theory explained virtually none of experienced productivity growth, the empirical researchers reported that their "finding" that technical change was responsible for 80 (or 85 or 75) percent of experienced productivity growth.
>
> —Nelson and Winter (1982)[543]

The last sentence focuses the present discussion: real progress is made through technological change. Beyond economists, many have adopted that worldview. In the United States today, for example, there is widespread debate about technological competitiveness. The new paradigm also responds to unease about how well the conventional paradigm catches the broad impacts of transportation development.

The functions of transportation loomed large in writing about the space economy, writing that accelerated during the 1950s with the development of location theory, quantitative geography, and regional science.[544]

Let's try to be clear. The conventional paradigm, view, or wisdom is not being thrown out. It applies pretty well to a very short run world with fixed production functions. It is a valuable guide to the optimal use of resources in such a world. Rather than being thrown out, it is being put in perspective. While valuable, it fails to help when trying to understand transportation or economic progress (or decline).

This change in priority toward ideas that are more supportive of economic progress was sharply illustrated during conversations in connection with a USDOT policy study. A discussion participant remarked, "... all that is available is the views of 1960s economists and engineering economists, and they are wrong. What's really important is structural change." "Wrong" isn't exactly the correct word. Economists stuck in the 1960s are not wrong in the terms of their reasoning, it's that the reasoning doesn't fit the situation as we now understand it.

The quote from Nelson and Winter amounts to an assertion that the conventional economics paradigm omits a lot. It doesn't say what is omitted, and it does not extend to our interest—how transportation improvements fit into the modern paradigm.

26.4 Status of Production

The idea displayed in Figure 26.2 suggests that prior to 1900, the system was growing using a certain set of technologies. It began to shift to another set of technologies at about 1900. Associated with each set is a range of development appropriate to that set. At about that same time, Andersson (1986) proposed a similar scheme.

To simplify the discussion that follows, we will use the word "production" as metaphor—it stands for things that transportation influences, such as industrial production, availability of resources, consumption, and social opportunities.

We need some way to catch the relation between production and the status of transportation, so we offer a disequilibrium paradigm, as shown in Figure 26.3. No time dimension is shown; we add this in descriptive verbiage.

Starting at point *a*, production is in equilibrium with system status. A strong nonlinearity in the improvement of the system drives production out of equilibrium at point *b*, and equilibrium is achieved at point *c*. From that point, equilibrium may shift to the northeast or to the southeast. If the latter, there is another phase shift at point *d*. Figure 26.2 conjectures about the temporal relations between the status of a system and production.

The temporal trace of the reduced costs of service is perhaps suggestive of the relations of the curves in the figure above (costs since railroad deregulation). The unit cost curve

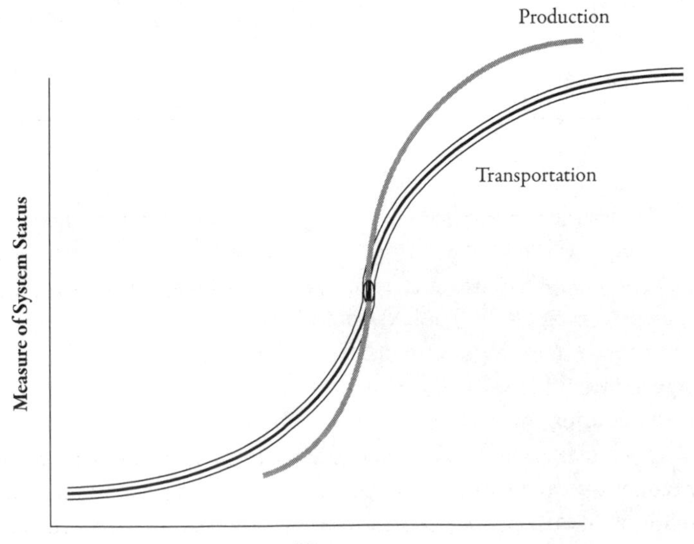

FIGURE 26.2 Relations Between the Status of Transportation and the Status of Production

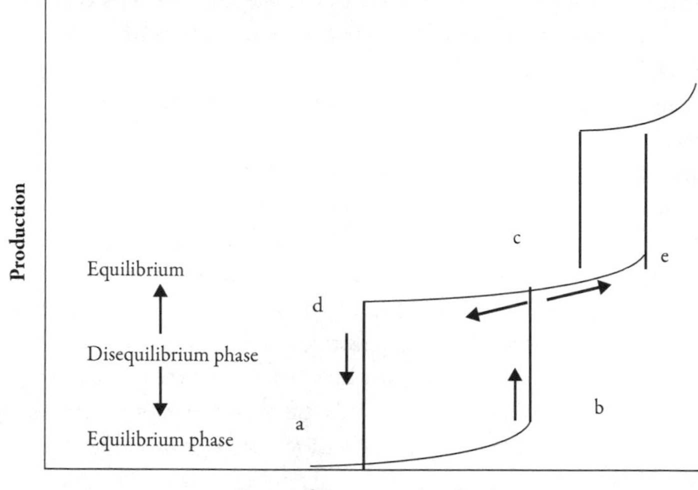

FIGURE 26.3 The Disequilibrium Paradigm

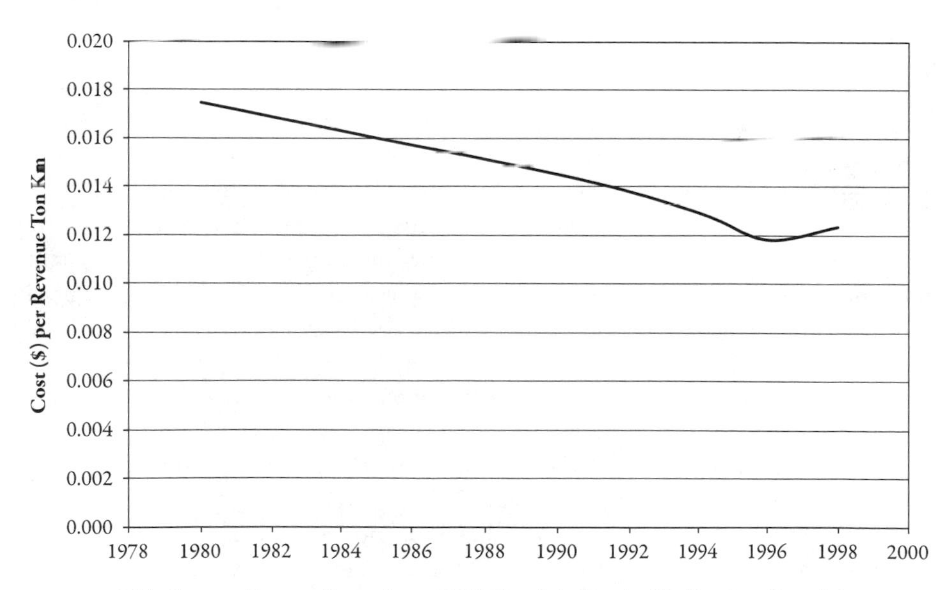

FIGURE 26.4 Cost per Revenue Ton-mile on US Railroads in Nominal Dollars (unadjusted for inflation)

typically has a (reverse) J-shape, as shown in Figure 26.4. Cost reductions are rapid early on, and then taper.

The question of the mechanisms relating the status of transportation to the status of production requires a two-stage analysis. First, we know that major gains in productivity, the creation of new products, the creation of new activities, and so on, have their major proximate cause in innovation. Second, we need to ask what causes (permits) innovations. A transportation explanation works very well. As Adam Smith remarked in 1776, transportation makes new resources available to the economy, enables economies of scale and scope, enlarges market areas, and is, generally, permissive of new activities. New activities follow from innovation.[545]

Our work on transportation technology explains the mechanisms through which transportation technology changes (or, in the future, can be changed). What is the mechanism that links transportation improvements to development?

Knowing that development is mainly driven by technological advances, thinking centered on such advances. The view that emerged is this: We should view transportation (and communications) innovations as energizing innovation and technology development generally. One may think of latent innovations. That is, at any point in time, there are many techniques that could bloom as usable innovations. The interactive development of transportation enables these latent innovations to be shaped for markets and to be tested in markets. Those that are successful drive a round of development. It is the burst-blooming of latent innovations that accelerates development as a new transportation system begins to be deployed.

We may also think of innovations that begin to appear as transportation service spreads. These are induced by the availability of new services.

26.5 Companion Innovation

There has been a lot of thinking about rounds or waves or cycles of economic development. Schumpeter proposed that they are innovation driven. Mensch (1979) identifies bursts of innovations and associates these with waves. Graham and Senge (1980) have developed some ideas about the mechanisms driving innovations. The process we imagine is this:

An improvement in transportation triggers a burst of innovations. Some of these were "on hold" because previous transportation constrained their introduction to markets (latent innovations). Some are new because improved transportation permits their being imagined (induced). This creates both a constructive and destructive situation. There is structural change, the old is replaced by the new.

The burst of innovations drives development. But once the transportation improvement is about half-deployed, there is a development downturn. Graham and Senge provide one explanation for such a downturn. We would add that the slowing of productivity gains from the transportation improvement plays a role.

A pent-up market for a new system and the increasing dysfunctions of the existing one trigger new transportation services and a new wave of economic development. We associate some innovations with transportation development.[546]

We do not claim, of course, that transportation steers all technological progress. There have been waves of chemical industry and electrical developments, for example, that only loosely relate to transportation.

Still, improved transportation technologies enhance services, enabling people and things to move more easily, while communications increase. Consumers benefit from more choices and more information about those choices. Consumers also gain opportunities to pursue other activities (such as recreation). Producers can replace low-grade inputs with higher quality resources as transportation increases the size of input markets. Increases in the geographical scale of markets enable reorganization of business practices to achieve

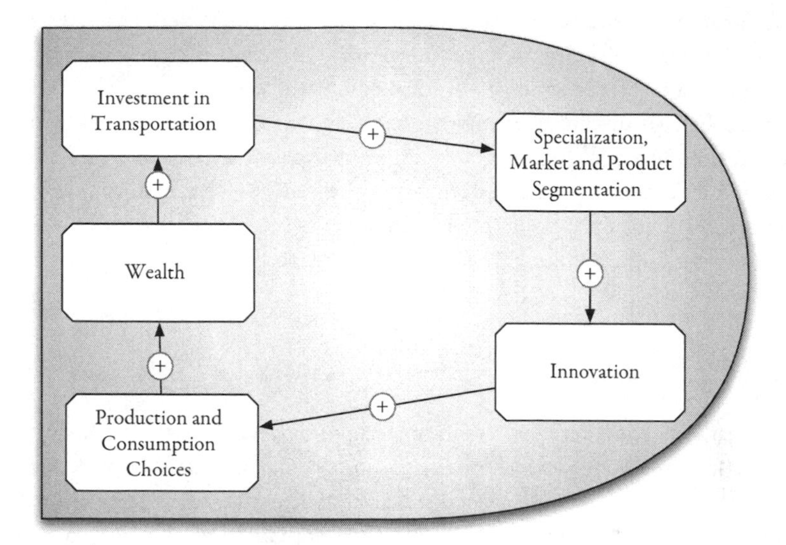

FIGURE 26.5 A General Model of Innovation and Technology Development in Transportation

economies of scale. As Adam Smith noted in 1776, the division of labor is limited by the extent of the market. As the market expands, division of labor, specialization, increases. More market niches are created, and more specialized tools and materials for production are created.

This specialization of production enables, and even requires, innovation in the technology used to exploit these new niches. These companion innovations result from innovative actions spurred by transportation development.

Further, as transportation and communication technologies are deployed, ideas diffuse across the networks more and more rapidly. Ideas that once took weeks to cross the Atlantic, and years to take hold, now take seconds to cross, and days or hours to take hold.

The process generalized in Figure 26.5 has been repeated numerous times through the transportation experience. The canal era, the road and wagon era, the ocean liner era, the railroad era, the automobile era, the air travel era, along with their associated communication innovations (semaphore, telegraph, telephone, television, Internet) have led to radical changes in politics, warfare, capital accumulation, and culture, as well as changing the natural environment in which human society is embedded.

Transportation and communication advances not only enable us to do old things faster (which is adequately captured in consumer's surplus and land value measures), but also enables us to do old things in new ways, and more important, to do new things. But this is only true of real advances, those that shrink the real and perceived distances across space.

How does the companion innovation concept enter the debate about transportation externalities? The literature on the external effects of transportation investment concentrates on negative externalities because, as Verhoef puts it in the case of highways "... we are not able to detect any significant external benefits of road transportation...."[547] The presumption seems to be that positive benefits are internalized, say, in wage rates or tariffs for service.

The companion innovation concept says that the no-significant-external-benefits conclusion is incorrect. The provision of transportation services triggers new technology development and improved productivity throughout the economy. Willeke (1993) comments that the usual treatment of externalities is static in character, and that one has to consider the dynamics of change in order to imagine external benefits. We agree. We would also point out that innovation is an externality in the same way that noise or air pollution is. No market is made for the output. Though innovators try to capture profits with patents, likely they cannot capture all the gains of their innovation.

26.6 Thrusts for World Power

How does transportation development influence nations attaining world power?

The question is posed for several reasons. For one thing, we observe that the transportation experience can be discussed rather effectively without much reference to the great social and political changes taking place as the experience unfolded. There were great social and political changes from, say, 1600 through the middle nineteenth century, but as we examined the transportation experience during that period, the changes were more in the background than up-front in the story. The question "How did changes in the sociopolitical milieu affect transportation?" seemed to require no special attention; broad-brush background statements would do.

There is also the question of the reverse causality: How did transportation development affect social and political changes? Some results of transportation development were the specialization of production; increases in the tools of production, income, and wealth (and changes in the holding of wealth); conflicts among social classes and institutions and adjustments of political power; and more. While the subject is vast, we limit ourselves to a discussion of an "attaining world power" version of the question and subject.

Modelski and Thompson (1996) date the modern world from about 1500, but point out that it was prefaced by a considerable period of gestation, starting at about year 1000. During that 500-year period, population doubled in both China and Europe, and there was much technology transfer from China to Europe. This dating is similar to that used in our discussion of the emergence of the modern world. There was growth of cities; coastal, road, and river transport; and trade in Western Europe prior to the circa 1600 date we used to mark the emergence of the modern world.

In this model, the first cycle of world power begins in the 1400s and 1500s when the Portuguese designed and built new types of ships and began to explore the Atlantic and, then, more widely. This was a sea development cycle, a cycle having to do with ship designs, exploration and the exploitation of native populations, church influence, and world power based on ship transportation. Both the Iberian countries of Spain and Portugal and the Northern Italian city-states of Venice and Genoa made thrusts for world power.

The Kondratieff cycle (or K-waves) allows us to organize our understanding of history. Kondratieff proposed long waves (of about 50 years) that track with the birth, growth, and maturity of clusters of major technologies, our S-curves. Historians and historiographers who analyze these things suggest K-waves dating back to the North Sung dynasty of China, dominated by leading sectors. Some embed these waves within even longer cycles of about

TABLE 26.1

Long Waves in Social Development

Long Cycles	World Powers	Date	K-Waves	Global leading sectors
LC1	Northern Sung	930	K1	Printing and paper
		990	K2	National market
LC2	Southern Sung	1060	K3	Fiscal framework
		1120	K4	Maritime trade expansion
LC3	(Genoa)	1190	K5	Champagne Fairs
		1250	K6	Black Sea trade
LC4	(Venice)	1300	K7	Galley fleets
		1350	K8	Pepper
LC5	Portugal	1420	K9	Guinea gold
		1492	K10	Spices
LC6	Dutch Republic	1540	K11	Baltic trade
		1580	K12	Asian trade
LC7	Britain I	1640	K13	American plantations
		1680	K14	Amerasian trade
LC8	Britain II	1740	K15	Cotton, iron
		1792	K16	Railroads
LC9	USA	1850	K17	Electric power, steel
		1914	K18	Electronics, motor vehicles
LC10		1973	K19	Information industries
		2026	K20	

Source: Modelski and Thompson (1996)

100 years. During each long cycle there is a dominant "world power," where the world is defined initially as Eurasian.

Rather than discuss subsequent cycles, Table 26.1 shows one interpretation of these cycles.

One would think that transportation developments would affect the capture (and loss) of world power. Transportation developments would affect the tools for attaining power and the loci where power is held. The challenge now is to say something insightful that links the information in the table to what we know about transportation development.

One can divide transportation developments into sea development and land development. In the early days, sea transportation was the tool for world power (Portugal, Spain, North Italy). The landside consolidation of power was not so much transportation enabled. It turned on the exertion of power by the nobility, the emerging trading classes, and the church.

The importance of land transportation in securing a power base began to be important in the Netherlands (early 1600s) and in subsequent struggles for power on the Continent. An exceptional situation was Britain I (early 1700s), for at that time the internal transportation system in Britain didn't amount to much, compared to, say, France. With that exception, it appears that the requisite for a power base increasingly became a strong integrated economy and the transportation that goes with it. The world reach of sea power was an extension of that power base.

With respect to establishing a land power base, the land powers of Europe (Spain, France, and, later, Germany) were preoccupied with controlling the landmass of Europe. The United States and England were not so concerned with land power challenges; they were on the peripheries. They devoted their energies to the global system and at a global scale. That was the case for England beginning in the early 1800s; the United States wasn't concerned with global power until the Spanish-American War of 1898.

Challenges for power on the European landmass continued to the end of the Cold War (with some extension in the various Balkan wars of the 1990s); we also see the continued emergence of power on the peripheries, example, Arabia and the Persian Gulf. Japan offers a partial illustration of the latter. By about 1900, Japan had emerged as a power on the periphery. It first made local scale challenges to Russia (ca. 1900) and then China (1930s) for power on land. Its world scale power challenge was in the 1940s. Today, Japan's power challenge is economic rather than military, and it is viewed on the decline compared with China.

Each successive successful world power has been larger, more powerful, wealthier, and so on, than those that went before, just as the critical global wars have increased in scope. Also, in each phase, the management of world power involved more complicated strategic associations with partners and thus management relations.

26.7 Transportation and Economic Development

Ontogeny recapitulates phylogeny.

—Ernst Haeckel

The statement, a largely discredited theory of biology, implies that when developing from an immature to mature state, an organism goes through stages reflecting the evolutionary development of its ancestors. In transportation and economic development, it would imply that before developing electric trains, a new country would deploy steam trains, or before deploying wireless, a country would need land lines. Fortunately we can technologically short-circuit deployment so that evolution need not recapitulate in development. Still, there is a sense that at least some paths must be retraced. When looking at rapidly developing China, for instance, we see motorization and the construction of massive new highways to emulate more developed countries, rather than forging a new path.

We understand how an observer in a nation might compare their area with, say, Switzerland and conclude: We are less developed when the states of transportation, nutrition, education, national income, and other things are compared. Such conclusions may seem straightforward and a point of departure for discussions and analysis. Even so, we think that such comparisons and the developed/less developed language confuses the issues.

For one thing, all nations, regions, and smaller areas are developed, given their historic paths, natural resource endowments, the tools available, and other attributes, including cultural and aspirational differences. The perspective from which we make that statement is this. It is a Euro-centered judgment. All nations have been touched by developments in Europe since year 1600 and have been developing. Four hundred years later, we say all are developed. They are certainly quite different from what they were in the year 1600.

Critics would be quick to point out that current status leaves much to be desired: given the attributes of undeveloped nations, a development status could be achieved that is closer to the status of Switzerland. Therefore, such nations are less developed than they could and should be.

Reasoning more or less like that, world-scale organizations, such as the World Bank, regional investment organizations, and individual nations have undertaken the stimulation of development. Essentially, the effort has been to provide attributes of capital, know-how, and services, such as educational and health services, that is, provide the infrastructure for development.

Although agency records suggest that development efforts have been successful—nations identified as less developed are better off with development efforts than they would have been without the efforts—that's a value-laden and contested judgment. Hirschman (1981) uses the word "trespassing" to summarize the argument He points out that governments have been destabilized, old ways of life changed, and so on. Demand has been stimulated more than production and balance of payments worsened. It is also pointed out that the gap between the developed and the less developed has increased. Within regions and nations, the equity impacts of development have been questioned.

We said that all places are developed given their historic paths, natural resource endowments, tools, and other attributes. Agencies promoting development take historic paths and natural resource endowments are givens; development efforts have focused on tools as instruments to build on those givens. Transportation has been considered as a tool for economic development, and it has loomed large in development assistance activities.

We have five things to say about the transportation tool.

1. The waves of transportation (and communication) development that we have examined—rail, transit, autos, air, and so on—were not just played out in the developed nations. Once the technology was standardized, they were deployed everywhere feasible. Encoded in standards, institutional formats, and equipment, know-how was readily transferred. Capital was quite mobile. So, for example, the round of rapid railroad deployment beginning about 1840 was a worldwide round.

   Recall that Stephenson's son, John, worked in South America for a time before joining his father's locomotive manufacturing business. Theodore Vail worked in South America building transit systems before being recalled to head AT&T.

   So if we had been involved at the time when development programs for the less developed countries (LDCs) were created, we might have argued that transportation should not be a centerpiece of development programs. We would have argued that transportation had already been tried. Those trials have resulted in a facade or layer of development. That's it. Not much more is to be expected.

   We would not have taken that position as an absolute, for surely distortions in the delivery system left unmet needs for investment, institution building, and other items.

2. Discussing transportation and production, we would have made much of ways production opportunities unfolded with the dynamics of system deployment and

development. Today's systems fit the production opportunities at places where they were birthed, rapidly installed, and extensively deployed. The less developed nations didn't participate in that development very much because the systems didn't "fit" their situations.

Suppose we have made too much of the "didn't fit" reasoning, deployment wasn't tried hard enough, or constraints now removed thwarted deployment and/or other things have changed. But even if we improved transportation in an LDC, that wouldn't be enough to induce development. The competitive developed nations might no longer have transportation advantages, but advantages such as agglomeration economies would still be held.

3. If we had been critics of development, we would have objected to projects oriented to decreasing costs of providing services that worked well for the production complexes of the developed nations. Rather, we would have argued that the priority should go to finding the services that matched latent production complexes in the less developed nations. We would have been very much critics of the use of transportation to stimulate demand and therefore increase the rate of growth of GDP.
4. Recall also that there is historic path dependence working within transportation. The transportation systems we have are not the best of all possible systems. They represent designs that worked at the time they were innovated, and that were subsequently locked in by custom, economies of scale, and standards.

   The fourth thing that we would have done would be to point out that the transportation systems exported to the LDCs are foreign in both space and time. They are suited to the resources and factor prices of the LDCs only to the extent that they are similar to those of the United States and Europe. System designs reflect know-how of the times when systems were birthed rather than modern know-how, though modern know-how has been used to improve systems.

   Also, the systems are cursed by many dysfunctions. If one insists on further deploying those systems, they should be cleaned of dysfunctions prior to deployment.
5. Observing that the production impacts of today's transportation modes in the LDCs are bound to be limited because the system technologies don't fit the situations, we see the LDCs needing processes of both transportation and production innovation directed to their production possibilities. By production possibilities, we mean possibilities defined on the resources and conditions of the LDCs.

The powerful lure of imitation has directed attention from innovations. Excuses claiming capital shortages were offered. (An excuse rather than a reason, for successful systems create the capital necessary to grow.)

Possible counter examples to our view are the East Asian experiences since 1960. Japan, Singapore, Hong Kong, Taiwan, and South Korea are in the "developed" category now. In these cases, there has been production expansion with transportation development mainly following. (See, for instance, the discussion of high-speed rail in Japan in Chapter 28).

Actually, of course, sharp changes in bulk, neo-bulk, and container maritime transport were quite supportive of developments in Asia. But land transportation followed rather than stimulated.

Considering those developments, low labor factor prices helped at first. Also, the Japanese read Schumpeter—after World War II, the first two English-language economics books translated into Japanese were Schumpeter's. The Japanese learned the importance of Schumpeter's (1934) "gales of creative destruction" and "carrying out of new combinations" as the edge of competitiveness and economic growth. They were also sensitive to the achievement of scale economies and economies of agglomeration.

26.8 Discussion

We can think of transportation providing a social contract with society. When there are strains on the social contract, government policy may be forged to ease the strains. We must deal with the substance of expectations. The question is: What does the transportation side of the contract say? What is transportation supposed to do for society?

To summarize, transportation enables thrusts for state or world power. For instance, the transcontinental railroads of the 1800s in the United States and French roads in the 1700s consolidated state power. With respect to world power, improved ships and navigation techniques enabled Iberian conquests beginning in the 1400s, Dutch efforts in the 1600s, and English accomplishments beginning in the 1700s. Transportation makes for more efficient production in a production function context. Transportation changes the comparative advantage of places. The mechanism here is land (transportation) rent. As land rents change, spatial patterns of production and consumption change. (It changes relative accessibility.) Transportation brings new resources into the economy. That is, market and supply areas are increased. (It increases absolute accessibility.) Transportation enables specialization of production and consumption, a consequence of larger supply and market areas, which opens opportunities for innovation and productivity improvements.

To the extent that our ideas about the emerging evolutionary paradigm are correct, they directly confront conventional wisdom about transportation management. We invest in and control transportation to lower costs and/or improve services. The focus is on making existing systems work better. One cannot deny the value of making systems work better, for efficiency is always desirable. Considering the maturity of systems, however, one should not be optimistic about the gains to be achieved. What's the confrontation? It is that conventional wisdom leads us to do things that are not very important; major technological improvements are important and should be sought.

Activities resulting from conventional wisdom are costly. The opportunity cost of doing this rather than that may be great. As conventional wisdom tells us, for instance, we could improve services by introducing ramp meters and advanced traffic management systems on some highways. That would take a lot of effort; perhaps that effort would be better spent seeking new technology and services, or enacting new policies.

One might claim that conventional procedures are satisfactory. After all, concern with elasticity of demand will focus attention on new things waiting to be birthed. Unfortunately, that is not true. Concern is with what can be seen and measured. Conventional procedures tend to favor the old and block the new.

There are many other implications. Of special interest is the developing regions or nations problem. They have invested in conventional systems to "catch up." That has merit, for it allows those areas to participate in the current round of development. Even so, that strategy can be questioned, for those areas that developed earlier have the lead in development. The developing world must run faster and faster just to keep up. Perhaps it would be better to pursue an alternative path rather than copying the well-worn path of the developed world. That might provide an opportunity to surpass rather than merely catch the developed world.

9 Wave Five: Modern Times

Development by which societies today meet their needs, without compromising the ability of future generations to meet their own needs.

—DEFINITION OF SUSTAINABILITY BY GRO HARLEM BRUNDTLAND (1987), *Report of the World Commission on Environment and Development*

27 Energy and Environment

27.1 Man and Nature

ENERGY AND ENVIRONMENTAL concerns are not new. In the first edition, George Perkins Marsh's (1850) *Man and Nature* pointed out how uses of the land affected microclimates and watersheds.[548] Already, Marsh was stressing "the dangers of imprudence and the necessity of caution." His concerns extended to damage to biological communities, and he mentioned the ways the Suez Canal would mix waters from alien seas. Jevons's late 1800s concerns about the exhaustion of coal resources illustrate another aspect of environmental questions, as does Thomas Malthus's earlier concerns about the inevitability of overpopulation. These references provide examples of what seem to be three major views of the environmental sustainability problem: (1) overpopulation, (2) natural resource exhaustion, and (3) damage to the environment.

In the opening quote, Gro Harlem Brundtland defined sustainability. While we dislike the term "sustainability" because it implies a static world, we think we understand the sentiment. We take it to mean a condition where the use of resources can support a healthy ecological complex into far distant futures. In our view, that shouldn't bar shifts from one resource to another as technologies and factor prices change. Actually, such shifts can aid attaining a sustainable condition. For instance, fuel shifts have decreased the carbon intensity of energy consumption (decreasing CO_2 per unit of consumption) and acid precipitation. The recent increase in natural gas fuel (in place of coal) has helped reduce US CO_2 output measurably.

It is widely felt that tomorrow's transportation choices must accept energy and environmental constraints (real or perceived). That there are constraints is undeniable, although it is countered by the argument that poverty, nutrition, and other sweeping problems ought not to be subordinated to relatively minor and manageable energy and environmental problems.[549]

There is also the opportunity cost concern about shortsighted policies that have unintended effects and waste resources, especially by giving an illusion of progress and diverting attention away from richer directions for actions.

The authors recognize that debates and actions of these types are common and regard them as overhead cost in an intellectually active society. The amount of that cost and its increase may be of issue, of course.

Energy and environmental issues bear especially on transportation, although they may have been eased by recent developments. Technology and price increases are making expensive, but extensive petroleum sources economically viable. The interrelated CO_2 emissions and climate change issues are being viewed more broadly. Intergovernmental Report on Climate Change says that "Climate change may be due to natural internal processes or external forces, or to the persistent anthropogenic changes in the composition of the atmosphere or in land use."[550] There is no longer the single issue of carbon fuel burning; it is recognized that land uses may affect temperature records, and there is room to consider forces that yielded the Medieval and Roman warm periods.[551]

27.2 Energy

27.2.1 RUNNING OUT

In 1984, the Reagan administration shredded 4.8 billion ration coupons being stored at a warehouse in Pueblo, Colorado.[552] These coupons had been prepared just in case federal gasoline rationing would be required, and were a response to the apparent fuel shortages of the 1970s. However, since deregulation (beginning in 1980 despite the efforts of President Jimmy Carter and continuing in the early 1980s under President Ronald Reagan), gasoline prices were falling and concerns about shortages disappeared.

Fuel shortages had been known before; World War II was an example. However, during the war, gasoline was being diverted from domestic to military use, and was largely unchallenged. The oil shortages of 1973–1974 were associated with Organization of Petroleum Exporting Countries (OPEC) oil embargo of October 17, 1973, which was a response to the Yom Kippur War when Israel's Arab neighbors attacked. The oil embargo shocked the system.

Oil is fungible, and OPEC members were not the only producers, so if OPEC stopped selling oil to certain parties, they might still sell to other countries, which could then substitute OPEC oil for their other sources, freeing those other sources to sell to the Western countries. But still, all the disruption and production cutbacks should at a minimum have driven prices up, thereby decreasing consumption.

In a well-functioning market, the other effect of such a price rise is to encourage new producers to enter the market. Oil that was previously too expensive might now be feasible. Well-functioning markets have a way of equilibrating shocks so that supply still equals demand, even if the price changes.

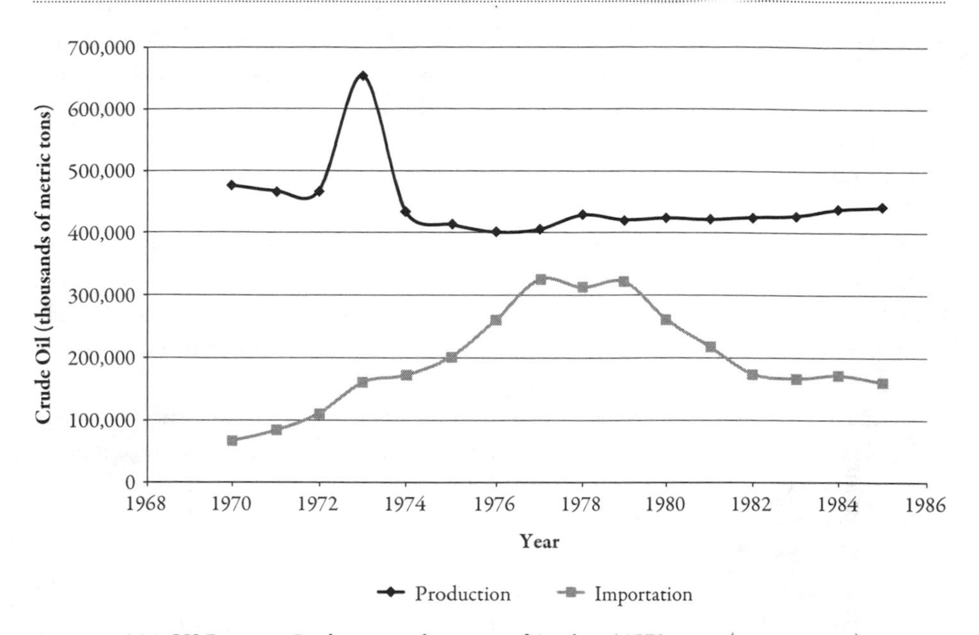

FIGURE 27.1 US Domestic Production and Imports of Crude Oil 1970–1985 (in metric tons)

So why were there shortages in the United States? Why were there lines at gas stations? The conventional explanation is that the embargo caused the shortage, but the evidence shows that domestic oil production dropped in 1974 (Figure 27.1), and oil imports increased. That is a very strange response to a price hike due to an international embargo. The better explanation for why domestic oil production dropped has to do with the federal policy response to the shortage. Just after the embargo, Congress passed, and President Richard Nixon signed, the Emergency Petroleum Allocation Act (EPAA) (which had been in the legislative pipeline for some time). This act instituted a two-tier price system, one tier for existing domestic oil reserves (whose prices were to be kept at pre-embargo levels), and a second tier for all other sources (whose prices were to be uncontrolled). The result was that domestic producers would sell little "old" oil at below market rates, and there was a contraction in domestic production until new oil fields (which would be priced at market rates) were brought on line, a process that takes several years. This law was in character with the era, which had seen other wage-price freezes. The Energy Policy and Conservation Act of 1975 extended EPAA and created a fixed maximum price for new oil. The Act also established the Corporate Average Fuel Efficiency (CAFÉ) standards, which are generally seen as successful.

The Iranian Revolution, followed by the hostage crisis of 1979, instituted a second period of shortage. Iranian oil was removed from the market, marking a 5 percent reduction in supplies to the United States Again, this would be expected to increase prices (which they did, but clearly regulated prices cannot respond as quickly as market prices). Normally, individuals would respond to price hikes by reducing consumption in the manner most efficacious to them. However, to prod the process, President Jimmy Carter proposed several conservation measures. These included a prohibition on the sale of gasoline during certain weekend hours, limiting thermostats in buildings to 18°C (65°F) for heating and 27°C (80°F) for cooling, and restricting non-essential lighting for advertising. Congress only approved the second measure. Nevertheless, because of a large change in the wholesale price of oil and caps on the retail price, shortages ensued.

The results of these policies follow the law of unintended consequences. New oil production boomed (while "old oil" was pulled from the market), which led to a boom in the Texas economy. Houston real estate especially took off. This boom lasted a few years into the 1980s, as people expected high oil prices to remain (markets hadn't equilibrated properly in the past, why should they now?). Ultimately it did crash, bringing their financiers, the newly deregulated (but still federally insured) savings and loan industry down with it.

With deregulation in the 1980s (especially the Petroleum Price and Allocation Decontrol Act of 1981), along with more fuel-efficient autos in the fleet and changes in consumption patterns, domestic oil production increased and imports of oil dropped. OPEC responded with production cuts to keep the price of oil high, but the effectiveness of this cartel was broken when Saudi Arabia increased production in late 1985. By 1986, oil prices had dropped significantly. In the 1990s, the fuel-efficiency of the US fleet decreased with the large increase in the use of trucks, especially minivans and sport utility vehicles, as passenger transportation. However, oil disruptions associated with the 1990 Iraqi invasion of Kuwait and the 2003 US invasion of Iraq did not result in shortages, though prices did rise in both periods.

27.2.2 IS TRANSPORTATION OBESE?

Transportation moves mass. To do so, energy is used to accelerate mass to cruise speed, up a hill, and so on. In theory, that is no matter because potential energy is recoverable when decelerating. In practice, of course, there is always energy loss when work is done. Energy is not recovered very well when decelerating; often it is wasted as heat when brakes are applied. For a given acceleration task, the amount of energy required depends on the mass to be accelerated. Because packages, containers, cars, and so on, of some sort are used to contain things to be moved, we need to be concerned with laden (gross) and unladen (tare) weights. Today's automobile weighs, for example, about 1,400 kg (3,000 lbs) empty (tare weight) and moves about 1.15 persons, say, 90 kg (200 lbs). A ten-wheel truck's tare weigh is about 9,000 kg (20,000 lbs) and the tare weight of a 120-ton rail car might be 25 tons. Sea-rail-truck containers have a 10:1 loaded/empty ratio, not bad comparatively, but containers must be carried on some vehicle. An inter-city passenger rail car might weigh 80 tons. If it weren't for acceleration-deceleration energy losses, gross weight wouldn't matter, except that systems with a high ratio of gross to laden weight suggest that work is being done and material being used unnecessarily.

Energy is required to overcome air resistance (Figure 27.2), rolling resistance (mechanical resistance), which includes the resistance to movement by bearings and seals, as well as resistance at the wheel-ground interface. Velocity dependent resistance results from the work done as loads shift, couplers rub, and so on. Note that air resistance begins to become important at velocities between (40 and 70 km/hr), and dominates at high velocities. This figure uses the Davis equations, which understate air resistance.

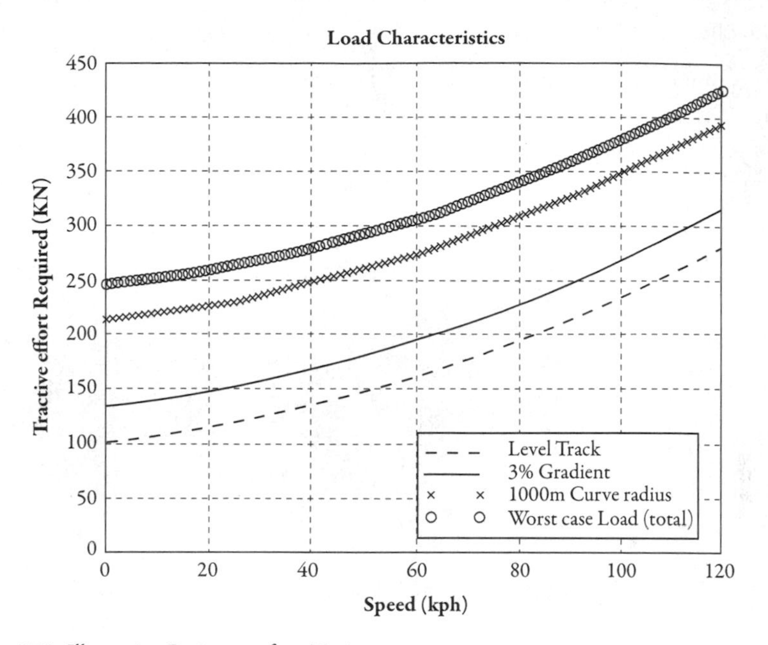

FIGURE 27.2 Illustrating Resistances for a Train
Source: Lyovic, 2000, Assumptions: see Table 27.1.

TABLE 27.1

Assumptions in Model, Figure 27.2		
Locomotive mass	198,000	kg
Number of locomotives	1	
Number of axles (locomotive)	6	
Wagon mass	140,000	kg
Number of wagons	80	
Number of axles (wagon)	4	
Total train mass	11,398,000	kg
Speed range	0–120	$km \cdot h^{-1}$
CSA of locomotive	11.3	m^2
Gradient range	0-3	%
Gravitational acceleration	9.8	$m \cdot s^{-2}$
Curve range (r)	1000–3500	m
Acceleration range (a)	0.02–0.1	$m \cdot s^{-2}$
Rotating mass factor range (x)	1.07–1.12	

27.2.3 ENERGY USE

Transportation is not the only user of energy. Petroleum comprises about 39 percent of all energy used in the United States, and transportation only used two-thirds of the petroleum (or about one-fourth of energy consumed in the US per year), as seen in Figure 27.3. The reasoning that development induces lots of transportation which consumes large amounts

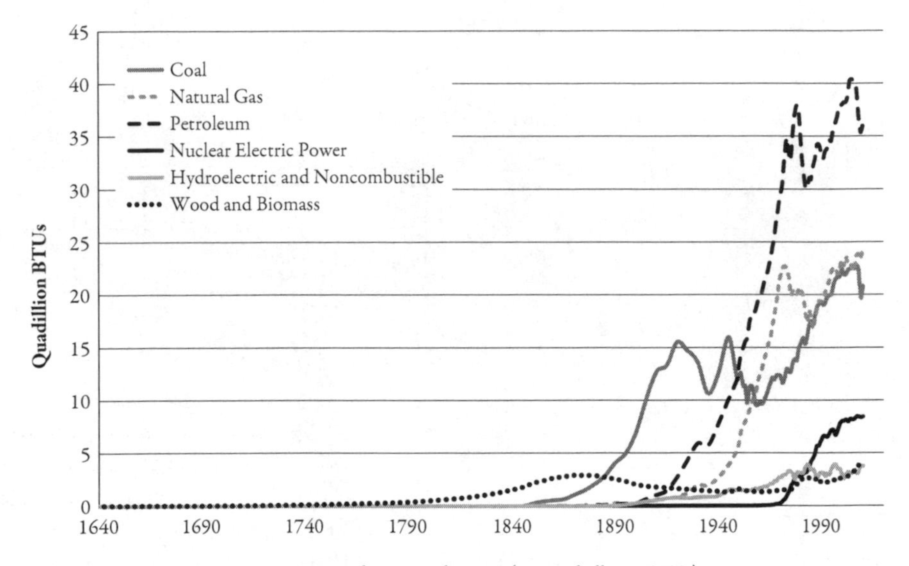

FIGURE 27.3 Energy Consumption in the United States (in quadrillion BTUs)

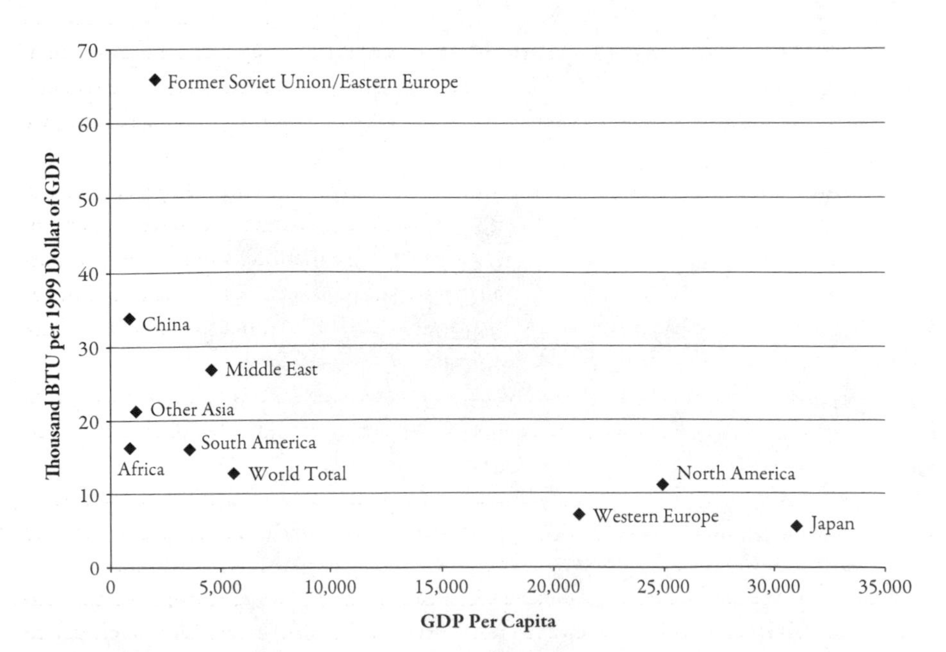

FIGURE 27.4 Relations Between Development and Energy Use (thousand BTUs per 1999 Dollar of GDP)

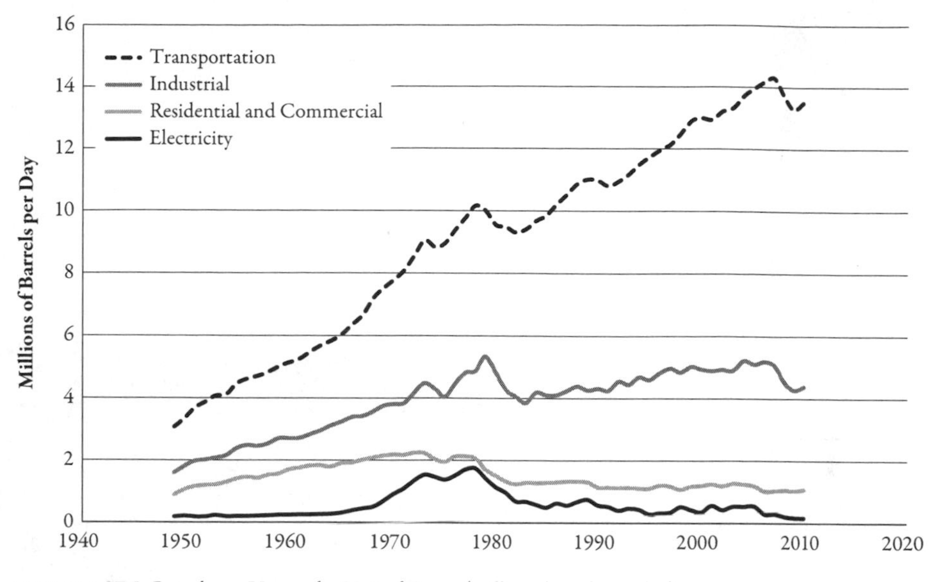

FIGURE 27.5 Petroleum Use in the United States (millions barrels per day)

of energy has to be tempered. Poorer countries are less efficient in their use of energy (or perhaps less efficient countries are poorer), as seen in Figure 27.4. Transportation mainly uses petroleum energy, of course. But even if we narrow the problem to petroleum energy use, it can't be cured by transportation focused actions alone. That's because about 33 percent of petroleum use is by non-transportation sectors, as seen in Figure 27.5.

Some say there is no urgency because a good many decades of petroleum energy are available, and alternative fuels, such as liquified natural gas (LNG), add more decades of business-as-usual. All we need to do is to begin to reduce emissions and be more fuel-efficient. This may be true. But not knowing either the severity of emissions problems, such as the CO_2 situation, or future political stability in petroleum resource-holding regions, we think it's past time to begin to seek new development paths for transportation.

Some hold the view that non-transportation energy use will be managed by transition to a nuclear, solar, or hydrogen society. It is argued that the public will discover and accept advances in nuclear technology. (Things get bad enough, something will happen.) Transportation may shift to electric power (either with fuel cells or batteries). Absent looking at other things, it may be a default strategy. Futurists speak of a "hydrogen economy," where electricity is generated by hydrogen-powered fuel cells.

Almost three-quarters of transportation energy use is by auto and truck. But curing auto energy use won't achieve sustainability because 16.2 quads of energy use would remain to be treated (Figure 27.6).

Light trucks have two axles and four tires. They consume about 60 percent of total truck energy consumption. As we know, many light trucks are used for car-like purposes. Yet they are regulated like trucks and so face fewer energy-efficiency regulations than autos.

Increases in automobile energy use have stabilized. Inter-city bus and rail have concentrated their services and this has increased their energy efficiency. There isn't much left to thin out, so the outlook is poor.

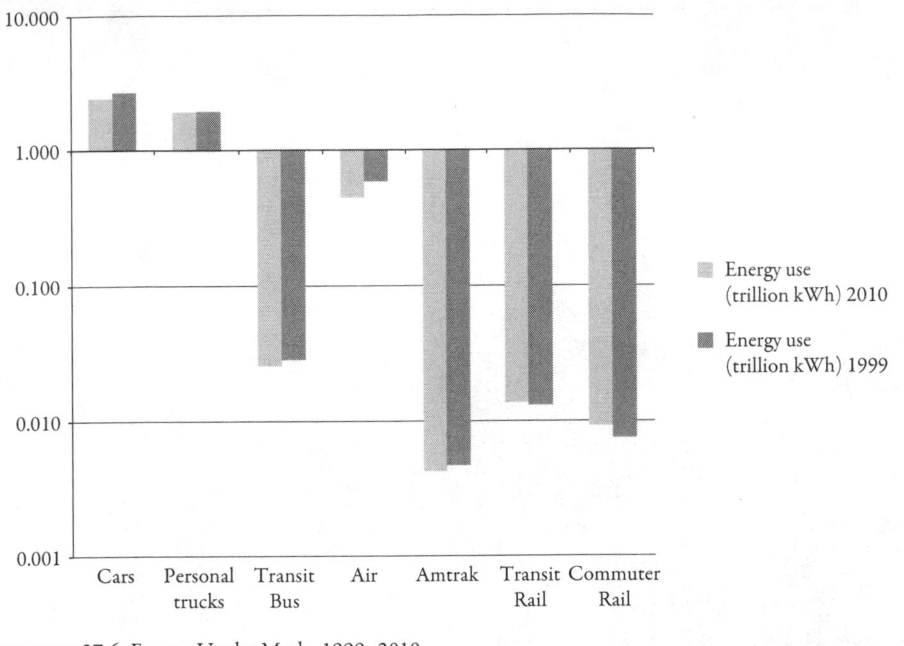

FIGURE 27.6 Energy Use by Mode, 1999–2010
Source: Transportation Energy Data Book

The urban rail, bus, and auto modes do not differ very much in their energy intensities. The outlook for improvements is pretty good for the automobile, at least technically. The outlook is not good for the transit modes under the present policy of expanding services to thinner markets, aiming to achieve spatial coverage goals rather than ridership goals. Commercial air transportation uses about 2.5 quads per year, and the new aircraft entering the fleet will continue improvements in their fuel efficiency.

Personal trucks are energy gulpers compared to the other passenger modes (Figure 27.7).

We need to pay special attention to the big users: automobiles and small trucks, as shown in Figure 27.6. What about the elasticity of demand as energy costs change? Unsurprisingly, higher gasoline prices result in the purchase of more fuel-efficient cars while the amount of auto use increases with lower gasoline prices.[553]

Inter-city trucks are fuel gulpers, as seen in Figure 27.8, but have improved by about 14 percent since 1970. It's said that about one-half of freight truck travel is in urban areas, and those trucks must be very energy inefficient because, in general, loads are light.

What about the diversion of freight from trucks to the other freight modes, especially railroads? This has been examined quite a bit, and the opportunity for inter-city freight diversion from trucks to other modes is limited because of the shipment size and service difference among the modes. However, Trailer (Container) on Freight Car (TOFC/COFC) is diverting freight from trucks partly because of fuel cost savings. The problem is that this diversion is limited to dense freight movement corridors, such as Los Angeles–Chicago, because it is in such corridors that the railroads can offer good service.

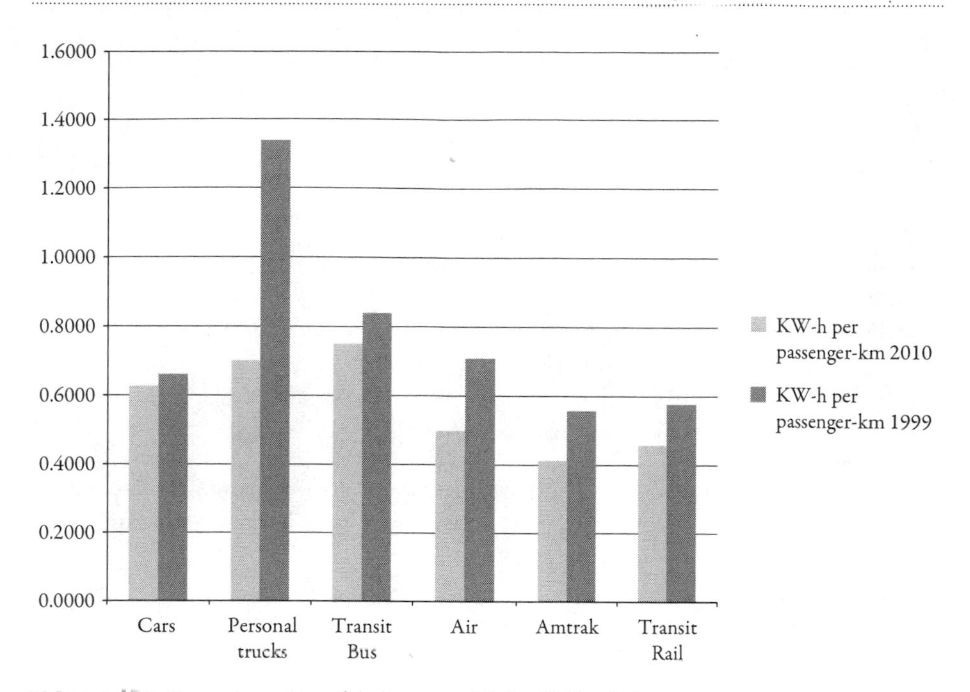

FIGURE 27.7 Energy Intensities of the Passenger Modes, 1999–2010
Source: Transportation Energy Data Book

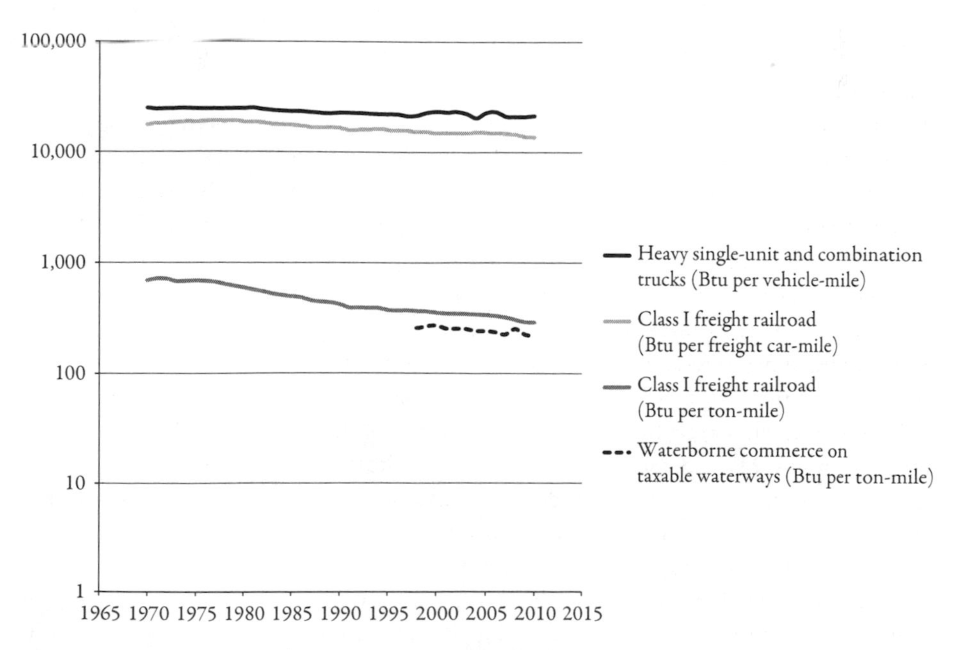

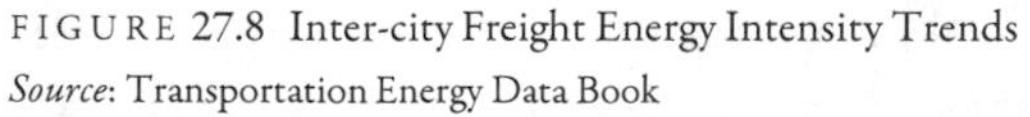
FIGURE 27.8 Inter-city Freight Energy Intensity Trends
Source: Transportation Energy Data Book

27.3 Environment

27.3.1 IS TRANSPORTATION SUSTAINABLE?

Uneasiness exists that the present transportation development path is not "sustainable." The word is used in many contexts, we hear:

- Transportation is gulping finite resources, petroleum in particular, and that cannot continue forever.
- Transportation induced air and water pollution, soil degradation, and ecological insults are or will soon be causing irreversible harm.
- Transportation services result in the conversion of farmland to urban uses and damage to parks in degrees that can't continue.

Even more broadly, the modern transportation-based society is consuming resources of all types at rates that can't be continued. The Sept/Oct 1994 *TR News* placed sustainable transportation first in its list of issues and says that transportation policy is incompatible with a healthy world environment. Whatever their shape, these questions are important and deserve concern and action.

Thoughtful debates about these issues range from analytic discussions about the truth of claims through adjustment mechanisms of an economic or technological substitution sort to calls for drastic policy constraints on consumption. While the discussion to follow will involve technological options, it will not attempt to illuminate these debates.

Nor will we argue a reactive stance: the "something will come along when things get bad enough" school of thought. Lessons from history do give this school of thought some creditability. For instance, Thomas Malthus's message on the inevitability of overpopulation has been countered by the tapering of population growth in the industrialized nations, as well as by development of more productive crops. The acute wintertime coal smoke pollution problems of London were largely countered by shifts in fuels. Soil erosion on the Carolina Piedmont has decreased following shifts in the location of cotton production. Even though history may blunt pessimism, we think the "something will come along" game is very chancy and costly. Not every environmental disaster has been averted; one need only look at some civilizations that have collapsed.[554] Moreover, that "something" is not dropped like manna from heaven; it requires explicit design, policy decisions, market shifts or behavioral changes.

We do not find "sin now and regret later" discussions of interest. As mentioned, Jevons (1866) pointed out that at then rates of increases of English coal production, coal resources would soon be exhausted. He said, ". . . we have to make the choice between brief greatness and longer continued mediocrity." It didn't work out quite that way, of course. At any rate, there seem to be those who think that mediocrity is inevitable, and "hunker down/get used to lowered expectations" is appropriate policy.

Instead of commenting on today's debates and actions, this chapter takes the view that energy consumption is both the problem and the opportunity. It is the problem because it wastes many types of resources, and emissions tie to energy conversion. It is an opportunity because achieving reductions in energy use could well improve transportation services.

There is more. Recognizing that energy use is one of an array of air, soil, water, nutrition, and other problems, Hollander (2003) makes the point that poverty is the real environmental problem, while affluence permits adjustments to problems and dealing with them. Transportation enters because it enables economic and social affluence and adjusting to changing circumstances. Transportation by creating wealth enables solutions to the problems it creates.

Also, population increases, resource use changes, and adjustments in economic and social activities will ask for improvements in transportation's flexibility and efficiency.

Postponing further treatment at this level of generality, we focus on the topical issue of greenhouse emissions, CO_2 in particular, because of the tie between CO_2 emissions and fossil fuel use.

27.3.2 CLIMATE CHANGE

Concerned about climatic impacts of greenhouse gases (sometimes called global warming), the United Nation has held conferences and appointed study groups. Many nations have committed to a Framework Convention on Climate Change (FCCC). Among other things, it calls for developing technologies, practices, and processes that control, reduce, or prevent anthropogenic emissions of greenhouse gases. Technology transfer is encouraged, especially to developing nations. For a variety of reasons, including doubts about whether global warming is actually under way, the role played by anthropogenic emissions, and the economic cost of reducing greenhouse gases, there is more talk than action.

For transportation, the gas of main concern is CO_2. As illustrated in Figure 27.9, CO_2 in the atmosphere is increasing. While there are a number of greenhouse gases, we simplify our discussion by reasoning: transportation uses energy, which yields CO_2; therefore, we should examine reducing energy use in transportation. A 20–80 percent reduction in emissions is

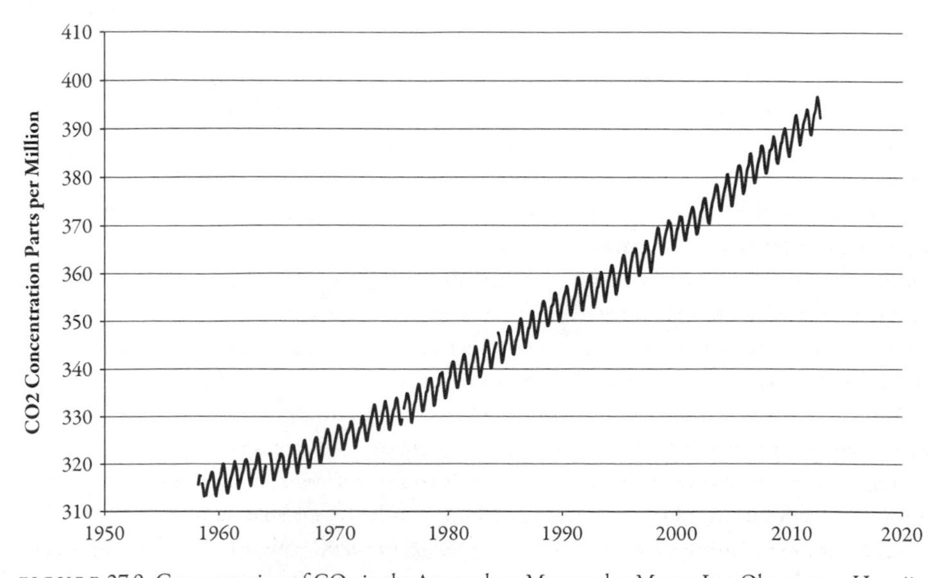

FIGURE 27.9 Concentration of CO_2 in the Atmosphere Measured at Mauna Loa Observatory, Hawaii

discussed. Can that be done in a reasonable way? Some argue for the need to adapt to prevent future damages.[555] This question is another reason for our focus on energy and CO_2.

27.3.3 AIR POLLUTION

Tetraethyl lead was added to petroleum in the 1920s to reduce knocking, improve octane ratings, increasing both power and fuel economy. Unfortunately, the lead when combusted would become an airborne pollutant, and when breathed in, would result in higher lead concentrations in the bloodstream. This was bad for health, and in children, could limit development, lead to lower IQs, and engender more anti-social behavior. It was also damaging to the catalytic converter, which aimed at reducing other pollutants. Lead was not banned in the United States until the 1970s (automakers and refiners were opponents), first no new leaded gasoline cars were produced, later (1995) lead was no longer sold. The Soviet Union banned leaded fuels in cities in 1967.[556] After the ban on leaded gasoline, IQs in the United States rose, consistent with, and possibly explaining, the Flynn Effect showing long-term increases in IQ,[557] though the causality here is still a speculative hypothesis.

It is easy to document that numerous counterproductive actions have been taken in the energy and environmental arenas. More constructive public policy requires much better information than is available; an ability to set priorities and make trade-offs, and a style that seeks satisfactory-workable paths for improvements, as contrasted to the setting of standards that "hold actors' feet to the fire." We ought to have better information. We ought to view achieving safety, cleaner air, and reduced energy consumption as some of the many ethical needs of society. But to say that is not to say very much. The literature on the subject is not very helpful for it "curses the darkness" rather than "lights a candle."

What are some lines of policy reasoning? One line of policy reasoning might go this way:

> We are heavily committed to policy and programs; there are associated heavy political and emotional commitments that cannot be changed easily. Therefore, for a period of time we must let things go along as they are. When people become aware that things aren't going well, then more efficacious policies might be put in place.

We are not happy with that line of reasoning, although we hear it often. Can we think of something better?

Consider clean air. The air pollution emissions problem has been stated as a health imperative. The problem then becomes a how-to-achieve question and the broad issue is whether to use a command and control style or charge those who pollute. Presently, with a few exceptions, a command-and-control style is in use. Perhaps a cost-based line of policy reasoning and policy formation might be useful. Such policies have been developed and applied within some niches (market trading of emissions from certain large generators, for instance).

It might be helpful if the policy debate would re-address the nature of the clean air imperative and how achieving it might conflict with other ethical needs or imperatives such as making a living, social interaction, housing, and so on. While switching from command and control to a decentralized market-based approach may be reasonable (and at least should be considered), we are not optimistic about it in the short term. The notion of an imperative defines and gives such priority to policies that they are not open to discussion.

TABLE 27.2

EPA Criteria Pollution Standards

Pollutant	Standard Value
Carbon Monoxide (CO)	
8-hour Average	$9 ppm$
1-hour Average	$35 ppm$
Lead (Pb)	
Quarterly Average	$0.15 \mu g/m^3$
Nitrogen Dioxide (NO_2)	
Annual Arithmetic Mean	$0.053 ppm$
Ozone (O_3)	
1-hour Average	$0.075 ppm$
Particulate ($PM_{2.5}$) Particles with diameters ≤ 2.5 μm	
Annual Arithmetic Mean	$15 \mu g/m^3$
24-hour Average	$35 \mu g/m^3$
Particulate (PM_{10}) Particles with diameters ≤ 10 μm	
24-hour Average	$150 \mu g/m^3$
Sulfur Dioxide (SO_2)	
24-hour Average	$0.075 ppm$
3-hour Average	$0.50 ppm$

Source: US EPA National Ambient Air Quality Standards (NAAQS) (2012)[558]

Hydrocarbons, carbon monoxide (CO), and ozone (O_3) are associated with health problems, and automobiles are primary producers of CO. Automobiles appear to produce something like 60 percent of ozone precursors. From the violation of standards stance, ozone is the big problem.

The federal standard for airsheds is given in Table 27.2 and these have been tightened over the years, both as the science and ability to measure has improved, and understanding of health effects has changed, and as the ability to comply with the standards has improved. California has much more restrictive standards. For instance, the San Francisco–Oakland area, with strong Pacific breezes, is usually in compliance with federal standards, yet it violates state standards from time to time.

Also with respect to the vehicle fleet, old cars are the bad actors. A policy would be in order to scrap all pre-1975 (or 1985, or 1995) cars, often considered "super-emitters," and give their owners new cars (see also Section 27.5.4). California is trying to phase in zero emission vehicles (electric vehicles or fuel cells), which have zero tailpipe emissions (but of course run on electricity generated somewhere). But the policy has met great resistance and has been stalled more than a decade. In part, the automakers complain about the inability to manufacture them, but are probably more concerned about poor consumer reception if the vehicles were not as good as (less powerful, shorter range, less reliable) conventional internal combustion engine powered vehicles, which have had a century to be perfected. This is discussed below (see Section 27.5.2).

One of the strategies many large cities have adopted to combat congestion and pollution is rationing of the use of automobiles. Athens, Greece, for example, adopted odd-even rationing (you were permitted to drive based on the license plate number, odd numbers could drive on odd days). The wealthy bought additional vehicles to skirt the regulation.[559] The less wealthy might buy second license plates.

27.4 The Automobile Vilified

27.4.1 A TAX ON SIN?

In the early days in the United States, the automobile was regarded as a "rich man's toy," just as today when Humvees or Priuses are viewed as status signifiers. Views of the automobile as a luxury have affected policy. In Europe, this resulted in a steeply graduated levy on engine displacement; only the rich had large cars. Early in 1989, Daniel E. Koshland, Jr., professor of biochemistry at Berkeley, proposed a tax on large engines: "A Tax on Sin: The Six-Cylinder."[560] A tax on sin, "... where sin is defined on the basis of national policy rather than personal peccadillo..." Economist George Stigler replied:[561]

> Daniel E. Koshland, Jr. (Editorial, 20 Jan., p. 281), proposes a proportional or progressive tax on automobiles on the basis of their fuel consumption. The benefits he lists are numerous: smaller deficits in the federal budget and foreign aid, cleaner air, and better care of the needy (this last a fine example of double-counting).
>
> The same argument calls for progressive taxation of dwelling units: they too use fuel; and to paraphrase Koshland on automobiles, most rooms in larger homes have less than one occupant. He appropriately remarks that if this kind of policy becomes widely accepted, it could be extended to other areas (room temperatures? illumination? travel?).
>
> Koshland's editorial presents by example his distinction between "national policy" and "personal peccadillo." Could he have confused the two?

We have no sharp understanding of why autos and auto travel have long been regarded as conspicuous consumption more so than other goods or services. We suspect that it has something to do with breaking up of old arrangements and the creation of new ones.[562]

27.4.2 THE AUTOMOBILE AS DRUG

Let's consider the automobile and the environment. Newman and Kenworthy (1989) popularized the term "automobile dependence," analogous to drug dependence. By neat association of the automobile with the evils of drugs, the auto too, because of its environmental and social effects, becomes evil in their nominally scientific, public health oriented worldview. As distasteful as it may seem, it seems that the drug problem and the environmental problem have much in common. Each moved from complexity to simplicity, each is treated in the moral imperative style of policy making. If the automobile is evil transportation, its alternatives must be good (the enemy of thine enemy is thy friend).

27.5 Responses

The ills described in this chapter—too much energy consumption, energy market instability, too much emissions—are all addressed by improved fuel efficiency. It is a vehicle matter, but change in the fuel for automobiles is constrained by other components of the system fuel supply. This sharply limits what can be done. Disjoint decision making, lack of incentives, and rules fashioned decades ago help prevent this from happening. For instance, the Corporate Average Fuel Economy (CAFÉ) standards, which have turned out to be relatively successful, developed after the fuel shortfalls of the 1970s, were applied to cars. Yet more than half of all US "cars" are now trucks (pickups, sport utility vehicles).

To move a person one km requires that we move along a ton or more of tare weight (the auto), and the movement of one ton of freight also requires, at best, that we move a ton of equipment. That is resource demanding. It constrains what we can expect to achieve from energy conservation efforts. It makes the services more expensive than they need to be. Just the room it takes to move tare weight is demanding of facility investment.

The system is locked in by design decisions made in earlier days. Modern vehicle equipment is descended from beefed up wagons and buggies. It is the result of incremental change. No other kinds of changes can be easily made. The reduction in the weight of automobiles has had safety and ride quality costs, and has hardly reduced the costs of vehicles. This section examines several strategies to reduce automobile fuel consumption: very small cars, electric vehicles, and scrappage schemes.

27.5.1 VERY SMALL CARS

Consider the collection of problems involving pollution, energy use, and access to the auto system. They flow from predominant design characteristics. As a result of those characteristics, there is a lot of mass in vehicles.

So let's intervene in the system to reduce mass (see Section 27.2.2).[563] With current technology, a passenger vehicle can be built for about one-half the cost of a compact car. It would weigh about 225 kg (500 lbs), achieve 80 km/liter (200 mpg), and have high performance, say, 0 to 95 km/hr (60 mph) in 6 seconds.[564]

Amory and Hunter Lovins at the Snowmass Institute have developed a concept for a standard volume, but low mass "hypercar" that, by using advanced materials, combustion devices, and energy storage, achieves 38 km/liter (90 mpg). They believe this automobile can be made available at a price acceptable to the market by early in the twenty-first century, and over the long term, efficiency can be doubled.[565] While the product technology information put forth sounds reasonable, there is an enormous assumption that somehow production for mass markets will greatly decrease production costs.

But perhaps the problem is scoping. Do we need high performance from a vehicle most of the time? Most travel is local, and on short trips we don't need powerful acceleration or to carry around the equipment needed for such power. Very small low-speed vehicles have less performance than regular cars, but come with much less cost.[566]

The general idea of specialized vehicles, roads, and operations is beginning to find markets. Disney considered it for the Celebration, New Town in Florida, and others at the Research Triangle Institute in North Carolina have as well, for example. Golf carts are widely used in Southern US retirement communities as off-road transportation, but they are also used on low speed streets, and some golf carts begin to resemble neighborhood vehicles. Laws need to be revised and networks developed to encourage them.[567] A match exists between the idea and planning notions of new urbanist villages.

So why hasn't the market pulled small vehicle development and sales except in small market niches? We suppose that current protocols for facility design are a limiting factor.[568] Several vehicle manufacturers have inquired, and the idea has been introduced to the Society of Automotive Engineers (SAE) community. We sense that the "sticky" question in deployment is changing road designs, and have given such designs attention (Figure 27.10).

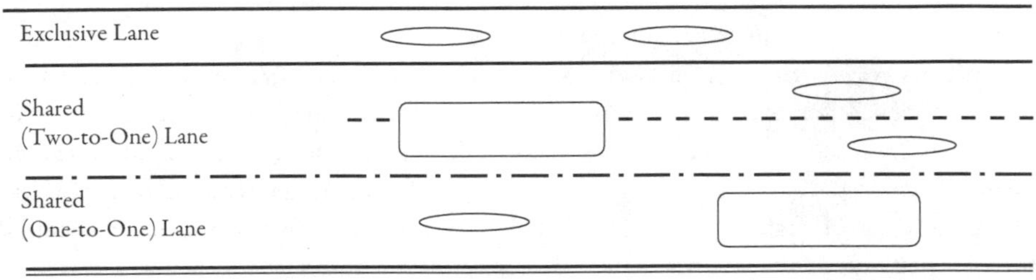

Exclusive Lane

Shared
(Two-to-One) Lane

Shared
(One-to-One) Lane

FIGURE 27.10 Accommodating Commuter Cars on Existing Roads

FIGURE 27.11 Electric Vehicle, Kyoto Japan

With small amounts of energy use, energy and pollution problems are sharply reduced. The small cross section suggests more efficient road use and parking, yielding major changes in congestion problems. The high performance version could be used for commuting with salutary effects on accessibility to jobs and residential sites. The low performance version might be used for neighborhood travel. Low cost and simple to operate, it would ease access (to the system) problems for the poor and elderly. It would reduce insult to residential areas.

27.5.2 EV REDUX

> It doesn't notice when you turn down your thermostat and drive a hybrid car. These actions simply spread the pain over a few centuries, the bat of an eyelash as far as the earth is concerned, and leave the end result exactly the same: all the fossil fuel that used to be in the ground is now in the air, and none is left to burn.
>
> —Robert McLaughlin, "The Earth Doesn't Care if You Drive a Hybrid," in *American Scholar*[569]

While popular at the beginning of the twentieth century (see Section 7.2), interest in electric vehicles waned for the decades from the 1920s until the 1970s. The Energy Crisis of the early 1970s revived attention. In 1976, the Electric and Hybrid Vehicle Research Development and Demonstration Act provided modest support. The electric utility industry, through the Electric Power Research Institute, also began support, seeing that EVs could become important for their public utility members.

By the 1990s, the market was getting more ready, with both energy and environmental concerns used as rationales. California implemented (but later deferred) policies requiring Zero Emission Vehicles on the road. GM's *Impact*, designed by Paul MacCready, entered market with high hopes. Later dubbed the *EV1*, it plied the roads of California for a few years before being killed by GM, as documented in the film *Who Killed the Electric Car?*, which considered oil companies, GM, politicians, and consumers as among the suspects. The main problem that challenged EVs in the first decade of the twentieth century, range, had not really been solved by the last decade of that century.

GM reentered the EV market in 2011 with the Chevy *Volt*, while Nissan entered with the *Leaf* and Honda with the *Fit*. (It is probably worth noting that GM and Chrysler, two of three largest remaining US automakers, were briefly nationalized following the economic shock of 2008. Ford remained private. Federal policies on emissions and fuel efficiency were ratcheted up in this period.)

New companies were formed to market high-end EVs. Tesla Motors, founded by entrepreneur Elon Musk (who also founded SpaceX and PayPal), sold about 2650 Model S cars in 2012, about the same number it sold in the first 2 months of 2013,[570] following on their more expensive Tesla Roadster. Sales of other EVs were also in the low thousands (estimated EV sales were at about 0.2 percent market share in 2012), while the US market was about 13 to 14 million cars and light trucks per year. That has not stopped Musk and others from predicting that EVs will be the dominant vehicle in the future, Musk has claimed most new cars will be EVs within twenty years.

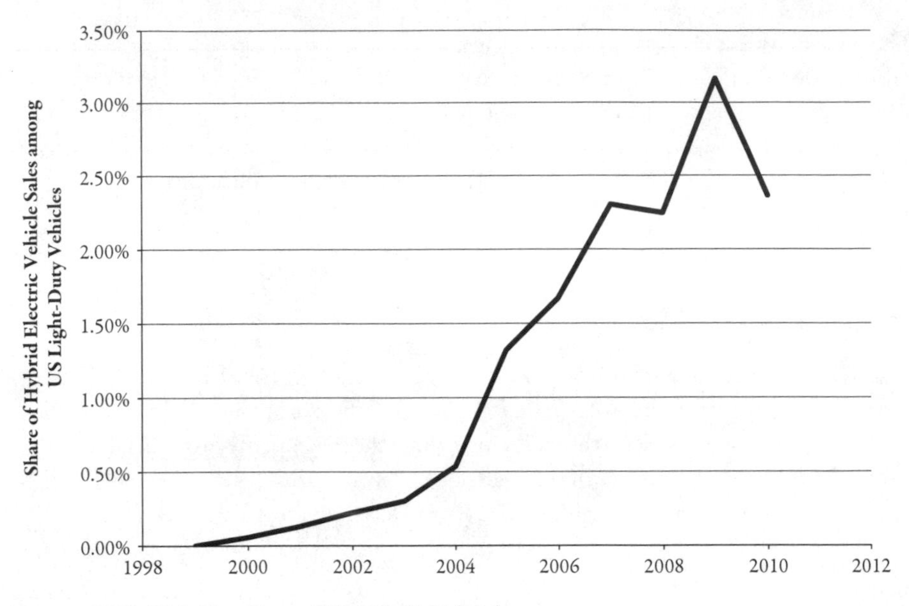

FIGURE 27.12 Hybrids as Share of US Light Vehicle Sales

To give a sense of the state of the technology, the Honda Fit had an energy efficiency of 29 kWh per 100 miles (18 kWh per 100 km) (lower is better) combined city and highway, which the EPA scores at 118 miles per gallon equivalent (50 km/liter) (higher is better), with an 82 mile (132 km) range.[571] As of 2012, the US government provides what might be characterized as an "infant industry" subsidy of $2,500 tax credits for plug-in EVs, and has in the past provided other subsidies for fuel efficient vehicles. Many states, and other countries, provide additional subsidies.

In the early 2000s, hybrid-electric vehicles (HEVs) started to become popular (see Figures 27.12 and 27.13). These vehicles had both an internal combustion engine and were electric powered, overcoming the range concerns as the electricity would be used on city streets, and the Internal Combustion Engine (ICE) could recharge the battery.

The main market remained gasoline-powered vehicles, which became much more efficient over time. Unfortunately for the environment, that efficiency was used up by producing vehicles with greater horsepower, rather than vehicles with better fuel economy. Fuel economy at 1985 levels of vehicle performance would be up to 30 percent better.[572]

In the United States, electric vehicles are a social status symbol, a way to wear your concern for the environment as a badge. Owners of EVs once painted the word "Electric" on their cars. With HEVs, the design of the vehicle itself was a slightly more subtle signifier (known to all who matter) that the individual cared more about the environment than you did (e.g., a Toyota *Prius* or a Honda *Insight*). As social markers, hybrids also become fodder for the culture wars, as noted in this section's opening quote.

Other electric vehicles, two-wheelers in the developing world, especially China, have been enormously successful from a market perspective.[573] These provide an upgrade path between the inexpensive, but speed-limited bicycle and the much more expensive, and rationed, automobile.

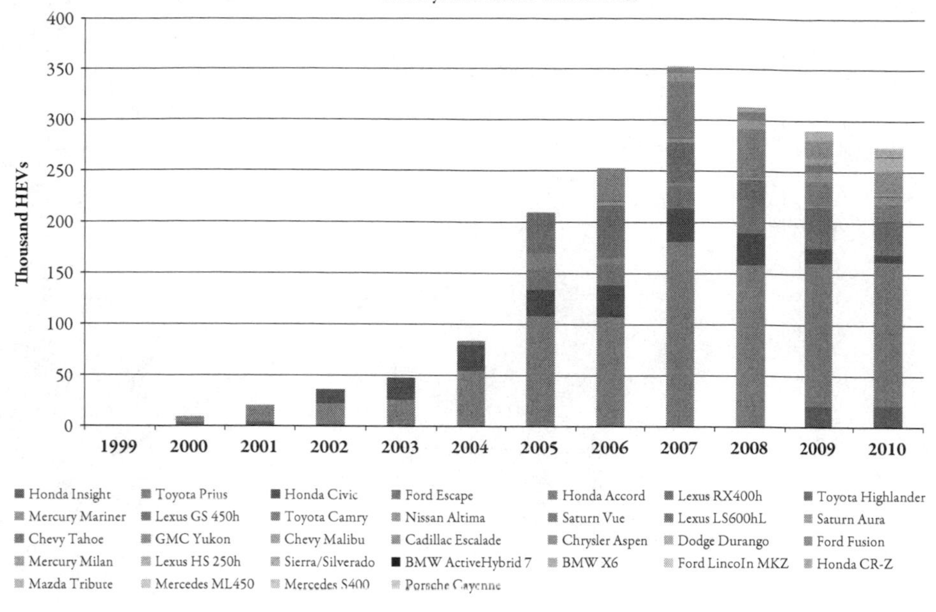

FIGURE 27.13 US Hybrid Electric Vehicle Sales

On-board batteries are only one possible technology for supplying electric cars with energy. Others include ultra-capacitors, fuel cells, which directly transform chemical fuels into electricity with oxygen, or another oxidizing agent, for on-board generation, or cables or wireless electric transmission from the network to the vehicle.[574] In the late 1990s, there was a lot of excitement about the prospects of hydrogen fuel cells.[575] But while costs have been driven down, they have yet to be cost-effective comparable with batteries, much less internal combustion engines, and the most optimistic thing you can say is the time frame of the promoters has been stretched.

27.5.3 IMPLICATIONS OF EVS ON FINANCING

If EVs or other non-petroleum-based energy sources become widely adopted, the primary source funding US highways since its 1919 debut, the gas tax, will come to an end. The United States, in contrast with many other countries, hypothecates motor fuel taxes to pay for transportation.

Imagine all gasoline vehicle users pay for all transportation costs. Imagine total expenses are $100,000,000 and the total number of users are 1,000,000, and all gasoline-powered cars get 30 MPG. In that case, if all vehicles are gasoline powered, the gas tax will be $0.30/gallon, in line with current costs. Now imagine, only half of all cars pay the gas tax, the tax jumps to $0.60 to cover costs, still quite tolerable, but as the gas tax rises, the number of gasoline-powered cars should be expected to fall.

Figure 27.14 shows the expected gas tax based on the above assumptions with a varying number of gasoline-powered cars on the road. Note especially that this is a log-log scale. At 50,000 cars with gasoline engines (95% non-gasoline powered), the tax jumps to $6.00 per

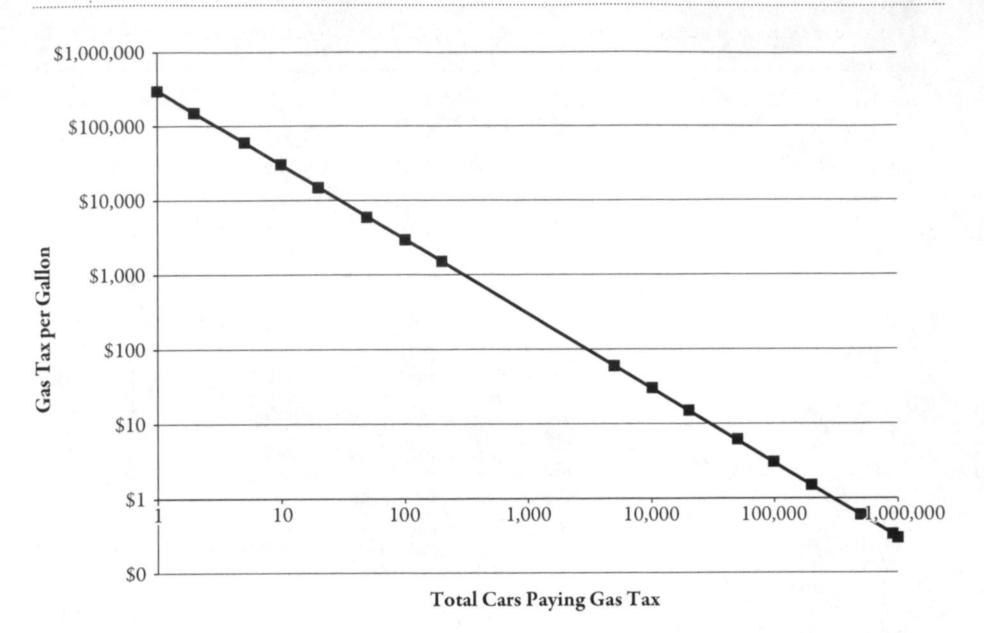

FIGURE 27.14 Continued Reliance on the Gas Tax for Funding with Shrinking Share of Petroleum Usage

gallon (European levels), but the last car has to pay $300,000 per gallon. The move away from the gas tax is a positive feedback system that will accelerate. A replacement will be required. We have discussed elsewhere some possibilities, with direct road pricing the most likely.

To be clear, there is no obvious reason to move from the gas tax before it is necessary; the tax is administratively very efficient, and accomplished the basic policy ends of raising funds from drivers roughly in proportion to use while discouraging gasoline consumption. But it will at some point become necessary if alternative fuels are adopted.

27.5.4 SCRAPPAGE SCHEMES

The UK Scrappage Scheme and US Cash for Clunkers were policies adopted during the Great Recession of 2008–2009 to help stimulate auto sales while improving average air quality by getting super-emitters off the road. The policy gave car purchasers money toward purchase of a new car when they turned in an older car with a worse fuel economy.

It has been estimated that the worst 10 percent of vehicles are responsible for 60 percent of emissions.[576] (This is similar to the Pareto Principle or 80-20 rule, which says that 80 percent of effects result from 20 percent of the causes). So if those cars could be removed from the road, air quality would improve greatly. The plans gave cash incentives to people trading in old cars for new cars, if the new car had better fuel economy. The idea was first proposed in the United Kingdom, and, as we have mentioned before, ideas are light baggage; it quickly traveled across the pond to the United States, where it was implemented.

The assessments of the programs are at best mixed. The US Cash for Clunkers program did induce new vehicle purchases in July and August of 2009, as intended, but these purchases were offset by fewer purchases in later months.[577] Fuel economy did increase, as the program required, but by less than 1 mile per gallon.[578] The cost-effectiveness of the program from an environmental perspective has been challenged, estimated to cost between $92 and $450 per ton of CO_2 avoided (at best 5 times market value).[579]

27.6 Discussion

Safety, congestion, energy, and environmental problems have been long with us in transportation, but environmental costs are garnering a larger share of attention.

An array of varied activities exist, some of which are leading to results already (or being) implemented: for example, more efficient diesel engines, better truck aerodynamics, and more energy efficient aircraft. Some of the activities are intended to "ready" technologies for use in the future, for example, improved batteries, inertial energy storage devices, use of lighter weight materials in vehicles, and fuel cells.

An optimistic scenario says that over a period of decades we can move into a future that has the properties of the polished present. We say that is optimistic because higher energy costs make otherwise relatively expensive technologies feasible. If these alternative low-mass, low-energy vehicles can be implemented across modes, the energy-for-moving-things-around problem, as well as the air pollution and CO_2 problems, will have been at least partially countered, but there remain other issues that beg improvements in transportation, example, congestion, avoiding decreases in productivity, and enabling settlement and social and economic adjustments responding to population growth and migration and pressures from disparities in economic and social progress, as well as shifts steered by changes in resource conditions and new activities.

Freight Railways worldwide are fundamentally profitable business and passenger railways worldwide are fundamentally unprofitable businesses.

—HENRY POSNER III,

Chairman of the Railroad Development Corporation *Trains*, September 2012, p. 8

28 Higher Speed Rail: Hubs and Spokes

28.1 A Magic Bullet?

AFTER FIVE YEARS of construction, and coinciding with the 1964 Tokyo Olympics, the first Japanese high-speed rail (HSR) system, the Shinkansen (new train) or bullet train opened. While the dates coincided, they were not mere coincidence; both events were aimed to globally promote the image of a modern Japan.[580]

In 1967, a second line, the Sanyo Shinkansen was begun. By 1970, Japan authorized a national Shinkansen network. Construction plans for five additional lines and basic plans for twelve others were approved in 1973, but despite the approvals, the cost of the five lines (five trillion yen, or about fifty billion dollars) combined with the oil shock and recessions delayed the lines until 1989. High petroleum prices, which increase the relative demand for non-petroleum-based transportation such as high-speed rail, also increased the construction cost for HSR and reduced available revenue, thus delaying construction.

Many of the new Japanese lines combine narrow gauge and wider gauge lines on the same structures, allowing both conventional and advanced technologies to use them. While the hybrid technology limits the speed of the bullet train on these routes, it permits later upgrades. As with the birth of the Shinkansen some thirty four years earlier, the 1998 Winter Olympics in Nagano, Japan, were a target for the opening of a rail line extension.

Within Japan, high-speed rail has confronted the break-up and privatization of the rail system, begun in 1987. The restructuring aims to achieve more efficient methods to ensure profitability in the passenger rail sector.

The Japanese have continuously improved all aspects of their system over the years.[581] Major improvements since the first opening in 1964 include the introduction of computerized crew training systems, double-decker cars to expand capacity, the use of regenerative

FIGURE 28.1 Japanese Bullet Train (Shinkansen)

brakes to conserve energy, lowering the weight and increasing strength of cars, the use of electronics in mechanical system management, mechanization of track maintenance, the introduction of tilt trains, the incorporation of aerodynamic considerations in train design.[582] As a result of these improvements, travel time from Tokyo to Shin Osaka has dropped by 90 minutes from 4 hours in 1964, to 2 hours and 30 minutes, achieving a speed of 220 km/hr (138 miles/hr). The current bullet train is shown in Figure 28.1.

This chapter considers the reinvention of passenger railroads. In the United States, the passenger railroad is thought to be dying; yet significant investment elsewhere in the world seeks to fight the trend. Strategies include redesign, new rights-of-way, and restructuring by privatization.

28.2 Reinvention by Redesign

High-speed rail is touted as a way to reinvigorate rail demand by reinventing the technology. There are two main technologies for high-speed rail: maglev and enhanced conventional.

Maglev, or magnetic levitation, has been operational on a 30 km (19 mile) section in Shanghai, China, since the start of 2004, reaching a peak speed over 400 km/hr (250 miles/hr) (the fastest system on a test track has been recorded in Japan at 581 km/h [361 miles/hr]). There are also small systems in place in Aichi, Japan, and Daejeon, South Korea, associated with fairs. Maglev differs from railroads as we know them, sharing the concept of cars being trained together and a track, but using completely different technologies for propulsion. However, the other aspects of system building, from land acquisition to managing stations and scheduling services, will inevitably copy logic from railroads. China had planned to expand the Shanghai system developed by the German organization Transrapid to Hangzhou, but its cost, and fears about radiation, constrained those plans, and it is now suspended and unlikely to be built, as a conventional high-speed line has been built in the corridor.

The Transrapid organization is seeking to fill three market niches with Maglev technology: airport to downtown, regional or commuter rail, and long-distance inter-city rail. However, with only one real system in operation, its future is unclear. It seems likely to be a technological success. But just because the engineers can do something, does not mean society should do it. If only because of its novelty, the economic analysis of this technology has yet to be resolved. However, as time progresses, if costs can be brought down due to scale economies of various sorts, maglev holds some promise.

In contrast to maglev, rail has been reinvented in places through enhanced conventional service. These lines include technological improvements in the train as well as more conducive right-of-way conditions (dedicated, straighter right-of-way but potentially steeper grades, since the rails don't serve heavier freight trains). The operating high-speed rail lines have, to date, followed this less technologically risky (but perhaps less rewarding) strategy. Serving high volume, nearby, inter-city markets, examples of this enhanced conventional service include the Shinkansen between Tokyo and Osaka in Japan and the Train å Grande Vitesse (TGV) between Paris and Lyon in France, both of which have been further expanded, in the French case to a trans-European network. The French TGV achieves an operating speed of 300 km/hr, with maximum test speeds of 515 km/hr in 1990. The Chinese HSR system has been rapidly constructed, but beset with corruption and other problems, leading to a fatal crash in 2011.[583]

28.3 Prelude to Reinvention

During most of the twentieth century, US (and other) railroads lost passenger traffic to autos, buses, and air service. Loss to the highway modes began in the 1930s, especially in short-distance markets. In the 1950s, those losses continued, joined by losses to air service in longer distance markets.

Most of the rail properties did not manage these losses gracefully. Passenger traffic had long been a source of institutional pride, as well as a source of revenue. Contrary to conventional wisdom, management worked hard to stem the tide of losses. It purchased new equipment and created improved services. The famous long-distance trains of the 1950s were the result, such as those operated by the Santa Fe. There were also efforts in shorter distance markets, with the Rock Island's Rockets providing an example.

Why did we say the losses were not managed gracefully? Management, in our judgment, did not recognize the depth of forces driving the loss of markets. It invested at a time when graceful disinvestment (rationalization) should have been undertaken. Management did not examine possible residual market niches where efforts to create and preserve markets might have been successful.

The Pennsylvania Railroad was an exception to the latter statement. Long a leader in the New York–Washington corridor, it saw service hard pressed in that corridor by auto, bus, and air services. Distances were such, however, that high-speed rail might be quite viable in the market niche. Seeing that, the Pennsylvania Railroad explored options for advanced technology, maglev services that could have been quite successful had not the federal government co-opted high-speed train programs.

Despite the decline of passenger rail service for over a half century, intercity high-speed ground transportation has seen renewed interest in the United States. America's National Passenger Rail Corporation, Amtrak, was formed in 1971 by assuming the passenger services of the commercial railroads, which were left to focus on freight. Despite a charter to be profitable, Amtrak has never seen black ink system-wide, and really only has hopes in the Northeast corridor (Boston–New York–Philadelphia–Baltimore–Washington). The organization is perpetually in danger of being shut down or radically scaled back; however, it provides service in many Congressional districts, giving it political cover.

In the Northeast corridor, Amtrak deployed the Acela service in 2001, which travels at a peak of 240 km/hour (150 miles/hr). However, the actual scheduled time is 2 hours and 44 minutes from Penn Station in Manhattan to Union Station in Washington. Thus the effective speed for this 368 km (230 mile) trip, including stops, acceleration, and deceleration, is 134 km/hr (83 miles/hr)—somewhat faster than an automobile, more so in congested periods. Figure 28.2 gives some information about rail passenger use in the United States; a rough comparison with population suggests at best 0.092 trips per capita per year nationally. Overall, Amtrak ridership is up 37.3 percent for the first decade of the 2000s, rising from 20.9 million in 2000 to 28.7 million in 2010, compared with a 9.7 percent population growth nationally. Ridership rose in most cities; however New York City declined, some of which might be due to the spike in 2001 after aviation was shut down, and some of which may be due to switching between Amtrak and commuter rail modes. New York remains Amtrak's best market, with about one trip per capita per year, or more than ten times the national average (notably, many of those trips are made not by New York residents, but by visitors to New York).

28.4 Conception of Reinvention

When discussing the life cycle of a technology, we noted three phases: birth, growth, and maturity. At any time, renewal or reinvention is a possibility, but there is a long mating dance before pregnancy. In the case of US HSR, this mating dance has gone on for five decades.

The deterioration of passenger train service in the Northeast corridor (Washington-Boston) and air traffic congestion became relevant in Congress circa 1963 when Senator Claiborne Pell of Rhode Island introduced a bill asking that Shinkansen-type trains be run in the corridor.[584] The bill was resisted by the administration in the absence of cost and market studies. However, after a good amount of pushing and pulling, the High-speed Ground Transportation (HSGT) Act of 1965 emerged. It provided funds to begin upgrading the corridor (there were rather different problems in the Washington–New York and New York–Boston segments) and for government research and development.

The status of work was reviewed annually in the reports on high-speed ground transportation.[585] The competitiveness of high-speed ground transportation was reviewed

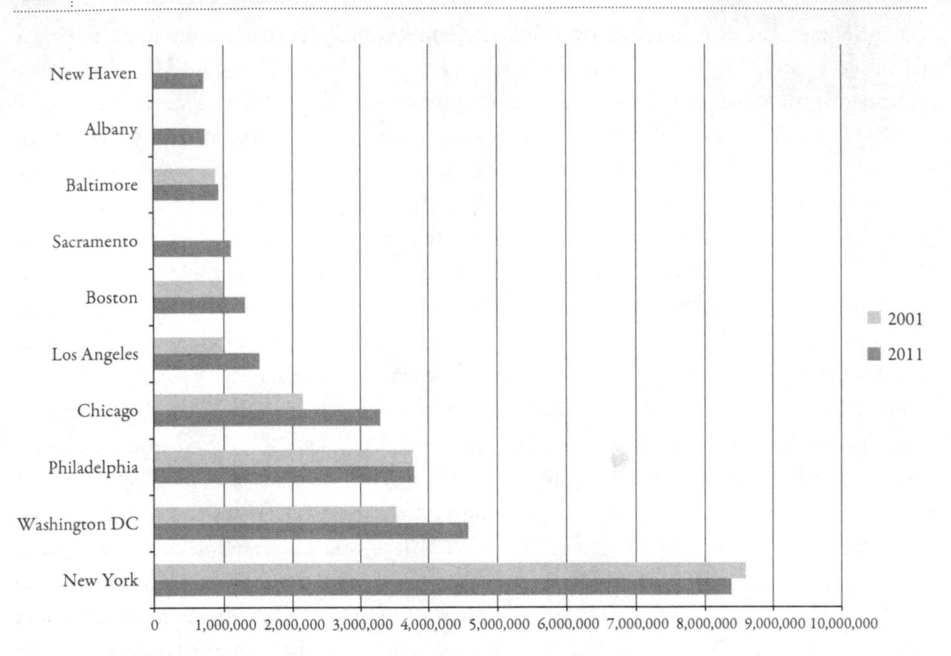

FIGURE 28.2 Amtrak Boardings and Alightings at 10 Busiest Stations
Source: Bureau of Transportation Statistics National Transportation Statistics

by a USFRA and Amtrak study of upgraded service in city-pairs and the first chapter of that report provides a useful summary.[586] NASA supported a series of studies by McDonnell-Douglas, Lockheed, and Stanford University in low, medium, and high-density markets.

The references above provide information through the early 1970s. There followed a period of several years when interest in corridors seemed to wane. The High-speed Ground Transportation Act programs were defunded—the Act remained for a time, but no money was provided. Amtrak continued its corridor upgrading. The USDOT phased out the work it was doing on maglev and other high-speed technologies. NASA studies of short haul air transport had not yielded technologies with great promise, and their program was reduced, although the tilt-rotor aircraft has received attention recently.

Agency wisdom at the federal level was influenced by the National Research Council, which in 1976 recommended that conventional speed rail systems be monitored, with the prospect of deploying high-speed (250 km/hr) systems in promising markets in the mid-term.[587] If markets evolve, then consider very high-speed systems (480 km/hr) in the long term. The Committee report just mentioned reviewed a number of corridors where there was interest in service, but the report was not addressed to nor known to those promoting those corridors.

Although agency funding was reduced to zero, some interest continued in Congress, mainly because Congress was approached for funds to support local interest in systems. The result was a report by the Congressional Office of Technology Assessment,[588] which emphasized that there was no clear-cut case for maglev or other varieties of high-speed service, and that data should be obtained and carefully evaluated on a case-by-case basis.

Even though the OTA report was not optimistic, federal interest was renewed in 1989 when news of high temperature superconductors came along, and Congress asked for an analysis of maglev transportation systems. (Actually, and in spite of what was asserted, such superconductors would have only a modest impact on maglev costs). The result was a USFRA report calling for "further analysis."[589] Other reports have been prepared with groups working directly with Congress. Both the Corps of Engineers and the US Department of Energy[590] joined the National Maglev Initiative. As roles evolved, it appeared that Argonne wanted to be a leader of technical work, with USDOT evaluating and funding local efforts. The Corps of Engineers was to be left without a role. The 1991 Intermodal Surface Transportation Efficiency Act (ISTEA) funded maglev research and development, but funding was sharply reduced in 1994 and disappeared in 1995. The reason given was: "it's too expensive."

Yet, like a ghoul from a horror movie, it would not die, and high-speed rail reappeared in later funding bills (TEA-21). Building on deployments in France and Japan, as well as the Acela line in the Northeast, preliminary planning was undertaken to bring high-speed rail transportation to the rest of the Unied States. After the election of President Obama in 2008, $8 billion was authorized for HSR construction in the American Recovery and Reinvestment Act of 2009 (the "Stimulus" bill). A few projects have seen construction authorized; most are modest improvements on existing Amtrak lines, and some money has been dedicated to the California HSR. Since then, the subsequent Congress zeroed out new appropriations for HSR, and three state governors canceled authorized projects in Wisconsin, Ohio, and Florida.

28.5 Hubs and Spokes

The network architecture of high-speed rail lines has tended to be in a hub-and-spoke pattern, connecting a hub city (e.g., Paris, Madrid, Tokyo) to secondary cities in tree-like architecture, with occasional crossing links, typically at both lower speed, lower frequency, and lower cost of construction. As these systems were designed nationally, and the largest city is often the capital (as in Paris, Madrid, and Tokyo), which is also (roughly) centrally located, it is no surprise that the hub was based where it was. Germany has fewer very high speed links (faster than 300 km/h), and a flatter (less-hubbed) network, perhaps reflecting its strong federalism, relative decentralization into a multi-polar urban structure and late formation into a nation-state. Italy has centered its hub in Milan, the largest metropolitan area in the country.

The reason for the hub-and-spoke architecture is to achieve economies of density in track usage and network effects at the hub city, which enable frequent service to multiple destinations. Multiple paths between origins and destinations would diffuse the network effects and result in less frequent service, and therefore reduce demand. The hub-and-spoke architecture, while benefiting the network as a whole when demand is insufficient to enable frequent point-to-point service, clearly benefit the hub cities the most, as they gain from all the incoming flows, which create additional demand, and thus greater service. In air transportation, airlines often use hub-and-spoke networks, and if they have a large market share at a hub airport, will use that advantage to charge a premium for travel, thereby capturing some, if not all, of the benefits of being located in a hub airport city.

As used here, a *hub* is a center of activity, from which multiple *spokes* (links connecting the hub with other locations) emanate. On a network with a tree structure, the primary hub is the point from which the maximum number of spokes emerge. There may be secondary and tertiary hubs on the network as well.

The proposed US system, such as it is, has no well-thought out national architecture. There were a number of independent proposals that have been drawn on a single map.

- The existing Northeast corridor, the only US claim to HSR, is part of the national "plan," though it received the least funding from the American Recovery and Reinvestment Act of 2009. The Northeast has the most developed network with semi-high-speed rail (Acela) running from Boston through New York to Philadelphia, Baltimore, and Washington. This could be described as a *New York Hub* (though it has not been pitched as such), with current non-high-speed lines from New York emanating in particular to Albany and then to Rochester and Buffalo or to Montreal, and spurs from New Haven to Burlington, from Philadelphia to Harrisburg and Pittsburgh, from Washington south to Richmond and Raleigh, and from Boston to Portland and Brunswick, Maine, all of which have been proposed for upgrade to high speed.

  Amtrak has separately proposed to upgrade the Northeast Corridor, with a $140 billion proposal, which would include major tunnels under New York, Philadelphia, and Baltimore. One version of increased access into New York City, a proposed commuter rail tunnel under the Hudson River called "Access to the Region's Core" or (ARC), was canceled by New Jersey governor Chris Christie in 2010.
- The proposed California Corridor, is based on a mainline that runs from San Francisco, through California's Central Valley to Los Angeles, with extensions to Sacramento and San Diego. The long-term vision of the national program has a line from Las Vegas to Los Angeles. With all of the commuter rail already in the Los Angeles region, the network could more accurately be described as the *Los Angeles Hub*. Even the Sacramento line is more oriented to Los Angeles than San Francisco, despite the distance. This Corridor was given bonding authority by California voters under 2008 Proposition 1A, which required matching funds from the federal government and private sector. It is still unclear whether construction will start, much less be completed, as environmental reviews and lack of Congressional support threaten the project. California Governor Jerry Brown is strongly behind the project.
- The most coherent of the new proposals is the *Chicago Hub*, which as its name suggests, hubs traffic from other Midwestern cities into Chicago. This proposal achieved agreement from all of the regional governors, and with a Chicago-based administration in the White House, not surprisingly received a large share of the recent federal allocations ($2.6 Billion). The Milwaukee to Minneapolis segment was notably canceled by Wisconsin governor Scott Walker. A related system centered on

Columbus connecting Cleveland and Cincinnati was canceled by Ohio governor John Kasich.

- The proposed Florida high-speed rail system runs from Miami though West Broward, West Palm Beach, to Orlando, Lakeland, terminating in Tampa with about 10 stations planned. Proposed additional extensions connecting Fort Myers, Jacksonville, and Tallahassee and Pensacola have also been drawn on maps, but these are further into the future. This could be described as an *Orlando Hub*. Though Miami is a larger metropolitan area than Orlando, the branching structure is naturally geographically based in Orlando due to its centrality on the Florida peninsula, as well as its central location vis-a-vis tourist traffic. Tourist traffic is important to this line, as stops at Disney and Port Canaveral have been included. It is anticipated the line will carry 2 million travelers yearly (5,500 per day on 12–18 round trips), and is 324 miles in length in total. With 10 stations, there is an average of 32 miles between stations, which will bear nuisance costs, and 10 station areas, which will see accessibility benefits. The line is anticipated to run along the I-4 and I-95 corridors for significant stretches, so those areas already see some accessibility benefits (at on-ramps and off-ramps) and nuisance costs (between interchanges). The Florida Corridor was killed by Governor Jeb Bush and reinstated by voters, only to be killed again by Governor Rick Scott. Some proposals for private passenger rail service in Florida have emerged.
- The Northwest region, or *Seattle Hub*, connects Vancouver, Canada, with Salem, Oregon.
- The South-Central region, once dubbed the Texas Triangle, and now the Texas T-Bone, may be described as a *Dallas Hub*, connecting San Antonio, Austin, Houston, New Orleans, Oklahoma City, Little Rock, and Memphis, among others. Similar proposals in the 1990s were defeated due to lobbying by Texas-based Southwest Airlines. Proposals are floating for a private passenger rail line from Dallas to Houston.
- The Southeast region is probably best described as an *Atlanta Hub*, as Atlanta is the key interchange in the region (hubbing traffic from Savannah, Jacksonville, Birmingham, Chattanooga, Nashville, Charlotte, and Raleigh), and the largest metropolitan area. There is also a line from Raleigh through Columbia to Savannah, bypassing Atlanta, which is helpful for long-distance train travelers from the Northeast going to Florida, but might not have much local demand.
- The Gulf Coast Corridor, or *New Orleans Hub*, connects Houston to Mobile and Atlanta. This is an official FRA corridor, but seems on a slower track than many of the others, not receiving funding in the most recent round.
- The long-term program includes a line from Phoenix to Tucson (a *Phoenix Hub*), and from Denver to El Paso (a *Denver Hub*), but these are both isolated corridors, indicated on the long-term vision, without any likelihood for construction in the short term. Describing these as hubs stretches the meaning of the term, but those are the primary cities on the respective networks, and are the only cities on their networks with significant feeder public transit. While these local spokes do not show on the national high-speed rail network, they still exist, and support the use of the term for these locales.

Several cities tie together multiple hub networks, these include New Orleans (connecting the Dallas and Atlanta networks as the hub of the Gulf Coast Corridor), Raleigh (connecting the Atlanta and New York networks), Louisville (connecting the Chicago and Atlanta networks), and Kansas City (connecting the Dallas and Chicago networks). Those with an eye to drawing networks would easily conceive of links (not yet on the books) connecting Memphis, Nashville, and Knoxville in Tennessee, or Pittsburgh and Cleveland or Columbus. The unofficial advocacy group, US High-Speed Rail Association, has the most comprehensive network plans, including staging, which includes many of these and other links.

These hub networks in the Federal High-Speed Intercity Passenger Rail Program include the top 47 metropolitan areas of the United States (and many smaller ones); the largest city not on the network is Salt Lake City, Utah, at 48, with just over 1 million people in the metro area.[591] There is, however, a private proposal for lines serving the Mountain West, including Salt Lake. These are proposed as maglev, noting the difficulty of high-speed conventional rail through the Rocky Mountains. The hubs themselves are metro areas ranked 1, 2, 3, 4, 9, 15, and 27.[592] The US High-Speed Rail Association network includes even more cities.

The political genius of the proposed inter-city passenger proposal is that it includes lines in all but eight of the fifty states.[593] This is a practice learned in transportation from previous national packages, the Interstate Highway System (with miles in all 50 states, including special routes in Alaska, Hawaii, and Puerto Rico) and Amtrak (nearly so), helping ensure strong support in Congress, particularly the Senate.

Network architecture matters a lot, not just in accessibility, but in user travel time. Hatoko and Nakagawa (2007) compare the Swiss railway network and the Japan network, and conclude the mesh-like network with precision timing architecture in Switzerland better serves its population than the hub and spoke mainline system in Japan.

28.6 Economic Effects of High-speed Rail Systems

> The spatial impacts of the new lines will be complex. They will favour the large central cities they connect, especially their urban cores, and this may threaten the position of more peripheral cities.
>
> —(Hall 2009)

> [T]he wider economic benefits of high-speed rail are difficult to detect, as they are swamped by external factors," but are likely to be larger in more central locations than more peripheral locations.
>
> —(Preston and Wall 2008)

Examination of local land uses around international high-speed rail stations suggests that were it not for commuter traffic, the effects on land use will not necessarily be localized near the station, the way they would with a public transit station. Downtown stations, if they were to see land use benefits, should see higher local densities, higher local rents, and the construction of air rights over the station and local yards.

Eurostar is a heavily used high-speed rail line connecting London and Paris, serving 9.2 million passengers per year. Gare du Nord in Paris, which serves Eurostar, has local land uses largely indistinguishable from other areas of Paris; St. Pancras in London similarly. Ebbsfleet International Rail Station and Ashford International Rail Station are surrounded by surface parking lots. The results in Taiwan are similarly insignificant (Andersson et al. 2010).

Tokaido Shinkansen, connecting Tokyo and Osaka and serving 151 million passengers annually, is an order of magnitude more successful. The densities around stations on this line are visibly higher, but still air rights are partially, but not fully developed, indicating limits to how valuable the land is, even in Tokyo. Shin-Osaka station is adjacent to surface parking lots.

The development effects are not local (unlike public transit stations), which is not surprising since if they are serving long-distance travel they are also serving less frequent travel, and as a consequence the advantages of being local to the station are weaker. Where they share space with local transit system hubs, the effects would be difficult to disentangle.

There is no grounded empirical work to date on the economic development impacts of high-speed rail in the United States, since such services do not exist. Little has been written from objective (as opposed to vested) sources; the effects elsewhere in the world are minimal or nonexistent.[594]

28.7 Nuisance Effects of Proposed High-speed Rail Lines

High-speed rail, while providing potential benefits at the nodes, guarantees costs along the lines. Evidence from hedonic price studies show that each additional decibel of noise reduces home value by 0.62 percent.[595] Using the methodology in Levinson et al. (1997), the noise per train and the number of trains per hour determine a noise exposure forecast. Applying the noise exposure forecast to the number of houses affected by each level of noise, and summing over all of the houses, and multiplying by the value of each house, gives the economic noise damages associated with the trains. So, for instance, for a project running 20 trains per hour at 241 km/h through an area with 1,000 housing units per square kilometer, each with a value of $250,000, would produce a total noise damage cost per kilometer of track of $1.975 million, a not insignificant cost. For a line of 500 km, this would be a system noise cost of nearly $1 billion. These relationships are nonlinear; even one train per hour would produce a total cost of $269 million. Running 20 trains at an average speed of 350 km/h would produce a cost of $1.5 billion.

The noise damages can be avoided if preventive measures are adopted. These include acquiring a much wider right-of-way so there is no housing near the tracks, or noise walls. Whether those costs are less expensive than accepting damages depends on the circumstances.

28.8 Reinvention by Restructuring

An alternative route for reinvention of passenger rail is privatization. A decision taken in 1987 divided the Japanese National Railways into six regional passenger companies and a

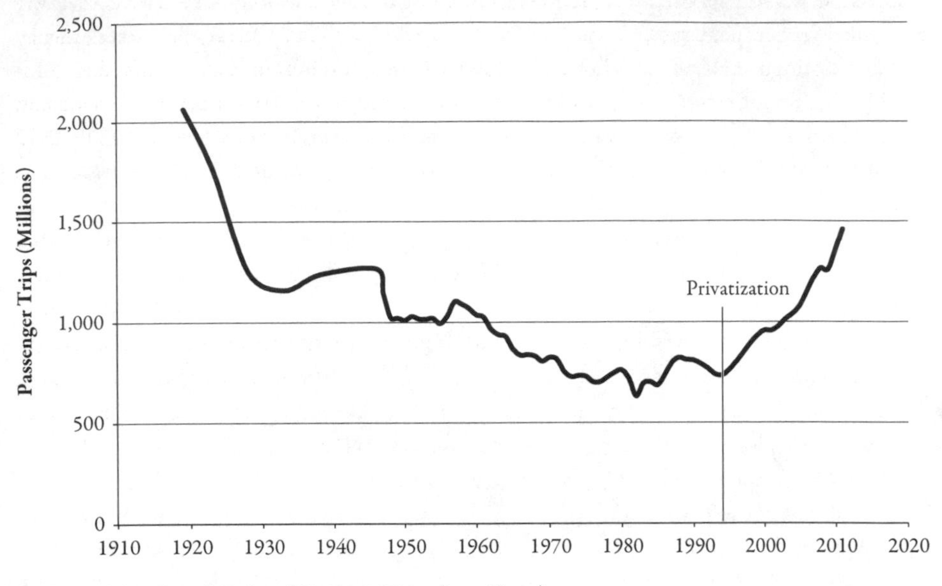

FIGURE 28.3 British Railroad Passenger Trips (in millions)

freight company. The British separated the ownership of the track from the trains. Private firms, including well-known brands such as Virgin, operated trains on specific routes, with scheduling and pricing freedom. Privatization has resulted in an increase in passenger trips (see Figure 28.3).

Mismanagement of the RailTrack organization, which operated the track for the carriers, caused an increase in the number of crashes. While the idea of having a separate track from carrier works in other sectors (roads vs. trucks, airports vs. airlines), trains and tracks are more integrated technologically (tracks steer trains, roads don't steer trucks), so the management tasks are more intricate if they are separated.[596] This should not cause the concept to be dismissed immediately, but should suggest caution.

Following on the experience (but not the success) of RailTrack, the central government in the United Kingdom required that Transport for London, the organization that runs the subway (Underground) system in London, reorganize the Tube using Public Private Partnerships, wherein the infrastructure would be maintained by private companies, though ownership would stay public. Christian Wolmar, author of the excellent history *Down the Tube*, says:

> One could hardly write the script as fiction. On the very day that Gordon Brown is teetering on the edge of oblivion and the House of Lords, one of his cherished projects, the London Underground PPP is breathing its last. The news that Transport for London is going to be taking over Tube Lines and running the contracts to maintain the Tube leaked out on the very day that voters were going to the polls. Since Metronet has already gone to join Railtrack, various franchises and the Strategic Rail Authority in the big dustbin of failed organisations, the demise of Tube Lines effectively means that the PPP joins this infamous group.[597]

Recent proposals for rail in England suggest going back to the "Big Four" British railway companies that resulted from the 1921 Railways Act, each railway (Great Western Railway [GWR], London, Midland and Scottish Railway [LMS], London and North Eastern Railway [LNER], Southern Railway [SR]) would be vertically integrated (operating trains and tracks) in a territory. These were consolidated into British Railways in 1948. In any case, despite the problems with vertical separation, passenger trips post-privatization have risen notably.

Because of its success with rail privatization, and rising rail demand, the United Kingdom is giving serious consideration to what is being called High Speed 2 (HS2), the country's second high-speed line after HS1, connecting a newly remodeled St. Pancras Station in London to the Channel Tunnel, which would functionally replace the West Coast Main Line. Major issues include where to connect to London, how direct to connect to Heathrow Airport, which cities in England are on the line, as opposed to on spurs, and where does it go in Scotland.

28.9 Discussion

This chapter reviewed the state of passenger rail planning and operations globally, with special focus on the United States.

Unlike traditional railroads, whose emergence was at least in part due to a forceful entrepreneur, example George Stephenson, high-speed ground transportation has been a product of central planning in Japan, France, and the United States, Operationally, the systems are largely adapted from conventional rail systems, with similar labor organization and ownership in Japan and France and similar architectures in many other respects.

Despite a great deal of legislative effort and lobbying by the large engineering interests, it is doubtful that without considerable subsidy high-speed rail could be constructed, much less profitable in the United States. In contrast to assertions of "operating" profitability (which conveniently ignores the very high capital costs), HSR has in all cases required government subsidy. It is clear a free market would never develop HSR. That leaves open the question of whether government should.

The conditions in Europe and Japan during the conception and birthing stages differ significantly from most parts of the United States. Land uses are denser and cities are closer together

A key distinction is that the regulated transportation sectors in Japan and Europe prevented competition from air travel to the same degree as in the United States when the HSR lines were planned and deployed. Thus the market for high-speed rail probably appeared more promising than in a deregulated environment. Had air travel been deregulated and privatized at the time, the decision to proceed with high-speed rail, particularly in Europe, may have been different. A ride on many of Europe's high-speed inter-city lines finds nearly empty cars; this is due to recent deregulation of air travel. As an illustration of this, Southwest Airlines was a major opponent of high-speed rail in Texas.[598] Another distinction is the willingness of European governments to engage in subsidies and tolerate what Gerondeau terms "a catastrophic level of debt."[599]

Real constraints on the growth of the highway and air travel systems exist. Widely cited are congestion, or capacity limits. Airports have limited capacity to serve aircraft in peak times, as do highways. In a priced system, this would result in higher user charges, but in an unpriced system, there are simply queues formed. High-speed rail, which has potentially very high capacity on its fixed corridors, offers the promise of relieving congestion on the other systems. In Europe and Japan, with important, though declining, conventional rail services, its extension and adaptation to a higher speed technology was a more obvious choice than in America.

In the United States, congestion/capacity problems are less severe than in Japan and Europe. Moreover, conventional passenger rail has long been a less important mode than the other two. Further, the high-volume, short-distance markets for which rail is best suited are less common in the United States. For these reasons, high-speed rail remains in a birthing or pre-birthing stage in the United States. Other cited complaints against the air and highway modes are their externalities: pollution, noise, accidents, and so forth. While it cannot be argued that either air or highway modes have internalized their externalities, it also cannot be argued that from a systems perspective high-speed rail does not create problems of its own.[600] Vibration is one issue.[601] Access to rail generally requires vehicle trips. These vehicle trips generate pollution of their own, for instance the most severe pollution comes from the so-called "cold start," or running a car before it is warmed up. It is difficult to establish how much of the potential demand on high-speed rail is diverted from other modes and how much is induced travel. While induced travel may expand the economy, it certainly does not mitigate externalities.

In short, while the conditions were favorable for the development of HSR in Europe and Japan, they are clearly less so in the United States.

The US plans generally call for a set of barely interconnected hub-and-spoke networks. There is sometimes a danger of a planner falling in love with his map. There is no danger here; even the same agencies have random maps. It seems as no one cares where the lines actually go, so long as they are high-speed rail.

The issue of *opportunity costs* is seldom mentioned. The United States carries a greater share of freight by rail than Europe. Converting rights-of-way into passenger only (which is required for HSR) may cost some of that freight share. Any money spent on HSR cannot be spent on something else.

The evidence from US transit systems shows that lines have two major impacts. There are positive accessibility benefits near stations, but there are negative nuisance effects along the lines themselves. High-speed lines are unlikely to have local accessibility benefits separate from connecting local transit lines because there is little advantage for most people or businesses to locate near a line used infrequently (unlike public transit). However, they may have more widespread metropolitan level effects (cities on the HSR network, and especially at the hub of an HSR network, may prosper at the expense of those off the network). They will retain, and perhaps worse, have much higher, nuisance effects. A previous study of the full costs of high-speed rail in California[602] showed that the noise and vibration costs along the line would be quite significant.

If high-speed rail lines can create larger effective regions, that might affect the distribution of who wins and loses from such infrastructure. The magnitude of agglomeration economies is uncertain (and certainly location-specific), but presents the best case that can be made in favor of HSR in the United States.

That said, remember that real HSR (not the short-term improvements to get to 90 or 110 mph, which may or may not be a good thing, but are certainly not HSR) is a long-term deployment, so it needs to be compared with cars and airplanes ten or twenty or thirty years hence. Cars are getting better from both an environmental perspective and from the perspective of automation technologies. Self-driving vehicles (described in Chapter 30) should be able to attain relatively high speeds (though certainly not HSR speeds in mixed traffic). Further, they may move less material per passenger than HSR (trains are heavy), and so may net less environmental impact if electrically powered. Aviation is improving as well, both in terms of its environmental impacts and its efficiency. Socially constructed problems like aviation security or congestion can be solved for far less money than is required for any one high-speed rail line.

While its technological advantages over conventional rail are obvious, as with all rail modes, there is a significant amount of inflexibility associated with the system design. The high-speed networks are limited, and the rails require specialized vehicles. HSR lacks the point-to-point convenience of the auto and the speed of the airplane on long trips. Compared with the greater flexibility afforded the untracked air travel system or the ubiquitous highway system, high-speed rail faces serious difficulties. As noted earlier, in mature systems, the benefits of new infrastructure in an already well-served area are elusive.

It will be interesting to observe the progression of the Japanese and European high-speed rail systems from growth to maturity, and to compare this with the earlier history of conventional rail. A notable chapter in the story is opening in Japan, where there is stagnation in the number of passengers served and deregulation is concentrating services on the more profitable routes.[603] Whether high-speed rail is a new story, or simply the final chapter to the history of conventional passenger rail, waits to be seen.

The Internet treats censorship as a malfunction and routes around it.

—JOHN PERRY BARLOW

29 Internet

29.1 Routing Around

SEVERAL SIGNIFICANT TELECOMMUNICATIONS networks have been deployed since the telephone, including broadcast and cable television (which remains largely one-way, though this may change with the advent of broadband cable modem service and video-on-demand), cellular telephone (including several distinct technology waves), and the Internet. All three have been very important in shaping activity patterns; the latter two have some important parallels with transportation as well. The Internet may be most interesting because of its potential.

What most distinguishes the Internet from the plain old telephone service (POTS) provided by AT&T and the regional Bell operating companies is the use of packets for communication rather than keeping an entire circuit open for two-way communications. The analogy that the Internet is the car/highway system to the POTS train has been made before and has some merit. Like the car, Internet packets are small and discrete, have a variety of origins and destinations, and share space on the network. Like a train, (especially when we only have a single track), telephone requires a circuit between the origin and destination to be kept open—no other traffic can use that circuit (no other trains are allowed on that track). The analogy of course can be stretched too far, and there important differences. In cars, the intelligence (one hopes) lies with the driver, while with the Internet, packets are given direction by routers (as if the traffic signal told you which direction to go, not simply whether to go or not). Another analogy between the Internet and containerization might be more appropriate, as packets are more like freight than people. If a packet (or freight container) between Minneapolis and Tokyo must go by way of London, so be it—no one will complain if it is the low-cost route.[604] A passenger making that trip would find it terribly inefficient.

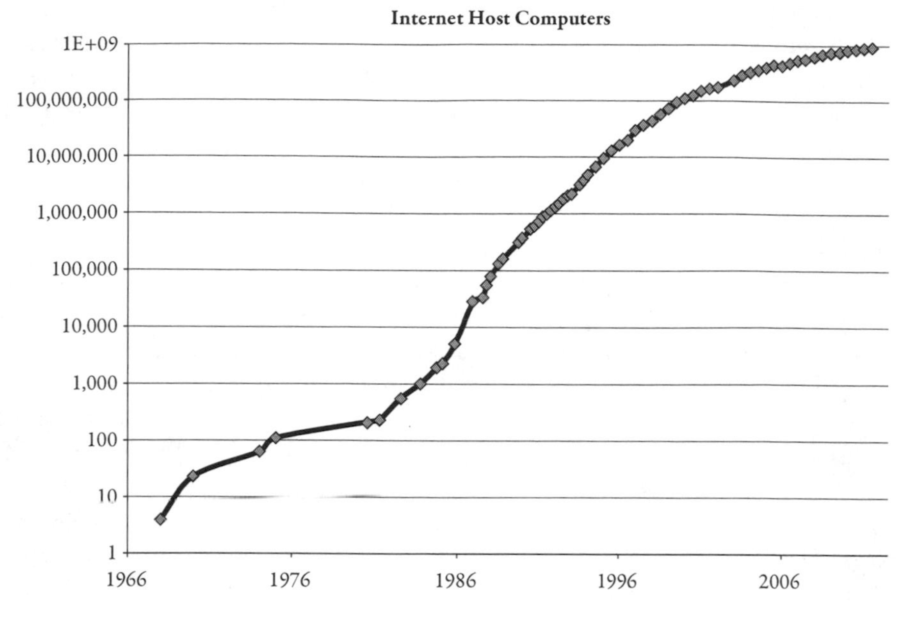

FIGURE 29.1 Internet Host Computers, 1969–2012

Another key facet of the Internet architecture is that it is distributed; disable a link and the system can route around it. This is similar to a robust highway network, and distinct from a hub-and-spoke network with a few critical points.

The Internet was developed in the 1960s with funding from the Department of Defense's Advanced Research Projects Administration, and was originally called ARPAnet—which was deployed in 1969.[605] Figure 29.1 shows the growth of the Internet from 1969 to 2002, using the number of computer hosts connected to the Internet as the metric of size. Other measures are number of users, amount of traffic, and so on, but those are less reliable than this infrastructure-based measure. They show the same basic trends. The S-curve phenomenon is (we expect) at play here as well. In the first edition of this book, we wrote: "we just have yet to see the slow down in new hosts." More recent evidence finds that slowing, at least a slowing in the rate of growth. Still, the maximum number of hosts is unknown. One can imagine, of course, one host for every personal computer (which would presumably max out at one (or two or three) per person), but then one for every cell phone, every car, and every appliance are not unreasonable speculations. Whether every electric outlet gets one may seem far-fetched, but the idea of a world where every electronic device has an address is plausible. That would give a very large number of potential host computers indeed.

The Internet has, despite significant government involvement in its creation, remained largely unregulated. Some attempts at content regulation (the Communications Decency Act in the United States) have fortunately been unimplementable. Other attempts at using the legal system to copyright violations have had more success, the shutdown of the music trading service Napster, for instance. But for every Napster brought down, a more robust network (in this case a peer-to-peer network with distributed indexes) such as Gnutella, or then BitTorrent, takes its place. Censorship remains in some countries—Saudi Arabia and China come to mind—but the ability to get data on the Internet may outstrip the government's ability to censor. Other problems remain, and don't see an easy solution. In 2004, Bill Gates was quoted as saying "two years from now spam will be solved."[606] It has yet to be.

The content regulation on the Internet has analogy in transportation in terms of the transportation of prohibited goods (e.g., drug trafficking), which have been very hard to restrict. While the penalties are severe if caught, getting caught is a low probability event, especially given the low staffing levels of inspection services. Guardians do not have the time (or in general, the legal authority) to open every suitcase, every car trunk, and so on, except in commercial aviation, where scanners are used, at considerable delay to passengers.

The protocols for connection on the Internet and on other communications technologies are clear analogs to protocols and standards in transportation. Trains must be on the same gauge as the track, and one railroad's tracks must be like others if transportation is to occur. Truck widths and weights are another example, and must be compatible with the road. Trailers must have an interface with the truck. More recently, with containerization and the logistics revolution, as well as intelligent transportation systems, information technology is bringing a new layer to transportation.

We discuss the Internet for several reasons. One is comparisons between the history and deployment of communications and transportation. We have discussed these in some earlier chapters (from the post office, through the telegraph and telephone), and think it is important theme to continue. There is also the unanswered question of the relationship between Internet service and transportation. Attention has been paid to telecommuting, and "clicks versus bricks" as shopping moves online, but already we suspect more. For instance, 3-D printing may suggest changes in transportation and production. One can purchase a replicator ($2,800), download an electronic design for a job, and produce a 3D product. It is certainly plausible to imagine a near future in which freight transportation ships more unfinished or raw materials, the Internet sends digital blueprints, and production of many things is local.

29.2 Network Neutrality

A major argument recently has been whether (and how) the Internet should be neutral. An advocacy group asserts "Net Neutrality ensures that all users can access the content or run the applications and devices of their choice. With Net Neutrality, the network's only job is to move data—not choose which data to privilege with higher quality service. Net Neutrality prevents the companies that control the wires from discriminating against content based on its source or ownership."[607]

This argument has been enmeshed in a public debate as providers of Internet connections seek to charge the senders of data for data transmitted over their wires to their customers. Some in favor of neutrality argue the customers have already paid an access charge for Internet service, and should be able to receive bits of data from anywhere connected to the Internet, with no differentiation in the quality of service those bits receive. An example of this argument can be seen in the following quote:

"So imagine if turnpikes charged when you got on the road, and then again when you got off. This is exactly what some of the telcos are trying to do with Internet access. They see that Internet access is a commodity and decreasing revenue from the cash-cow that is circuit based voice service and are looking for new sources of revenue. . . ."[608] Similarly, an author satirically imagines "Transportation Service Providers" charging for sidewalks in front of people's houses.[609] Again in that vein: "Think of the pipes and wires that you use to go online as a sidewalk. The question is whether the sidewalk should get a cut of the value of the conversations that you have as you walk along."[610]

The notion of Internet network neutrality garnered attention after the publication of an article by Wu (2003). Networks, he argues, are not like typical businesses that discriminate, and the consequences would be much more severe. That paper suggested legislation to ensure network neutrality.

This set of arguments is not new; it has played itself out in the development of earlier networks. Evidence about network neutrality, or the lack thereof, in the context of transportation networks follows in the next section. Historically, no other network has been strictly neutral.[611] Rather, "neutrality" that permits differentiated quality of service and different prices has been closer to practice.[612] Opponents of network neutrality, funded by large telecommunications companies (in contrast to the large web content companies on the other side of the debate), argue that the Internet should not be regulated.[613] This logic, too, has arisen in the transport sector, which has swung between unregulated and strictly regulated over the course of time. The issues of common carriage, price discrimination, exclusion and rivalry, and transaction costs are discussed in turn.

29.3 Common Carriage

Government regulation has long been an important element of the provision of transportation service, as government permission has been required to construct networks or network elements (e.g., bridges, locks, ports, turnpikes, canals), and in some cases public subsidy provided (most notably land grants and powers of eminent domain associated with railroads, in other cases the public purchase of shares or bonds of private network providers). While common carriers have been identified in English Common Law since the fourteenth century, regulation has since the late 1800s assumed an especially important role, particularly for the then largely mature railroad industry, and many networks become rate-regulated. In the United States, the Interstate Commerce Commission was established in 1887. A Supreme Court case a few decades later noted:

> The list of rate-regulated occupations is not too long to be here given. It includes canals, waterways, and booms; bridges and ferries; wharves, docks, elevators, and stockyards; telegraph, telephone, electric, gas, and oil lines; turnpikes, railroads, and the various forms of common carriers, including express and cabs. To this should be added the case of the innkeeper (as to which no American case has been found where the constitutional question as to the right to fix his rates has been considered), the confessedly close case of the irrigation ditches for distributing water *(189 U. S. 439)*, and the toll mill acts. This, of course, does not include the case of condemnation for

> governmental purposes or for roads and ways where no question of rates is involved. There may be other instances not found, but it is believed that the foregoing numeration exhausts the list of what has heretofore been treated as a public business justifying the exercise of price-fixing power against persons or corporations.[614]

Bouvier's Law Dictionary provides a long description of the rights and duties, and liabilities of common carriers.[615] Common carriers are paid in advance, and in return must deliver passengers safely to their destinations. They also in general could not charge more than a posted fare. They could of course discount. As any modern traveler knows, when visiting a hotel room and looking at the rates posted on the back of the door, there is usually a large discount between actual prices and maximum prices. There is also a minimum quality of service guarantee (e.g., "not to overload the coach with either passengers or luggage" assures a limit to the congestion on the vehicle) associated with common carriage.

Common carriage is a much less stringent notion than strict definitions of network neutrality. It does not prohibit offering a higher quality of service. Examples date from the earliest dates of common carriers providing first, second, and third class services. Most widely known are passenger railroads and airlines, which, while common carriers, differentiate quality of service for different fares. This is discussed in more detail in the section below. Other differentiators include means of payment (the willingness to take credit and open accounts rather than requiring cash prior to delivery as had been required), which helped United Parcel Service differentiate itself from the US Post Office, and speed of delivery, as when one train or bus service is faster than another (e.g., express vs. local), but both are provided by the same firm under the same common carrier obligations.

The common carriage definition given above applies to carriers, but similar rules may apply to the physical network itself; turnpikes don't discriminate between trucks based on their content (provided their content is not hazardous, is legal, is driven by a licensed driver, meets weight and size restrictions, and is carried on a legal-sized truck that has paid its taxes and fees). However, the turnpike has not historically guaranteed a speed (though this may be changing, as discussed below). Turnpikes also charge different types of trucks different tolls. At one point, California had seventeen different rates for trucks at their toll bridge crossings, a number reduced to six only because the automated vehicle classification system the California Department of Transportation developed to administer electronic tolling could not distinguish many of the categories. It should be noted that railroads did engage in value of service pricing.

Noam (1994) further discusses the history of common carriage and comes to the conclusion that common carriage was unlikely to survive in the telecommunications sector, particularly as telecommunications firms became systems integrators and entered the content arena.

29.4 Price Discrimination

Price discrimination is the idea that suppliers charge more to some customers than others. While suppliers do this to maximize profit, it may also be efficiency enhancing, especially if price discrimination is voluntarily selected by the consumer and coupled with service differentiation. One means for doing this is by differentiating quality. Under common carriage, the same vehicle can be divided into different classes of service (first class, second class). Although the classes would depart and arrive in their town at simultaneously (on a given vehicle), first class passengers alight the vehicle first, and get better service along the way, a service selected consumers clearly value otherwise they would not pay the premium. As noted above, even common carriers differentiate quality of service by providing faster speeds for passengers or freight who pay a premium.

A second means for doing this is to differentiate customers based on their willingness to pay, example when buses give discounts to students (or seniors) to ride the exact same bus at the exact same time. In public transit, this is often pitched with an equity rationale, but cinemas provide discounts to the same groups with no such pretense.

There are numerous variations and combinations of these two techniques. An example is Priceline, where approximately equivalent services have lower prices, in exchange for less certainty in advance about the quality of service being provided (to go from Minneapolis to Chicago, you might change planes in St. Louis, a significantly out-of-the-way transfer point). Similarly HOT lanes (see Section 22.4.6) differentiate roads by quality and willingness to pay.

Another transportation example is the Paris Metro, which historically (until the mid-1980s) had first class and second class cars; the first class cars would charge twice as much, and passengers would likely get a seat, while in second class, passengers might be standing in very crowded cars. The system is self-regulating; if the first class gets too crowded, people stop paying a premium for the better quality of service.[616] If second class gets too crowded, some people switch to first class. Odlyzko suggested an analog for this for the Internet. Band width would be broken into several channels and different prices. Each channel would make a best effort to move its packets, The more channels, the more closely each individual's value of time would be approximated.

An important issue that arises is that users are averse to drowning in a sea of small charges, and many prefer flat rate packages rather than thinking about each decision on a case-by-case basis. This is rational for users, who have limited time and intellectual resources to devote to optimizing small decisions and would prefer to satisfice. User frustration with airlines who, in what is called *yield management*, micro-manage pricing so that two travelers in adjacent and equivalent seats pay hundreds (or thousands) of dollars difference in fares, or with cell-phone company rate plans, is notorious.

Non-price discrimination has an invidious history in transportation, most infamously in parts of the United States the requirement that blacks go to the back of the bus. The challenge to this, the Montgomery Bus Boycott of 1955–1956, initiated by Rosa Parks, ultimately launched the US Civil Rights Movement.

29.5 Enclosure, Exclusion, and Rivalry

By converting some existing roadway capacity to toll lanes, when they were previously *free* (not paid for directly), the amount of free capacity is reduced, and those users are generally worse off (except for those with a very high value of time, who now find the better service outweighs the additional cost). The perception of equity of new toll lanes depends on

TABLE 29.1

Types of Goods

	Excludable	*Non-Excludable*
Rivalrous	Private (e.g., congesting limited access highway)	Congesting (e.g., congested city street)
Non-Rivalrous	Club (e.g., roads within a gated subdivision)	Public (e.g., uncongested city street)

whether those lanes were converted lanes or they are new construction. Tolling additional capacity is not resisted nearly as much as tolling what had been seen as free.

This echoes arguments put forward by telecommunications companies that additional services that provide additional choices should be allowed, at higher fees, that do not harm existing users. The risk, in both transportation and telecommunications, is that these existing services somehow come at the cost of existing services, which might be allowed to degrade in some fashion, thereby encouraging consumers to upgrade service just to stay in place. In some transportation systems, bypasses put paying customers at the front of the queue for bottlenecks, rather than having everyone queue together (first come–first serve). While this appears to be additional capacity, that capacity is on a non-critical section of infrastructure. A notable example of this is in airport security lines in the United States, where at many airports, first class passengers can jump the queue to clear security faster because they paid the airline more for their ticket (even though they did not pay more for security, as the security surcharge is a fixed tax per passenger).

Perception of inequity associated with changing social relationships can be the source of significant grievances; there were many riots in the Turnpike era from the late 1600s to the mid-1800s. Between 1839 and 1844, to cite one example from history, there were a series of what were called *Rebecca Riots* in Wales against the construction of gates and imposition of tolls across roads, that resulted in significant destruction of the toll collection apparatus and buildings, and in one case, murder of a toll collector, by "Rebecca and her daughters," men dressed in women's clothing to hide their identities.[617] In November 2004, there was a riot involving over 1,000 police and significantly more rioters in Jieyang City in Guangdong Province, China, when a woman was assaulted by toll collectors after complaining about the toll imposed on her motorbike.[618]

As shown in Table 29.1, economists define four types of good, based on whether the good is excludable or not, and whether the good is rivalrous or not. Goods that are both excludable (I can charge you for it directly) and rivalrous (my use prevents or interferes with yours) are called *private*, while goods that are neither excludable nor rivalrous are *public*, the classic example of which is national defense. The difficulty in drawing analogies is that roads fall into all four boxes. Private goods need not be privately owned, though typically they are in most sectors of a capitalistic market economy.

Road financing differs for several of these categories. In Europe and parts of the United States, inter-city limited-access highways are paid for with tolls charged to users. In many parts of the United States, local streets are built by developers, and in some cases are maintained by local homeowners associations, who tax themselves for the privilege. City streets are paid for with a combination of local property taxes and state gas taxes, while state roads and "free" interstate highways are paid for by state and federal gas taxes. Homeowners are often responsible for the cost of sidewalks, and are certainly responsible for their maintenance (snow removal in winter) despite their being public property. This harkens back to the way roads were "financed"; the road was simply a right of passage across private property, and was the obligation of the property owner to maintain.[619]

Establishment of property rights in cyberspace, as the enclosure of common grounds and roads in real space, is likely to be a contentious issue. What should be a free commons, what publicly controlled and charged for, and what privately controlled? The rights and expectations of the institutional arrangement in contracts between users and monopolistic providers remain unsettled.

29.6 Transaction Costs

An efficiency argument against network discrimination considers that the cost of differentiated pricing may outweigh the benefits, because of the transaction costs of implementing toll collection.

With pricing in transportation, an entire infrastructure of toll collection and user differentiation needs to be established, and this is not costless. The MnPass HOT Lane system in the Twin Cities does not recover system operating costs (and may never recover capital costs), while the London Congestion Charge, which now charges the vehicle owner 8 pounds per day to travel in the center of London, before the recent toll increase had enforcement and collection costs of about 67 percent of operating revenue. These examples compare with a less than 1 percent collection cost loss associated with gas taxes.[620]

The degree to which the costs of collection outweigh the benefits of differentiation depends on the technology of collection, and how much interference it imposes on all packets to read each packet and route it on higher speed or slower speed paths, based on its origin or fare class.

We have long recognized that different types of freight have different priorities (overnight, two-day, and ground are among the choices for shipping); and in transportation, the transportation profession has slowly begun to recognize that same kind of differentiation applies to drivers. Different drivers have different values of time at different times; of the day. Thus the ability to pay a premium and travel at a better level of service during peak times provides a service not generally available, but which may add value to the system and improve overall welfare. There is no reason to believe that different packets don't have different value to different users at different times; some are more urgent than others. If that differentiation can be achieved with a minimum of transaction costs, a net welfare improvement can be obtained. Whether that is the case is an empirical, not a moral, argument.

The implications of this are several; high collection costs for discrimination and per use charging are only valuable if the benefits of the discrimination outweigh the costs. In many examples in transportation, the costs of price discrimination and per-use charging outweigh the benefits by suppressing use and increasing costs.[621] While this of itself does not necessarily favor a regulatory approach to network neutrality, it suggests that providers should be reluctant to engage in non-neutral behavior, as that would increases their costs and reduce their demand (and perhaps profit).

29.7 Learning—Over-engineering—Prioritization

In communications, former head of the University of Minnesota Digital Technology Center Andrew Odlyzko has described the life-cycle process as "learning—over-engineering—prioritization." While this maps to our life-cycle process of birth—growth—maturity, and the individual facility process of design—build—operate, it has particular connotations. First, the S-curve of supply growth precedes the S-curve of demand growth.

A key aspect is the learning enables us to do things better. This results in a drop in cost over time. As with any system, early adopters pay more for access to the highest quality service, and we have the familiar economies of scale and density processes that costs drop as more and more users join the system.

Due to over-engineering, we have excess capacity in early stages. On a U-shaped cost curve, the left side is dominated by fixed costs and the right side by variable costs. Without congestion, large infrastructures (both transportation and communication) are dominated by fixed costs; with congestion, they are dominated by variable costs. We are more likely to see major transportation facilities on the right side of the cost curve, as it is easier to lay new wire than new road, rail, or airport capacity.

Value of time is also different, communication falling somewhere between passenger and freight transportation. Delaying the message is not the same as delaying the person, but messages usually have higher priority than goods. Yet communications are now getting so fast that we can get real-time videoconferencing or movies "on-demand" with a minimum of latency, even movies that can be delivered faster than in real-time (so you can begin watching it, and download the rest, so even if you are disconnected later on you can see the whole thing). This differs from physical items or people, which are generally thought of as delivered or not, not being partially delivered.

In the end, the final stage is prioritization, as demand eventually catches up with supply, and some congestion occurs in the peaks. Prioritization, managing demand and giving precedence to some over others, needs to occur. This might be rule-based (e.g., real-time video has preference over movies), or price-based (the cost of transmission increases in the peak, and whichever bits people want to pay for get first dibs).

As with transportation, in general there is structural separation between content/cargo and carrier. There are conflicts, among them those discussed above in the section on network neutrality about carriers privileging their own content over that provided by others. When supply is adequate, this doesn't create problems. If supply is scarce, such privileging has economic consequences: which movie service would you subscribe to, the one that is real-time, or the one with a 1-hour delay?

29.8 Discussion

Many of the analogies that are used about transportation in the public debate over network neutrality are misleading at best. The Internet, like most networks, can be divided into sections: first mile, last mile, and linehaul or backbone. The firms that provide these services are different. Each is a potential bottleneck along the trip from content to consumer. While most capacity is unused most of the time, it is the peak times that are of interest, when differentiation matters most.

A key difference is that local streets are expected to be slower and serve less traffic. On the "last mile" of the Internet, this creates a bottleneck. In other words, the "last mile" of roads is seldom rivalrous (except perhaps for on-street parking), while the "last mile" of the Internet clearly exhibits rivalry at selected times and locations. Removing this bottleneck by expanding capacity to serve selected traffic can be paid for either by a use charge on either the content providers (who may pass it on to their specific customers), or the content consumers, or an access charge levied on all traffic. A use charge falls on those who benefit. An access charge, while smaller per user, falls both on users who benefit and on users who do not. Claiming that the access charge is inherently fairer (an implicit argument of some network neutrality advocates) is disingenuous.

While homeowners do pay property taxes to support local streets to access the wider network (just as a consumer subscribes to an ISP to access the wider Internet), and shippers pay property taxes on their end, carriers pay gas taxes for all road use in between as well (including taxes on gas consumed on the "access" roads), and tolls if they are using tolled facilities.

In transportation, it is customary that the producer pays the shipper for the cost of shipment (a cost which is then passed on to the recipient/consumer). The shipper pays for the truck and labor, and fuel, and pays fuel taxes on the entire trip. The shipper is partially subsidized by local property owners (including producers and consumers), who pay the infrastructure cost of local roads (the "first mile" and "last mile") through property taxes. Shippers then pass this transportation cost on to the consumers who already paid an access charge.

No one source pays for the entire road network, and no one, undifferentiated charge should be expected to pay for Internet service, which has associated with it both fixed and variable costs, features congestion at time, and for which a per-use charging infrastructure is costly to implement. For highway networks, it has been shown that there is a cost-minimizing split between state and local government shares of highway finance, with neither all-state nor all-local funding necessarily the most efficient.[622] Although the data for private telecommunications networks are not available in the same way as for public highways, one imagines that there is some appropriate split between charging producers and consumers for the cost of different services, and that split may involve both per-use charges and flat rate access charges, depending on the circumstances.

While much transportation has fallen under the common carrier rubric, there is not a mode of transportation that strictly adhered to a notion of neutrality that prohibited both price and quality of service differentiation. Some network operators were more intrusive than others in understanding their customers' shipments, their characteristics, and their ability to pay. Moreover, different networks offer similar but non-identical services (travel by air vs. passenger rail, for instance). In the beginning, successful networks try to remain simple. As maturity sets in, ensuring profits requires differentiating between customers, who have both different willingness to pay and desire differentiation in quality of service. While perceived as inequitable or inefficient, product and price differentiation may also be necessary when users of the system impose different demands on it.

The network service providers, especially those who have a monopoly or share an oligopoly for service, would like to overcharge and undersupply to maximize profits. In the absence of significant competition and regulation, they will be able to succeed. The difficulty in regulation is finding an appropriate balance between the needs of the consumer and the needs of the regulated provider, without stifling innovation.

Internet businesses also remind us of the airline and railroad sectors. Several aspects are: the boom and bust cycle, the large fixed and relatively low marginal costs, and the winner-take-all aspect.

Network industries are subject to large swings in profitability. Why? We can call it the empty seat phenomenon. When an airplane takes off with an empty seat, the airline is leaving money on the table (or the ground). The marginal cost of operating the plane with an additional passenger is almost nil, given the plane is taking off anyway. It would cost but a few dollars in fuel and meals to have had one extra passenger. When load factors (percent of seats occupied) are high, so are profits. But when load factors drop just a little bit, it implies that demand is softening. Airlines are faced with an unappealing dilemma to try to retain or restore profits. They could try to raise fares to get additional revenue from the remaining passengers, but this risks chasing away more passengers. They could try to cut fares to encourage additional passengers, but this lowers revenue from the passengers they would have anyway, as some who would have paid the higher fare now pay the lower fare. They can try to cut costs and service, but this potentially costs revenue as well. Airlines of course do all three, and hope that by intelligently price discriminating (giving discounts only to marginal travelers, not to inelastic business passengers) they can recover. And, of course, they can wait until demand picks up due to external circumstances. Trains and other time-sensitive transportation industries have the same pressures. Internet backbone companies, those that carry long-distance Internet communications, are in a similar situation. Unless their wires are brimming with traffic, like the airlines taking off with an empty seat, they are leaving money on the table. The wires are there, costing money (in terms of paying back the lenders or paying off shareholders), whether or not they carry traffic. Unlike manufactured goods, this capacity cannot be stored.

A second phenomenon is the lumpiness of network industries. The capacity in networks comes in discrete lumps: an airplane (which carries 150 or 300 passengers per hour), a wire which carries million of bits per second. Acquiring an airplane or a wire takes time, while there may be a few extra planes lying about, reconditioning them for service is not instantaneous, and orders for new planes take years. Similarly, fiber optic cables take time to lay down.

Third, profits during the good years attract new entrants. When times are good and demand is growing, network industries are profitable. These profits provide a signal to others that there are excess profits in this industry, and it is a good field to enter. By excess profits, we are not implying a moral judgment, simply that profits are above market averages for an investment of apparently the same risk. So new companies enter the market. In the airline industry, buy and recondition a few planes and you can enter the industry, rent a gate at another airport and you can enter a new market. Every so often, a host of new carriers try to make it in the airline industry. A few survive, a few are acquired, and

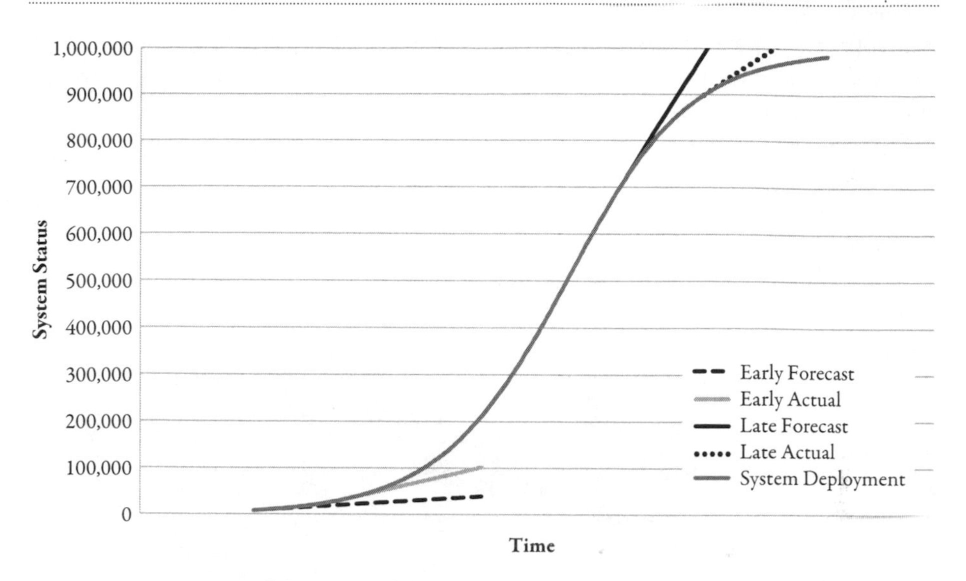

FIGURE 29.2 S-Curve and the Danger of Extrapolation

others go bankrupt when the market turns bad as the economy moves through cycles. Similarly, since telecommunications deregulation in the 1980s, the United States has seen many new entrants, first in the long distance telephone and then the Internet markets. Firms saw potential excess profits and high growth rates. Each assumed that it would capture the lion's share of the new market and built new capacity accordingly. When they found they were not the only company with this idea, but had a score of competitors, prices had to be lowered to a point where they no longer covered the high capital costs (though generally remained above the operating or marginal costs). Thus there is a situation with excess capacity. Over time, as information exchange explodes, this capacity may become fully utilized, especially if technologies like video (or better yet, 3-D projections) over Internet take off. But in the meantime, debts have to be paid and companies lack the resources to do so. Hence, we see numerous bankruptcies in this sector (WorldCom, the parent of MCI, and Global Crossing, but two of many). Failing companies try to play with demand, as noted above, but most eventually cut costs, reducing purchases from suppliers. They may even play with the books, leading to accounting scandals. This reverberates to equipment manufacturers.

The reason for overshoot can be understood by returning to the S-curve. Assumed forecasts are made by extrapolating previous results, which is how many businesses and investors operate, as shown in Figure 29.2. In early years (birthing and early growth) the rate of growth each year is greater than the previous year. But in late growth and maturity, growth is slower than the previous year.

As connecting or linking technologies, communications and transportation have much similarity, as well as competitive and complementary relations. Work now is mainly concerned with substitution of communications for transportation. An obvious extension is to complementary relations. Communications improvements are already pulling freight transportation services in new ways. A quick analysis says that improved communications and control are enabling such things as "just in time" inventory policies and real-time traffic control. Deeper analysis is needed, for much more than inventory and traffic policy may be involved. We need to know about changed ways of doing business, productivity gains, and the development and use of new production technologies.

[If such drivers have no faster alternative route], Those are the people who I would encourage to change jobs or change houses.... The City of Edina needs to build some arterials.
—Mn/DOT Ramp Meter Chief Engineer (November 28, 1999, *Minneapolis Star Tribune*)

30 Technology: Hard and Soft

30.1 Metering Motoring

WE DEFINE TECHNOLOGY as a way of doing things. A technology may be hard, and embodied in physical tools, or soft, and embodied in protocols. Usually these are linked. Highways, for example, have a technology for traffic control that is embodied in signal systems, and so on, and also a set of rules for drivers. At a much broader scale, one may think of the physical makeup of a mode as a technology, with a companion set of soft technologies embodied in institutions, regulations, standards, and so on.[623] Here is an example that introduces current technology development efforts.

Ramp meters, traffic signals posted on freeway entrance ramps, seek to regulate the flow of traffic entering the freeway. They serve two main purposes; first, they limit the number of vehicles trying to merge simultaneously, smoothing traffic flow (and reducing crashes); second, they keep the total number of vehicles on the freeway trying to simultaneously use a critical bottleneck just below a threshold (capacity), so that freeway flow doesn't exceed capacity, and thereby avoiding queueing. In and of themselves, those are both reasonable goals for managing a mature system, and most travelers readily accept traffic lights in other contexts. Yet somehow, in the Twin Cities of Minneapolis and Saint Paul, Minnesota, ramp meters became the transportation issue of 2000.

The reasons are clear in retrospect, but may not have been in advance. As can be seen in Figure 30.1, ramp meters were slowly deployed in the 1970s and 1980s, and became much more widespread in the 1990s. As road capacity was built out, additional roads became more and more difficult to build, not only in monetary cost, but also in political will. The leadership of the Minnesota Department of Transportation (Mn/DOT) viewed ramp meters as a way of stretching the system slightly further, eking out a small capacity improvement

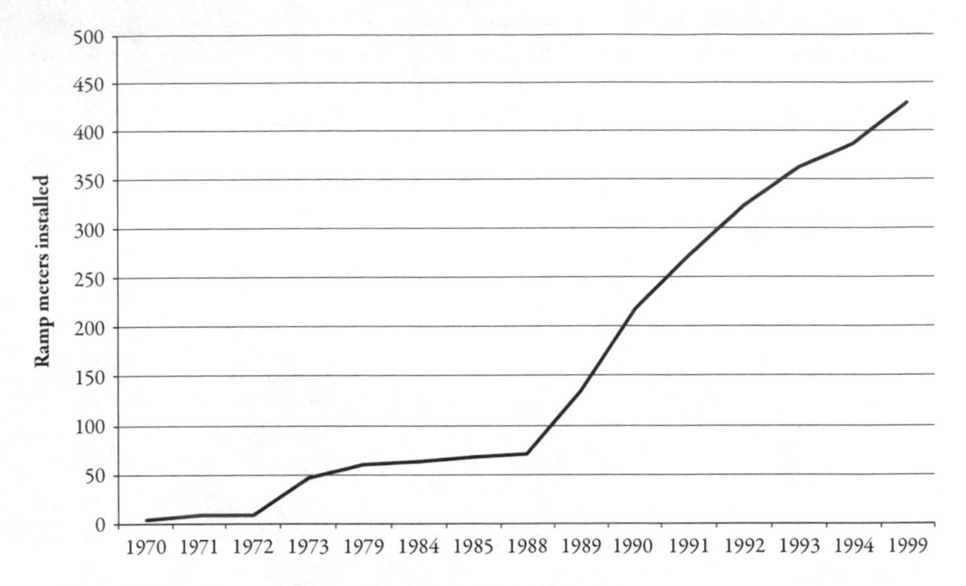

FIGURE 30.1 Deployment of Ramp Meters in the Twin Cities

and a significant speed improvement at a cost much below that of adding lanes to the freeway.

Yet the Twin Cities continued to grow, as did peak-hour travel demand. The primary effect of ramp meters is to move delay from the freeway to the entrance ramp. By the late 1990s, some commuters experienced long delays at some ramps, in cases upward of 20 minutes. In 1999, Dick Day, a state senator from Owatonna, Minnesota, a rural community outside the (metered) metropolitan area, pushed a "Freedom to Drive" package. This package called for shutting off all of the ramp meters, allowing all cars to use HOV lanes, and establishing the left lane as a passing-only lane. (Day claims to drive 70,000 miles a year, which averages to over 3 hours a day in his car—the reader can assess whether this is reasonable or hyperbole).

Day was able to obtain press for his initiative, and in November 1999, the *Minneapolis Star-Tribune*, the state's largest newspaper, printed a large Sunday, front-page piece on ramp meters. (The opening quote of the chapter is from that article.) Discussions with the engineers reveal several things. First, they were certain metering was the right thing, and they believed that shutting off the meters would be "catastrophic." Second, they were indifferent to the fact that some drivers had long commutes so that others would have shorter commutes.[624] They did not see ramp delay as an important metric. Rather, if the freeway flows were higher with than without meters, and at higher speeds, they knew they were reducing total delay (if more total travel is using the freeway, then there is less total travel on alternative slower routes). Third, they were highly resistant to outside analysis, probably because of distrust that the outcome would be different from their own.

Nevertheless, to avoid the threatened shutdown, Mn/DOT commissioned three separate University of Minnesota studies to evaluate meters.[625] One might suggest that these studies were a holding strategy, essentially telling the state legislature "see we are studying this—please go away." However, those studies did not involve shutting down the ramp meter operations; rather, they would conduct computer simulations to examine operations with and without meters, compare metering approaches from a number of cities, and examine empirical data. Despite these studies, in May 2002, the Minnesota state legislature insisted on a shutdown experiment, which would last at least 4 weeks. A large consulting firm[626] was hired to conduct the study. Many of Mn/DOT's ramp meter engineers were excluded from the study process (their biases and lack of political acumen having been demonstrated), as were the university researchers (who were funded by Mn/DOT and therefore tainted by association). Traffic data were collected before the shut off period, and then the meters were to be shut off for a period of at least 4 weeks to conduct the study in October 2000. Because of weather, the study was extended a few more weeks. Due to the lack of catastrophe, the study was extended a few more, since it was clear that Mn/DOT could not return to the old metering strategy, and no new strategy was obvious. Eventually the meters were turned on (December 2000), but running at their fastest rate, so that queues would not get too long. Over time, a new strategy was developed to cap maximum waits at the ramps at 4 minutes.

Dick Day was not entirely satisfied, and Mn/DOT staff was unhappy with the shift in their worldview, but the residents of the Twin Cities seem happier with the system than before.

Ramp meters are but one of many "new" technologies collectively dubbed "intelligent transportation systems" (ITS). The new technologies vary from simple operational improvements (e.g., ramp meters, freeway service patrols [tow trucks that quickly service disabled vehicles]) to more sophisticated information systems (in-vehicle navigation systems that tell drivers the shortest path given real-time congestion information, or bus stops that tell you to the minute when the next bus will arrive) to control systems (adaptive cruise control that will follow the speed of the vehicle ahead, automated highway systems that will control all movement of vehicles). While some of these technologies have already been deployed, many have not, and won't be for another decade, if ever. Yet already controversies emerge.

This chapter will treat the origins and status of ITS technology development and implementation efforts. It will remark on the roles of standards. ITS is receiving attention, and it provides a technology example for considerations of innovation and the generation of imaginative actions. Before talking about ITS, the question might be asked: Do we see "Twin Cities" stories as we visit the modes and efforts for their improvement?

"Well, in a way," is one answer. Recalling the within mode structure of systems, there are activities in equipment, facilities, and operations sectors. Recalling behaviors over the life cycle, marginal improvements to entrenched ways of doing things is the rule at mature stages in the lifecycle. Recalling that advances are made by importing new tools, it is not surprising that information and communications technologies are being introduced into transportation efforts.

But there is more on the technology shelf. There are improvements in physical and human resource management, materials, lubrication, and endless other things. So we see fly-by-wire aircraft, automated monitoring and control of ships, widespread monitoring and control in the logistics aspects of freight movements, monitoring and control of freight trains and locomotives, application of smart cards to enable electronic toll collection and fare payments on transit, and many other activities. Improvement rather than replacement is the rule, and technology delays the onset of diminishing returns.

"Well, no," is another answer. That's partly because markets are better defined in some non-ITS situations and there are competitive pressures in some other situations: firms operating ships and aircraft, managers of competitive ports, trucking firms investing in location finding and communication devices. The champions of technology improvement command the payoffs, the rents. That's not the case with some ITS technologies and services.

30.2 Magic Motorways

The dream of what we now call intelligent transportation systems first gained widespread attention at the 1939 New York World's Fair. In particular General Motors' *Highways and Horizons* exhibit, better known as *Futurama*, pictured the world of a large metropolis in 1960, containing completely grade-separated freeways running some type of automated highway system.

As part of the 1939 World Fair and the Futurama display, GM commissioned a film *To New Horizons*, which is a paean to progress. It emphasized "New places to go and new means of getting there."[627]

This Norman Bel Geddes (1940) vision of the future may have been too soon (obviously 1960 came and went and the Interstates were barely under construction, much less an automated highway system), yet the vision remains. He documented this in his book *Magic Motorways*.[628]

Bel Geddes started as a set designer on Broadway, and migrated to industrial design. He is also known for his advocacy of art deco streamlining, as described in *Horizons*.[629] (Fittingly, his daughter, Barbara Bel Geddes, was an actress best known as playing matriarch Miss Ellie on the popular primetime oil-industry soap opera *Dallas*.)

In *Magic Motorways*, Norman Bel Geddes made a great number of suggestions to improve the efficiency of then existing auto-truck-highway system. The modern road-builder should be concerned with safety, comfort, speed, and economy. The existing system, he noted, emerged from cow-paths and the like, the result of subsequent incremental improvements, but no systematic design. If all roads were designed like the best roads, the death toll would be reduced by 80 percent. Even the best road of the day, the Long Island Motor Parkway, had curves that were too sharp, hills that were too steep, and was too narrow.

He complained about truck weigh stations at every state line as an infringement of Interstate commerce. He further complained about speed traps, noting that in Connecticut, example, traffic constables were rewarded proportionate to the revenue they brought in from speeders, which was unfair to tourists. In Toledo, he reports, red lights were turned off at night because motorists were being held up when they stopped. Bel Geddes endorsed one-way street conversion, noting that volume and speeds both increased by 20 percent in Philadelphia when Chestnut, Walnut, and Market streets were converted.

He argued that motorways should connect cities but not enter the actual concentration points. He suggests that perhaps there were too many private cars in congested areas of cities.

Cars that did not have business on a street should not be on the street. This kind of pattern is only achievable with a highly hierarchic street network. He wanted road building and management to be under the auspices of higher levels of government to improve quality and network consistencies. "When road building depends on local and personal whims, a four-lane highway is likely to be sent swooping through a mess of haberdasheries and then, without explanation, to peter out at the town or county line into a narrow macadam road." He reported a plan of architect Ernest Flagg that grade-separated pedestrians, fast vehicles, and slow vehicles.

Technology, too, could be a factor. To identify one example (that has yet to be resolved some 70 years later): Why must drivers dim their brights manually when facing an oncoming car? Surely this could be automated.

There were several implicit objectives of this vision (illustrated in Figures 30.2, 30.3, 30.4). One was to promote cars and highways as the vision of the future. Clearly, that much came to pass. While there are buses in the Futurama exhibit, trains and streetcars were downplayed. Another removes the driver from control of the vehicle. This was seen as advantageous because drivers are the cause of most crashes. Further machines respond faster than humans, and so can follow more closely at higher speeds while still remaining safe. A third was the separation of pedestrians from vehicles. The elevated sidewalks (or skyways) are an idea borrowed from other visions of the future, and have both safety and efficiency aspects, though urban designers often see separation as problematic.

Notably, BPR chief Thomas MacDonald "thought Bel Geddes a crackpot; the last thing the country needed was fourteen-lane bands of concrete crisscrossing the hinterlands."[630]

FIGURE 30.2 Interior View of General Motors Highway & Horizons Exhibit

FIGURE 30.3 Postcard View of Futurama

FIGURE 30.4 Two 14-Lane Express Highways Cross in 1960

30.3 Automated Highways: The Clash between Technology and Deployment

In August 1997, at DEMO 97 in San Diego, California, researchers of automated highway systems (AHS) demonstrated a system on the reversible high occupancy vehicle (HOV) express lanes of I-15. As shown in Figure 30.5, specially equipped cars were able to follow at high speeds, without driver intervention, at a fixed 6.4 m (20 ft) following distance using advanced communication technologies (and some low-tech in-road magnets to help with lane guidance). While the demonstration was a technical success, funding for research in this area was promptly cut, and the National Automated Highway System Consortium, which sponsored the demonstration, was disbanded. That is not to suggest it was a waste; much was learned.

First, the possibilities of deployment of such a system (which relies on intelligent highways and inter-vehicle communication) are much more difficult than deployment of autonomous intelligent vehicles, sometimes referred to as self-driving vehicles (SDVs). While SDVs are harder to engineer, they can operate in mixed traffic without upgrading both the infrastructure and all other vehicles. AHS suffered from a "chicken and egg problem" *par excellence*. Who would buy an expensive AHS vehicle with no routes, and who would build new a network of AHS lanes with no vehicles? Where the intelligence in a network technology lies is a critical question that needs to be answered. We have seen various mixes throughout the history of transportation, from smart vehicles (cars and their drivers—we hope) and dumb links (simple roadway), to dumb vehicles (trains that don't have steering) and smart links (the tracks steer the train at junctions). However, it is clear that compatibility with existing systems is a crucial question. While you can drive

FIGURE 30.5 I-15 Demo of Automated Highway Systems

a car over railroad tracks and get somewhere (though it may be a bumpy ride), it would be difficult to take a train and drive on streets (it would keep going straight until it hit something).

Second, much knowledge was gained about the specific technologies, which ones work, which don't, and which can be adapted to other technology paths, for example intelligent vehicles.

Most participants in ITS work would probably agree with these conclusions. The interpretations we will make now require modest to major stretches. Even so, we make them because we think they may lie behind the faltering of the ITS national program and applications such as the Twin Cities ramp metering.

It is a modest stretch to speculate that the inherently disjoint nature of the structures of the auto-highway and truck-highway systems is a major shaper of technology improvement. In a conversation with some auto manufacturers, we found that they have a policy of internalizing all improvements in the vehicle. "We cannot expect the highway people to do anything." (We were discussing vehicle guidance systems. The company will use expensive inertial devices to track location rather than off-board fixed facilities. GPS has significantly helped in this regard, but this infrastructure was provided by the military, not for civilian transportation, and was only reluctantly opened up.) We find especially in the highway sector that because vehicles, vehicle trips, and highways are produced separately (unlike in railroads, where a single organization specifies vehicles, operates them, and controls the track), highly disjoint behavior restricts development options. This diminished expectation of infrastructure, and especially integration, is the reason heard most often for compartmentalizing development efforts to drivers (navigation systems), automobiles (intelligent cruise control and vehicle braking), and highways (traffic flow sensors and ramp metering).

It is a bit more of a stretch to ask: "Why can't one count on others?" The mismatch of institutional norms (which may create adversarial relations) and controls (political or market) might be a partial answer, and lack of motive for joint efforts another. Glory, status, and feel-good rewards motivate only so much. The motive that counts is making the profits or capturing the rents, saving time, or what have you. "What is mine is mine" is at work. This point about who captures the rents has already been made, but it bears underlining in the institutional context.

Stretching some more, there is a clash of cultures. To illustrate, ramp metering aims to make best use of agency resources and minimize total delay. Perhaps drivers see this as inequitable because some are stuck in traffic while others whiz by. Is the agency limiting my use of common property (the road)? Some see it one way, others another.

30.4 Intelligent Vehicles: The Rise of the Robots

Automation in the automobile sector resulted in horses being replaced by engines. On the production side, labor is already being replaced by machinery and robots, the hours of labor per car is steadily declining over time. Brynjolfsson and McAfee (2011) relay a possibly apocryphal story: "Ford CEO Henry Ford II and United Automobile Workers president Walter Reuther are jointly touring a modern auto plant. Ford jokingly jabs at Reuther: 'Walter, how are you going to get these robots to pay UAW dues?' Not missing a beat, Reuther responds: 'Henry, how are you going to get them to buy your cars?' "

We may eventually see drivers replaced with robots. In late 2010, Google announced that it had been secretly testing autonomous vehicles in traffic. After hiring the members of the successful teams from the 2007 DARPA Urban Challenge, including Sebastian Thrun, they developed a small fleet of autonomous vehicles, which they drove for over 100,000 miles (160,000 km) in traffic in the San Francisco Bay Area (as of 2012, the number is over 500,000 miles [800,000 km]). While the cars still had drivers ready to take the wheel, the tests were successful (the cars themselves caused no crashes while in autonomous mode, one was rear-ended, and later, the car got into a fender-bender when a human driver was in control).

Why robot cars matter:

1. Safety: Cars would be safe if only there weren't drivers behind the wheel. Driverless cars seldom get distracted or tired, have really fast perception-reaction times, know exactly how hard to break, and can communicate (potentially) with vehicles around them with Mobile Ad Hoc Networks.[631] But this improves not only vehicle safety, it improves the safety and environment for pedestrians and bicyclists.
2. Capacity: Robot cars can follow other driverless cars at a significantly reduced distance, and can stay within much narrower lanes with greater accuracy. Capacity at bottlenecks should improve, both in throughput per lane and the number of lanes per unit roadwidth. These cars still need to go somewhere, so we need capacity on city streets as well as freeways, but we save space on parking (see Section 15.5), and lane width everywhere. If we can reduce lane width, and have adequate capacity, we can reduce paved area and still see higher throughput. Most road space is not used most of the time.
3. Vehicle diversity: Narrow and specialized cars are now more feasible with computers driving and increased overall safety. Especially if we move to cloud commuting (as below), we can have greater variety, and more precision in the fleet, with the right size car for the job.
4. Travel behavior: If the cost of traveling per trip declines (drivers need to exert less effort, and lose less effective time, since they can do something else), we would expect more trips (my taxi can take me wherever) and longer trips and more trips by robocar.
5. Land use: If acceptable trip distances increases, we would expect a greater spread of origins and destinations (pejoratively, sprawl), just as commuter trains enable exurban living or living in a different city.
6. Parking: my car can drop me off at the front door, and go fairly remotely to park, so we don't need to devote valuable space to surface parking or parking ramps (garages) (we still need space, it is just far away). Searching for parking is also less critical. On-street parking can be abolished.
7. Transportation disadvantaged: Children, the physically challenged, and others who cannot or should not drive, are now enabled. Parents, friends, and siblings need not shuttle children around, the vehicle can do that by itself. The differences between

transit and private vehicles begin to collapse. We can serious consider giving passes to driverless taxis for the poor, since costs should drop with lower labor costs, and if the point below holds, paratransit services become much less expensive as well.

8. Reduced auto ownership: Cloud commuting becomes possible. People no longer need to own a car, they can instead subscribe to a car-sharing service.

   "Cloud commuting" is the idea that cars would be provided from a giant pool operated by remote organizations (in the "cloud"). The organization would dispatch a vehicle that drives to the customer on demand and in short order, which then delivers the customer to the destination. The vehicle would have each customer's preferences pre-loaded (seat position, computing ability, audio environment). The customer benefits by not tying up capital in vehicles, nor having to worry about maintaining or fueling vehicles. The fleet is used more efficiently, each vehicle would annually travel two or three times more than current vehicles, so the fleet would turn over faster and be more modern. Fewer vehicles overall would be needed. It is likely customers would need to pay for this service (either as a subscription or a per-use basis). One imagines that stores might subsidize transportation, as might employers, as benefits for the customers or staff, just as they subsidize parking now.

   We have car-sharing services, which require large networks of members to be valuable (so the member can be near a vehicle, and there will be vehicles available). In the United States, there are several companies, including the once stock exchange–listed (though not profitable) Zipcar providing these services. With greater value, such a service should have more members, and thus more vehicle choice and shorter wait times. In short, there is a potential magic bullet to be had (see Chapter 10).

The story above is a story of a future, one where we can look back at how technology changed transportation. Robot cars will, of course, allow us to do the same thing (driving to and from) better, replacing the easily distractible attention of the driver with the perfect attention of a computer. In that respect, it polishes the existing system.

But robots are also a new beginnings, a transformation that will allow us to do new things. No longer will only licensed drivers have auto mobility, children and the disabled can access cars as needed. No longer will families need to own a fleet of vehicles, each can call the right-sized vehicles on demand. No longer will drivers lose an hour or two per day traveling, while travel will not be timeless (until the Transporter from *Star Trek* is invented), travel will be much more productive, as other things can be done. No longer will cities be swamped with parking lots, as robot cars can go away and park themselves where land is abundant. No longer will people need to maintain cars, that cost can be centralized. If the cost of travel is lower, we would expect more of it. This change will not be necessarily welcomed, but it is a logical progression and presents one possible future.

At the time of this writing, the technology is not quite ready. Finding the right formula for mix of vehicles, location of vehicles, amount of time people will wait when summoning vehicle before arrival, and so on, will be important experiments in this era. Google is petitioning states to change their laws to permit on-road driverless vehicle testing;[632] California, Nevada, and Florida have so far gone along.

30.5 Planning Technology

The issue isn't whether technology is incorporated in planning processes, but rather how it is incorporated. The urban transportation planning process (see Chapter 24), which was developed for a growing mode (highways) and is now applied to mature ones (highways and transit), doesn't explicitly recognize technology. Rather it takes technology as a given and fails to consider change in technology. At best, some kludges to test new technology are incorporated as network elements and demand shifts, often computed with "post-processors."

Fortunately, the urban transportation planning process is not the only one. Other planning efforts consider technology in different ways as the life cycle progresses. As the predominant technology emerges, hard and soft technologies receive much attention. They are used as building blocks for system designs. Here the issues are how widely innovators scope for building blocks and the decisions made as designs are frozen by standards. During this period the process is not ergodic. Each decision made by "historical accident" will not be "smoothed out" by future actions. A history-dependent path is created.

Technology is also an agent for planning as a system grows to maturity. Hard and soft technological standards constrain the planning process. They also serve as objectives, for planning seeks to put those technologies in place in some desirable way.

At maturity, one issue is operations; other issues have to do with productivity declines, market channeling, repair and maintenance. Mensch (1979) uses the word "pseudo-innovations" for related technology efforts. That word recognizes that there is not much to be accomplished by tweaking a mature system. Similarly, Tyler Cowen identifies a *Great Stagnation* at play.[633] Our view is negative, but not as negative as Mensch. For one thing, we begin to see some cross-component work, such as ITS. There is also a willingness to bring in some "outside the system" technologies, such as those being incorporated in railroads control systems.

Because of the division of labor between design and policy, planners often assume that technology considerations are being managed elsewhere. There are good reasons for division of labor. The large number of people involved in technology fields and the large literature suggest that technology is being taken care of elsewhere. We can understand the planner's assumption. However, the reader of this text will recognize that it is, of course, an unacceptable assumption for transportation.

30.6 Standards and Orthostandards

There are two related meanings for the word "standard" that are of interest to transportation professionals. The first has to do with compatibility. The idea traces from interchangeable parts; we want a standard to ensure that part A from one manufacturer can fit with part B from a second manufacturer. It is contrasted with "custom." The second has to do with quality or performance, for example achieving a level of service (LOS) standard, which we will call "orthostandards" (meaning "correct standard") to avoid confusion. We may want to ensure that we meet the orthostandard that the road operates at LOS C or better, and will use that orthostandard to decide how many lanes the road should have.

Both types of standards serve to ensure consistent design; though one can be thought of as hard and the other soft, each will translate into the other. Standards are established as systems grow. Often they are simply ratifying whatever decision the first innovator made. An example is George Stephenson's use of what became the standard rail gauge, which remains after nearly two centuries.

The use of classification and standards follows the rationalist-scientific paradigm of the nineteenth and twentieth centuries. However, the extension and ubiquity of normative orthostandards, beyond what is necessary for inter-system compatibility, is more difficult to understand, and here are some thoughts.

1. One has to have orthostandards in order to classify, so orthostandards resulted from the urge to classify.
2. Experts have superior knowledge and can prescribe what is best.
3. While it is true that orthostandards do not fit every case exactly, they fit fairly well and offer great efficiencies in processes of planning and management.
4. Standards and orthostandards support articulation goals. (In the case of roads, standardized driving environments [e.g., signs] are essential.)
5. Without orthostandards, people would cheat (e.g., authority high in the hierarchy can control dishonesty at lower levels in the hierarchy).
6. Orthostandards support the common carrier, common law, public utility ethic of similar service to citizens in similar circumstances.
7. Consumer protection requires standards and orthostandards (e.g., for safety).
8. Standards and orthostandards reduce construction costs.

Conversations with USDOT staff mainly involve (2), (5), (7), or (8), depending on the subject. But it is difficult to get conversations going. Most engineers and planners seem to take orthostandards as a matter of course, caring about "how" and never worrying "why?"

Transportation professionals are very committee-oriented, and there exist a number of committees to establish all sorts of transportation standards and orthostandards, from uniformity in traffic control devices to what factors to include when determining the value of highway assets. Most of these committees deal with known problems, and are simply ratifying existing procedures, and regularizing them, or choosing among competing alternatives.

However, there is one notable example of trying to put the cart before the horse, so to speak. The transportation community has spent hundreds of thousands of hours and several forests worth of trees to document the National ITS Architecture, defined by its advocates as "a common framework for planning, defining, and integrating intelligent transportation systems." They note that "it is a mature product that reflects the contributions of a broad cross-section of the ITS community (transportation practitioners, systems engineers, system developers, technology specialists, consultants, etc.)."[634]

However, there is very little to show for this effort. The few ITS applications in place developed organically (like the Twin Cities ramp metering system, local freeway service patrols, automakers' in-vehicle navigation systems like On-Star, or Google's self-driving vehicle) rather than as a result of this process. The most successful ITS application, electronic toll collection, has produced transponders that are incompatible between regions, largely because (1) they made decisions and grew independently, (2) they felt no reason to pay more than lip service to the architecture, and (3) the cost of coordination outweighed the benefits of compatibility. However, over time, one expects that compatible transponders will be developed, so that one can use the same device to pay tolls in Maine and California. Either one of the technologies will be selected, and phased in to the other region; or the groups will get together to standardize on a next generation technology that is better, and deploy that over time. ETC transponders and "smart cards" are likely to face a convergent path, as the wireless transmission of money has applications far beyond paying tolls.

Stepping back, we discussed ITS at the beginning of the chapter. ITS was funded for several reasons, improving transportation probably being secondary to helping large defense firms convert to civilian sectors as the Cold War ended. It is no surprise that the defense industry applied their top-down systems-engineering approach that was successful in the weapons-making business to the more target-oriented transportation sector.

An alternative approach would have allowed the technologies to emerge bottom-up, and then work toward compatibility and standardization among successful systems. This organic approach is more in keeping with the history of successful technologies in both transportation (there was no National Railroad Architecture, at least not until the era of rationalization, and then it was largely ignored) as well as communications, where the Internet developed without a master plan, but rather a series of decisions (politely called "requests for comments") made when they were needed.

30.7 Standards as Economic Decisions

Economic analysis is applied in highly constrained circumstances. The existence of agreed-upon standards constrain the analysis context. Given the standards, what does analysis say?

Consider the Interstate. Decisions were made about the product. There were a lot of layers to those decisions. They started with the free, limited access design decision, then laid out the general location of freeway links, involved the formulas for the distribution of funding among the states, and, finally, focused on the level of service to be provided. There was much economic content in each decision, but the decisions were regarded as standards somehow detached from economic decisions.

The level of service decision was important. A level of service class C was to be provided on urban freeways (B on rural), and that was targeted on the 30th highest traffic hour for the design year. Although set as a design orthostandard by AASHTO, that was very much an economic decision. There were equity-economic decisions mixed in. All cities got the same service level; the value of time was set low, and the value of time was the same for all members of the population. From discussions with AASHO leaders, the decisions were driven by the question: How far will the available money stretch? Urban facilities are expensive to construct relative to rural ones, and that's the reason for the rural-urban difference.

The idea we wish to transmit is this: Orthostandard-setting substitutes for economic analysis. There remains room for economic decisions within the context of the standards that had been set. These residual decisions were important, but small compared to the big standards setting decisions. We find this a truism in many sectors, in addition to highways and the public sector generally.

30.8 Discussion

Professionals often believe (and are affirmed by management consultants) that they are engaging in strategic planning when they are doing nothing of the sort. They are merely producing pseudo-innovations. Recommendations to undertake more radical technology planning fall on deaf ears; change threatens. Beyond the ways technology has been incorporated in planning, we suggest two additional strategies.

One is begin to incorporate system-changing technologies (in the building block, birthing style) as systems age. The objective would be renewal, one system rising from the ashes of another. A strong argument can be made for this strategy. We strive to create competing technologies once most of the "goodies" have been extracted from the old mode. More important, we know how to do this because the historical record is clear, we just emulate the past.

A more desirable strategy would be to smoothly create new options. This requires forging an inquiring style of planning that continually considers and envisions new futures and creates paths toward those futures. Society can then control its future by choosing among alternatives. This dynamic planning would create new hooks for the future. An object-oriented approach, to borrow some jargon from computer science, would allow one layer to be removed and replaced without worrying about others. At the micro-level, this can be as simple as ensuring that technologies are extendable. At the macro-level, this involves considering option value in decisions. We know transportation consumes land in specific ways, ground transportation in long strips, sea and air transportation in large, relatively compact ports. Preserving right-of-way, ensuring that capacity is available, is one way of preserving future options. The problem, of course, is spending present dollars for future possibilities, which must be discounted.

It is urgent to deal with strategic-technology questions and opportunities. We have mature systems, and further investment has low, if not negative, returns. The failure of transportation to improve does not bode well for economic and social development.[635] We should use planning to create situations in which there is wise inquiring about new, enhanced futures.

Technology improvements are not new, and advances in ITS-like things pose planning opportunities. The opportunities presented are close-to-the-heart actions by traffic engineers to improve traffic flow. Planning ways of thinking might bring these out of niche applications to networks. For instance, the automated highway idea is consistent with the highway providers dogma: concentrate resources and achieve economies of scale.

That is soft and general, and to make it understandable the authors would be tempted to use examples of some things to be done. There is a trap here that we have faced in many discussions: The medium (examples) are taken as the message (what to do). Yet the message is not that the technology forms used as examples are the things to do—the message is to explore likely options and let the public choose.

10 Beyond the Life-cycle

Every discussion of duty has two parts. One part deals with the question of the supreme good. The other with the rules that should guide our ordinary behavior.

—CICERO, *On Moral Duty*

31 Policy and Its Lacunas

31.1 The Experiential Policy Model

WHY DID THE system policies and rules that we have discussed emerge? In the spirit of William of Occam, our best explanation is that just as experiences follow from policies, policies follow from experiences. This experiential policy model is simple and explains a lot, but puzzles remain.

Policies divide nicely into (1) those based on social and political consensus and established and implemented by law or social custom and (2) those created by the modes for their own purposes and embedded in modal practice. The first category is what most "policy analysts" think about when the word policy arises, but the second is important as well, but is more often in the domain of the "management consultant." We talk about both here.

Policies reshape experience, and the process continues. We visualize this process in Figure 31.1. As we have seen, the experience in one mode frames policy making for other subsequent modes.

Understanding the transportation experience is the key to understanding policy and the Life-cycle of technologies. An operative word is "learning": we have learned from the transportation experience, and learning has yielded rules (guidelines, regulations, etc.) that tell us how to create, deploy, and operate systems. We take those rules to be policies. We need to address:

- Where they originate?
- Who enforces them?
- What is tacit and what is explicit?
- How well do they work? and
- How are they changed?

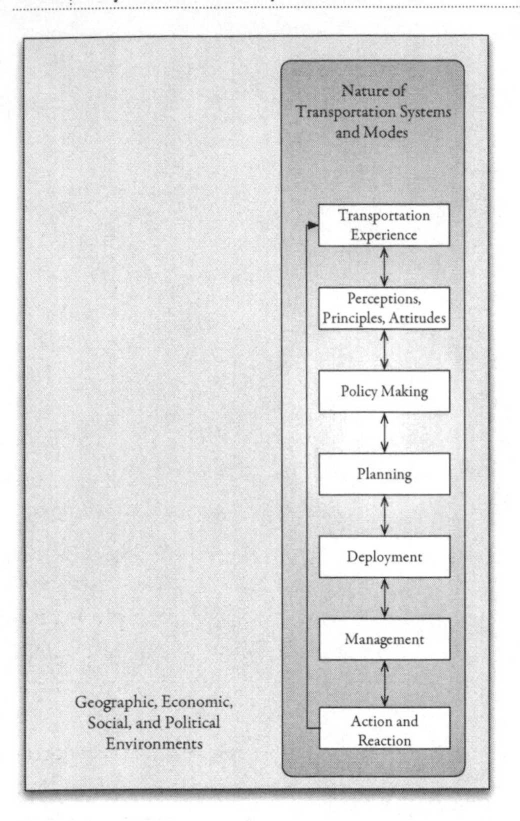

FIGURE 31.1 Experiential Policy Model

The unfolding of the transportation experience organized our discussions; it told us how policy protocols are modeled on experience. We spent a good bit of time on railroads because they provide a mother logic that applies to other modes. But railroads are not the only mother logic; the maritime experience affected other modes, as surely the automobile/highway system will affect future modes. Each mode has its own "internal logic," as well as "borrowed logics" it has adopted from other modes.

31.2 Policy Models

The introduction of this chapter gave a terse statement of the authors' point of departure, Figure 31.1, the experiential policy model.

The transportation experience is embedded not only in geographic, economic, social, and political environments; it is also corralled by the limits of technological structure and the nature of specific modes. The nature of transportation and the greater environment, collated with the transportation experience, gives rise to perceptions, principles, and attitudes. Those beliefs generate a layer of policies (both government and private) that translate into actions. Action and reaction indicate that the modes adjust performance to cope with problems. Those actions shape and re-shape the transportation experience.

Our view is not conventional.

Like the comparison of the life-cycle model of the deployment (and evolution) of technologies and networks to the Newtonian model, the experiential model we propose can similarly be compared to conventional models operating in a more fixed environment. For instance, one modern version of classic "rational" policy analysis model is described in Patton and Sawicki (1993). It suggests a very orderly process of six basic steps: "(1) defining the problem, (2) establishing evaluation criteria, (3) identifying alternative policies, (4) evaluating alternative policies, (5) displaying and distinguishing among policies, and (6) monitoring policy outcomes. Rather than a rigid-lock step approach, the process involves feedback and iteration among the six activities." It is normative, describing what the policy analyst should do. But it is ahistorical and aspatial. It does not tell us where policies actually come from. It does not tell us what values are important to the players. It does not explain why choice sets are constrained (either artificially or economically).

Other policy models are descriptive, and tend to be politically oriented, with actors (variously elites, interest groups, organizations, bureaucracies, self-interested actors) who have some power attempting to pull strings and competing for legitimacy. Decisions are generally incremental, as it is difficult to make non-incremental changes.

The experiential policy model says that perceptions, principles, and attitudes are forged from experience interacting with the nature of transportation systems. For example, principles bearing on the organization of systems, as well as public and private sector roles, stem from experiences when systems were deployed. Policies relating transportation investment to economic development result from past development experiences.

Our experiential policy model considers the formation of principles from the transportation experience as part of the system (or endogenous). It is subject to the objection that experiences other than transportation ones are bound to bear on transportation experiences and attitudes. That is a valid objection and must be true, yet we have a strong response to it.

One aspect of the response is that the transportation experience has so permeated all social and economic experience that we should not think of purely "outside transportation" experiences. The deep and economy-wide impacts of the railroad experience on institutional forms, financing, government activities, regional economic organization, and so on, illustrate the impacts as well as the transfer of transportation experiences. As we point out elsewhere in the book, national industrial policies have roots in transportation experiences. The explicit recognition of embedded policies is another important way the experiential model differs from the conventional model.

The experiential model explains the structure and performance characteristics of systems to which the policy is applied. The characteristics of systems create the need for policy studies and condition their results. The characteristics may be thought of as providing stages for daily and annual debates about regulatory, funding, pricing, and investment legislation.

31.3 The Subjects of Policy

31.3.1 1978 NATIONAL TRANSPORTATION POLICY STUDY COMMISSION

Another way to approach issues is to classify them by their associations with inputs to systems, systems per se, or outputs from systems. Consider the list below from the National Transportation Policy Study Commission's 1978 *Special Report No. 1*.

1. Federal Economic Regulatory Reform
2. Air Carrier Regulation
3. Motor Carrier Entry
4. Rail Abandonment
5. Standard Highway Rules and Regulations
6. Public versus Private Ownership of Transportation
7. Proliferation of Government Agencies in Transportation
8. Consolidation of Transportation Regulation Agencies
9. Federal Transportation Planning Assistance
10. Federal Subsidies
11. Modal/Intermodal Trust Funds
12. Block Grants to State and Local Governments
13. Maritime Trade Support
14. Waterway User Charges
15. Financing Urban Mass Transportation
16. Maintenance, Repair, and Upgrading of Highway Facilities
17. Transportation Industry Capital Formation
18. Coal Slurry Pipelines
19. Energy Conservation
20. Transportation and the Environment
21. Highway Accident Reduction
22. Labor-Management Relations
23. Stimulation of Employment Through Transportation Facility Construction
24. Regional and Community Development Through Transportation Policy
25. Mobility Rights.

About ten of the items belong to the inputs and outputs classes, and the remaining issues deal with the running or operations of the systems. Many of the systems operations issues bear on intergovernmental relations. Another group has to do with efficiency to be obtained through deregulation and/or assurance of a "level playing field" for competition.

Perhaps the only item from the 1978 list no longer relevant is coal slurry pipelines (though other pipelines are still problematic), while several others have been largely resolved for the time with deregulation, namely motor carrier entry and rail abandonment. Some items would be framed differently, but at the core remain the same issues. Everything else from this 1978 list is still a policy topic today.

On review of this list (and considering the many issues into which they divide), we note both narrow definitions and lack of attention to outputs. Issue 24 does have an output orientation. It refers to development. But by and large the social and economic purposes that transportation serves are not stressed. It is implicit that making transportation efficient and controlling ills is all that is needed. Agreeing with Dupuit (1844), who said, "The ultimate aim of a means of communication must be to reduce not the costs of transport, but the costs of production," we think policy ought to pay more attention to outputs. His insight, stated conventionally that transportation is a derived demand, is too often lost in transportation debates.

31.3.2 2008 NATIONAL SURFACE TRANSPORTATION POLICY AND REVENUE STUDY COMMISSION

Meet the new boss, same as the old boss."

—The Who, *Won't Get Fooled Again* (1971)

Transportation policy issues in the new millennium, as identified by the 2008 US National Surface Transportation Policy and Revenue Study Commission, a blue ribbon commission authorized by Congress, resemble those of decades earlier. Almost everything it says about the problems are correct. Where it fails is in solutions, which while not necessarily wrong, are very much of a "more of the same" character and will not enable any kind of innovation.

It asks "Why is Transportation Important?" and identifies five primary reasons:

1. Making Goods More Convenient and Accessible
2. Improving International Competitiveness
3. Developing Markets Within the United States
4. Enhancing Personal Mobility
5. Determining the Nation's Energy Use
6. Impacting Health and Safety.

Each of the subsequent chapters asks a question:

1. What Are the Future Demands on the Surface Transportation System?
2. How Does Our System Function Today?
3. What Are the Long-Term Capital Investment Needs of the System?
4. What Revenue Sources Are Available for Financing Surface Transportation Improvements?
5. Are Program and Institutional Reforms Instrumental to Achieving Our National Vision?

Better management, streamlining programs, improving focus, instituting pricing, and adequate funding are all necessary for the legacy surface transportation systems. And as a short-term policy where everyone is resigned to living in and using a mature system, it is rational. However, the commission billed itself as *Transportation for Tomorrow*, and it is more about preserving the Transportation of Yesterday. In addition to there being no technological fixes for the problems identified (aside from mention of electronic toll collection, the report could have been written 30 years earlier with slightly different numbers), there is no thought given to institutional fixes, affecting who delivers transportation services.

31.4 Interaction with the Transportation Experience

The statements in the lists above are general. How do they generate specific issues? We think the transportation experience interacts with general ideas to yield specific issues. How does that work?

The experience-yields-policy part of the Figure 31.1 suggests a closed system moving along a trajectory. But the environment changes, and it is the transportation experience plus the changed environment that trigger issues. These become policy issues if the situation is in conflict with one or more of the items on the list.

We should recognize that government involvement in transportation is large and long-standing. Also, there has been feedback from the transportation experience to the general functions of governments. The transportation experience thus has affected governments and ideas of what governments should do in a rather broad fashion. The relation is not one-way: transportation has affected the development of government roles; government roles have affected transportation. It is also worth noting that sometimes general policy applied to transportation is taken to be transportation policy, for instance accounting rules or EPA rules.

Also, we should keep in mind that the transportation experience is not broadly known. Individuals know fragments of it, depending on how they are positioned within a mode. The recent experience is better known than historical experience. The longer experience extends to many of today's topics: toll roads, defense, value capture, subsidy for maritime transportation, transportation and land use relations, deregulation, network rationalization, etc. However, since that experience is not widely known, it is sometimes ignored when framing policy.

31.5 Doing Better

The many social programs created or expanded in the 1960s provided windows for the uses of social science knowledge. What was known could be used to cure housing problems, fix education, solve urban problems, and do many other things. But for many reasons, not all of which were related to adequacy of knowledge, applications of the social sciences did not solve problems, and a considerable disenchantment set in. In 1970, social scientist, Nixon administration official, and future senator Daniel Moynihan pressed a change of view; his notion of "benign neglect" became a working rule.[636] As a result of perceived failures, researchers have drawn back from claiming values for applications. In addition, current published work gives short shrift to applications, and methodological contributions are emphasized.

The 1960s rush to apply social science knowledge through policy didn't affect transportation very much. However, economics-based policy has long found an applications window in transportation, and was not rejected in the 1970s. Work by researchers pressing for deregulation was well received and led to the transportation deregulation movement of the late 1970s and early 1980s. In Western Europe and elsewhere (though largely not in the United States), policy that recommended privatization of government transportation assets was implemented in the 1980s and 1990s. Today, congestion pricing represents an application of knowledge directly to policy.

There is a long tradition of the study of government structure and behavior, in political science and law in particular. One result is that policy problems are thought to result from poor government structure, and their cure is to change its configuration. For example, it was felt that the regulation of the airlines was dysfunctional, and the disbanding of the Civil Aeronautics Board (CAB) was the result, a radical change in structure.

The dysfunctional government structure view has much currency in Washington. Indeed, the curing of government structural dysfunctions is often a motive for national transportation policy studies. It is a motive for the reorganizations that occur frequently in agencies.

The work of the Buchanan School has introduced economic reasoning into structural/behavioral work.[637] It employs the well-known "free rider" insight that economists use in discussing the structure of markets and market failures. In political situations, the rational individual would not exert effort toward an end desired by a large political group because that individual would share in the end, even if no personal action were taken. In a large group, the action of one individual is inconsequential. This reasoning helps explain why special interest groups are so powerful. The intensity concept used by political scientists is similar to the free rider notion, and political scientists emphasize how the geographical organization of political power in the United States aggravates the intensity problem. We note that the "free rider" analogy comes directly from transportation.

31.6 Beyond the Supreme Good

We have just suggested some models used to guide policy thinking. Just prior, we discussed the purposes of policy: Why do we create and implement rules to control systems? If there are dysfunctions, something is wrong, and policies seek to manage or correct wrongs. Our brief remark on improving what systems do that is worth doing introduced another aspect of the purposes of policies. They seek improvements going beyond correcting wrongs.

This section will not rehash those remarks. It seeks to emphasize the ways equity and efficiency issues are seen in the minds of beholders and the roles they play in debates.

Imagine being at a conference table or in a public meeting where policy is at issue. You wish to understand what is being said and why. What we have said thus far is helpful. People do come from a vision of what's wrong, and they apply their calculi of how things work to make suggestions. They work toward the "supreme good." They also frame their suggestions (or counter-suggestions to what others have said) in light of their fragment of the transportation experience. If they are working, say, on airport design in response to congestion, that experience frames their views of the air transportation system and of transportation in general.

But it seems that there is something more, something beyond the "supreme good" suggested by Cicero in the chapter's opening, and the influence of experience. Some rules are telling them exactly what to say.

By and large, we would reject the implication that personal or institutional interest, greed, or something like that is shaping remarks. Rather, we think that they are making equity and efficiency trade-offs (see Figure 31.2). Those trade-offs guide "ordinary behavior."

Horizontal equity has a spatial or among-places content; vertical equity has an interpersonal or among-actors of different classes context. Efficiency at either micro or macro levels has to be traded off against equity of different types.

Returning to our imaginary conference table or public hearing, we hear an owner-operator trucker demanding that policies be imposed to reduce the cost of short-term chassis rentals or a small airport operator demanding facility investment subsidies. In our

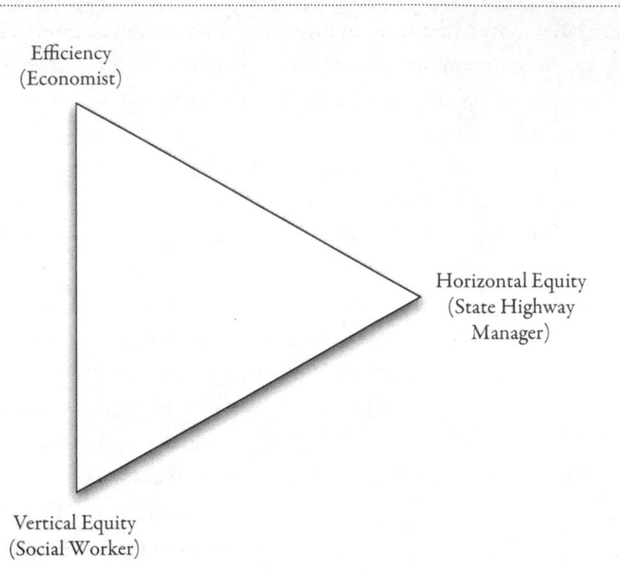

FIGURE 31.2 Efficiency and Equity Space

experience, greed isn't the reason for such demands, although at first glance it may seem to be; rather, it is rules about equity and efficiency. (However, to be less Pollyannaish, people may choose to emphasize the dimension of equity that aligns with their personal interests.)

31.7 Discussion: Application of Policy to Highways

Turnpike and other early road experiences were building blocks for twentieth-century road program governance, construction, and financing. In addition to the transfer of railroad regulatory experiences, which were applied to the trucking industry, the railroad experience affected construction because railroad engineering was the training experience for many civil engineers.

But while the railroads integrated rolling stock and track in the same organization (if in separate branches), the highway system differs. This has posed a great number of system difficulties. However, the disjoint control of trucks (owned by trucking firms) and pavements (owned by governmental road agencies) has created a number of extra costs that proper management of the system might avoid. Pavements are rated for different loads of trucks; roads are restricted to 5-ton, 7-ton, 9-ton, and 10-ton axle weight trucks. Shipments across this network are constrained by the lowest weight limit permitted on the roads to be used (or risk violation—though weight enforcement off the Interstate highways is very sparse). Some roads should be upgraded, some trucks should have more axles, but the disjoint nature of the control makes this coordination difficult. A major solution to these problems lies in rethinking highway financing. The ability to charge truckers different amounts for different roads would put the proper incentives for behavior back in the system.[638]

Economists have been arguing this for several decades, and the policy community, which has some working examples and the promise of modern revenue collection technologies, is finally absorbing it. A second solution, to improve materials to the point that they are "too cheap to meter," that is, so that they are sufficiently strong that it doesn't matter the load using them (within reason), is the analog to building your way out of congestion. Laying pavements with near zero variable (per use) costs may be technically possibly, but their up-front fixed (one time) costs are likely to be very high.

The objective of this book was not to say that certain products (intermodalism, separate truck highways, a Big Dig) are the answer to problems. The discussion was to illustrate how system thinking offers ways to consider policy options. No discussion of how to proceed was stated. The policy question is: How do we forge policies so that system behavior manages problems? That is, we wish that transportation systems had capabilities to learn, reorganize, and renew themselves. They should be testing markets and changing the services they are able to deliver. They should be using technology as an instrument for change. They should have capability to react to problems in full ways.

No large complex system has those capabilities in a high degree. We do not know if transportation systems as a class are worse or better than other complex systems. The disjoint nature of decision making by facility suppliers and system operators (users) is one Devil, as we have stressed over and over again. We think that the urban auto and truck highway systems are "worst case" transportation systems, closely followed by the air system. Within the urban highway system, we think that fixed facilities are the "worst case" component. They limit the options available.

What must we do? First, we need to recognize that problems are rooted in structure and behavior; second, we need to ask how structure and behavior might be changed; and, third, we need policies supporting the actions needed to change structure and behavior.

The question of how structure and behavior might be changed is answered in part by consideration of actions not taken when the systems are developed. Taking some of those actions are an option. Actions should be consistent with social trends (specialize the systems) and problem management (e.g., sharp reductions in energy use, urban congestion, and environmental impacts)

What kinds of policies are needed to support change? System histories answer that question:

- Policy should support innovative individuals and institutions exploring niche-specific markets.
- Policy should give priority to new combinations. The old will enter those combinations, for instance, the extensive right-of-way of the highway system. Lots of new technologies might enter, such as developments in electronics.
- Policy should recognize resource constraints and ecological realities.
- Policy should also recognize that successful transformations will track social and economic development trends.

We think the four general statements just made should guide policy, but are well aware that most policy makers would want something more specific. Being somewhat more specific and paying attention to only the highway system, which is the main focus of transportation policy, one might suggest rethinking the scopes of federal, state, and local interest in the financing and provision of facilities. We think that exercise would result in a much-reduced federal presence in the highway arena. However, there are other considerations. The second consideration is the federal interest in productivity increases and economic growth. That, as we see it, gives the federal government a role in efforts to improve services and the policies to support change just discussed. The third and final consideration is federal interest in health (safety, clean air). Again, we return to the policies to support change just listed.

Hughes (1989) has pointed out that "mature systems suffocate nascent ones." Might the development path we are on suffocate the concepts presented here?

As just remarked, many policy makers want specifics, and one way to generate suggestions is to think about markets. What about high priority freight, small shipments of bulk freight, collector-distributor freight, passengers, and so on?

One may think about existing or emerging technologies that might enter as building blocks in new endeavors. There is much more than the new propulsion, information, and computer technologies so often mentioned. For example, existing tunneling and pumping (for air pressure reduction) technologies suggest vehicles in tunnels moving in a partial vacuum. Applications might be found in passenger and priority freight markets. In addition to savings in tunnel excavation costs (vehicles act less like pistons in pumps and this simplifies needs for cross-over and ventilation tunnels), energy savings from reduced air resistance would reduce the cost of high-speed services.

But there are existing systems for which we have not been able to think of promising redesign schemes. For example, about the only way major energy use reductions can be achieved in petroleum pipelines are by reducing pumping pressure and/or increasing the diameters of pipes (reducing turbulence). Higher energy prices might lead to such actions, but will not open a new pathway for development.

To make a difference in transportation, one has to fiddle with system designs in market niches. All it takes is imagination. Exercising a little imagination, one can think of many system designs that build from what we have and meet social, physical, and other constraints. Applying the market niche constraint reduces that number. One might say that all that is missing is the will and energy to try some things out. But that's not the case, for one needs an overall environment that values innovation and rewards the required risks. The lack of such an environment is part of the problem. The lack of imagination is, too. Because while the world abounds in imaginative people, imagining design changes in transportation requires stretching system designs.

He who has imagination without learning has wings with no feet.

—JOSEPH JOUBERT

32 Speculations

32.1 Capstone

THIS CHAPTER AIMS to place a capstone atop our explorations of the transportation experience. However, perhaps "capstone" claims too much; "speculations" might be a better word. A simple statement of our objective is: "Learn to do better." That's quite a charge, for it bears on the abilities of organized societies to improve and adjust to changing conditions.

What policies are needed to steer systems in more appropriate ways, that is, do better? In Chapter 10, we described the "magic bullet": Demand creates economies of scale that either improve quality or reduce cost, thereby increasing demand. This is true at a macroscopic scale, and describes how technologies are successfully deployed. But moving from within the life-cycle of a technology to the next technology's life-cycle is harder. There are many kinds of systems and all are complex, so there is no "silver bullet" for that technology shifting problem. No single action fits all cases.

What sets transportation and communications systems apart from others are their connecting-enabling roles and the ways they pervade so much of life. They are universal input industries (part of social infrastructure). These features, we think, give special importance to our task, for doing better with these systems enables social and economic development. At the same time, the everywhere-available character of these systems, coupled with lack of recent development, places many aspects of their problems and opportunities out-of-sight and out-of-mind.

We said "learn" because one must learn the rules in order to use, bend, circumvent, finesse, or break them, as well as to have a sense about which ways to go.

32.2 Policy Debates Matter

We have explored policy and policy making as it has been sifted, sorted, and judged by political scientists and by transportation policy wonks, such as James Q. Wilson, John Hazard, David Jones, and Alan Altshuler. This exploration told us that transportation policy is a game that, with a little effort, anyone can play.

The professional should understand issues, varied concerns, and work effectively in contexts where issues are debated and decisions made, so this learning about sizing up and participation seems, *ipso facto*, valuable.

But we must also recognize that many, if not most, debates are about marginalia or miss the point because of poor process specification. Does one's effectiveness in such debates have merit? "Yes," is the answer to that question when one takes the view that much progress takes place in small steps, and even though the debate may be "off the mark" a bit, it's about moving in the right direction.

Consider this example of such a debate. The local mercantilism policy theme often gets simplified to job creation. Cost per job, the incidence of costs, and equity issues are debated. Those matters are important if, say, the debate is about tax reduction and other inducements for the location of a firm in an area. If the action at debate is transportation enhancement, a similar debate may occur about the jobs created to build and operate something, say, an airport. This misses the point. Such jobs may represent a cost rather than a benefit to the economy.[639]

But one could argue that jobs created for facility operations have some positive correlation to the effectiveness of the facility—if a new airport creates jobs (above what jobs there would have been without the new airport), it implies that the airport has increased activities in its service area and something good must be going on.

Yet that argument is quite fuzzy. For one thing, decreased efficiency may require an increased number of jobs, or increased airport efficiency may disconnect airport jobs from job creation in the service area. Other things may be said, but we have made our point. The example debate suffers from poor process specification and argument about airport jobs is argument at the margin of the development process.

To reconsider the question, "Does one's effectiveness in debates such as that just illustrated have merit?" We said "yes" before, but could have said "no" or "it depends on how the subject at debate is framed." We will strengthen our speculation by observing that there are opportunity costs when policies are debated and actions taken.

32.3 Pick Consequential Battles

Debating and taking actions on low-priority or misspecified issues may sap energy and attention from more fruitful endeavors. Perhaps one can tar with the opportunity-cost brush most of the debates responding to transportation energy and pollution problems and also debates focusing on transportation system management.

The Transportation Experience attends to the structures that characterize transportation (and similar public facility) systems, to their growth dynamics, and to the nature of decision-making within systems and between them and the rest of the world. We know that consequential changes in what systems do are rather rare, and are rarely a part of personal experiences.

We have learned that a consequential change in the capabilities of a system involves a change in system format. We know that a change in system format may occur in several ways. One way that format change occurs, for example, is when innovators put building blocks together into a new format (e.g., container shipping); another is when a change in a component forces change in other components (as jet aircraft did).

We know that a system format has a technological as well as an institutional aspect, and that the institutional aspect usually falls in place to match the technological format. (That is why we emphasize technology as the instrument for change.) That is not to say that there are no generic ways to understand institutions. After all, institutions result from learning, and they have behavioral characteristics.

We have learned that once the system format is established, decision making is more and more component and subcomponent constrained, and the likelihood that a change in a component will force system format change becomes less and less. As the scope of available decisions gets more and more locked in, the priority for kinds of decisions changes. Near the end of the growth cycle, we shift from deployment decisions to operations decisions.

This explains why system structure and behavior in transportation provide a context unlike the contexts in which many experienced technology managers and innovators have accumulated their deep understandings, and why mapping from their experiences to transportation is difficult.

That learning can help us place perspective on day-to-day experiences. It explains, for example, why many departments of transportation privilege technologies for system operations. It also explains the authors' opinions that the priority is not likely to lead to anything very consequential, even though there are lots of exciting and interesting technologies to be played with and used as building blocks.

Policy and planning ought to be inquiring, exploring activities, ever opening new options for social and economic choices. They should open pathways rather than mine-out existing ones. The word "consequential" implies that actions should offer a factor of two or more improvements in the human condition. Resources for analysis and implementation are small; apply them to things that will make a difference.

To strive to understand how and why questions, we have invested in seeking what the literature says. There are literatures on the history of science and technology, on the economics of invention and innovation, and on program or event-specific things: for instance, the Panama Canal.

The literature does have a lot to say about what, how, and why, but the information is limited to parts of systems or to the economist's atomistic system. When we ask that something be consequential in the context of large system behavior, the relevance of the literature is sharply limited. The literature rarely analyzes large systems, such as transportation, as a whole.

People do analyze things of consequence to large systems—example, we know the Ford Model-T story very well. What is missing is the connection of the analysis to system behavior, constraints, payoffs, and so on.

Where does that leave us? It requires that we add to what the literature and our own personal experiences say. We urge the reader to make a liar of Georg Hegel, "What history teaches us is that men have never learned anything from it."

Consequential action, as we have defined it, demands thinking about the nature of transportation and its functions. In turn, that thinking suggests interesting work. We cannot resist the comment that conventional action hardly takes any thinking at all.

32.4 Conventional Wisdom versus Consequential Action

Over the years, Garrison has spent many, many hours and days in meetings with "experts." Here are some examples: Not long ago, he met with a group of about twenty, including some policy analysts from the White House, to work out what, if anything, should be done at this time to begin to substitute methanol for gasoline. About the same time, he met with a group of about forty, mostly scientists, to recommend "breakthrough technology" opportunities for energy conservation. He has been in lots of meetings to talk about technologies for highway traffic operations.

Experiences such as these have accumulated into the hundreds over the years; topics have ranged from bus design through regional economic development to earth resources sensors.

What do these experiences teach that might be useful to planning? In particular, is there anything that suggests more consequential planning concepts and methods, as opposed to methods that help us do more than deploy a given system marginally better?

The answer to this question is "not much," although the experiences do provide a point of departure from which we have tried to craft ways of thinking.

Each experience is different, of course. Even, so, they tend to be at two extremes. In the first are situations where there is unquestioning consensus, and the meeting achieves smooth conformity. An example is an experience several years ago dealing with how to incorporate energy conservation experiences into transportation system management (TSM) protocols. The consensus was that TSM is a good thing to do; protocols already adopted were just right; fine-tuning was the problem. Dullsville.

It is true that if something consequential is going on, then conforming and doing better is worthwhile. But we desire to learn how to shift development paths and achieve more consequential outcomes, and cannot learn that from conforming behavior.

Here, Garrison's past ideas conflict with his present ideas. Suppose we had been asked in the late 1940s whether the idea of needs studies was a good one. The answer would have been, sure—all the states ought to do such studies because it will make the highway program a lot better. Needs studies were viewed as a consequential thing to do. In the long run, of course, needs studies locked in ways of thinking and highway development.

In a situation like that today, we would want to do consequential things, but not in a style that locks in inquiring planning.

Garrison's experiences on the National Academy of Engineering Committee on Transportation (NAECOT) advisory committee to the USDOT were at another extreme. The committee didn't conform to conventional wisdom at all. Nothing was taken for granted. The advisory committee comprised individuals who had been very successful in technology innovation and management situations. Members came mainly from private sector, high-tech environments. Each person could speak from personal experience, and previous successes gave authority to what they had to say.

Suggestions made by committee members were far from conventional ones. Compared to conventional suggestions, they had an out-of-left-field flavor. At the same time, suggestions obviously had a lot of insight and knowledge backing them up.

The committee's work was not very effective for two reasons:

1. There was a misfit between the successful (and exciting) experiences brought to the table and the inherent nature of the transportation system. There was lots of talent at the table, evidenced by the major accomplishments of the actors, but the talent just didn't fit transportation.
2. The reason that successful experiences and rich insights could not be transferred to transportation was because while committee-members had good intuition on what might be done, they did not have systematic ideas about how and why things might be done; they had rich insights that couldn't be transferred. It takes how and why knowledge to transfer insights from one field to another.

Other experiences cluster toward either the TSM-energy planning or DOT advisory committee extremes. By definition, most people's thinking conforms to conventional wisdom (often reflecting government programs; if DOT says it, therefore it is important).

Similar experiences are formed when general technology, policy, planning, or other topics are the subject. Such conversations are often impression-based, and, in contrast to talking with experts, tend not to be deep. "I think. . . ." dominates, with little authority-experience to back up the "I think." Often, even the insights and hunches aren't very good.

As a result of these experiences, we conclude:

- That conventional, conforming behavior is wrong headed;
- That unless "experts" have a track record backing up their insights (as was the case on the DOT advisory committee), most "I think" statements are to be disregarded;
- Where the speaker's record says one should pay attention to insights, they can't be used very well without good understanding of why and how questions.

32.5 Interest Groups Are Socially Constructed (Where You Stand Depends on Where You Sit)

We have also undertaken the daring and little attempted task of looking beyond what some sages say. We examined the full modern transportation experience. We asked what the railroad experience says about the logic of policy, and then saw that logic working in other modes.

Why did we explore in this fashion? We have already said that one has to learn the rules, and exploration of the full experience fleshes out the rules. It ought to be apparent that there is much more to the rules than what is found in the current literature and topical debates. Perhaps Friedrich Nietzsche's remark is appropriate, "the surest way to corrupt our youth is to teach them to hold in higher esteem those who think alike than those who think differently," and we have sought to interest the student in thinking differently and questioning those who think alike.

Those who think alike may be thought of as holding conventional wisdom; those who think differently are not so taken by conventional wisdom. To illustrate this point, consider the rather large literature on decision making relative to policy. That literature emphasizes interest groups and the structures within which interest groups seek their ends. It's rich in the sense that trade-offs, side payments, and other such interactive and behavioral dimensions are introduced.

Without denying the usefulness of the literature, we would raise that point that interest groups are not "God given." Their roots have to do with what we have called the national political ethic, transportation systems structures, particular development paths, and other things. We think that understanding the evolution and shaping of interest groups is rich from a causal point of view, for that says how and why the die is cast. How interest groups behave doesn't reach so much to the heart of the matter. That's not conventional wisdom. Interest groups are socially constructed, and how and why are first questions.

To illustrate that speculation, consider the difference between inland water policy debates and highway policy debates. The stage (the die, to reuse the word) tells us the US Army Corps of Engineers is very much a player on waterways matters at the national level, while the highway establishment is relatively a lesser player at that level and more of a player at state and local levels. The explanations for the differences involve the French engineering/statist style of the Corps, and many other things.

Consider the groups that support and oppose road pricing. Where do they lie on the conventional political spectrum? (Recall that social constructions are shaped by technological evolution.)

32.6 No Decision is "Truth" (Many Roads Lead to Rome)

Another speculation is conveyed by the "there are many roads to Rome" expression. We saw the difference between English, French, and US policy styles, and one can imagine how styles in other nations relate to those differences. Each style has its advantages and dysfunctions and these may be debated. However, the bottom line is that many things turn out more or less the same, regardless of the road taken. Many roads lead to Rome. There is no single correct way to proceed, conventional wisdom notwithstanding.

32.7 Linear Reasoning Yields Conventional Wisdom

To this point we have discussed conventional wisdom at a very broad level. Conventional wisdom assumes a mythic level. A first glance interpretation of why this is true might go this way:

- First, something seems obvious, therefore it's true. (When we ride a bus, there are more riders on a bus than in a car; therefore, a bus is more energy efficient than a car.)
- Second, we are lazy, and don't bother to go beyond the obvious. (What about the times when there aren't so many riders on the bus? Because only a few people are

acting as observers, it follows that low ridership is less likely to be observed than high ridership. What about dead-heading, thin collector/distributor service, etc?)

- One can add the underscoring of the obvious by advocates and the inherent difficulty of rebutting the obvious by those opposed to an issue.

While this line of thinking may apply to cases, we think it oversimplifies the situation and is pessimistic about the ability of individuals to think about things. Ordinary life experiences say "you can fool some of the people only some of the time."

The calculus of reasoning yields conventional wisdom. We reason, using symbols, metonymy, and metaphor, to turn complex situations into common sense. This obviousness leads to "ends justifies means" programs and to incarnations of moral imperatives. "Good versus evil" is easy to see. That's how automobiles and highways became peccable; mass transit became good and worth achieving, no matter what.

What's the bottom line? Metaphors are useful, but they can trap us. For instance, the system life-cycle is useful, but we should be wary of the inference that maturity is inevitable and that maturity halts development.

32.8 Policy and Design Would Be Improved by Bridging the Soft and Hard

Should hard or soft technologies be emphasized? Which aspect of technology, the hard or the soft, is instrumental in leading off change? It is important to have a clear answer to that question if we wish to use technology as a planning instrument.

Consider how the rail system came about. The steam engine, coal fuel, tramways, and so on, came together into a workable system format. That workable format called for soft operations, financial, legal, and institutional technologies. That took about fifteen years, and when the hard and soft were in place the system was ready for deployment and the changes in technology that deployment required.

Such a discussion is neat, but we do not regard it as sufficiently deep to serve our search for causality and critical instruments. The reason is that it does not deal with what was behind the first step. What caused the building blocks of the steam engine, tramways, and so forth, to be melded into a system? Those building blocks were ripe for use; the technology was ready. But they were melded into a system by wise innovators, in particular situations, and for particular social and economic purposes.

That reaching for "why" says the soft side is of interest, along with the building blocks or tools. Indeed, the first question is how and why people do innovative things.

This observation changes the question. It is not a matter of emphasizing either the hard or the soft side of a technology. Interest must be on the process through which the hard and soft are birthed, the innovation process. That's where to start.

It is not news that there is a considerable chasm between thinking in the social sciences and allied fields and thinking in the physical and biological sciences and allied fields. It is also not news that the policy world divides into what for shorthand we will call hard and soft. Energy, aerospace, defense, and health policies are largely framed by hard thinking; welfare, education, trade, international relations, antitrust, and some other policies by soft thinking.

The chasm is illustrated by energy policy. By and large, it is dominated by hard thinking, and the creation of more efficient technologies is the objective of policy. Soft policy mainly addresses constraints on demand.

Bridging or merging the soft and the hard would vastly improve policy and policy-making processes. That point is somewhat obvious, and it fleshes out very well when we remember that progress is achieved by design and through design. That's just saying what we have said elsewhere using expressions such as "putting old things together in new ways." It's true that most of our discussion has used hard language to describe creating the new. That's an easy way to transmit images. At the same time, we have striven to include the soft sides of designs.

It needs to be emphasized that the "progress by design" notion isn't just a notion held by in hard arenas. Sanger and Levin (1992) make the point that administrative and policy innovations follow from the "assemblage of old stuff in new ways." These are words very similar to those we use. They go on to say that the sequence "aim-fire-ready" is the one that works, and we see this observation close to our emphasis on trying things out.

What about achieving bridges between the hard and soft worlds? Many would say that economics permeates policy in both worlds and that is the bridge that needs to be strengthened. That's true; yet we have some problems with the economics emphasis. For one thing, that connection has been present for a long time and the hard and soft are still worlds apart, so to speak. (Indeed, both benefit-cost and consumer surplus ideas evolved in transportation contexts.) Economics can be said to serve a "checklist" role for testing the economic feasibility of hard proposals. That's a limited role, and there is no counterpart—a checklist for the hard implications of soft thinking. The creation of designs would provide venues for merging hard and soft thinking.

32.9 Ideas Are Light Baggage

> Next to the originator of a good sentence is the first quoter of it.
>
> —Ralph Waldo Emerson, *Quotation and Originality*

> Good artists copy, great artists steal.
>
> —Pablo Picasso (attributed but unsourced)

> One of the surest tests [of the superiority or inferiority of a poet] is the way in which a poet borrows. Immature poets imitate; mature poets steal; bad poets deface what they take, and good poets make it into something better, or at least something different. The good poet welds his theft into a whole of feeling which is unique, utterly different than that from which it is torn; the bad poet throws it into something which has no cohesion. A good poet will usually borrow from authors remote in time, or alien in language, or diverse in interest.
>
> —T. S. Eliot, "Philip Massinger," *The Sacred Wood*, New York: Bartleby.com, 2000.

Without innovation, nothing would happen. Without imitation and replication, the innovation would stay localized. Imitation is a critical component of the development process. As important as the initial creation is the decision to "copy" or "steal" it. There is a major difference between "copying." which is simply adopting the external form, and "stealing," assimilating the underlying logic. Steve Jobs, co-founder of Apple Computer, embraced this quote, arguing in essence, "If there were a great idea, why wouldn't you steal it and make it your own?" Jobs also said "Real artists ship," meaning that the product must actually be delivered, not simply reside in the minds of the engineer.

Ideas are light baggage. News about good ideas can spread fast and far. Within a few years of the development of the first steam railway in England, imitators were developing others in England, on the Continent of Europe, and in the United States.

As ideas are imitated, they may be re-engineered or reverse engineered, resulting in different implementations. The new instances are recognizable as deriving from the original, but they are no longer exactly the same. American steam railways evolved to take into account American circumstances.

Chicago Area Transportation Study–rooted urban transportation planning methods were adopted worldwide in about a decade. Private sector-like arrangements for government activities are spreading. Tolls in Singapore are emulated in Rome, Milan, Stockholm, and London. The technology for the manufacture of gunpowder seems to have migrated to Europe before Marco Polo's travels. The list is long.

But there is the problem of counterexamples. Ideas about road technology and road programs didn't seem to migrate from France to England very well, though canal technology did. The English read the fate of canals and toll roads versus railroads very quickly, but that reading didn't seem to migrate to the United States very rapidly.

So the notion that ideas are light baggage needs some tightening and interpretation. Can we interpret the notion by examining the carry-over to later times of ideas from the English pre-railroad experience?

The equity issues associated with toll road development and use were mentioned in our discussion. Were those issues light baggage? In a short time frame, they certainly were. The evolution of railroad commodity pricing and Parliamentary trains responded to those issues. In a long time frame, however, those ideas have not proved light baggage.

We didn't mention environmental and property rights issues very much in our discussion. To augment the discussion, we note that the first act enabling road relocation carefully specified that no gardens, lawns, or structures should be taken for road use. Parnell (1837) was concerned about such trespassing.

Our conclusions about environmental and property issues are similar to the equity-toll issues. In the short run, ideas are light baggage; in the long run, they are lost. Experience of earlier debates seems to be forgotten. Few anticipated the vigor of environmental debates, and no one seems to remember how previous debates were resolved.

32.10 Transportation Provides a Mother Logic for Society

We said that the railroads provided a mother logic for modern transportation systems. The transportation experience provides a mother logic useful beyond the transportation context, a logic useful when interpreting political attitudes, institutions, and many other things about societies. A simple reason for this is that transportation systems are large, touch on all of life, and are omnipresent. Caveats are needed. We said "a" mother logic, rather than "the." Also, we should think of the realizations of systems as a result of interactions between systems and society. As systems were developed, they were a venue for learning, shaping, and framing.

The Pennsylvania Railroad as a model for the modern decentralized corporation was mentioned. For the rail mode and other modes, a long list of other such models may be identified—use of job descriptions, provision of health care for workers, retirement schemes, rule establishment and rule following, the union movement, designated inspectors, standards setting institutions, and so on.

Our discussion made much of local entrepreneurship in the interest of local competitiveness and the evolution of constrained capitalism. Are there other themes at this broad level?

Alexis de Tocqueville stressed the tension between the individual and community in American life, and transportation was certainly a stage for the play of those themes because local mercantilism required a community base and community actions were needed, for example, for the provision of road systems. For this reason, one might suppose that community was shaped by transportation.

Yet transportation (and communications) decreased the tyranny of space and the sense of place as a base for action. As Webber put it, we now live in a "no place" world.[640] The sets of interactions of individuals and institutions span long distances, and transportation and communications have yielded placeless community. It has weakened traditional community based on place. At the same time, it may have strengthened community based on common interests. The transportation-communications-community relation has changed over time.

32.11 Transportation Regulatory Agencies Provided a Mother Logic for Regulatory Government

Recall that while government planned and financed transportation improvements in the 1800s, the experience was unsatisfactory because of misuse of funds, shoddy work, and other ills. This experience translated into a suspicion of big business and played a role in the move to constrain capitalism.

It seems to have been widely understood at the time that the fault was partly with government. Governments were small, consisted of many elected persons with few skills beyond politics, tainted by bribery, and had other characteristics limiting abilities to understand, monitor, and control private organizations.

Among other things, this situation and the transportation experience laid the basis for the progressive movement that swelled in the early 1900s. (It also swelled because of the increasing interest in "scientific" management.) Governments were to have expert bureaucrats on board equipped to run government in an efficient and equitable way. Government procurement rules, civil service laws, personnel programs, and so on, followed. The implementation of progressive government was speeded by expanding road programs (and public health programs) where the meaning of expertise was clear and diffused costs and benefits asked for equity.

Transportation problems and programs seeded and nurtured government by professional bureaucrats. (A point made earlier about government's learning about regulation.)

How did the transportation experience steer today's intergovernmental relations? We speculate that there was a lot of steering, even though the relation often seems fuzzy. Consider the growth of federal power. The up-the-hierarchy movement of government power in transportation began with the nation setting port charges (favoring US ships in the coastal trades) in the early nineteenth century, Army Corps of Engineers programs in the mid-nineteenth century, and the creation of the Interstate Commerce Commission in the 1880s. Power moved up the hierarchy in other programs subsequently. Did the transportation experience influence those programs? We can find cases where the answer seems "Yes," such as the use of federal grants to steer local programs. But the link seems weak in other cases.

Along with power moving upward, there has been an increase in the size of the federal establishment. Today, the US DOT has some 63,577 employees.[641] There is also the state and local government partnership character of programs. Developing early, the transportation experience, especially the highway experience, must have affected partnerships that came along later, but "how" poses a larger question. The answer to the "how" question is fuzzy in part because the transportation experience was a very diverse one, example, the Corps of Engineers program development experience was very different from the highway program experience.

32.12 Transportation Opportunities Enable Innovation in the Economy

For completeness we need to include the companion innovation speculation. Transportation improvements open opportunities for innovation. The pursuit of those opportunities steers development generally.

32.13 Transportation Is Bound by a Social Contract

We have said that government policy formation and implementation has, in part, a "default" character, and we noted government actions in instances where the embedded policies of the railroads could not meet the needs of the properties or manage their problems. We argued that government actions are also taken when system services fall short of social expectations, that is, when the social contract is broken.

To deal with transportation's relations with society, it is useful to think of a social contract. Society is one member to the contract. Members of the population and social and economic institutions hear about and/or experience and learn what transportation can do; expectations develop. Service providers test the market, and they gain a sense of what is expected of them. In a sense a kind of contract emerges, a social contract, and problems arise when the parties to the contract no longer honor it.

For an example of a contract in trouble, today's highway establishment is upset because the public is no longer providing the money to continue with "business as usual." The public is unhappy, too—facility providers are no longer providing a better road system. The social contract has been violated.

Speaking especially to highways, David Jones[642] put the matter this way:

> Transportation development is a contractual phenomenon. At its core is a negotiated agreement about what to build and how to pay for it. This agreement has the character of a social contract. Transportation development can proceed when there is agreement on what to build and who shall pay. It falters when such agreement erodes.
>
> Transportation development is also a technological phenomenon. Progress is made as improvements in technology are introduced, perfected, and built up into systems. The introduction of successive technological improvements—developmental progress—hinges on successful adjustment of prevailing agreements on what to build and how to pay for it. In turn, sustained development hinges on renewing that agreement in successive rounds of investment. Each round of investment reopens the question: What to build next and how to pay for the next increment to be added. Without renegotiation, consensus erodes, and development falters.

And Jones sees political actors as negotiating successive rounds of development investments. We like Jones's discussion, but would not want to restrict the notion of social contract to instances where the contract is formally negotiated or those where there are successive rounds of investment. We think of a contract's existing if what is being done is acceptable to both parties. Things are copasetic if the "if it ain't broke, don't fix it" rule is holding.

32.14 The Life-cycle Dynamic of Systems Strains Social Contracts

For example, as long as a system is in a deployment phase, service improves and expectations increase. Growth by deployment results in access to ever increasing markets and social opportunities. But once deployment is complete, service providers cannot deliver expected improvements: growth slows; profits return to normal; all customers are served. At this point, companion innovation slows down until the next technology comes along.

The social contract implies that society will serve the needs of its members. With respect to the physical facilities used by the public, needs are recognized by observing how users behave. The physical dimensions of needs are then defined through analysis. Geared to efficiency, safety, and other goals, the effort is made to provide for needs in desirable ways.

We see cars consuming numerous gallons of gasoline per day, we count the traffic on a highway route, or we observe how travelers check in at an airport. The professional regards such behaviors as needs, and strives to define them in physical terms. The latter, the definition in physical terms, is often stated as the need: we need an M meter-wide pavement of P centimeter-thick Portland cement concrete on route R; at time T, we need N open ticket counters for flight XXX. Highway planners today are observing the growth of the suburbs. Tomorrow's needs are for suburban facilities.

While the definition of physical facility needs is less than perfect, there is no doubt that there has been steady progress in responding to needs defined this way. Compared to other wealthy nations, the United States does a pretty good job of meeting needs.

Unlike needs, the notion of wants has not been well operationalized. Wants is a vague idea. While needs may be limited, wants are practically insatiable.

In the mythical, purely atomistic world of the economist, there is market clearing. The goods or services that consumers want get turned into needs (things purchased) through a calculus that balances preferences, disposable income, supply and demand relations in aggregate markets, and prices.

How well do wants get translated to needs? That depends on lots of things, with the level of disposable income being critical. The rich can get what they want; the poor do not do as well.

Taking one step back from a static market situation, how well do the goods and services supplied track on changing wants?

In the atomistic world there are innovators developing new products to replace the old. There is a constant exploring of ways to meet old and new wants. Successful in the market, new products or services displace the old. The new, responded to in the market as a need, matches wants better than the old. Something needed before is no longer needed. The market tells me there is little or no need for buggy whips. Now, there is a need for GPS-based in-vehicle navigation systems.

At least that's the way it is supposed to work in the private sector.

In the public sector, the lack of fit between the ever-changing wants of publics and needs as defined by suppliers gets treated through political markets. That process works to some extent. It has some distortions, of course. Publics with intense interest in some want can get that turned into a general need. For instance, there are those who intensely want American jobs and bottoms in coastal trade. Therefore, the nation needs Jones Act protection.

The private sector-style innovator exploring new ways to meet wants is generally absent in the public sector. When an activity is identified, interest is mainly in more efficient ways to meet wants. It isn't in new products or services that displace the old in the fashion that we imagine in the private sector.

The planning style suggested is aimed at stimulating innovation in the public sector—transportation in particular. The innovation of concern is market- or wants-exploring.

We have suggested that the private sector provides a model for that kind of innovation, but that's arguable. The private sector is less than perfect. Some private sector managers have little use for thinking about new things; others (like the reasonable man in the Shaw quote at the head of the chapter on innovation—Section 4.10) accept the world as it is and concentrate on playing the cards they hold. And some let others innovate, assuming that they can produce a look-a-like that's better. Let others take the risk; after all, the pioneer is the one with the arrow in his back.

In the early days of the life-cycle of systems, two questions are explored. The first concerns supply: "What technology designs are feasible?" The second relates to demand: "What do publics want?" New systems meet wants rather than needs, as real needs have already been satisfied by older systems. The social contract says when satisfying new "wants" don't unsatisfy old "needs." This is in line with the notion of undertaking only Pareto improvements, those that make some people better off without making anyone else worse off. The problem arises when "ensuring steadily improving conditions" or "ensuring that everyone achieves proportionate gains" appear as needs rather than wants.

32.15 Transportation Tomorrow Will Resemble Today

We have mainly been looking back. We have seen how the policy, planning, deployment, and management tasks changed as modes ran out their life-cycle dynamics. We also looked back at ideas. We have also seen how techniques have evolved. Looking back has an element of looking around, for we look back using today's perspectives—we interpret what-used-to-be using today's perspective.

Looking around, we see that much has been accomplished. Most everywhere, systems are safer, more reliable and accessible, cleaner, and so forth. That's in spite of the growth in demand and the many who predicted disasters just around the corner because of shortfalls of capacity.

Looking around says a good bit about looking ahead, for tomorrow we will be reacting to today's ideas and problem statements, institutions and policies, and social and economic trends, as well as the running-out of the life-cycle dynamics of systems. Assuming that progress similar to that we are making today continues, tomorrow will be a polished today. (Assuming away the possibilities for war, depression, violent social conflicts, disease, and sudden ecological disasters.)

A major reason that tomorrow will be much like today is that the turf is occupied by legacy systems. There are the physical systems—highways, transit, airports, and so on. Institutional and financial arrangements are there, too—metropolitan planning organizations, congestion management agencies, fuel taxes, state agencies, and so on. And there is an endless list of things ranging from legal arrangements to protocols for construction contracts and the job descriptions of transportation planners and workers. Legacy systems change slowly.

Institutional inertia bears on the continued life of legacy systems. But we think social contracts are the strong glue that says that legacy systems will continue into the future. It's decided to do something a certain way, and that's the contract. Consider the every five year or so surface transportation bill as a social contract. It will stick in ways, regardless of what Congress does. First there is the cost of contracting—getting the legislation. It takes years, much effort, and lots of horse-trading (side contracts). The bill occupies the intellectual turf as the solution, and other solutions have to find a place to spawn, before pushing aside the current wisdom. Although it may not be so great in the case of a single pork-barrel-laden surface transportation bill, most contracts have ethical implications if they are broken—people get hurt.

The social contract is often described in terms such as those applying to jobs, rights to education and welfare, and security. It really is much broader than that, for it shapes the accepted views of problems and solutions; it grandfathers things.

We are observing that actions are constrained by contracts. However, a naysayer might comment that contracts do get broken (e.g., as corporations downsize) and that Congress is engaged in massive contract breaking (downsizing and devolution)—welfare, education, environment, and so on. That's true, and it underscores the cost of contract breaking, as well as the role of imperatives when contracts are broken. Competition is the imperative in the marketplace, and the national debt is the culprit in the government case.

We wonder if the use of a highly metaphoric language in transportation is a dysfunction. We speak of arterials, bypasses, transportation-housing balance, and many other things using metaphors. Does this limit the scope of planning actions and affect the ways that tasks for planning are identified?

The disjoint character of systems is another dysfunction. It constrains the way systems are conceptualized and thus constrains the ideas for improving planning. The situation is worse, however, because turf is occupied by ideas of limited effectiveness. They translate to techniques that are similar to hammers. If that is what one has, everything looks like a nail. In this kind of world:

> ... research is involved in a positive feedback spiral where increasingly inefficient research becomes the training ground for each new generation of research workers.
>
> —Forrester and Brown (1975)

Finally, life-cycle-based processes play a large role in defining the stage and thus how ideas are shaped and techniques used. With respect to the life-cycle, one presumption is that things will go along as they are: maturity and stasis continue. Emphasis on transportation management is one result. Among other things, there is fine detail mining-out of opportunities—such as opportunities to develop HOV lanes or park and ride lots. Another feature of maturity is that institutions are very conservative and risk adverse, as well as being very process oriented.

At the same time, achieving scale economies in actions is critical. That's partly to reduce risk. It is also a requirement for being competitive. Facilities such as the new Denver airport, the Channel tunnel, and high-speed inter-city trains are very risky, and a critical question is: "Will they achieve necessary scale?" Within urban areas, people movers and rail transit investment have faced the scale question.

There is the option of breaking the tyranny of the existing life-cycle. That could happen as the result of inventors and innovators creating new technologies that break the dependence on old systems. An alternative would be to implement planning and policy to break the tyranny.

What is surprising is the number of project starts made even when it was already clear that a superior system was replacing the system being planned. This was the case for many canal projects in the United States. Investments were made even though there were ample hints of the potential of railroad competition. The competition dynamic says that plans ought to consider what the competition will do, as well as considering strategies for decline in the face of competition. The inability to break away from legacy systems places a curse on posterity.

32.16 All Nations Are Both Developed and Undeveloped Concerning Transportation

> For we already live in the openness experiment, and have for two hundred years. It is called the Enlightenment—with "light" both a core word and a key concept in our turn away from 4,000 years of feudalism. All of the great enlightenment arenas—markets, science and democracy—flourish in direct proportion to how much their players (consumers, scientists and voters) know, in order to make good decisions. To whatever extent these arenas get clogged by secrecy, they fail.
>
> —David Brin[643]

David Brin argues that the Enlightenment experiment has been about positive sum games in open-competitive economic markets, science, and democracy. Free movement of people and ideas is a critical enabler of open science, open markets, and open democracy. That movement is made "free" by policies, such as those that allow migration or eliminate tariffs, but it is also made freer and lowered in cost by new transportation and communication investments that allow more to be done with less, more travel in less time or at lower cost.

In a similar vein, Ferguson (2011) credits six "killer apps," or social developments that set Western Europe and the Americas on a faster growth path than the rest of the world: competition, science, property, medicine, consumption, and work. Peter Gordon has suggested adding positive agglomeration and networking economies to the list,[644] which are of course enabled by the relative location of activities to each other and the transportation between them.

Although there are gradations, it's useful to speak of three types of nations: developed, developing, and undeveloped (following the maturity, growth, and birthing stages). From a transportation perspective, all nations are developed nations. That seems to counter ordinary experience in undeveloped and developing nations, where service isn't of high quality or everywhere available. In what sense could this be true?

The modern transportation systems were birthed in the "developed" world environment, energized development through companion innovation processes, and were deployed as development pushed and pulled deployment. At that time, they were deployed in the undeveloped world. They were pushed and pulled by the same processes. For example, there were early railroads in Africa and South America where development opportunities called for them. The difference between the developed and the underdeveloped nations is that the undeveloped nations experienced Western-style development at the fringe, so to speak. The companion innovations that bloomed as modern transportation was created and deployed fit the Western situation very well. They took hold only in limited ways in other places.

The economic development programs that emphasize deployment of the systems successful in the developed nations in the undeveloped nations don't much make sense. Wilfred Owen argues equity as a basis for subsidized deployment. It isn't fair and just for the undeveloped nations not to have good highway and other services. That argument has merit. After all, their deployment has already been tried with limited success. What's needed is the development of services suited to the situations in the undeveloped nations.

On the other hand, one could rightly view all nations as undeveloped in a transportation sense. That follows from observing that modern systems are not so modern. They were developed using once-modern tools to fit once-current circumstances, and they are obsolete today.

Following Ferguson, the social developments in the developed nations cumulatively fed on each other and made progress occur faster. Following the life-cycle model, though, as those developments play themselves out, the benefits from them experience diminishing returns, allowing less developed nations to catch up if they emulate (with short-cuts) the development path of the mature economies.

Another useful quote is from science fiction writer William Gibson: "The future is already here—it's just not very evenly distributed."[645] One example is M-Pesa, found in Afghanistan, Kenya, South Africa, and Tanzania, which allows sending cash from one mobile phone to another, as a means for avoiding highway robbery. This innovation is driven by needs in the developed rather than developing worlds. Similarly, development in the less and more developed countries differ more in distribution than existence of transportation technologies and networks.

32.17. Policy Should Create Environments for Innovators to Improve Service

In the first paragraph of this chapter, it was said that our objective is to do better. Toward doing better, the discussion above sought to dig deeper. For example, it was said that we ought to go beyond how interest groups behave and be concerned about their structures and origins. It was said that the use of metaphors has an upside and a downside. Running through *The Transportation Experience*, there was effort to understand policy in a generic fashion.

Also, we sought to scope widely; for example, we considered both mode-embedded and government policies, and we recognized logic-copying, learning from experiences, and so on. We scoped to considerations of national points of view about the roles of government. Growing out of the examination of life-cycles and the mature status of today's systems, there was effort to scope to the energizing, redesigning, or rebirthing of systems.

The development of more efficacious policies surely will be aided by our digging deeper and scoping widely.

Scoping to energizing adds a dimension to policy, and our final speculation is that policy should create environments within which organizational and individual innovators can search for system-scoped service improvements. All the speculations bear on this task. For example, an environment ought to encourage the merging of hard and soft ideas, it ought to provide for rich problem and opportunity statements. We will not discuss the environment for innovation further because that topic goes beyond our policy speculations.

Turning now to dysfunctions, recall that a large part of the improvements in the financial health and productivity of the railroads has been achieved by treating dysfunctions accumulated during the era of regulation. A similar sort of thing might be imagined for urban transportation. For instance, the railroads constrained their common carrier responsibilities, changed pricing to a modified Ramsey (inverse elasticity) scheme, and rationalized network and institutional structures. Urban transportation systems might adopt similar changes.

Some obvious dysfunctions in transportation service reside at interfaces: interfaces between local and regional roads, interfaces between transit lines, interfaces between airports and rail stations and local transportation services. There are failures to balance costs and prices—leading to over-consumption and under-production of transportation services. There are also negative externalities such as air pollution and noise. A long list could be made of such dysfunctions. Such dysfunctions are targets for plans and actions.

Planning for changes of this sort would represent an expansion of existing activities. A sticky aspect is that some social contracts might have to be modified.

However, there is one policy question that should be mentioned. There is renewed interest in proactive policy (as opposed to what might be called enabling policy). Tools include subsidy, tax advantages, regulations, investment in demonstrations and research and development, and the use of government procurement to create markets. How proactive should policy be?

Considering national policy, our bias is very much toward emphasizing enabling more than enacting policy. Our view is that there are lots of latent ideas that would bloom if allowed. Something that has to be pulled, say, by subsidy is probably not a very good idea. Also, national-level programs suffer from one-size-fits-all, an emphasis on big programs, and other ills.

Let's put this another way. There are lots of enthusiasms out there. The niche market venues for trying out lots of things would seem to be at the local level, where local mercantilism is a motivating force. That suggests proactive local policies, and that is fine because risks, payoffs, measures of progress, objectives, and so on, would be under close scrutiny. This close coupling isn't present at the national level.

One may object to the authors' position by claiming that some things, such as research and development and massing large amounts of capital, might be most efficiently done at the national level. And one might add the notion that the national government could best recognize and utilize creative talent. As we have said, however, progress has come from new arrangements of old things in unsatisfied market niches. Research and development to improve systems and finding capital for their deployment become salient only after something worth doing is found.

A second objection follows from the "more than one road may lead to Rome" observation—a national proactive stance is another road. That could be true. But it seems chancy at this point in time because national government activities in transportation have not nurtured the development of the traditions, insights, and talents that a national proactive program would require. One could say that they aren't present at the local level either, so the issue becomes that of where learning best takes place: in local laboratories or by national direction. Experience will show the way.

11 Afterwords: Reflections on Transportation Experiences

33 The Fall and Rise of the I-35W Mississippi River Bridge

by David Levinson

C: STARTING DATE Wednesday, August 1, 2007, at 6 hours, 5 minutes and 38 seconds pm

V1: Metro, fifty cars there's report of some sort of collapse in the construction zone. North of University [Avenue]

V2: 437

V3: 551 I will be 10-8 [available for incidents] down here, she's on her way doing triple A now

V4: 224

V5: Metro, West Metro's going to be 10-33 [alarm sounding, emergency] at this time, all cars, fifty cars, we have a bridge collapse, the River Bridge over the Mississippi River Bridge is down

V5: 2500 do you copy? [siren in background]

V6: 80 metro 60 cars being routed metro

V5: We will need southbound closed and northbound, both sides are down

V5: 2500 do you copy?

V7: In route 169 to 97

V8: 10-4 [OK, I acknowledge] 1806

C: Ending Date Wednesday, August 1 at 6 hours, 6 minutes and 53 seconds pm

Note: C: is the computer time stamp, Vn: are the different voices heard on the recording. The 10 codes have been defined in [brackets].[646]

33.1 Introduction

After spending 10 months on sabbatical in London studying the co-evolution of transport and land use, on Tuesday, July 31, 2007, my family and I returned home to Minneapolis. We had many things to do to restart a household, among them shopping. Relatives were in town for a conference, and they had rented a minivan. The next day we went to a warehouse club (Costco) to stock up on basic stores, filling the back of our vehicle with typical American middle class goods (paper towels, diapers, etc.). While we traveled to the store on what any online mapping service would suggest is the shortest path, on the return at about 3:00 PM we took West River Parkway instead of the highway to avoid traffic and have a more scenic view. That route runs along the Mississippi River, and passes immediately below the I-35W Bridge. I did not look up at the Bridge from below as we drove under it.

At 6:05 PM, CDT, August 1, 2007, the I-35W Mississippi River Bridge famously collapsed.

By 6:10 PM, I and the world knew about the collapse from watching both local news and CNN. People from around the world contacted me wishing well.

I didn't know any of the 13 dead or 145 injured at the time, and was as surprised as anyone at the collapse, having driven under and on the Bridge many times. We discovered that it was an eight-lane truss arch bridge that had opened in 1967, and carried about 140,000 vehicles a day. Everyone in the Greater Minneapolis–St. Paul region has their own story, some heroic, most mundane. Everyone, though, remembers it.

The Regional Traffic Management Center (RTMC) in Roseville, Minnesota, received news instantly, and had video cameras in the area, which they quickly pointed in the direction where once stood the Bridge. The recording of their audio inputs are transcribed in the opening quote. In the stream of random and mundane information coming into the center, communications were received about the collapse.

The next morning (August 2, 2007, 8:24 AM) Paul Levy of the local *Star Tribune* newspaper reported an article with the headline (one which varied across the day and week) *4 Dead, 79 Injured, 20 Missing after Dozens of Vehicles Plummet into River*.[647]

As with any tragedy, information as of August 2 was incomplete. People were missing, some of whom were found alive, others dead. The estimates of injured went up as better counts were made.

In the days following, I received some seventeen media contacts asking about the traffic effects. My structural engineering colleagues received many, many more. As researchers, my transportation colleagues and I quickly proposed studies to examine the consequences of the collapse.

Users take infrastructure for granted. From the roots "infra," meaning below or underneath, and "structure," meaning building or assemblage, infrastructure is by its very nature not obvious, and is often hidden in plain sight. Yet its absence is noticed. Americans seldom complain about lack of on-demand electricity (blackouts), natural gas, or water, but often complain about lack of on-demand transportation capacity, which we call congestion. When construction or events close routes, so you cannot get from here to there, the complaints rise. But when infrastructure fails unexpectedly, it engenders shock rather than complaint.

Why did the Bridge collapse? And what does it say about the state of infrastructure in the United States, and for that matter, the developed world?

In a one-time event, blame is a useless exercise. My blaming you will not produce better future outcomes. But in a world with signals and repeated games, blame can lead people to behave better in the future, and the prospect of being blamed for failure may encourage behavior to avoid failure. Too much blame for failure, and insufficient reward for success, will lead to risk-averse outcomes. In some arenas—conventional finance, structural engineering—risk-aversion is probably a good idea. They provide a lattice on which the rest of society depends to accomplish their own work. The gains from innovation are likely small, the losses higher. In other areas, for example war, risk-aversion on the part of the weaker army may ensure defeat.

To answer the question of "why the Bridge collapsed," one can look to physics, and blame gravity. One can look to structural engineering and blame undersized gusset plates. Or one can look to construction engineering practices and blame overloading. Or one can look to traffic engineering, and blame the need for all those people to get from A to B. Or one can look to politics, and ask why the Bridge, which had (different) known problems, had not already been repaired. And so on. The layers of blame are worth exploring.

The collapse of the Bridge illustrates several different kinds of networks in action. The bridge itself was a structural network, a connection of steel and concrete elements designed (but in the end failing) to transmit force safely from the Bridge deck to the ground. The Bridge was a link in the transportation network, an element of the limited access US Interstate Highway System enabling people to travel by car from point to point without stopping. The news of the collapse of the Bridge was transmitted over communications networks (both electronic and social); it was a quickly transmitted piece of information.

33.2 Structure

Bridges are designed to overcome gravity. They take travelers over a trench, river, or chasm of some kind to reduce the costs of travel. In their absence, travelers would need to descend and ford a river, take a ferry, or make some other less convenient accommodation. Bridges are networks, sometimes simple, sometimes complex, for transmitting forces from the air to the ground. These networks may be of stone, concrete, wood, steel, or other materials. The network elements are connected in various ways.

The I-35W Bridge was constructed as part of the Interstate Highway System. It was not the first crossing of the Mississippi River in the City of Minneapolis; one can see many other crossings from the photos and maps. Immediately upstream we find the oldest extant crossing, the curved Stone Arch Bridge, dating from 1883, which originally brought trains of the Great Northern Railway across the Saint Anthony Falls from Old St. Anthony on the east bank of the River to the Mill District on the West Bank, and now acts as a pedestrian crossing. Immediately downstream is the 10th Avenue Bridge, opened in 1929, and still carrying vehicles. The first river crossing in Minneapolis was the 1855 Hennepin Avenue Bridge, a tolled suspension bridge, which lasted at least twenty years.

In principle, engineers know (or knew) how to build long-lasting structures, that with proper maintenance could last centuries. In fact, the Pons Fabricius in Rome was originally constructed in 62 BCE, more than 2,000 years ago, and has remained in continuous use. So something went wrong on I-35W for it to last only forty years, and something has gone wrong in civil engineering practice if we are designing bridges to only last fifty years.

The National Transportation Safety Board, the federal government agency for investigating failures, engaged in an extensive one-year study of the collapse. Inadequately sized gusset plates, sheets of steel that connect truss members, beams, girders, and columns in bridges and other structures, were the proximate cause. While the gusset places were too thin for the design, they were not so thin that the Bridge collapsed earlier. The Bridge was undergoing some construction at the time of the collapse, only two of the four lanes were open to traffic, while the others were being resurfaced. It was the combination of the undersized gusset plate with increased weight of the Bridge over time (due to things like pavement resurfacings), and in particular, the loading of construction materials on the Bridge, above the gusset plate that day that was the proverbial "straw that broke the camel's back." Once one gusset place cracked and could not support the loads, cascading failures led to the collapse.[648] The Bridge was fracture critical or "non-load-path-redundant," meaning that once one critical element failed, there was no redundant element to take the load.

Tom Fisher says that fracture-critical design has four characteristics: lack of redundancy, interconnectedness, efficiency, and sensitivity to stress.[649] Beyond that, it has long been known the Bridge was structurally deficient, and it had been investigated for other possible failure modes. A report by my late colleague Bob Dexter is interesting in that it said,[650] "As a result, MnDOT does not need to prematurely replace this bridge because of fatigue cracking, avoiding the high costs associated with such a large project." The report was correct as far as it went, since fatigue cracking was not the source of failure. It did not identify the problems with the gusset plates, nor did any inspections after construction.

The United States still has about 18,000 fracture critical bridges.[651] Some 465 have similar designs to the I-35W Bridge. There are about 72,500 structurally deficient bridges, according to USDOT, out of about 600,000 bridges.[652] Another 80,000 are functionally obsolete, which does not imply a bridge safety problem, but means they are not to standard, for instance with narrow lanes, or are under-capacity for demand.

Bridge failures on the Interstate are not as uncommon as one might think. Table 33.1 shows significant Interstate bridge failures and their causes.

Other bridges have been closed before failure, and repaired or replaced. The Sherman Minton Bridge across the Ohio River was closed in 2011 after cracks were discovered, and repaired. While some causes seem to be acts of nature (earthquakes) or difficult to predict (barge collision, truck explosion), good design will defend against even those failures, at least to a point. The trade-off inherent in all design is the amount of failure to be accepted. Will we accept one Interstate bridge failure in the United States every day (no), every month (no), every year (no), every decade (yes), or every century (yes)? Nine failures in twenty seven years indicates about one every three years is somehow acceptable.

Each higher standard is increasingly expensive. At some point, money spent on reducing fatalities by making ever safer bridges outweighs the same money spent on reducing

TABLE 33.1

Interstate Bridge Failures and Their Causes

Location	Date	Facility	Proximate cause
Tampa Bay, FL	May 9, 1980	I-275 ship collision	
Greenwich, CT	June 28, 1983	I-95	Metal corrosion, fatigue
Oakland, CA	October 17, 1989	Bay Bridge	Earthquake
Oakland, CA	October 17, 1989	I-880	Earthquake
Milwaukee, WI	December 13, 2000	I-794	Weather, traffic?
Webbers Falls, OK	May 26, 2002	I-40	Barge collision
Bridgeport, CT	March 2003	I-95	Car-truck fire
Oakland, CA	April 29, 2007	MacArthur Maze	Truck explosion
Minneapolis, MN	August 1, 2007	I-35W	Design, construction

deaths some other way (e.g., increasing traffic safety or reducing air pollution). For instance, a billion dollars annually spent on reducing expected fatalities from bridge collapses by one person per year is 200 times more than would be spent reducing traffic fatalities (where the "statistical value of life" is on the order of $5–6 million per person), and would be a misallocation of resources from a safety perspective. We can of course potentially add the costs of infrastructure replacement avoided, but we currently spend more per life saved on safety in structures than on safety in traffic. As with aviation crashes, bridge collapses are highly visible and are perceived as more common than they really are.

33.3 Communication

The I-35W Bridge collapse occurred before the advent of Twitter, when there were only 50 million users of Facebook (as of July 2013 there were over 1.1. billion users, and growth in user numbers seems to be leveling off). I joined Facebook on November 13, 2004, so they tell me, when Facebook had fewer than 1 million users,[653] but it was pretty much useless to me until late 2008 when enough people I knew were on to make it interesting to check in. And though I added 24 "friends" in 2007, I never posted. It did not even occur to me to update my Facebook status, which would likely be the first place many Twin Citians would go today in such an event. I did update my blog the next day.

Yet the news traveled fast. TV, radio, on-road variable message signs, phone calls, e-mails—all helped transmit this knowledge. We have evidence on how the news traveled by looking at traffic counts. Figure 33.1 shows the difference in counts between August 1 and a week earlier, July 25, which are otherwise similar days. As noted, behavior changed quickly that night, traffic counts were lower systemwide, but especially upstream and downstream of the collapse. In contrast, the best long-distance alternatives (Mn100 and I-35E) saw upticks in traffic.

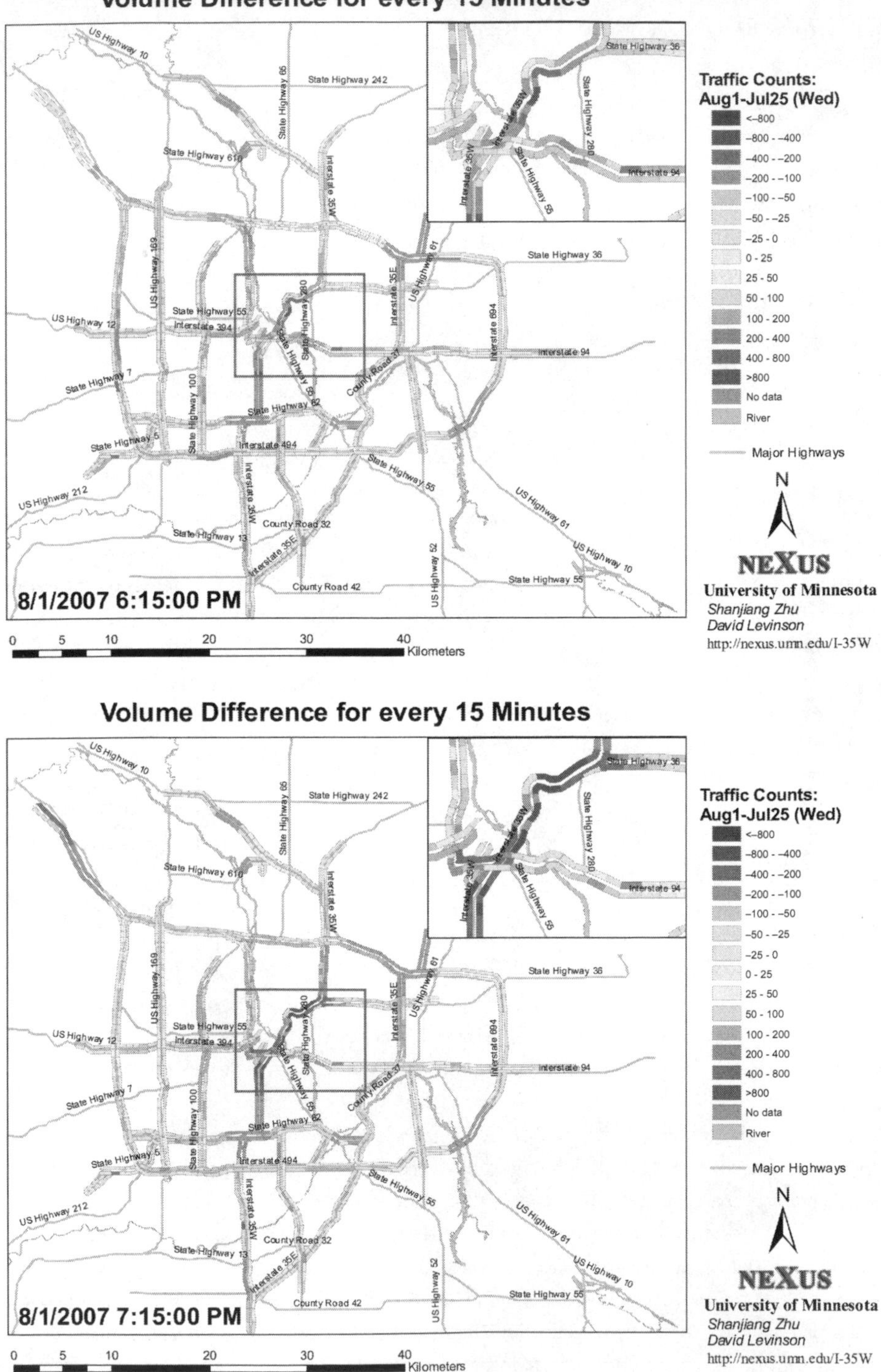

FIGURE 33.1 News of the Bridge Collapse Traveled Fast. Parallel (substitute) routes saw an increase in traffic. Upstream and downstream links saw a decrease in traffic.

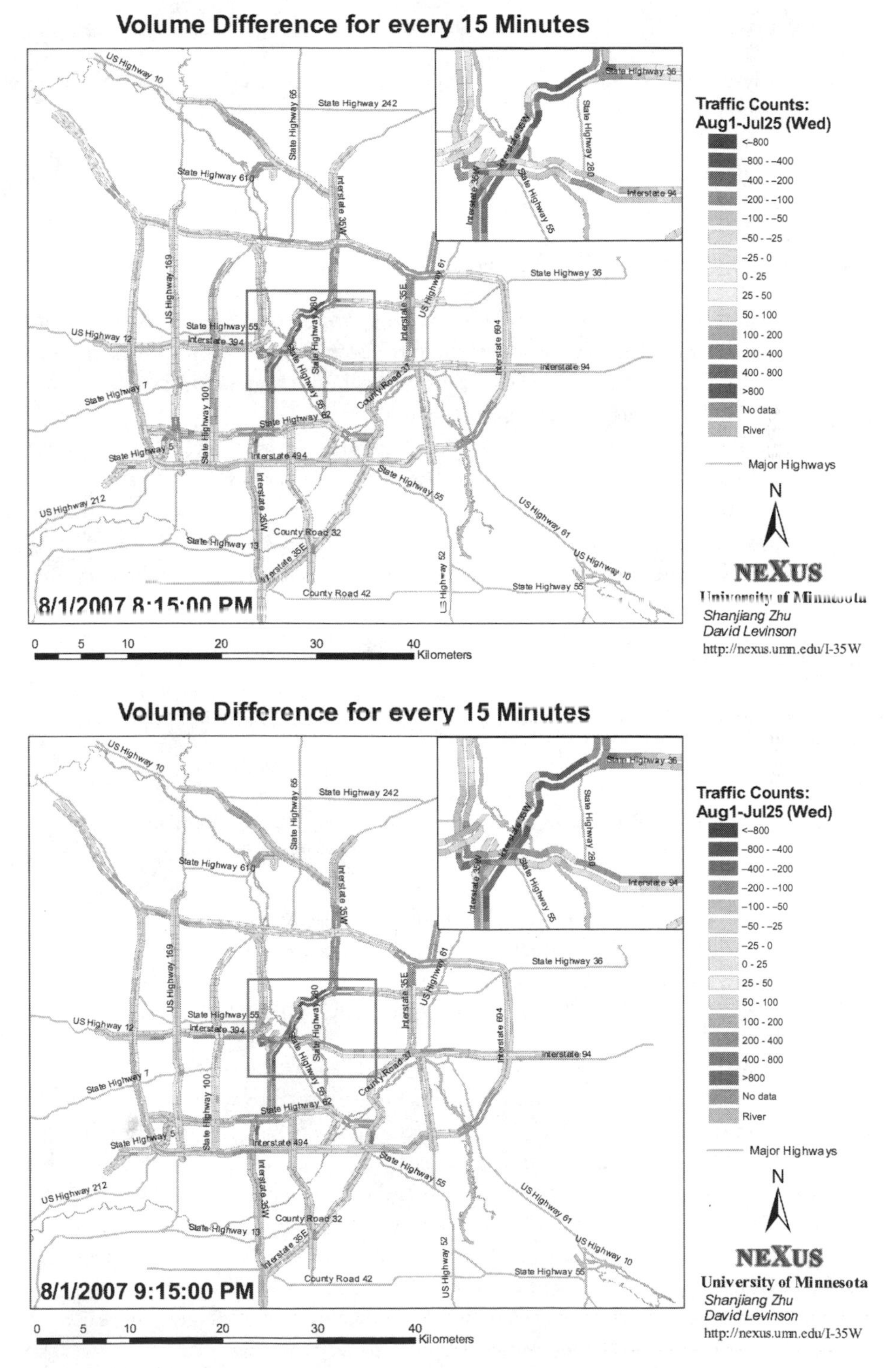

FIGURE 33.1 (*Cont.*)

33.4 Politics

Some will say the Bridge collapse was not about money. Throwing money at the Bridge would not have kept it from falling. Others note money could have bought:

- more inspections;
- a structural (finite-element) model of the Bridge;
- better, faster repairs;
- the ability to replace the Bridge sooner.

Money could have been spent more wisely. More important, money is always a constraint on decision making at MnDOT. As was noted in the *Star Tribune*: "Phone call put brakes on bridge repair: Plans to reinforce the Bridge were well underway when the project came to a screeching halt in January amid concerns about safety and cost."[654]

Governor Tim Pawlenty had already vetoed a legislature-passed increase in the gas tax that could have raised money to repair bridges like this one. The latest vetoed gas tax would not have solved this problem, but previous taxes that were not passed (due in part to Pawlenty's previous veto threat) may have, had the money been spent on this kind of thing. The gas tax had not been raised in Minnesota since 1988, and thus its purchasing power had diminished significantly, while the network was expanded and aged, and traffic levels increased. Pawlenty's campaign took pride in this veto, posting a clipping from the *Star Tribune* on its website:

> WEDNESDAY, MAY 16, 2007 STAR TRIBUNE: Pawlenty vetoes gas tax, income tax bills
>
> By Patricia Lopez, *Star Tribune*—Gov. Tim Pawlenty struck swiftly and with strong language Tuesday to veto a gasoline tax increase and an income-tax-for-property-tax swap that were at the heart of the DFL's agenda for the session. —DFLers accused him of protecting the state's richest 1 percent—those who would have borne most of the income tax increase, which would pay for the proposed property tax relief—at the expense of everyone else. But they conceded that some of their top objectives are fast sliding out of reach." — `http://timpawlenty.blogspot.com/` PREPARED BY PAWLENTY FOR GOVERNOR, PO BOX 21887, EAGAN, MN 55121[655]

Gas taxes in the United States and Minnesota are dedicated to transportation (and in some cases just to roads). The 2008 Minnesota gas tax bill phased in an increase of the gas tax by 8.5 cents a gallon by 2014. Of that, 3.5 cents of the gas tax increase was dedicated to paying the debt service on $2 billion in road and bridge bonds. The bill borrowed $1 billion in 2009–2010, with $600 million earmarked for repairing or replacing the state's thirteen most dangerous bridges. In addition, the bill increased the sales tax in the seven-county metro area by 0.25 percent for transit. It also increased license tab fees on newly purchased cars and trucks (1.25 percent on sale of new cars, drops 10 percent per year).

The bill was passed by the legislature, but vetoed by Governor Tim Pawlenty. The governor had run on a "no new taxes" pledge, and clearly had political aspirations. He was frequently mentioned as a possible vice presidential running mate for the 2008 GOP

TABLE 33.2

Political Fallout from Gas Tax Increase

Representative (District)	Party Endorsement?	Primary Victor	General Election Victor
Bud Heidgerken[656] (13A-Freeport)	Retired		Paul Anderson (R)
Rod Hamilton[657] (22B-Mountain Lake)	Rod Hamilton	Rod Hamilton	Rod Hamilton (R)
Ron Erhardt[658] (41A-Edina)	Keith Downey	Keith Downey	Keith Downey (R)
Neil Peterson (41B-Bloomington)	Jan Schneider	Jan Schneider	Paul Rosenthal (DFL)
Jim Abeler (48B-Anoka)	No endorsement	Jim Abeler	Jim Abeler (R)
Kathy Tingelstad (49B-Andover)	Retired		Jerry Newton (DFL)

Note: R indicates Republican, DFL indicates Democratic-Farmer-Labor.[659]

presidential nominee, John McCain, and was hosting the GOP convention in St. Paul, Minnesota, that year. In the event, the governor of Alaska, who had more foreign policy experience, got the nod as the VP candidate, Pawlenty continued to look toward higher office, and was for a time a candidate in the 2012 GOP nominating process, before dropping out for lack of support and will.

The *Override Six* are Republicans who voted with the DFL to override Governor Pawlenty's veto of the gas tax bill. Four of them lost their seats due to Primary challenges, while the Republicans lost two of those seats to the DFL in the 2008 general election. This leads to the rule that voting in favor of a gas tax increase can be dangerous to your political health, if you are a Republican.

Carol Molnau was the state's lieutenant governor and the MnDOT commissioner. While she had been confirmed in the first Pawlenty administration, when she was reappointed, the DFL legislature did not confirm her, and her appointment expired in February 2008. That can be directly tied with dissatisfaction with her and the governor's performance dealing with the Bridge. Notably she was not the Department's point person with the media in days, weeks, and months following the collapse. She was replaced by Tom Sorel, a federal civil servant who had worked on the Bridge replacement process.

The political problem is deeper than just the fate of a few politicians, though. It is a classic problem in transportation funding. Ribbon cuttings on new projects are much more attractive to politicians (and newspapers and TV news) than maintaining what we have. People are also more interested in road surface than the underlying structure. Yet pavement failure, while bad, is not nearly as bad as structural failure. "Failure" in the traffic level of service sense (LOS "F"), while economically costly and personally annoying, and perhaps leading to more (or at least different) crashes, does not have anywhere near the same connotation as structural collapse.

The competing uses of funds are ultimately political decisions. Should money be spent for bread and circuses (e.g., football and baseball stadiums) rather than genuinely productive infrastructure? Five years later, should money be spent on new bridges with added capacity (e.g., the Saint Croix River Crossing in Stillwater) while over 1,000 structurally deficient bridges remain in Minnesota[660]?

33.5 Economic Effects

At the request of MnDOT, shortly after the collapse, we estimated the Twin Cities seven-county region daily vehicle hours of travel with and without the Bridge, using a planning model under two assumptions. The first kept the trip table fixed. This means that people did not change the number of trips, or destinations, in response to the Bridge failure. This should give an upper bound to the effects of the Bridge failure. The second allowed trip destinations to vary (though keeping the number of trips fixed). This provides more of a lower bound of the effects. Clearly, some people can switch destinations, or avoid trips altogether, if the cost of their previous destinations are now too high. On the other hand, not everyone can do so. The exact number of people who change destinations is not something we can easily know.

Note, these are direct model outputs, so while the precision is high, the accuracy is not nearly as high as implied by the precision. We monetize these numbers using values of time from MnDOT[661] of auto $12.63/hour and truck $20.41, and we assumed 80 percent auto and 20 percent truck, giving a composite value of time of $14.19.

I believe the MnDOT Value of Time for Trucks is very low; our estimates[662] put the number at closer to $50 per hour. If we used that, we would get a composite value of time of $20.14.

The results in Table 33.4 are, of course, estimates. However, the number is large and positive, which we expect. And the numbers lead us to conclude that letting bridges fall down is bad public policy—which most of us already knew. The number does have uses aside from (rhetorically) beating people over the head; it tells us, for instance, how much we should reward contractors for early completion.

The problem is that those who benefit from the Bridge (or lose from the absence of the Bridge) differ from those who pay for it, and are responsible for maintaining it. If presented

TABLE 33.3

Descriptions of Scenarios

Scenario	Time	Trip Table	Planning Network
0 (Base)	Before	N.A.	Complete network
1	After	Variable	Crippled network
2	After	Fixed	Crippled network
3	After	Variable	Crippled network with upgrades
4	After	Fixed	Crippled network with upgrades

TABLE 33.4

Early Estimates of Economic Cost of Bridge Collapse

	Scenario 0 (Base)	Scenario 1	Scenario 2
Daily VHT (10^6 veh.hrs)	1.122	1.131	1.134
Daily economic loss (VHT)		8,980	12,013
Daily economic loss ($)		$127,390	$170,425

TABLE 33.5

Measures of Effectiveness of Alternative Scenarios

	Scenario 0	Scenario 1	Scenario 2	Scenario 3	Scenario 4
Daily VHT (10^6 veh.hrs)	1.427	1.432 (0.35%)	1.442 (1.09%)	1.431 (0.31%)	1.441 (1.00%)
Daily VKT (10^6 veh.kms)	86.53	86.27 (−0.31%)	86.58 (0.05%)	86.27 (−0.30%)	86.58 (0.06%)
Daily economic loss ($)	N.A.	71,466	220,198	62,408	203,409
Ave trip length (kms)	18.82	18.76 (−0.31%)	18.83 (0.05%)	18.76 (−0.30%)	18.83 (0.06%)
Ave trip time (mins)	18.61	18.68 (0.35%)	18.82 (1.09%)	18.67 (0.31%)	18.8 (1.00%)

Note: Numbers in parentheses indicate the percentage change of a measure as compared to its counterpart in the base scenario.

with the choice of paying and keeping the Bridge up and not paying and letting it fall, most users would have gladly paid more than was required to keep the Bridge up.

We conducted a more thorough analysis later (Table 33.5),[663] considering upgrades to alternative routes, such as restriping I-94 to add a lane and upgrading Mn280, with somewhat lower results.

33.6 Traffic

Travel behavior changes after network disruption as well as after the replacement of disrupted links are not well understood. Table 33.6 shows the number of river crossing trips by type of facility before and after the collapse and the reopening. We discover 46,000 lost trips daily after the collapse (nearly a third of what the I-35W Bridge had carried), and 20,000 found trips after the new bridge opened. Those lost trips may not have been made, or more likely, found different destinations not requiring a river crossing. This provides additional evidence to the phenomenon of induced demand (see Section 24.4.4).

Gains from the Bridge for three peak periods re-estimated with accurate (observed) travel times (but a fixed and not-observed OD matrix), shown in Table 33.7, were on the order of $70,000 per day (somewhat below our initial low all-day estimates ($127,000), far below

TABLE 33.6

Traffic Lost and Found

	Collapse 1-Aug-2007			Reopen 18-Sep-2008		
Bridge	Before	After	Change	Before Reopen	After Reopen	Change
I-35W	140000	0	−100.00%	0	120349	
Arterial total	152311	197566	29.70%	169983	95895	−43.60%
Freeway total	572274	481040	−15.90%	488717	583127	19.30%
Total	724585	678606	−6.30%	658700	679022	3.10%

TABLE 33.7

Change in Vehicle Hours Before and After Bridge Reopening

	"Daily" VHT[664]	Daily Difference VHT	Value[665]
Phase1	1.48110^6		
Phase2	1.47610^6	5000	$70,950
Phase3	1.47810^6	3000	$42,570

MnDOT's, and also below the contractor early completion bonus). Much of that gain is lost once the I-94 bridge lane disappears, as consistent with the original results. $42,000 annualizes to about $15 million in benefit, for a $250 million bridge (which pays off in about 23 years at 3% interest). The I-94 lane restriping paid off in a matter of a month.

33.7 New Bridge

The replacement bridge cost $251 million, funded almost entirely by the federal government. We can debate whether the federal government should have paid for it (it was originally built with 90 percent federal contributions, 10 percent state, but matches recently are much more balanced), since most traffic using the Bridge both originated in and is destined for Minnesota. With Minnesota congressman James Oberstar then chair of the House Transportation Committee, there was plenty of political support behind this.

The replacement bridge was hurried, completed by September 2008, several months ahead of the original schedule. This is good; a lack of a bridge costs the economy. (Xie and Levinson estimated between $127,000 and $170,000 per day, MnDOT estimated $400,000 per day). The contractor received $200,000 per day bonus for early completion. So perhaps in an economic sense, too much was paid to complete it a few months early.

Rebuilding a collapsed bridge is of course a crisis, but it is also an opportunity to do something interesting. Rushing designs may mean that ideas are missed. What was built is a functional bridge, and there are state-of-the-art real-time structural health monitoring systems installed, so I have no fear driving over it. It was also ensured that the Bridge would be compatible with any future light-rail transit lines, though none is planned for this bridge (and how they would transition from the center of the Bridge to any reasonable destination is extremely unclear). But could more have been done?

The snow removal and icing problem was not deeply considered. Minnesota is famous for its winters. The previous I-35W bridge had installed a de-icing system (in response to earlier crashes), which had been speculated to be responsible for corrosion of the structure. While the NTSB did not find that, de-icing chemicals do have environmental consequences.

A solution not considered was air rights. A bridge over the Mississippi is expensive. But imagine having a two- or three-story office building hanging from below, and/or built above the highway. The views from the River are fantastic. It would not impair other views of the river especially much, and would generate a significant amount of revenue to pay for reconstruction. One example would be the historic London Bridge, which had houses and stores along the side, encroaching on travelways. Obviously it would increase the initial construction cost, and perhaps time, but that would be amply repaid over the long term. That structure would further have shielded the roadway from ice and snow, reducing road snow clearance costs and crashes. There are better ways to combine transportation arteries with development opportunities, and creative design can show the way.

The bridge opened in the early darkness of September 18, 2008. A parade of first responders, and then a bulge of traffic all hoping to be the first (and none succeeding) went across. Soon the Bridge was attracting 120,000 vehicles per day, measurably off the pre-collapse levels.

33.8 Policy Implications

Unlike bridges, transportation networks are seldom "fracture critical." While the Interstate Highway System did what it could to sever local streets and channelize traffic onto fewer, larger, limited access links to achieve economies of scale and higher speeds and throughputs and the expense of redundancy, there was enough remaining redundancy to ensure that this one bridge collapse would not have devastating transportation implications. Had the I-94 bridge collapsed instead, I am sure the consequences would have been much worse, as the alternative paths are not as convenient. But again, the Twin Cities would have muddled through.

Several lessons can be drawn. If there is redundancy, maintaining road operations during construction may not be necessary, and may needlessly delay construction while exploding its cost. There have been projects in Minnesota, notably Mn36, where closing the road entirely in order to speed construction has been successful. Rebuilding a bridge while keeping it open is like doing brain surgery on oneself. In principle it is possible, but why? The length of bridge construction in other examples versus the less than one year from design to opening that this bridge took is instructive.

Actual effects were much less than forecast. People are quite adaptable in their travel patterns, especially if there are alternatives routes and destinations.

As a consequence of the Bridge collapse, many other bridges were repaired more quickly than they otherwise would have been. The list of recent Bridge closures in the region below is illustrative.

1. August 1, 2007, I-35W;
2. January 2008, Hastings Bridge;
3. March 20, 2008, St. Cloud Bridge;
4. March 26, 2008, University of Minnesota Pedestrian Bridge;
5. April 25, 2008, Lowry Avenue Bridge;
6. May 6, 2008, Blatnik Bridge (I-535) in Duluth;
7. June 4, 2008, MnDOT barricades Hwy. 43 Bridge over Mississippi River at Winona;
8. July 2012 Plymouth Avenue Bridge.

We see three successes:

- Emergency response
- Immediate traffic restoration
- Quick completion of bridge with design/build.

and we see three failures:

- Bridge collapse
- Removal of successful, low-cost restoration measures
- Overpaying for bridge.

Overall, Americans seem good at short-term tactics but poor at longer term strategy. This needs to be rectified. While there were positive outcomes, including the passage of the gas tax, and increased attention to maintenance, there may have been some over-reaction in terms of replacement of bridges.

While there was some accountability, some key personnel remained in place. Firings and resignations need to be used more often in the aftermath of events like this; some people should assume responsibility.

We need to address the question of the appropriate role of politics in infrastructure decisions. We don't expect politicians to make recommendations about which electric power cables are replaced. Why are they so involved in maintaining existing roads? (The construction of new roads is, on the other hand, more obviously political in nature.) Should MnDOT and other similar agencies move toward a public utility model to de-politicize?

We are not good at dealing with low-probability, high-consequence decisions. We are not good at assessing the value of either.

Unless something is done, failure will be more frequent. The Interstate system is aging and nearing the end of useful life for many components.

In *The Transportation Experience*, we identify a set of strategies for maturity:

- Abandonment
- Cash cow (using resources for something else)
- Maintenance and rehabilitation
- Replacement.

Abandonment of the Interstate system as a whole seems premature, and while a handful of selected urban links have been closed, these are very much the exception. The cash cow strategy is in fact what has been employed for the past several decades, as the gains from the built infrastructure exceeded by far the amount that was reinvested in it. This is of course economically correct; infrastructure should be fully exploited to its capacity. But this needs to be coupled with appropriate levels of maintenance and rehabilitation, and ultimately planned replacement; otherwise, to mix our farm metaphors, the goose that lays the golden egg will die.

The "Weak Bridge" sign that one sees all over England is not terribly reassuring, and indicates an unwillingness to recapitalize the network. In the US context, there should be more money for transportation maintenance and rehabilitation; we are eroding our physical capital. If there isn't—we should spend our money more carefully, taking care of the existing systems and users first, and engaging in graceful abandonments as necessary, but not building new infrastructure that will require long-term maintenance without any means for doing so. The mantra "Fix it first" has been suggested for this strategy, and I agree. Unless something is done, the problem of decaying infrastructure will only get worse (the physical world being subject to entropy and all). Bridges do not repair themselves.

The Real Question is not to explain our sorry reality, but to improve it.

—AUGUST LÖSCH

34 The Transportation Professional and Transportation Policy (The Design of a Life)

Lösch (1967) *Economics of Location* is one of the great books of the twentieth century. It's must reading for anyone in planning, location economics, transportation, or geography. Written during the 1930s and 1940s in Germany by an outsider to National Socialism and the university community, the words "sorry reality" run deep.

These afterwords will address a most difficult question: What is the reader to do with the understandings of policy absorbed from *The Transportation Experience*; how should those understandings affect the reader's work? Because each reader differs and there are many paths through life, there is a limit to how far we can work the question. We can address how understandings might be used, with their use depending on the student and the opportunities that unfold.

Toward the end of the words, Garrison comments on the interactions in policy affairs and about the world as he would like it—a world where options are explored and new paths taken.

34.1 The Sorry Reality

One sorry reality is that transportation systems are not controllable in important ways. They get started; they grow and develop; and then sail in the sunset of obsolete stasis. Subject to responding to markets, systems are self-organizing and self-steering, and embedded policies play a dominant role in shaping the ways they are organized and the performance they achieve.

At fine detail, society can ease barriers to deployment (and subvent deployment via subsidy), control the behavior of system actors when they blatantly violate social norms, force fixes on externalities, and fine-tune deployment decisions to places and local circumstances. These are tuning policies rather than controlling policies.

It seems to us that understanding the problem of control is a major contribution of *The Transportation Experience*. Control is a non-trivial matter. It deserves our attention as a transportation question. And we see the transportation question as one of a larger set of questions about the social control of technology. The question of social control is usually phrased in more limited and less deep ways, and understanding the transportation situation gives us some general, non-trivial insights.

Maturity of systems yields two aspects. One is that systems are not yielding on- and off-system improvements in the way that we have known them in the past. Transportation's external benefits and reorganizational changes have been exploited. Indeed, in some cases, service improvements are negative.

The other element of maturity is the limited options available to managers and to policy makers. The options are a sorry lot because they are tinkering or tuning ones. Managers can seek to improve product quality and differentiate standardized services to minor variations in markets. With the objective of lowering costs, they strive to get the scale of production just right; they search for lower factor prices, example, by using non-union labor. Fine-tuning pricing policy, market channeling, and product differentiation become priorities.

The maturity of transportation systems deserves our attention as such, and also because of similarities among transportation systems and other maturing activities. Understanding the transportation situation yields some general, non-trivial knowledge.

A fourth aspect of sorry reality emerges from considering what we found when we examined how policy is studied, debated, made, and implemented. We saw iron triangles, policy being made from "pure" technology considerations, exercise of many-against-the-few and few-against-the-many power, and ends-justifies-means actions.

Focusing on these aspects of reality, Darling sees policy as "witchcraft." We would not go that far. Given the options available and the circumstances in which policy is made, actors impose their rationalities on systems. If that leads to gross subsidies for coal transport over Crows Pass, unlike Darling, we judge those subsidies as rational. If policy makers in Los Angeles compare streetcars versus buses and opt for streetcars, we judge that decision as rational within the environment it was made.

34.2 But to Improve It

To assist the discussion of the improvement of our sorry reality, shorthand expressions will be useful; three words will be used: control, maturity, and rationality. Control has to do with intervening in the apparent inevitability and irreversibility of the life-cycle. Maturity describes the present state of affairs. And rationality has to do with ways we make judgments and decisions.

Taking life-cycle and control matters first, the usual prescription is to live with the tyranny of the life-cycle, adopt an incrementalist muddling through prescription. There's the body of concepts in micro-economics. The situation is taken as is, production functions are fixed. Optimization is at issue.

We accept the need for incrementalism, but point out that incremental actions consistent with embedded policies and the movement of a system along its life-cycle are one thing, and incremental actions yielding consequential improvements are another. Fulton, Stephenson and Pease, McLean, and the others who created auto, air, transit, and other systems worked incrementally. In the transportation situation, muddling through yields only squabbling over marginal things.

Ayres (1984) offers another prescription. He would have us track how-goes-it for activities and implement control when activities are in the adolescent stage in the cycle. The instrument of control would be innovation, and at adolescence the manager-planner would make investments in R&D so as to obsolete the maturing product. In this scheme, systems avoid maturity. New products are put in the pipeline to obsolete old products just when old products begin to mature. There is action of this type in the private sector.

Ayres terms this strategy, policy, or activity "indicative" planning. In his world, intellectual or innovation resources would be deployed here and there as needed to renew-replace activities before they mature.

Elsewhere, Ayres and Steger (1985) argue that new technologies are changing the name of the game, offering the possibility for small-scale production. Thus the standardization and production at large scale that lock in mature activities no longer are needed.

We like Ayres's insights, but have problems with his indicative planning. It demands knowledge and insights, and it suggests the centralized allocation of resources for innovation. The latter is risky for democratic societies.

Some policy wonks and business types talk about industrial planning, using Japan's MITI as a model. That debate lacks Ayres' insights. Its essence is how to live with mature activities, and a close examination of MITI's activities gives only lessons on how to gain comparative advantage in mature activities.

To this point we have spoken to the life-cycle and maturity as we see it in history; it is irreversible and inevitable. History says more, for new products do come along and obsolete the old. All we have to do is wait, something will happen.

While we admit it is likely that "something" will happen, we give this strategy little coin. As we have seen, there is lots of happenstance to what happens. There is no reason to believe that what will happen will have desirable properties compared to the universe of all things that might happen.

Furthermore, while this strategy may work reasonably well in an atomistic world, in today's locked-in big-system world it may be a long time before anything comes along.

In years to come, the reader will be hearing a good bit about long waves (on the order of 50- to 80-year cycles) in the economy—recession, depression, recovery, recession.... Observers have called attention to these waves for about fifty years, but lacking a theory and a convincing empirical base, most (such as Samuelson) have called them "science fiction."

But improved data and some articulate causal arguments have emerged. Understandings of the relations among investment, productivity, and innovation are at the core of the causal argument.[666] Innovation waves are stimulated by depressions. The depressions reflect adjustments, as investments are made obsolete. That's not a pretty picture. We are not happy that life rides on waves beyond our control. We are not attracted to a "wait till things change and something comes along" strategy.

Kant's perception of ethical behavior is one we find useful. It takes a book to state it, but it boils down to "never use others to serve one's purposes." Another version is to avoid imposing one's calculus of values and rationality on others.

Kant's perception is often violated in transportation affairs and in life generally. For example, we often hear transportation planners saying, "We have got to get land use under control in order to make transportation work." We hear that in the light rail debate: it will work just fine if we control land uses. That's bothersome. A transportation person is trying to get people to do things they would not otherwise do to make transportation work. It's pushing our collective failure to develop the right kinds of transportation on someone else.

Another case involves opportunity costs. Society implements lots of policies that attempt improvements by throwing money at problems—save the cities with transit, smart highways will manage congestion, solve the air pollution problem by degrading combustion efficiency, and so on. We suspect those are not very good ways to go, but in the spirit of Kant, we limit ourselves to saying that and do not interfere when others proceed in that direction.

But that puts us in a bind because opportunity costs are so high—we could be using resources of attention, energy, and money in better ways.

Finally, there are ethical-equity dimensions. There is the here and now question of vertical and horizontal equity. In our minds, though, the now and tomorrow equity question overrides. Are we doing things that are inequitable to future generations? We hear some urbanists say the public ought to invest in heavy rail. True, it won't work very well now, but think of how wonderful it will be for future generations. That's a mind-bending way of thinking.

34.3 Design of a Life

What is the reader to do? Does deep knowledge of our sorry reality immobilize us? Are problems so tough that nothing can be done?

Garrison, who has been involved in many things and claims to have done no harm, says, "I manage by wearing two hats. I wear one hat when judging whether or not to enter ongoing affairs. I wear a second hat when debating goals, programs, opportunities, the allocation of intellectual resources, and such."

There are thousands of policy debates ongoing at any time. For reasons of where one happens to be or whom one happens to know, the professional has the choice of entering the debate or sitting out the debate. The question boils down to when to enter and when to sit out—subject, of course, to limits on energy, attention, and time. The answer to the question has to be tempered by judgments of the probability of something happening.

One question to ask of an opportunity is the extent to which it addresses making a mature system work better. We know, for example, that anything that better matches a mature product to markets is an improvement. We can be alert to whether the opportunity is concerned with on-system or off-system payoffs. Will there be a direct confrontation with low-likelihood-of-changing embedded policy? What magnitudes of payoffs are involved?

Armed with understandings from *The Transportation Experience*, the reader ought to be able to work with efficiency and taste.

There are ethical questions that bear on the enter/sit-out decision. As already remarked, we hesitate to impose our values on others, so we sit out when decision makers choose differently than we would. But at the same time, we are quick to speak out when we see those in power using others for their purposes. We have found "That's immoral because . . ." statements effective in many circumstances.

The "What to do about the state of systems?" question is central. Aware of the systems problem, we have some ideas about desirable systems attributes and how change might be achieved. Indeed, we have some ideas about specific designs that might be desirable.

But not wanting to impose our will on others, we talk and write using a widely scoped view of our circumstances. We emphasize process rather than products, verbs rather than nouns. As a result, we hope readers better understand their circumstances, and that those who are in the right place at the right time try out some designs that might be system changing. We hope scarce resources are not dissipated on unfruitful efforts.

The fact of the matter is that strategy hasn't worked very well. People want specific, what-to-do guides—build this, build that.

That's what we want, too. But we want specifics to be found in a world where folks holding rich understandings of transportation and using integrative ideas identify many options. That's a lot more than just the few things we can imagine. In such a world, society could choose from a menu of options that open rather than close opportunities for future generations.

We would like a learning, self-organizing world in place of an accepting, reacting world.

These last remarks are drawn from the designs of our respective lives as transportationists. With them, we leave the design of a life as a transportation consumer or professional as an exercise for the reader.

If history were taught in the form of stories, it would never be forgotten.

—RUDYARD KIPLING

35 Commencement

35.1 Introduction

THE TRANSPORTATION EXPERIENCE explores the genesis of transportation systems; the roles that policy plays as systems are planned, innovated, deployed, and reach maturity; and how outcomes might be improved. While the territory to traverse is vast, underlying themes that characterize system development and implementation facilitate our journey. In a sense, transportation systems matters are simple. They just seem complex because games are played on diverse stages, with many actors, and with issues that appear under different names.

The Transportation Experience emphasizes the American and British experiences since the beginning of the industrial revolution. The American or British experiences are hardly unique. They have roots in Western Europe, and each country is but one stage for the playing out of themes common to all places. And while much of the transportation system in Europe and North America is mature (if not senescent), most of the rest of the world is still planning, developing, and deploying. The accomplishments and mistakes of the more developed countries generate lessons that may be applied to places where networks remain nascent or adolescent.

To begin at an arbitrary point in time and place leaves unanswered the question of how experiences before that point in time shaped beginnings. Sometimes Western Europe is considered the locus for the emergence of what we call the modern world. We know that there were beginnings, or the resources for beginnings, in many places. China, in particular, demonstrated the capacity to organize knowledge, resources, and technologies for large-scale public works and transportation activities. It developed navigation instruments, defense walls, great canals, roads, and bridges. In the 1300s, China was a major maritime power, using large ships and considerable organizational capability. China imposed its will in Southeast Asia and as far away as Africa. The advantages to be had from transportation and trade were there. Additionally, the development of knowledge was well advanced in China compared to Europe. Yet in the following centuries, Europe came to dominate technology and progress.

However, it is unclear the extent to which precursor experiences in and outside Europe affected the developments birthed in Western Europe. Ideas are light baggage, and through contacts during the Crusades and through travel and trade, Western Europe might well have borrowed ideas.

35.2 Quest

> History doesn't repeat itself, but it does rhyme.
>
> —Mark Twain

There are a number of ways of organizing the text. It could be based modally, telling the story of each mode in turn. It could be a giant time line, giving the history of transportation from when humans first walked on two legs to the present.[667] It could use the "life-cycle" paradigm of birth, growth, maturity, and senility, and describe the modes in parallel (but out of chronological sequence). It could order by "structure," considering infrastructure, equipment, and operations as our basic organizing scheme. It could distinguish between urban, rural, and inter-city transportation, and passenger and freight transportation, giving us a 3 x 2 matrix. It could distinguish between nodes (ports, airports, terminals, intersections) and links (roads, rails) and services (transit, freight, airlines, etc.). It could be organized by "supply chain," considering inputs, process, and outputs.

The first edition of this book organized modally and thematically. In this second edition, we have instead tried to remain more chronological in the histories of the modes, treating modes in parallel. We organize the second edition text into five waves, roughly fifty- to sixty-year periods. That said, we intersperse our discussion of the waves of history with discussion of phases of development. These phases recur in each mode's development, so their placement is many respects arbitrary. But we discuss birth, growth, maturity, the life-cycle process, and what we call meta-cycles, the process by which life-cycles recapitulate with new technologies after each wave.

These discussions counter some conventional wisdom. Most think of each mode as having a unique history and status, and each is regarded as the private playground of experts and agencies holding unique knowledge. However, we argue that while modes have an appearance of uniqueness, patterns repeat and repeat: system policies, structures, and behaviors are a generic design on varying modal cloth. The illusion of uniqueness proves no more than myopic.

As our discussion proceeds, themes such as these emerge:

- Systems are built from experiences. At the dawn of a system, experiences are mainly transferred from other, older modes.

- Policies mirror the intrinsic characteristics of systems and the interplay of those characteristics with deployment problems. This theme overlaps with the previous theme. The words "intrinsic characteristics" refer to the structure, behavior and performance of systems, and this theme notes that these characteristics affect policy.
- Rules for behavior may be strictly embedded in system organizations and protocols or, at another extreme, in governments. The question of appropriate loci and shared power is long-standing, and it has mainly been answered on pragmatic grounds.

35.3 Structure

A transportation system can be usefully viewed as having a triad structure:

- Fixed facilities such as airports and airway navigation facilities or railroads and terminals. On the soft side, there are institutions that match these.
- Operations involving many kinds of institutions and protocols, as well as hard technologies, such as traffic lights.
- Equipment and its production, care, and feeding: locomotives, airplanes, liners, shipping firms, automobile dealers, repair shops, insurance companies, and so on.

Each of these system components has associated institutions, as stated. There are governmental and private institutions, as well as professional associations. Each has specialized financing, management, and fiscal arrangements. To a large extent, policy, planning, and management are scoped by components.

Taking railroads as an example, at a first cut there are interacting locomotives and cars (equipment component), routes and yards (fixed facility component), and control systems (operations component). Looking with somewhat wider scope, we see firms producing equipment, constructing and maintaining routes and terminals, and providing services. Expanding still further, we see suppliers to those activities, professional organizations, the Association of American Railroads, federal and state government activities, the operating officers association, and so on.

We may also remark that the triad structure has a *unitary character*. That's observing its "same everywhere" character. Transportation systems are unitary systems mainly because operations over networks require standardization. There are other reasons noted in the text. This unitary structure affects behavior. Standards enforce highly predictable behavior, and standards are valued as an instrument of control.

A second feature of behavior is its *reactive disjointness*. Actors in each component of the triad monitor the states of other components and adjust their affairs to fit. For instance, a topic of concern today is how to adjust highway facilities to fit increasing use and consequent congestion. We seek to strengthen pavements to accommodate larger and heavier trucks. The air traffic control system is being improved to accommodate growing traffic. These actors (planners, policy wonks, engineers) are Newtonian, adjusting speeds on existing trajectories, but constrained from finding new paths.

The unitary character of transportation, together with the disjoint, nature of system structure, places sharp limits on images of what planning should be and what it can do. Often, planning is component constrained, and its system impact may be limited; it may also be limited because unitary standards limit degrees of freedom. Planning strives to catch up with and adjust to developments elsewhere. Broadly, a system gets started and the predominant hard and soft technologies are frozen. From the point of view of *steerability*, the die is cast. System development moves along a predetermined path.

Bouladon (1967) provides a notion of inherent service capability and compares service capability to demand, shown in Figure 35.1. Bouladon suggests gap filling as a role for planning. This notion has been adopted in Japan (Figure 35.2). Innovative people imagine other kinds of gaps. In a sense, that is the definition of truly original thought, to identify a new dimension on which to characterize things. And once we have a new dimension, we can see where there is a product, and where there are gaps. Unfortunately, public policy strives to fill well-known gaps with subsidized services, rather than discovering new ways to think about the problem.

In the Figure 35.1, the left scale is a log scale, and the optimum utilization line indicates how the demand for trips decreases with distance. The x-axis is also a log scale. Plotting the service that a mode can provide, one sees some combinations of distances and demand volumes where there are gaps. For instance, the "too far to walk and too close to drive" market is only partially served. Most commuters don't have horses available, and bicycles don't work well for many individuals and in certain climates and terrains. The Segway, for instance, seeks to fill this gap and uses the idea in marketing. The "too close to fly, too far to drive" gap is the target for high-speed rail planning.

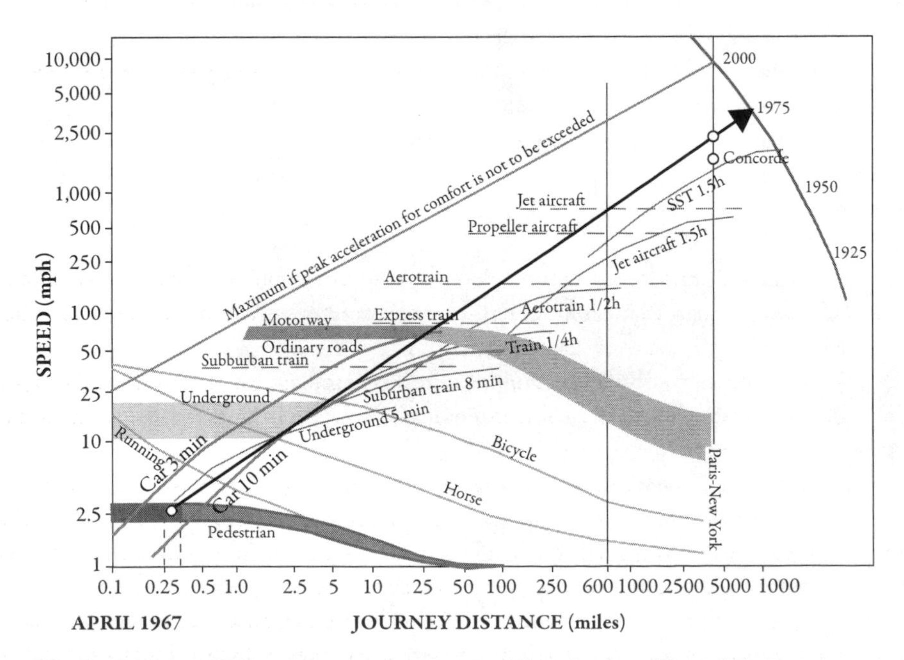

FIGURE 35.1 Transport Gaps: Density vs. Distance. Bouladon, G., adapted from: "The Transport Gaps," *Science Journal*, April 1967.

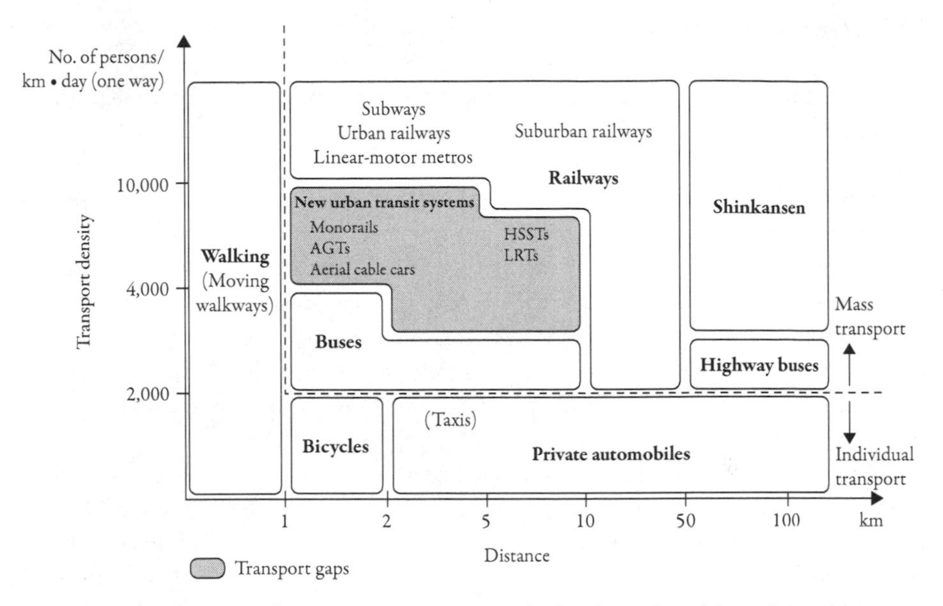

FIGURE 35.2 Adaptation of Transport Gaps to Japan. Nehashi, Akira, adapted from: "New Urban Transit Systems Reconsidered: A Better Transport Environment for the Next Century"

We think of the unitary, disjoint, and lack of steerability attributes of systems as dysfunctions. They suggest three roles for policy analysts, planners, and managers:

1. Accept the dysfunctions and work within the constraints they set.
2. Identify and fill transportation gaps.
3. Seek ways to truly steer systems by working around or breaking the tyranny of the dysfunctions.

35.4 Performance

There is another generic consideration bearing on transportation. It is what we like to refer to as the intrinsic character of transportation systems: how systems behave and how they perform.

We emphasize innovation and the life-cycle model throughout the book. Such S-shaped life-cycle curves characterize many features of transportation systems and other systems as well.

Most of the planning experience has been with stable, linear processes. Yet planning addresses changes, and for change to occur we must create and manage unstable, nonlinear processes. A system is unstable if a small perturbation from its past track sends it off in a changed direction. It is nonlinear if there are branch points or discontinuities.

We now assert that the planning (and control) of nonlinear, unstable processes is today's central transportation problem. *The Transportation Experience* discusses that assertion; it permeates our discussion.

35.5 Conduct

Transportation systems are an attractive topic for study because of the strong interrelations between policy and the nature of transportation. The policy story tells us what transportation is, does, and can or might do.

Policy may be defined as sets of formal and informal rules that control the innovation, construction, operation, financing, service provision, and other attributes of transportation systems. That is a vast subject area. There are policies for the testing of concrete pavements, for land taking, the funding of airports, the domain or scope of agency concerns and powers, controlling the range of products offered by equipment manufacturers, safety inspections, the subsidy of liner operators, and so on. Further, the limits on the subject area are not well defined. For example, policy and planning overlap, for in many ways planning is the application of policy.

This breadth is of concern because transportation professionals are expected to know the rules that apply to their environments. Knowing requires more than an ability to list policy rules to which work should adhere and where those rules are to be applied. Even so, policy courses historically have had a what-is-it thrust. Business schools offered policy courses dealing with the regulation (and deregulation) of commercial transportation. Where policy was treated in civil engineering departments, it emphasized highway funding and construction protocols. The results of such courses are but a limited snapshot of a descriptive sort.

The Transportation Experience seeks to provide deeper "knowing." It strives to help the reader understand how and why policies (rules) are developed and how to assist in forging policies to improve the functioning of systems.

At larger scope, the study of policy says how society has learned to create, deliver, and operate large, complicated systems that serve specific tasks very well. The insights extend beyond transportation because there are a number of large, complicated systems that share the structure of transportation systems. Transportation policy has lessons for all public facility systems.

In spite of those nice words about the bright side of the study of transportation policy, there is an overwhelming dark side. As a result of that dark side, some regard policy work as trivial, foolish, or counterproductive. It is often treated as a second-class subject, at best.

There are some quite perceptibly troublesome things. It is true that poor analysis gets published by claiming values for policy. Policy often results in using resources in unproductive ways. Often policy yields large cross-subsidies, and the ethical and social values of these are unclear. Policy often yields stasis rather than development.

One summary statement about the dark side compares policy making to sausage making, an activity that fastidious persons should avoid. Another is that the debate is at best petty and unseemly squabbles over marginalia. Writing in 1988, Howard Darling, the dean of Canadian policy analysts, had this to say:

> ... (Policy has been) ... transformed into a kind of witchcraft: a combination of incantatory spells, proprietary gifts, and rites of exorcism to banish the evil spirits whose presence was felt everywhere but defied any concrete identification.

Such statements often apply. Although Darling mainly had railroads in mind, his remark certainly applies to the debates about urban transportation. Some of the reasons that words such as Darling's apply are apparent and are not unique to transportation. In the United States there is a geographical organization of political power that affects the forging of policy and the distribution of gains and losses from policy actions. There is the "many against the few" consideration in decision making (and its reverse: the well-organized few against the many). These features of the political scene affect transportation.

Transportation systems have complex social and economic interrelations, and individuals and organizations have limited abilities to perceive their structure and performance. The general public's perceptions stem mainly from subjective user experience; traditional media coverage has scant depth. Policy needs are seen in response to the evil things that happened yesterday, and policy becomes, in Darling's words, "... rites of exorcism to banish evil spirits." We would not fault the public for wanting corrections for the problems it senses, and that is not the intent of these comments. Rather, it is to point out that professionals too often fail to respond to the signals the public sends. The public identifies an undesirable symptom (or states a goal), yet the professional is unable to provide an appropriate diagnosis and suggest a cure. Professionals lack effectiveness when they fail to perceive the nature of system processes and the relation of policy to those processes. There are many excuses for this.

One excuse is the disjoint structure of transportation systems—they divide into fixed facilities, equipment, and operations. The typical professional is a member of a component, and has limited system understanding. Cognition, need to know, and loyalty are component bound. There are highway experts, vehicle experts, traffic experts, transit experts, airport experts, logistics experts, and so on, but few transportation experts. Division of labor is a virtue, of course, and not everyone can or should hold deep knowledge across transportation. Even so, all should know enough to contribute to development or at least know enough to do no harm. There are also specialists from non-transportation fields who wish to have their say on policy. Regardless of the elegance of the home discipline, such experts have limited cognition.

Another excuse is the large time-span that characterizes the behavior of systems. Processes work out over decades, and the professional is informed by only that part of the process experienced. Altshuler (1979) remarked that highway and transit answers are preselected solutions to the urban transportation problem, and that is a comment on the experiences of those offering solutions. So even the professional engages in petty squabbles over marginalia—marginal because the debate is over whether to do a bit more or less of what went before, and petty because of narrow scope and thus consequence. Darling also used the word "unseemly." It is simply unseemly for the professional to debate from limited knowledge.

The paragraphs above identify excuses. There are no reasons that the transportationist cannot command policy, enact plans, and deploy systems. All that is needed is to broaden our cognition in a full, methodical way, and that is the essence of *The Transportation Experience.*

35.6 Imagination

Each modal chapter takes a proactive stance, focusing on development pathways. It proceeds by illustration. Each section illustrates how, with a little imagination, pathways were (or in the later chapters, might be) found.

To illustrate the pathways idea, recall that the urban interstates were built "on the cheap"—they were to provide capacity for some urban trips for a twenty-year period. At the time they were planned, no one said "what next." They were constructed with no room for more capacity within rights of way, and the social and economic costs of more freeways, if not already known at the time freeways were constructed, were soon apparent. We conclude that urban freeways did not open a pathway for managing urban mobility problems. More broadly, we question whether the transportation modes of the developed world provide development pathways for the less developed nations.

Concerns about scarcity of energy and land, rising costs and congestion, and the environment require that we seek alternative transportation development paths. We say alternative because we cannot fully know the shape or scale of the problems. We want to be able to go this way or that (and faster or slower) depending on how situations evolve and depending on what we learn as we proceed along paths. An associated idea is that we want flexible paths: flexible because some paths may be found to be more socially desirable than others and it would be useful to have alternatives available. Flexibility is also desired because situations may change: new options to be taken advantage of, problems not now recognized emerge.

It should go without saying that development paths should be sought that offer returns throughout society, are consistent with the built and social environments already in place, and build from resources already available, such as guideways, institutions and technologies, and so on. All kinds of constraints are set by the properties of the physical and biological worlds. It has been estimated (by Siemens engineers) that if the automobile had been developed as rapidly as semiconductor circuits, today's car would travel at 4,800 km/hr (3,000 miles/hr), weigh less than 60 g (2 oz), go 480,000 km (300,000 miles) on 4.5 liters (1 gallon) of fuel, and cost less than $3.00. Accomplishing that would require breeding smaller people so that they would fit into something smaller than a matchbox. That path doesn't seem to be feasible. (In defense of transportation, it has also been pointed out that cars don't crash nearly as often as software.)

Our view is that the behaviors of the systems are a consequence of birthing circumstances, of the ways the predominant technologies were defined, and the resulting structures of the systems. The systems are now rather mature. Problems recognized are symptoms of maturity. Striving to "fix" problems absent deep intervention in systems structures results only in superficial changes. Too often, we wind up "admiring" problems, like congestion or the national debt, instead of solving them.

Our discussion in *The Transportation Experience*, though lengthy, is hardly exhaustive, and the reader is invited to have fun imagining alternative development pathways that will exchange today's problems, both obvious and obscure, for new ones.

12 End Matter

Appendix: US Public Sector Transportation Institutions

Many federal, state, and local agencies have transportation responsibilities.

The United States Department of Transportation (USDOT) was created in 1967. The Office of the Secretary manages policy, overall administration, and international affairs. It is divided into modal administrations; the major ones are:

- Federal Aviation Administration (FAA): Promotes safety by issuing and enforcing regulations, and certifying aircraft machinery and pilots. It operates the airspace system: traffic control centers and instrument landing systems. The Airport and Airway Trust Fund was created in 1970. The FAA makes grants to over 3,000 airports for capital improvements. The fund also is used for capital investment in the air traffic control system and to defray a portion of its operating costs.
- Federal Highway Administration (FHWA) (formerly the Bureau of Public Roads): Administers the federal-aid highway program. The Highway Trust Fund (primarily from gas taxes) provides financing to states for the pay-as-you-go system.
- Federal Motor Carrier Safety Administration (FMCSA): Created in 1999, it regulates truck safety.
- Federal Railroad Administration (FRA): Charged with ensuring that the nation has a safe, efficient railroad system. It is now mainly involved in safety regulation.
- Federal Transit Administration (FTA): Provides planning and financial assistance to transit systems as per the Urban Mass Transit Act of 1964. It assists in the development of improved equipment, operation techniques, and methods; encourages planning and establishment of area-wide systems; provides assistance to state and local governments for the operation of systems. Funds are now available for both capital and operating projects.

- Maritime Administration (MARAD): Charged to develop a strong merchant marine system, it is involved in training, ship construction and operating subsidy programs, and bilateral maritime agreements.
- National Highway Traffic Safety Administration (NHTSA): Seeks to improve the safety of motor vehicles by creating and enforcing safety standards, and administers highway safety programs in cooperation with the states and local communities.

The Surface Transportation Board (STB) adjudicates disputes and regulates interstate surface transportation through various laws pertaining to the different modes of surface transportation. It is the successor to the Interstate Commerce Commission. The STB has independent decision-making power from USDOT, though it as administered within USDOT.

US Army Corps of Engineers administers navigable waterways. Inland waterway projects are about 50 percent funded by the Inland Waterway Trust Fund (user fuel impost), while since 1986 the Harbor Maintenance Trust Fund (surcharge on cargo value) provides for 40 percent of harbor maintenance costs. The remainder of funding is from general revenues.

States have roles dating from the early canal and railroad days. State highway departments were formed about 1910 and were largely concerned with the federal-aid system. Now most states have transportation departments involved with several modes; the situation varies from to state to state. State regulatory activities may or may not be lodged in the transportation departments. The states play a major funding role. They expend state and federal funds on state systems and often assist local governments. AASHTO (the organization of state DOT leaders) is a major player in the development of state and federal policies and programs and sets standards, goals, and so on.

Cities and counties traditionally have funded projects from user fees and general revenues. Now they depend more on federal and state funding for transit, highways, and airports.

Regional agencies have mainly planning and coordinating activities. Regional planning commissions (RPCs) and regional planning agencies (RPAs) were created as a result of the Housing Act of 1954. The metropolitan planning organizations (MPOÕs) were created as a result of the 1962 Highway Act that required states to form agencies to undertake planning. Planning and coordination now extend to FHWA, FAA airport, and FTA projects. Regional agencies prepare plans to serve as a guide for future projects, prepare transportation improvement programs (TIPs), and are encouraged to engage in transportation system management (TSM). TIPs and TSM documents identify projects scheduled for the current year, and the next two- and five-year periods.

Special districts formed by local initiative to provide special projects or services. Special districts often have taxing authority.

Notes

1. Nathan Read also invented a steam powered wagon, see 3.1

2. For a history of these and related events, see Sutcliffe (2005); for a history of the patent office, see Dobyns (1994)

3. Sutcliffe (2005) p.220

4. Reid (1840)

5. See Turnbull (1979)

6. Ward (2003)

7. See Johnson (2006)

8. See Arthur (2009) p. 82

9. There was also the Caledonian Canal (Commission established in 1803) across Scotland.

10. See Wood (2009) p. 361

11. See Taylor (1951) p. 34

12. See Taylor (1951) p. 37

13. See Taylor (1951) p. 48

14. See Taylor (1951) pp. 34, 53

15. From Niles Weekly Register IX (December 2, 1815) 238–239, quoted in Taylor (1951) p. 4

16. See Albion and Pope (1970)

17. See Durrenberger (1931)

18. See Taylor (1951) p. 10

19. See Santini (1988)

20. Competition between canals and railroads is discussed in Goodrich (1961); McCullough and Leuba (1973).

21. See Bartholomew and Metz (1989)
22. See Shallat (1994); Sherow (1990)
23. Fausset (2009)
24. See Ferejohn (1974)
25. See Wood (2009) p. 368
26. For instance, as a boy, Garrison learned the heads of navigation on the streams in Middle Tennessee from people who lived in the area. He didn't set out to do that. It was part of regional conversation and lore.
27. See Burton (1992)
28. Despite the lack of war affecting policy much, the Department of Defense has a long-standing concern about insufficient strategic sealift and insufficient construction capability.
29. See Jeans (1875)
30. See Howe (2007) p. 562
31. See Smiles (1858)
32. See Lienhard (2006)
33. Dabney (1993) cited in Sutcliffe (2005)
34. See Beasley (1988)
35. Interestingly, deer trails now follow human-built roads (Mech et al. 1980), and of course tend to avoid wolves. These create additional problems, such as vehicle-deer collisions (Danielson and Hubbard 1998).
36. See Scott (1818)
37. The Corps des Ponts et Chaussées compares in some ways with the US Bureau of Public Roads and its engineers. See Seely (1984).
38. Macadam pavements were adopted in the early 1800s, and road development continued until about 1860 when inter-city rail began to co-opt traffic. Between then and the development of the automobile, work continued on local and feeder roads.
39. See Singh (2008) p. B12 for mention of tolls during the Chola period (9th–13th century) of India. Also see Indian Analyst, The (2007a)
40. See Indian Analyst, The (2007b)
41. See Marshall (2008); Sun-Herald The (2004)
42. See Deloche and Walker (1993)
43. See Mishra (2000)
44. Among these routes were *Kuga-no-michi* (Northland Road), *Umitsu-michi* (East Sea Road), and *Nishi-no-michi* (West Road), linking the Yamato Region (present-day Nara prefecture) with Chikushi (present Fukuoka prefecture), which later became the San-yodo *Higashi-no-yamamichi* (East Mountain road).
45. Technically, "asphalt" refers to the binder which ties together the aggregate (rocks), so asphalt (or bituminous) concrete roads differ from Portland cement concrete roads in their binder, and perhaps in the composition of the aggregate, but both are concrete that have binders (asphalt, portland cement) and aggregate.
46. See Gerhold (1993a)
47. See Gerhold (1993a)
48. See Gerhold (1993b)
49. See Wood (2009) p. 459
50. See Gerhold (1993b)

51. See Wood (2009) p. 338
52. Source: Dartford Grammar School, 2003
53. See Albert (1972)
54. See Howe (2007) pp. 31, 40 and Taylor (1951) pp. 15–17, 33–35.
55. See Taylor (1951) p. 26
56. See Taylor (1951)
57. See Taylor (1951) p. 27
58. See Haines (1818) "Considerations on the Great Western Canal," p.11, cited in Wood (2009) p. 730
59. See Raitz and Thompson (1996)
60. See Howe (2007)
61. See Swift (2011)
62. Although a rail, not highway, application, Alexandria, Virginia, for instance, blocked the Baltimore and Ohio Railroad from making connections through the Shenandoah Valley in Virginia, in favor of their own rail connections.
63. See Jones (1990)
64. See Howe (2007) p. 557
65. Herodotus' famous inscription is on New York's James A. Farley Post Office, eventually named for a famed Postmaster General and political advisor to President Franklin Roosevelt. The Post Office opened in 1914 and employs a Beaux Arts design by architects McKim, Mead, and White. That Post Office is being converted into Moynihan Station, a new passenger concourse for the existing adjacent Penn Station, the busiest on the Amtrak system. The current Penn Station (under Madison Square Garden) is the sorry 1963 replacement for an older, 1910 Beaux Arts masterpiece also designed by McKim, Mead, and White.
66. See Howe (2007) p. 225
67. See Wood (2009) p. 478
68. See John (1998)
69. See Ferrara (1990)
70. See Parnell (1837)
71. See Arthur (2009)
72. See Basalla (1988)
73. See Arthur (2009)
74. See Arthur (2009) p. 89
75. See Schumpeter (1912)
76. See Hippel (1988)
77. Hippel (1988) examines the loci of innovative activity. Which actors in a system are motivated to innovate, what do they do? For a more general discussion and stress on the cultural context, see Pacey (1974).
78. See Johnson (2010)
79. See Kelly (2010)
80. See Tabarrok (2011) and Heller (2008)
81. See Heller (2008)
82. See Simon (1972)
83. Quoted in Kaufman (1997)
84. Hutcheon (1831) p. 37, quoted in Wood (2009) p. 631

85. See quote in Dillard (1990) and Arthur (2009) p. 79

86. See Gladwell et al. (2011)

87. See Von Hippel (1986)

88. Quoted in Smiles (1884)

89. See (Richardson 1980)

90. See Dickerson (1951)

91. See Wood (2009) p. 622

92. Cawston and Keane (1968) describes the history of many trading companies.

93. See Keay (1991)

94. Andrew Jackson and the Americans won the Battle of New Orleans despite the Treaty of Ghent having been signed weeks before.

95. See Zingales (2012)

96. See Konvitz (1978)

97. See Jackson (1983)

98. One metric ton equals 0.9842 imperial ton, so the difference is small.

99. See Odlyzko (1997)

100. See Howe (2007) p. 835

101. See Bowker and Star (2000) p. 77

102. There are many biographies of Isambard Kingdom Brunel, see for instance Vaughan (1991).

103. See Hay (1985)

104. No obvious technical reason existed for its conclusion that the 7-foot gauge was superior, given the fixing of wheels to axles and the problems of curving. Hilton (1990) discusses the work of the Commission and the evolution of various gauges.

105. We think of such a standard as an interface standard. Later, we will stress the critical importance of interface standards or policies. The British selected height clearance standards tighter than those used on the Continent and elsewhere. This has limited the attainability of economies of scale and created interface problems with the Continent.

106. Total cubic contents in "tons" of 100 cubic feet (2.8 m^3).

107. Liner coal consumption was not very efficient. At the time, seawater was used in the boilers, and it was thought that low pressure (ca. 0.2 MPa [30 psi]) should be used to reduce scaling in boiler tubes. As a rough approximation and for a given velocity, the power required to move a ship increases with the square of ship length; capacity increases with the cube of length. A large ship, carrying a lot of coal, was needed to evolve steady steaming.

108. See "The first industrial city: Manchester 17601830" in Hall (1998)

109. See Howe (2007) p. 562

110. See Sobel (1996)

111. Garfield (2002)

112. Wikipedia (`http://en.wikipedia.org/wiki/William_Huskisson`, accessed March 18, 2011) notes: "5 December 1821, when a carpenter, David Brook, was walking home from Leeds along the Middleton Railway in a blinding sleet storm. He failed to see or hear an approaching train ... and was fatally injured." —Richard Balkwill; John Marshall (1993). *The Guinness Book of Railway Facts and Feats* (6th ed.). Guinness. ISBN 0-85112-707-X. According to parish council records, a woman in Eaglescliffe, Teesside, thought to be a blind beggar,

was "killed by the steam machine on the railway" in 1827. "Corrections and Clarifications." *The Guardian*. June 21, 2008. Retrieved February 5, 2009.

113. See Howe (2007) p. 563

114. There are several excellent histories of transportation in England, Aldcroft and Freeman (1983), and the US, Ringwalt (1888).

115. See Wolmar (2010a)

116. See Smith (1990)

117. See Howe (2007) p. 563

118. See Taylor (1951)

119. See Taylor (1951) p. 92

120. See Ripley (1950)

121. See Wolmar (2010a) pp. 253–255

122. See Chandler (1979)

123. See Wellington (1902)

124. See Dahl and Lindblom (1953), on which this discussion of incrementalism is based.

125. Most state departments of transportation have no global highway transportation goals. They seek to maintain the roads in the fashion that their values say they should be maintained, given the priorities of their road classification and subject to the monies and other resources available.

126. See Churchman (1979)

127. See Flinn (1957)

128. See Blaise (2001)

129. See Sobel (1996)

130. See Lay (1992); Webb and Webb (1913)

131. For example, as a boy, Garrison lived about one-quarter of a mile (400 m) from the Natchez Trace, a walking way from Natchez, Mississippi, to the Nashville Basin and places to the north for traders who carried farm products on flatboats down the Mississippi and its tributaries to New Orleans. Now improved for much of the way as a parkway by the federal government, it bears no functional or design resemblance to the historic Natchez Trace. Levinson grew up near Columbia Pike (a one-time turnpike emanating from the District of Columbia), in Columbia, Maryland (which was named for a nearby post office, which was named for the turnpike, which was named for the District, which was named for Columbus). Columbia Pike is now part of US 29. He went to college at Georgia Tech on North Avenue (US 29) in Atlanta, and returned to Maryland to work just off Colesville Road (US 29) in Silver Spring. The road which runs from Ellicott City, Maryland, to Pensacola, Florida, is sometimes a two-lane rural route, sometimes an urban street, and elsewhere a freeway. With a little searching, most any reader could also tell a "growing up near" story.

132. Quote from: *Lancashire: Brief Historical and Descriptive Notes*, Grindon (1892)

133. See Wolmar (2010a) p. 270

134. See Gerondeau (1997)

135. See Sheriff (1997)

136. See Siddall (1974)

137. Quoted in Siddall (1974)

138. Notables have been involved in other transit incidents. In 1888, the Czar of Russia was on a train that derailed in Borki, killing sixteen members of his court, though he was unscathed.

Queen Victoria's staff checked the line for bombs placed by Fenian terrorists. See Wolmar (2010a) p. 245.

139. See Howe (2007) p. 816

140. See Wolmar (2010a) p. 134

141. See Folwell (1921) p. 369, cited in (p. 51) for a fuller discussion of Lord Gordon-Gordon.

142. See Levinson (2007) p. 59

143. See Grossman (2012)

144. Quoted from Dunn and Kemper (1919)

145. See Earle (1929)

146. US Patent 59,915 was issued for the Velocipede in 1866.

147. See Herlihy (2004)

148. See Schiffer et al. (1994)

149. In one version of the Greek myth, Phaeton, the son of Helios, drove very fast while pulling the chariot of the sun; however, he did set the earth ablaze in the process, until Zeus killed him.

150. Ford worked for Edison as chief engineer of the powerhouse of Detroit Edison (the local electric utility) but was already working on a two-cylinder gasoline powered car. See Schiffer et al. (1994) p. 43.

151. See Swift (2011)

152. See Brynjolfsson and McAfee (2011); Clark (2008)

153. See Nye (1992)

154. Schiffer et al. (1994) p.102

155. Cited in Schiffer et al. (1994) p. 105

156. See Schiffer et al. (1994) p. 119

157. See Schiffer et al. (1994) p. 140

158. Schiffer et al. (1994) p. 142

159. See Schiffer et al. (1994) p. 71

160. See Schiffer et al. (1994) p. 145. Note: Arms were broken when the engine started and the crank was not yet disengaged.

161. See Schiffer et al. (1994) p. 146

162. Schiffer et al. (1994) p. 85

163. See Duncan and Burns (2003)

164. See Schiffer et al. (1994) p. 89

165. See Swift (2011)

166. See Swift (2011)

167. See De Luca (2012)

168. See Potter (1891)

169. Impacts are not well documented, but for examples of the social impact of the automobile in rural areas, see Berger (1979).

170. See Swift (2011)

171. See De Luca (2012)

172. See US Department of Transportation, Federal Highway Administration (1977) and Preston (1991)

173. See McCullough (1992); Weingroff (2002)

174. Klapper (1978)

175. See Howe (2007) p. 20

176. http://en.wikipedia.org/wiki/Shilling

177. See Jones (1978) p. 134

178. See Washington (1902); White (1966)

179. This section is adapted from Levinson (2008).

180. According to the 1831 UK census, the area that is now London had 1.099 million people living there. See Vision of Britain (2006).

181. See Bogart (2007)

182. See Course (1962)

183. See Hoyle (1982)

184. The City of London, the original center of Greater London, is now an administrative district within London. The "box" keeping railway lines out of the center of London was penetrated at Victoria, Charing Cross, and the extension from Blackfriars to Farringdon.

185. Accounting in the early years of the Metropolitan Railway, especially prior to the Regulation of Railways Act of 1868, was a bit lax, and dividends were reportedly paid out of capital. To quote Jackson (1986) p. 38, describing the era of 1865, "It was . . . a house of cards, a precarious game in which the level of dividend was kept up at all costs, by finding money from somewhere, with no regard to sound accounting or financial rectitude."

186. See Simmons (1978) p. 121. This large number of Bills involves duplication both within firms (lines proposed one year and not approved might be proposed again with minor modifications) and between companies (substantially the same line proposed by more than one company).

187. For good overviews of the history of the New York subway see Cudahy (1995) and Hood (2004).

188. See Cudahy (1995) p. 31

189. See Cudahy (1995) p. 38

190. See Cudahy (1995) p. 103

191. King (2011)

192. See Hilton and Bier (1997)

193. See White (1966)

194. See Barrett (1983)

195. This section is adapted from Levinson (2008).

196. Cited in Course (1962)

197. Thorn (1876) p. 31, quoted in Simmons (1986)

198. See Jackson (1986) chapter 9

199. Croome and Jackson (1993) pp. 4, 45

200. See White (1963)

201. See Foxell (2005)

202. See Jackson (2006)

203. See Edwards and Pigram (1986)

204. Young (1808) p. 418

205. See Konvitz (1978)

206. See Mohring and Harwitz (1962)

207. See Walmsley and Perrett (1991) p. 127

208. See Knight and Trygg (1978)

209. See Garrison and Deakin (1992)

210. Quote from Mohring (1993)

211. E.g., Voith (1991) found that the accessibility value of commuter rail stations in Philadelphia was capitalized in home values. Pollakowski and Wachter (1990) have examined land value and transportation in Montgomery County, Maryland, and found that distance from downtown Washington, D.C., was negatively associated with a housing price index, as was distance to employment.

212. See Garrison et al. (1958)

213. See Howe (2007) p. 691 and John (1998) p. 68

214. See Stover (1995)

215. See John (1998), p. 88

216. See Standage (1998)

217. See Gordon (2002)

218. See Howe (2007) p. 854

219. The discussion of US telegraph messengers is based on Downey (2002).

220. Decreasing costs ask for Ramsey pricing. In the 1950s Garrison served on the Advisory Committee for the Highway Cost Allocation Study where there was focus on cost to build with the intent of decreasing marginal costs. At the same time, there was concern that increasing speeds on the Interstate came at increasing marginal costs to the vehicle operator.

221. The TV show *Mythbusters* (Ep. 34) suggests you will be safe from a direct shot if you are under 8 feet (2.67 m) of water.

222. See Hauer (1994)

223. See Levinson and El-Geneidy (2009) on circuity

224. See Mokyr (2004) p. 232

225. See Mokyr (2004) p. 231

226. See Ship Design Codes used by the US Maritime Commission at `http://shipbuildinghistory.com/history/merchantships/wwii/designcodes.htm`

227. dwts: dead weight tons, the tons a ship can lift. The dwts a ship can lift are the tons that will depress a ship from its unloaded to its loaded water marks. Dwts are usually long tons of 2,240 lbs.

228. Flags of convenience are not limited to ships. We see the same thing in urban taxicabs. In the Twin Cities, many taxis were registered in the City of St. Bonifacius because of convenient regulations and low registration fees (about 1/6 of the larger cities). The small town had one licensed taxi for every three residents. See Reckdahl (1999).

229. Pursuant to Section 902 of the Merchant Marine Act, 1936

230. See Branson (1998)

231. Garrison reports that William "Pat" Patterson (president of United Airlines from 1934 to 1966) tells the story: "Boeing could not make a decision on building the 707 and its price. I went to LA and ordered several DC 8s.... went to Seattle and went fishing for several days ... neither my company or Boeing could reach me. When the newspaper said Boeing was ready to build, I asked how much?"

232. See Botting (2001) for discussion of the history of the Zeppelin Company.

233. See Heppenheimer (1995)

234. The nickname Model-T of the air has been applied to other vehicles as well, notably Henry Ford's Flivver Plane (weighing a mere 350 lbs, 160 kg), or the widely used Piper Cub general aviation vehicle.

235. See Barrett (1987)

236. See Doig (2001)

237. This 35-mile radius of the Port Authority explains why the Tappan Zee Bridge is just beyond the 35-mile radius, to be under separate, New York governance.

238. See p. 29 Babson (1948)

239. See Levinson and Xie (2011) for a fuller discussion of first mover advantage in transportation.

240. Letter from Charles Moore, chairman of the US Commission of Fine Arts, to Andrew Mellon, secretary of the US Treasury, dated April 5, 1927, Minutes of Meeting of the Commission of Fine Arts, April 14 and 15, 1927, Exhibit H.

241. The Crown privatized exploration and settlement. It didn't create a NASA-like organization to explore and then build stations in space.

242. See Hillman (1968)

243. Early in system development, intervention requires (near) ubiquitous service. Later, reductions in service may require intervention.

244. Providers may be at the mercy of large service purchasers. Though they had natural monopoly characteristics (high fixed costs), there were multiple competing railroads in some markets (thus there was a natural monopoly without a monopoly), leading to cutthroat competition. Attempts to control this competition through pooling and rate making turned out to be a rough and unworkable game. Further, railroads needed protection from large shippers and express companies.

245. Systems are unable to implement marginal cost prices because marginal costs are less than average costs. Marginal cost pricing is socially desirable and ought to be implemented. To achieve this goal, social decisions have to be made about the additional resources required to cover system costs, i.e., funds over and above those obtained by charging marginal cost prices.

246. Desired results are in the minds of beholders, and they range all over the map.

247. There had been a number of well-publicized abuses through watering stock.

248. The German situation was quite different from that in France. The kingdoms and states authorized private railroad development, with the only early state railroad being in Bavaria. As a result, there were about 50 rail companies by 1850. Rail development was market driven, and standards varied according to what markets warranted and could pay for. Several railroad centers emerged in addition to Berlin.

249. See Smith (1990)

250. See De (1955); De Tocqueville (1831)

251. See Miller (1971)

252. See Kolko (1976)

253. See Hamer (1987)

254. See Wilson (1980)

255. See Fellmeth (1970)

256. The indicted companies were: National City Lines, Inc., American City Lines, Inc., Pacific City Lines, Inc., Standard Oil Company of California, Federal Engineering Corporation,

Phillips Petroleum Company, General Motors Corporation, Firestone Tire & Rubber Company, and Mack Manufacturing Corporation. The individuals indicted were: E. Roy Fitzgerald and Foster G. Beamsley of NCL; H. C. Grossman, GM; Henry C. Judd, Standard Oil of California; L. R. Jackson, Firestone Tire & Rubber; Frank B. Stradley, and A. M. Hughes, Phillips Petroleum.

257. See Yago (1984)

258. We have heard some Marxists say that suburbanization and automobilization artificially propped up the economy, postponing the collapse of capitalism. The conspiratorial interpretation of the NCL conspiracy is widely known and taken to be true. Indeed, we know of "urban experts" that take Yago (1984) to be the definitive work on urban transportation.

259. See Snell et al. (1974)

260. Snell says the conversion of the New York system in 1936, then the world's largest streetcar system, was due to GM. In this case the agent was the Omnibus Corporation. GM's involvement with Omnibus was later than 1936, however. There are other problems of fact in Snell's writing. For GM's "sweetness and light" side of the matter, see *The Truth About "American Ground Transport,"* General Motors (1974). Other writings critical of the conspiracy theory include Adler (1991), Bianco (1998), Slater (2003).

261. See Babson (1948) p. 22

262. See Jones (1985)

263. PCC-type cars provided service in Moscow and other Eastern Europe cities after World War II.

264. See Carlson et al. (1980)

265. Bion J. Arnold was another leader; his *Report on Transportation Facilities, City of San Francisco* (1913) is most interesting and illustrates the points we have made. (Arnold underestimated the impact of the automobile on transit.) The Kelker, De Leuw & Co. *Report and Recommendations on a Comprehensive Rapid Transit Plan for the City and County of Los Angeles* (1925) is another product typical of the times.

266. See Simon (1959)

267. The federal government took action requiring that the electric utilities divest their holdings in transit companies. The action was intended to encourage the competitive pricing of electricity. The action also responded to the feelings of many that the utilities were thwarting the conversion from electric rail to diesel buses.

268. There are some devil theories that run the other way; see Hamer (1976).

269. See Spengler (1930)

270. See Hillman (1979)

271. See Altshuler and Luberoff (2003) pp. 249–251

272. As of the first edition of *The Transportation Experience* (written c. summer 2003), www.asphaltisbetter.com and www.concreteisbest.com both remained available, but they are since taken but unused. The industry websites were largely unchanged between 2003 and 2013.

273. See Swift (2011)

274. See Barker and Gerhold (1995)

275. There have also been some special regional planning endeavors, such as the Appalachian Regional Commission and US Forest Service planning activities.

276. See Geddes (1940)

277. See Swift (2011)

278. See Weingroff (1997)
279. See Swift (2011)
280. See Jakle and Sculle (2008)
281. ICD-10, 1:274; see Bowker and Star (2000) p. 101
282. See Norton (2008)
283. See Norton (2008)
284. See Norton (2008)
285. Erskine was president of Studebaker from 1915 until 1933, when he resigned; shortly after, he committed suicide. The firm and he personally ran into financial difficulties, and his name was dropped from the Bureau.
286. See Rankin (1997). Also Robinson (2011) discusses the backgrounds of 51 influential early transportation professionals.
287. See Norton (2008) p. 158
288. See Norton (2008)
289. See Barnes (1965)
290. See Foster (1981)
291. See Schultz (1989). City engineering organizations are interesting. Good insight can be obtained from Phelps et al. (1978), *Public Works in Seattle*, which details the 100-year history of the Seattle Department.
292. See McClintock (1926)
293. See Eno (1939)
294. See Jones (1990)
295. See Swift (2011)
296. See James (1916)
297. See Norton (2008)
298. See Burnham (1961)
299. See Geddes (1940)
300. See Buechner (2004)
301. See Newbery and Santos (1999)
302. For an example of one history of local road financing, see Zhao et al. (2010)
303. See Viner (1926)
304. See Swift (2011)
305. See Bureau of Public Roads, U.D.O.A. and the New Hampshire State Highway Department (1927); Bureau of Public Roads et al. (1927); Bureau of Public Roads U.D.O.A. and the Vermont State Highway Department (1927)
306. See Blanchard (1919)
307. See Lind (1965); Swift (2011) for profiles of MacDonald
308. See Coke (1968)
309. Weiner (2008) text has been published in several places: as a commercial book and as articles. It is a very useful collection of information, but has two faults. In the main, it views transportation planning as something innovated after World War II. Also, it is Washington-centered in the extreme. The text says that brilliant Washington bureaucrats had flashes of insight about rules and regulations. That's nonsense, at best.
310. See Holmes and Lynch (1957)

311. As discussed in Holmes and Lynch (1957), the Bureau of Public Roads (1944) *Manual of Procedures for Home Interview Traffic Studies* treated interviews at dwelling units and at the roadside.

312. Here is the way the needs are calculated: The condition today is known. With the traffic growth projected into the future (particularly truck traffic), the pavement deteriorates to lower and lower PSRs. The decrease in the PSR is projected using information from the AASHO road test. If we know the number of projected 18 kip (kilo-pound, equivalent to 8,164 kilograms) equivalent standard axle loads (ESALs) on a pavement of a particular type, then we can estimate how its PSR will change. Needs may then be estimated. The pavement may need rebuilding or repair, etc.

313. See Office (1987)

314. Garrison served on the advisory panel for the federal study triggered by the Interstate Act, and there have been subsequent federal studies. Congress asked the study to consider benefit-based financing—pay according to benefits received—but it ended as an incremental study. There were views of benefits received that could not be reconciled.

315. There is also the issue of central or social planning, determining centrally the outcome of society, which evidence has shown doesn't work particularly well. That is planning over-reach. Just because one can't plan a society, doesn't mean one can't plan a canal, a road, or a subway. Just because one can plan that thing doesn't necessarily mean one should, as the plan says nothing of trade-offs, opportunities forgone by applying resources in a particular way.

316. See Konvitz (1985)

317. See St. Clair (1986)

318. See Hegemann and Peers (1922)

319. See Hénard (1904)

320. See Collins (2005)

321. See Burgess (1929)

322. See Hoyt (1939)

323. See Harris and Ullman (1945)

324. See Horwood and Boyce (1956)

325. See Fogelson (2003)

326. See Arthur (1990)

327. Layton (1971) reviews the social responsibility debate, which lasted into the 1940s.

328. See Wilson (1989)

329. See Reps (1992)

330. See e.g. Ford (1911)

331. See Altshuler (1965)

332. Here is an example. Several years ago, Garrison participated in work for the National Council on Public Works Improvement. From the way turf has been divided in the past, one would have expected civil engineering to dominate analysis. That was not the case. Why was it not the case? Why have others encroached on the traditional turf of public works engineers?

333. See Ford (1911); Olmsted Jr. (1910)

334. See Lewis (1916)

335. See Stine (1981)

336. Lewis held important positions. In addition to his textbook, he published widely and wrote with much force. He headed planning organizations. Because of his influence, the American Society of Civil Engineers established a city planning division in 1923, after establishing power, sanitary, irrigation, and highway divisions the previous year.

337. The urban planning student ought to examine an early edition of Nelson P. Lewis's textbook. The debt that current concepts of city planning owe to Lewis shows clearly when one examines Lewis's book.

338. It would seem (c. 2003) on the West Coast of the United States that "geek" is a compliment for an intelligent, technically proficient individual with social skills and "nerd" an insult (indicating a lack of social skills), the mirror of what is found on the East Coast. `http://people.mills.edu/spertus/Geek/`

339. See Robinson (1955)

340. See Spann (1988)

341. See Barrett (1987)

342. See Konvitz (1978)

343. See Shea (2010) p. 33

344. Crandall and Waverman (2000)

345. Did the FCC decision permit that service? The decision was written by Bernie Strassburg, chief of the FCC's Common Carrier Bureau. We have never reviewed it. It is said to be very dense and complex and to appear to permit many-to-many services. On close reading, the FCC's position was that the decision did permit that service, although that was not the intention of the decision. The fact is that the FX decision was one that received little attention at the time. FCC had bigger fish to fry in debates about TV and the radio spectra. FX was a minor thing in the stagnant telephone field; the Commissioners approved FX on staff recommendation and without discussion.

346. Data from Cellular Telephone Industries Association; see Shea (2010) p. 195

347. See Abernathy (1978)

348. See e.g. Bijker et al. (1987)

349. See Maguire et al. (2000)

350. See Latour (2005)

351. See Latour (1996)

352. See Gifford and Garrison (1993)

353. In addition to transportation systems, large technical systems include electrical and other energy systems, sewage, and communications. They are infrastructures that are ubiquitous, have homogeneous technologies, require special technical knowledge, and have other features in common. See Summerton (1994).

354. See Hayek (1989)

355. See David (1986)

356. Actually, it isn't that simple; see Lay (1992).

357. See Hilton (1990)

358. The term "Zombie Transportation" follows the lead of John Quiggin, who titled a book *Zombie Economics: How Dead Ideas Still Walk among Us*.

359. See Jomini (1841)

360. TEU: Twenty-foot container equivalent units 8′ x 8′ x 20′ (2.4 x 2.4 x 6 m)

361. See Cudahy (2006)

362. See Rosenstein (2000)

363. See Sauerbier (1956)

364. See Altonn (2003). Pfeiffer was from a seafaring family, and born in Fiji. He began working the docks at age 12. Garrison recounts "I once flew Washington-SF with the president of Matson (Bob Pfieffer, who later moved on to head A&B); he then headed the Maritime Research Board and I met some other maritime folks at dinners he organized. Pfeiffer was into stunt flying and had his own aircraft."

365. See Table 6.3 in Cudahy (2006)

366. In 1997, Neptune Orient Line purchased American President Lines.

367. See Tai and Hwang (2005)

368. See Wright (2011)

369. See American Association of Port Authorities (AAPA). 2004. "Public Port Facts." Industrial Information. http://www.aapa-ports.org/industryinfo/portfact.htm, accessed March 2004.

370. In elementary school in the late 1970s, Levinson remembers the 1:05 PM Concorde flying over Columbia, Maryland, on its way to Dulles Airport, heralding that after-lunch recess was over.

371. Garrison visited US Secretary of Transportation Alan Boyd in his office. He had a one-foot stack of reports and studies on why the US should not develop an SST program and a small "yes" stack. He remarked that in spite of staff advice he would establish a program—in his judgment the probability of high payoff mitigated the lukewarm and negative recommendations.

372. See Nova (1998)

373. On April 1, 2003, the first date of this writing, Air Canada declared bankruptcy. The day before, American Airlines announced an agreement with unions to renegotiate labor contracts to reduce costs. US Airways emerged from bankruptcy that day as well, while America's second largest airline, United, remained in bankruptcy, having failed to come to terms with its unions.

374. See Bureau of Transportation Statistics (2012) Table 3.16

375. See Nakicenovic (1988)

376. See Sigafoos (1983)

377. See Lynn and Reilly (2002)

378. See `http://en.wikipedia.org/wiki/Computer_reservation_system` for a list.

379. See Schneier (2006)

380. Adapted from text by Fallows (2010). Brin (1998) makes essentially the same point.

381. See Sunstein (2003)

382. Quoted from Schwartz (2012)

383. To this day, railroads question the need for nationalization and its results; it is a sore point with them that the public might have thought that they were not doing their job. Note: A later USRA (The United States Railroad Association) acquired bankrupt railroads in the 1970s, and was eventually sold to Conrail.

384. See the National Transportation Policy Report of 1961.

385. See Burt (1998)

386. An interesting thing about network revamping is its contingent character. A certain pattern of route ownership was in place at the start of the game. (Think of this, say, as the situation at about 1950 when there were about 50 major railroads.) The-run-of-the-game (think of this as

plays involving mergers, acquisitions, and abandonment) yielded a five or six large-firm system by the end of the twentieth century (depending on how you count). This is not the same end state as would have been achieved with a different starting arrangement. Considering all possible starts, we do not know if the realized end of the game compares well or poorly with possible ends. Also the plays might have been sequenced in many different ways. Some attempted plays were thwarted by regulatory actions, and the outcomes of some plays were tempered by railroad requirements, such as allowing running rights on trackage. The value of the plays available at any time depended on what had gone before. An Example: Suppose there are two routes between *A* and *B*; Route 1 is longer than Route 2, all other features are the same. Suppose as the "game is played" Route 1 gets lots of traffic because of a merger. Economies of traffic density give it lower costs than Route 2. The potentially superior (shorter) Route 2 is abandoned because of its relatively high costs. It just had bad luck and wasn't included in a merger that increased its traffic. Actually, this example applies to only a few cases.

387. See Burns (1958) for a guide to mergers and other changes in the decades following the 4R Act.

388. Due (1979); Due and Leever (1993) review the state rail plans that were available at that time.

389. See Goodwin (2002)

390. See Hankla (2001)

391. See Humphreville (2012)

392. See Aguilar, Julian. *Border Welcomes First Rail Line in More Than a Century*. Texas Tribune, 2012.

393. See Sellnow (2000); Sommer (2002)

394. *Norfolk and Western v. American Train Dispatchers Association*, 89–1027 and 89–1028

395. See Cupper (1990)

396. See Levinson (2001)

397. That rather murky sentence appeared on p. 683 of Carlton and Perloff (1990).

398. See Gifford (1984)

399. See MacKaye (1930)

400. See Bassett (1930)

401. See Swift (2011)

402. See the *Roadside Trilogy* of Jakle and Sculle for details on these changes: Jakle and Sculle (2002a, b); Jakle et al. (2002)

403. See Bureau of Public Roads (1955)

404. See Horwood and Boyce (1956)

405. See US Bureau of Public Roads (1939)

406. See Rose (1990)

407. See Weingroff (1996)

408. On this topic and the urban Interstate generally, see the full review by Schwartz (1976). Schwartz emphasizes financing, examples, and legislative history. Mertz has also posted an article at Mertz (n.d.).

409. See Seely (1987)

410. There are assertions that Eisenhower was surprised about the existence of the urban Interstate program, discussed in Mertz (n.d.), though this is disputed. He was not happy about it, though.

411. Schwartz (1976) p. 512

412. Altshuler (1965) p. 21

413. See e.g. Swift (2011) for discussion of Baltimore, Schrag (2004) for Washington, Issel (1999); Johnson (2009) for San Francisco, or Mohl (2004 2008) for more general discussions.

414. The debate between Moses and Jacobs is chronicled in Flint (2011).

415. Jacobs's major works include Jacobs (1961 1984 1992 2005); Jacobs et al. (1970)

416. See Weisberg (2009) p. 136

417. Central Artery/Tunnel Project Official Page: http://www.bigdig.com; see also http://www.boston.com/beyond_bigdig/. Also see the following link, which gives the details of Scheme Z in the history of the CA/T: http://libraries.mit.edu/rotch/artery/schemez_chron.htm Info regarding the open space created by the submerged central artery: http://enews.tufts.edu/stories/070102BigDigBonus.htm; Altshuler and Luberoff (2003); Hughes (2000).

418. See Korr (2002)

419. Quotes from http://www.iccyes.org/ See also http://www.iccfacts.org for the opponents' point of view.

420. See Di Caro (2012)

421. The detailed story of I-69 is told in Dellinger (2010).

422. See http://www.i69tour.org/cost_estimate.pdf

423. See Kozel (1997)

424. See News-Record (1996)

425. *Washington Post*, August 4, 1996.

426. See Newswire (1999)

427. See Journal (1999)

428. See Sullivan (2000)

429. See Meehan (2002)

430. See Kasarda and Lindsay (2011) for a discussion of airport-land use interactions.

431. See Meyer (2008)

432. See Dahlgren (1998)

433. See Spielberg and Shapiro (2000)

434. See Fielding and Klein (1993)

435. See Watt (2012)

436. See Daniels and Trebilcock (1996); Poole (1997); Poole Jr and Fixler Jr (1987)

437. See Albalate et al. (2009)

438. See Samuel and Poole (2005)

439. See Samuel et al. (2002)

440. See Fields et al. (2009); Poole and Samuel (2006); Poole and Sugimoto (1995); Staley and Moore (2009)

441. See Levinson (2002a)

442. See Albalate et al. (2009)

443. See Engel et al. (2006)

444. See Hillsman (1995)

445. See Saleth and Dinar (2004)

446. See Saleth and Dinar (1999)

447. See Stigler and Friedland (1962)

448. See Starkie (1988)
449. See Newbery and Santos (1999)
450. See Forkenbrock (2008)
451. See Zhang et al. (2009)
452. See Levinson and Odlyzko (2008a)
453. See Jenkins-Smith and Herron (2004)
454. See Pew Research Center for the People and the Press (2010)
455. See Garbarino and Lee (2003)
456. See Glaskowsky (1986)
457. See Association (1999)
458. See Winfrey (1968)
459. See Fawaz and Garrison (1994)
460. See Swift (2011)

461. The source of this quote is undetermined, but Wikiquote states: "According to a letter to the *Daily Telegraph* by Alistair Cooke on 2 November 2006, this sentiment originated with Loelia Ponsonby, one of the wives of 2nd Duke of Westminster who said 'Anybody seen in a bus over the age of 30 has been a failure in life.' In a letter published the next day, also in the *Daily Telegraph*, Hugo Vickers claims Loelia Ponsonby admitted to him that she had borrowed it from Brian Howard. There is no solid evidence that Margaret Thatcher ever quoted this statement with approval, or indeed shared the sentiment." `http://en.wikiquote.org/wiki/Margaret_Thatcher`. Loelia Ponsonby gives her name to a secretary in Ian Fleming's early *James Bond* novels.

462. See Danielson (1965) discussing Nelson (1959)
463. Quoting Danielson (1965) p. 158

464. Garrison served on a small committee to draft the turf-protecting, who could do what, agency agreement between HUD and the Department of Commerce. When Congress created the transit agency, it gave money to HUD to get it started. HUD managed to transfer the agency but keep the money.

465. Knowledgeable about transit, especially from contacts with George Krambles of the Chicago Transit Authority, Garrison followed early UMTA developments quite closely. Because UMTA was established at about the same time that the DOT was established and it was feared that the organizational problems of DOT would interfere with the take-off of UMTA programs, and people wanted to keep UMTA away from "transportation think," UMTA was housed in the Department of Housing and Urban Development (HUD) during its first year. Garrison participated in the study group that arranged the transfer to DOT and the conditions imposed on that transfer. During the first ten years of so of UMTAs life, he served on a variety of other advisory-study committees. These early UMTA days are not regarded as ones in which Garrison did constructive things.

466. But perhaps not so many as one finds in local governments. At the time, it was felt that special skill and knowledge were needed to operate transit properties.

467. We knew a few who managed to be "quick studies," but most didn't. We think this was because the agency didn't have a core who could give it appropriate instant-agency-culture, as well as political appointments at local levels.

468. The first demonstration grant made by UMTA went to Selma, Alabama, hardly a big city bail out.

469. BART Actual Weekday Ridership in 4th quarter of FY2003 was 291,011, in the fourth quarter of FY2010 it had risen to 342,055, while the number of stations also increased http://www.bart.gov/about/reports/indicators.asp (accessed September 16, 2003), http://www.bart.gov/docs/Quarterly_Exits_Q32011.pdf (accessed April 29, 2011).

470. MTC, 2011 http://www.mtc.ca.gov/maps_and_data/datamart/stats/baydemo.htm

471. One needs to be cautious as what constitutes a trip raises definitional questions about transfers and the accounting of linked vs. unlinked trips. In transit accounting, transferring from bus to rail may count as a trip. In travel demand modeling, going from origin to destination counts as a trip, even it if uses multiple modes.

472. San Francisco Bay Area Rapid Transit District Independent Auditor's Report, Management's Discussion and Analysis, and Basic Financial Statements for the Years Ended June 30, 2010, and 2009 http://www.bart.gov/docs/FY2010_financials.pdf

473. See Yamamoto (2003)

474. See Webber (1976)

475. See Schrag (2006)

476. See WMATA Vital Signs Report. A Scorecard of Metro's Key Performance Indicators (KPI) 2012 1st Quarter Results. http://wmata.com/about_metro/docs/Vital_Signs_May QTRLY 2012.pdf

477. District of Columbia, Total Nonfarm, Seasonally adjusted - SMS11000000000000001. http://data.bls.gov/cgi-bin/surveymost?sm+11

478. See Journey-To-Work Trends in the United States and its Major Metropolitan Areas 1960–1990 FHWA-EP-03-058, June 2003. http://ntl.bts.gov/DOCS/473.html.

479. See Chen (2007)

480. As Garrison did in Denver some years ago and Levinson has done in Minneapolis.

481. We hear that we tried new technology programs in the early days of UMTA and they were a failure. So new solutions don't exist. This turns the discussion to why UMTA technology programs failed, a discussion that isn't helpful when policy is under discussion in an urban area.

482. See Spanhake (2001)

483. See De Soto and Instituto Libertad y Democracia (Lima (1989); Klein et al. (1997); Richmond (2001)

484. See Rice and Rogers (1980)

485. See Rogers et al. (1979)

486. See for Universal Design (2011)

487. See Latour (1996)

488. See e.g. Anderson (1999); Burke (1979); Lowson (2005)

489. See Bernier and Seekins (1999)

490. See Heseltine and Silcock (1990); Kennedy (1995ab); Mackie et al. (1995); Nash (1993); White (1990 1997)

491. Passenger journeys on local bus services by metropolitan area status and country: Great Britain, annual from 1970. http://www.dft.gov.uk/statistics/tables/bus0103/

492. "If you are not paying for it, you're not the customer; you're the product being sold" was first posted on the MetaFilter online community by Andrew Lewis at 1:41 PM on August 26, 2010, and is believed to be the source of this sentiment. http://www.metafilter.com/95152/Userdriven-discontent#3256046.

493. See Denneen and Company (2009)

494. See Kilgore (2012)

495. See Federal Transit Adminsitration (2009)

496. See Foley (1990)

497. See Jevons (1866) p. 356

498. See Lewis (1916)

499. See Kuhn (1957)

500. The agency was called the Chicago Area Transportation Study (CATS) until it changed its name to CMAP—the Chicago Metropolitan Agency for Planning—in 2006 upon a merger with the Northeastern Illinois Planning Commission (NIPC).

501. See Beckmann et al. (1956); Boyce (2002)

502. See Ekelund and Hébert (1999)

503. See Wellington (1902)

504. See Lee Jr (1973)

505. See Churchman (1979)

506. See Armstrong et al. (1976); Garrison was involved in this study.

507. See Office of Technology Assessment (1983)

508. See Downs (1992)

509. Levinson worked at the Montgomery County Planning Department of the Maryland National Capital Parks and Planning Commission during this period (1989–1994).

510. See Brinkman (1982); Flyvbjerg et al. (1990); Wachs (1982, 1990)

511. See Button et al. (2010); Flyvbjerg (2007); Flyvbjerg et al. (2002 2005 2003); Kain (1990 1997); Pickrell (1992); Richmond (2001); Taylor et al. (2009)

512. See Parthasarathi and Levinson (2010)

513. See Barnes and Davis (2001); Chumak and Braaksma (1981); Levinson (1999); Zahavi (1974); Zahavi and Ryan (1980); Zahavi and Talvitie (1980)

514. See Prendergast and Williams (1981); Tanner (1981)

515. See Whyte (1993)

516. See Zahavi (1981)

517. See Sinclair (1935)

518. See Metz (2008); Millard-Ball and Schipper (2011a)

519. See http://www.google.com/publicdata/explore?ds=d5bncppjof8f9_&met_y=en_atm_co2e_pc&idim=country:USA&dl=en&hl=en&q=us+co2+emission+trends, accessed August 23, 2012

520. See Administration (2012)

521. See Sinclair (1994)

522. See Mokyr (2004) pp. 148–149

523. Researchers at the Institute for Applied Systems Analyses (IASA) in Austria have charted lots of data on American and European transportation systems, and the authors have also engaged in such exercises. See Batten and Johansson (1985); Garrison (1989); Garrison and Souleyrette (1996); Grubler (1990); Marchetti (1980); Nakicenovic (1988)

524. See Sperling and Gordon (2009) for a discussion of the rate and risks of Chinese motorization.

525. For discussion of some of the quality control problems, see Langfitt (2012). For discussion of China's environmental problems, see Smyth et al. (2008).

526. See Metz (2008); Millard-Ball and Schipper (2011b)
527. See Nakicenovic (1988)
528. See Batten and Johansson (1985)
529. The subject of network evolution is thoroughly discussed in Xie and Levinson (2012).
530. See Christaller (1966); Lösch (1967)
531. See Hirschman (1985)
532. See Aschauer (1989a,b 1990a,b 2000)
533. See Munnell and Cook (1990) as well as Nadiri and Mamuneas (1994)
534. See Aaron (1990); Slater (1992)
535. See LeHong and Fenn (2012)
536. See Garrison (1984)
537. See Porter (1990)
538. See Stigler (1951)
539. See Garrison and Souleyrette (1996); Isard (1942)
540. See Korotayev and Tsirel (2010)
541. See Mensch (1979) p. 131
542. See Mokyr (2004) pp. 232, 237, 277 and Mokyr (1993)
543. Nelson and Winter (1982) are referring to Hicks (1932); Schumpeter (1934); Solow (1957).
544. See Garrison et al. (1958)
545. See Garrison and Souleyrette (1996)
546. See Garrison and Souleyrette (1996)
547. See Verhoef (1994) p. 278
548. See Marsh (1885)
549. See e.g. Lomborg's position in his debate with Pope in Pope and Lomborg (2005).
550. See Intergovernmental Panel on Climate Change (2007)
551. See Sapart et al. (2012) who associate methane in ice cores with human activity.
552. See Tugwell (1988)
553. See Brons et al. (2008); Goodwin et al. (2004); Graham and Glaister (2002 2004)
554. See Diamond (2006)
555. See Kahn (2010)
556. See McNeill (2000)
557. See Nevin (2000)
558. See US Environmental Protection Agency (2012)
559. See McNeill (2000)
560. See Koshland (1989)
561. Stigler's reply is in *Science*, 3 March 1989.
562. See Berger (1979)
563. For some years, Garrison has explored redesign of personal highway transportation. Similar to the exploration of trucking options, the effort has been to build from existing technologies and institutions, as well as market niches. Designs have striven to specialize and optimize vehicles, roads, and operations. Neighborhood and commuter market niches have received priority.

564. A vehicle called the Tango holds similar promise; it is small and skinny but not light, primarily because of its batteries. However, because of limited production runs, it is being sold assembled for $200,000. It is hoped mass production would drive down costs.

565. See Institute (2003)

566. See RILEY (1994)

567. See Hunter-Zaworski (2012)

568. See Garrison and Ward (2000)

569. See http://theamericanscholar.org/what-the-earth-knows/

570. See King (2012). *Source*: December 2012 Plug-In Electric Vehicle Sales Report Card by Jay Cole, January 4, 2013 http://insideevs.com/december-2012-plug-in-electric-vehicle-sales-report-card/

571. See Blanco (2012)

572. See Sperling and Gordon (2009)

573. See Cherry and Cervero (2007)

574. See Sperling and Gordon (2009)

575. See Hoffmann (2002); Koppel (1999)

576. See Lawson (1993)

577. See Mian and Sufi (2012)

578. See Sivak (2009)

579. See Knittel (2009), Li et al. (2012)

580. SNCF, the French rail agency, also after many years of planning from 1966, opened the TGV in 1981 (nearly 20 years after Japan). The opening ceremonies were a significant event, being reported internationally, but not associated with a major showpiece such as a World's Fair or Olympics.

581. See Japanese Railway Engineering (1994)

582. Market segmentation in Japan has principally focused on the business travel market. The French have focused on business travelers. Pleasure travel is a secondary market, though many of the French extensions connect with vacation beaches on the Atlantic and Mediterranean. Friday evenings are the peak time for TGV. The system has lowered prices on long-distance travel to better compete with air service, and have even turned cities now within an hour of Paris (by TGV) into commuter bedroom communities, increasing its own market while restructuring land use.

583. See Osnos (2012)

584. Sometime in the 1960s, US Department of Commerce Undersecretary Plowman asked Garrison to visit Senator Pell. The Japanese National Railway representatives showed us a film and Pell said, "Lets build that."

585. See US Federal Railroad Administration (1987)

586. See US Federal Railroad Administration and National Railroad Passenger Corporation (1981)

587. See Federal Railroad Administration,Department of Transportation, US, Washington, D.C., *Northeast Corridor Improvement Project, Joint Annual Report*, 1976–1987

588. See Office of Technology Assessment (1983)

589. See US Federal Railroad Administration (1990)

590. See Johnson et al. (1989)

591. The next largest city not on the network is Honolulu, Hawaii, at 55.

592. New York (1), Los Angeles (2), Chicago (3), Dallas (4), Atlanta (9), Phoenix (12), Denver (21), and Orlando (27) http://en.wikipedia.org/wiki/Table_of_United_States_Metropolitan_Statistical_Areas.

593. Alaska, Hawaii, Idaho, Utah, Wyoming, North Dakota, and South Dakota are excluded from the Intercity Passenger Rail Program. However, North Dakota and South Dakota have been included in the Minnesota State plan (connecting to Sioux Falls and Fargo).

594. See Levinson (2012), which summarizes the literature.

595. See Levinson et al. (1997)

596. We would not go quite as far as Wolmar (2005), but there are clearly problems with vertical separation in the railroad sector.

597. See Wolmar (2010b)

598. See Krumm (1994)

599. See Gerondeau (1997) p. 123

600. See Chester and Horvath (2010) for a discussion of HSR externalities and time to payback.

601. See Krylov (2001)

602. See Levinson et al. (1997)

603. See Takeda and Mizuoka (2003)

604. See Taggart (1999) for the analogy.

605. See Berners-Lee and Fischetti (2000); Hafner and Lyon (1996)

606. See Shea (2010) p. 195

607. See Save the Internet (2006)

608. See Pulver (2006)

609. See Frankston (2006)

610. See Crawford (2006)

611. No network is neutral in the original Crawford (2006); Lessig and McChesney (2006); Wu (ND) sense of neutral.

612. This is the Berners-Lee (2006) sense of neutral. The *AT&T standard* (the conditions imposed upon AT&T during its merger with BellSouth) permits quality of service differentiation by the customer (consumers can pay more for higher bandwidth), but not the producer (a website cannot pay more to reach customers faster than other websites). That the AT&T standard also does not discriminate based on "source, ownership, or destination" is consistent with common carriage law, but may be seen as inadequate by advocates of strong neutrality, as it does not prevent differentiation by media type or willingness to pay. Federal Communications Commission (2006).

613. See Hands Off the Internet (2006)

614. See Lamar (1914)

615. See Bouvier (1856)

616. See Odlyzko (1999)

617. See Griffiths (1999)

618. See Markus (2004)

619. See Levinson (2002b)

620. See Hensher (2003); Wachs (2003)

621. See Levinson and Odlyzko (2008b)

622. See Levinson and Yerra (2002)

623. We have been in many situations where the mood is "technology is not the problem. The problem has to do with institutions, protocols, and things like that." We think such views misread the situation, for the focus is on soft technology. In addition, the problem may be misread. Sometimes, the problem is one of the mismatch between hard and soft technologies. Sometimes the hard technology seems to be the problem, and adjustments to soft technology have little to do with that problem.

624. The discussions revealed that many of the traffic engineers lived outside the metered area, and so could enter a freeflowing freeway and drive relatively unencumbered by traffic, which was sitting at freeway on-ramps.

625. The three university studies were published as Hourdakis and Michalopoulos (2002); Kotsialos and Papageorgiou (2004); Kwon et al. (2001); Levinson and Zhang (2004, 2006); Zhang (2010); Zhang and Levinson (2002, 2004, 2010)

626. From outside the state (and therefore seemingly incorruptible by future Mn/DOT consulting contracts) (Cambridge Systematics).

627. A 23-minute version of the video *To New Horizons* is available on Youtube, `http://www.youtube.com/watch?v=aIu6DTbYnog`.

628. See Geddes (1940)

629. See Geddes (1932)

630. See Swift (2011)

631. See Zou and Levinson (2003)

632. See Efrati (2012)

633. See Cowen (2011)

634. See Iteris (2012)

635. See Garrison (1985)

636. See Kihss (1970); Moynihan (1970)

637. See Buchanan and Tullock (1962)

638. See Small et al. (1989)

639. There is a subtle macro-economic point, that jobs represent a benefit if the construction worker would otherwise be unemployed, but not if the worker would otherwise be doing something else.

640. See Webber (1964)

641. However, of the 63,000 non–Coast Guard employees (according to the Office of Personnel Management 2001 Fact Book) (this number also predates the creation of the Transportation Security Administration), about 16,000 are in air traffic control and 15,000 are involved in airways facility jobs, so about 32,000 bureaucrats run the shop. That seems to be lots of people (though down over the past ten years). Holt's (1923) study of the Bureau of Public Roads said that about 24 employees managed the federal-state relation very well. (It might also be noted that the Department of Transportation, with a 79 percent white work force, is the whitest federal department.)

642. The source for this quote is Class Notes distributed c. 1985.

643. Brin, David (2012) "Contemplating Civilization: Its Rise, Fall, Rebuilding and Future"

644. Gordon, Peter (2012) "Thinking about Growth, the Seventh Killer App"

645. William Gibson said this on the NPR program *Fresh Air* (31 August 1993).

646. "10-codes" are widely used by police and emergency personnel in the US to convey information, and were first published in 1940, though there is great debate as to their

usefulness compared with plain speech. See Dispatch Magazine online for some discussion, http://www.911dispatch.com/info/tencode.html

647. See Levy (2007). The article can be found at: http://www.startribune.com/local/11593606.html

648. A summary of the NTSB report (2008) can be found at Roads&Bridges (2008)

649. See Fisher (2009)

650. See O'Connell et al. (2001)

651. See Wolf (2012)

652. See of US Department of Transportation Bureau of Transportation Statistics (2008)

653. See Foster (2012)

654. See Kennedy and McEnroe (2007)

655. This article can be accessed at http://timpawlenty.blogspot.com/2007/05/star-tribune-pawlenty-vetoes-gas-tax.html. Accessed April 19, 2012

656. Bud Heidgerken lost leadership position in Republican Party.

657. Rod Hamilton lost leadership position in Republican Party.

658. Ron Erhardt switched to the Independence Party but lost the general election. He ran in 2010 as a member of the DFL party, but lost to Kevin Staunton in the primary (Staunton lost in the general election to Kevin Downey). He was elected in 2012 as a member of the DFL, defeating Downey.

659. The Minnesota Democratic Farmer Labor Party is the local affiliate of the US Democratic Party, and was created as a merger in 1944 by Hubert Humphrey.

660. See Kimball (2011)

661. Current MnDOT values of time can be found at http://www.dot.state.mn.us/planning/program/benefitcost.html.

662. See Smalkoski and Levinson (2005)

663. See Xie and Levinson (2011)

664. "Daily" VHT comprises Morning + Mid-Day + Afternoon periods

665. Value at $14.19 Value of Time

666. See Marchetti (1980)

667. See e.g. Bruno (1993)

Bibliography

Aaron, Henry J. (1990). Discussion of "Why Is Infrastructure Important?," in Alicia Munnell (ed.). *Is There a Shortfall in Public Capital Investment*, Boston: Federal Reserve Bank of Boston, 51–63.

Abernathy, W. (1978). *The Productivity Dilemma*, Volume 35. Baltimore, MD: Johns Hopkins University Press.

Adler, M. (1985). Street Parking: The Case for Communal Property. *Logistics and Transportation Review 21*(4), 375–387.

Adler, S. (1991, September). The Transformation of the Pacific Electric Railway: Bradford Snell, Roger Rabbit, and the Politics of Transportation in Los Angeles. *Urban Affairs Quarterly 27*(1), 51–86.

Albalate, D., G. Bel, and X. Fageda (2009). Privatization and Regulatory Reform of Toll Motorways in Europe. *Governance 22*(2), 295–318.

Albert, W. (1972). *The Turnpike Road System in England: 1663–1840* . Cambridge: Cambridge University Press.

Albion, R., and J. Pope (1970). *The Rise of New York Port (1815–1860).* David & Charles.

Aldcroft, D., and M. Freeman (1983). *Transport in the Industrial Revolution*. Manchester: Manchester University Press.

Altonn, H. (2003, September 28). A & B's Longtime Leader Dies At 83: Robert J. Pfeiffer Led the Company Through Its Transition from Sugar to Development. *Honolulu Star-Bulletin*.

Altshuler, A. (1965). *The City Planning Process: A Political Analysis.*. Ithaca, NY: Cornell University Press.

Altshuler, A. (1979). *The Urban Transportation System: Politics and Policy Innovation.* Cambridge, MA: MIT Press.

Altshuler, A., and D. Luberoff (2003). *Mega-Projects: The Changing Politics of Urban Public Investment*. Washington, DC: Brookings Institution Press.

American Society of Mechanical Engineers and Regional Transit Authority, A. S. and R. T. Authority. St. Charles Avenue Streetcar Line, 1835. National Historic Mechanical Engineering Landmark brochure, `http://www.asme.org/history/brochures/h101.pdf`.

American Trucking Association (ATA). 1999. *ATA's Statement on the Passage of Legislation for the New Federal Motor Carrier Safety Administration*. Inside ATA. November 19. Alexandria, Va.: ATA. `http://www.truckline.com/insideata/press/111999_wbm_passage_of_bill.html`, accessed November 2003.

Anderson, J. (1999). A Review of the State of the Art of Personal Rapid Transit. *Journal of Advanced Transportation 34*(1), 3–29.

Andersson, Å. (1986). The Four Logistical Revolutions. *Papers of the Regional Science Association 59*(1), 1–12.

Andersson, D., O. Shyr, and J. Fu (2010). Does High-Speed Rail Accessibility Influence Residential Property Prices? Hedonic Estimates from Southern Taiwan. *Journal of Transport Geography 18*(1), 166–174.

Armstrong, E., M. Robinson, and S. Hoy (1976). *A History of Public Works in the United States, 1776–1976*. Chicago: Public Works Historical Society.

Arthur, W. (1990, February). Positive Feedbacks in the Economy. *Scientific American*, 92–99.

Arthur, W. (2009). *The Nature of Technology: What It Is and How It Evolves*. Free Press, New York.

Aschauer, D. (1989a). Does Public Capital Crowd Out Private Capital? *Journal of Monetary Economics 24*(2), 171–188.

Aschauer, D. (1989b). Is Public Expenditure Productive? *Journal of Monetary Economics 23*(2), 177–200.

Aschauer, D. A. (1990), "Why is Infrastructure Important", presented at the Federal Reserve Bank of Boston's Conference on "The Third Deficit: The Shortfall in Public Capital Investment", Harwich Port, Massachusetts, June.

Aschauer, D. (1990a). Highway Capacity and Economic Growth. *Economic Perspectives 14*(5), 4–24.

Aschauer, D. A. (1990b). "Why is Infrastructure Important," presented at the Federal Reserve Bank of Boston's Conference on *The Third Deficit: The Shortfall in Public Capital Investment*, Harwich Port, Massachusetts, June.

Aschauer, D. (2000). Public Capital and Economic Growth: Issues of Quantity, Finance, and Efficiency. *Economic Development and Cultural Change 48*(2), 391–406.

Ayres, R. (1984). *The Next Industrial Revolution: Reviving Industry through Innovation*. Cambridge, MA: Ballinger Publishing.

Ayres, R., and W. Steger (1985). Rejuvenating the Life Cycle Concept. *Journal of Business Strategy 6*(1), 66–76.

Babson, R. (1948). *Looking Ahead Fifty Years*. Harper.

Barker, T., and D. Gerhold (1995). *The Rise and Rise of Road Transport, 1700–1990*. Cambridge: Cambridge University Press.

Barnes, G., and G. Davis (2001). Land Use and Travel Choices in the Twin Cities, 1958–1990. Technical Report 6, University of Minnesota, Center for Transportation Studies.

Barnes, H. (1965). Man with Red and Green Eyes. Dutton Publishing (New York)

Barrett, P. (1983). *The Automobile and Urban Transit: The Formation of Public Policy in Chicago, 1900–1930*. Philadelphia: Temple University Press.

Barrett, P. (1987). Cities and Their Airports: Policy Formation, 1923–1952. *Journal of Urban History 14*, 112–137.

Bartholomew, A., and L. Metz (1989). *Delaware and Lehigh Canals*. Center for Canal History and Technology. Easton, Pennsylvania

Basalla, G. (1988). *The Evolution of Technology*. Cambridge: Cambridge University Press.

Bassett, E. (1930, February). The Freeway—A New Kind of Thoroughfare. *The American City*, 95.

Batten, D., and B. Johansson (1985). *Industrial Dynamics of the Building Sector: Substitution and Trade Specialization*. Economic Faces of the Building Sector, Stockholm: Swedish Council for Building Research.

Beasley, D. R. (1988). The Suppression of the Automobile: Skullduggery at the Crossroads. New York: Greenwood Press.

Becker, G. (1965). A Theory of the Allocation of Time. *The Economic Journal 75*(299), 493–517.

Beckmann, M., C. McGuire, and C. Winsten (1956). Studies in the Economics of Transportation. Technical report, New Haven, CT: Yale University Press.

Bejan, A., and S. Lorente (2011). The Constructal Law Origin of the Logistics S Curve. *Journal of Applied Physics 110*, 024901.

Berger, M. (1979). *The Devil Wagon in God's Country: The Automobile and Social Change in Rural America, 1893–1929*. Hamden, CT: Archon Books.

Berners-Lee, T. (2006, May). Neutrality of the Net. Weblog entry. May 2.

Berners-Lee, T., and M. Fischetti (2000). *Weaving the Web: The Original Design and Ultimate Destiny of the World Wide Web by Its Inventor*. New York: HarperCollins.

Bernier, B., and T. Seekins (1999). Rural Transportation Voucher Program for People with Disabilities: Three Case Studies. *Journal of Transportation and Statistics 2*(1), 61–70.

Bianco, M. (1998). 60 Minutes, and Roger Rabbit: Understanding Conspiracy Theory Explanations of the Decline of Urban Mass Transit. Portland, Oreg.: Center for Urban and Public Affairs, Portland State University, Discussion Paper 98–11.

Bijker, W., T. Hughes, and T. Pinch (1987). *The Social Construction of Technological Systems: New Directions in the Sociology and History of Technology*. Cambridge, MA: MIT Press.

Blaise, C. (2001). *Time Lord: Sir Sandford Fleming and the Creation of Standard Time*. New York: Pantheon Books.

Blanchard, A. (1919). *Highway Engineers' Handbook*. New York: John Wiley.

Blanco, S. (2012, June). Honda Fit EV rated at 118 MPGe with 82-mile range by EPA *UPDATE. *Autoblog*.

Bogart, D. (2007). Neighbors, Networks, and the Development of Transport Systems: Explaining the Diffusion of Turnpike Trusts in Eighteenth-Century England. *Journal of Urban Economics 61*, 238–262.

Botting, D. (2001). *Dr. Eckener's Dream Machine: The Great Zeppelin and the Dawn of Air Travel*. H. Holt. New York

Bouladon, G. (1967, April). The Transport Gaps. *Science Journal*, 41–46.

Bouvier, J. (1856). *A Law Dictionary Adapted to the Constitution and Laws of the United States of America and of the Several States of the American Union with References to the Civil and Other Systems of Foreign Law* (6 ed.), Volume 1. Philadelphia: Childs and Peterson.

Bowker, G. and S. Star (2000). *Sorting Things Out: Classification and its Consequences.* Cambridge, MA: MIT Press.

Boyce, D. (2002). Is the Sequential Travel Forecasting Paradigm Counterproductive? *ASCE Journal of Urban Planning and Development 128*(4), 169–183.

Boyce, D. (2007). An Account of a Road Network Design Method: Expressway Spacing, System Configuration and Economic Evaluation. In *Frastrukturprobleme bei Bevölkerungsrückgang*, pp. 131–159. Berlin: Berliner Wissenschafts-Verlag.

Branson, R. (1998, Dec 7.). Juan Trippe: Pilot of the Jet Age. *Time Magazine.*

Brin, D. (1998). *The Transparent Society: Will Technology Force Us to Choose between Privacy and Freedom?* New York, Basic Books.

Brinkman, P. (2003). *The Ethical Challenges and Professional Responses of Travel Demand Forecasters.* Ph. D. thesis, University of California at Berkeley, Department of City and Regional Planning.

Brons, M., P. Nijkamp, E. Pels, and P. Rietveld (2008). A Meta-analysis of the Price Elasticity of Gasoline Demand: A Sur Approach. *Energy Economics 30*(5), 2105–2122.

Bruno, L. (1993). *On the Move: A Chronicle of Advances in Transportation.* Detroit: Gale Research.

Brynjolfsson, E., and A. McAfee (2011). Race Against the Machine: How the Digital Revolution is Accelerating Innovation, Driving Productivity, and Irreversibly Transforming Employment and the Economy. *Amazon Digital Services.*

Buchanan, J., and G. Tullock (1962). *The Calculus of Consent: Logical Foundations of Constitutional Democracy*, Volume 100. Ann Arbor: University of Michigan Press.

Buechner, W. n.d. History of the Gasoline Tax. `http://www.artba.org/economics_research/reports/gas_tax_history.htm`, accessed March 2004.

Bureau of Public Roads (1955). General Location of the National System of Interstate Highways (Also Known as the Yellow Book). Technical report, US Department of Commerce.

Bureau of Public Roads, U. D. o. A., and the New Hampshire State Highway Department (1927). Report of a Survey of Transportation on the State Highways of New Hampshire. Technical report, Washington, DC.

Bureau of Public Roads, U. D. o. A., the Ohio Department of Highways, and P. Works (1927). Report of a Survey on the State Highway System of Ohio. Technical report, Washington, DC.

Bureau of Public Roads, U. D. o. A., and the Vermont State Highway Department (1927). Report of a Survey of Transportation on the State Highways of Vermont. Technical report, Washington, DC.

Bureau of Transportation Statistics, B. (2008). Transportation Statistics Annual Report. Technical report, US Department of Transportation.

Burgess, E. (1925). *The Growth of the City*, in R. Park, E. Burgess, and R. McKenzie (eds.), The City. Chicago: University of Chicago Press, 37–44.

Burke, C. G. (1979). Innovation and Public Policy: The Case of Personal Rapid Transit. Lexington, Mass: Lexington Books.

Burnham, J. (1961). The Gasoline Tax and the Automobile Revolution. *The Mississippi Valley Historical Review 48*(3), 435–459.

Burns, J. (1958). *Railroad Mergers and the Language of Unification*. New York: Quorum Books.

Burt, W. D. (1998, Fall). Was the Conrail Monopoly Necessary? *Journal of Transportation Law, Logistics and Policy*, 19–53.

Burton, A. (1992). *The Railway Builders*. London: Trafalgar Square Books.

Button, K., S. Doh, M. Hardy, J. Yuan, and X. Zhou (2010). The Accuracy of Transit System Ridership Forecasts and Capital Cost Estimates. *International Journal of Transport Economics (Rivista Internazionale de Economia dei Trasporti) 37*(2).

Byers, J. (1940). Selden Case, The. *Journal of the Patent Office Society 22*, 719.

Carlson, S., F. Schneider III, J. Bromley, and R. Jackson (1980). *PCC: The Car That Fought Back*. Glendale, CA: Interurban Press.

Carlton, D. W., and J. M. Perloff (1990). *Modern Industrial Organization*. Harper Collins Publishers. New York

Caro, R. (1974). *The Power Broker: Robert Moses and the Fall of New York*. New York: Knopf.

Cawston, G., and A. Keane (1968). *Early Chartered Companies: AD 1296–1858*. Number 140. Burt Franklin Publisher. New York

Chandler, A. (1979). *The Railroads: Pioneers in Modern Management*. New York: Arno Press.

Chen, W. (2007). Analysis of Rail Transit Project Selection Bias with an Incentive Approach. *Planning Theory 6*(1), 69–94.

Cherry, C., and R. Cervero (2007). Use Characteristics and Mode Choice Behavior of Electric Bike Users in China. *Transport Policy 14*(3), 247–257.

Chester, M., and A. Horvath (2010). Life-Cycle Assessment of High-Speed Rail: The Case of California. *Environmental Research Letters 5*(1), 014003.

Christaller, W. (1966). *Central Places in Southern Germany*. Englewood Cliffs, NJ: Prentice Hall.

Christensen, C. (1997). *The Innovator's Dilemma: When New Technologies Cause Great Firms to Fail*. Cambridge, MA: Harvard Business Press.

Christensen, C. (2012, November 3). A Capitalist's Dilemma, Whoever Wins on Tuesday. *New York Times*.

Chumak, A., and J. Braaksma (1981). Implications of the Travel-Time Budget for Urban Transportation Modeling in Canada. *Transportation Research Record* (794), 19–27.

Churchman, C. W. (1979). *The Systems Approach and Its Enemies*. Basic Books, New York.

Clark, G. (2008). *A Farewell to Alms: A Brief Economic History of the World*. Princeton, NJ: Princeton University Press.

Coke, J. G. (1968). Antecedents of Local Planning, in William I. Goodman and Eric C. Freund (eds.), *Principles and Practices of Urban Planning*. Washington, D.C.: International City Managers' Association.

Collins, C. (2005). *Werner Hegemann and the Search for Universal Urbanism*. WW, New York

Couclelis, H. (2004). Pizza over the Internet: E-commerce, the Fragmentation of Activity and the Tyranny of the Region. *Entrepreneurship & Regional Development 16*(1), 41–54.

Course, E. (1962). *London Railways*. B. T. Batsford Ltd. London

Cowen, T. (2011). *The Great Stagnation*. Penguin, New York

Crandall, R., and L. Waverman (2000). *Who Pays for Universal Service? When Telephone Subsidies Become Transparent*. Washington, DC: Brookings Institution Press.

Crawford, S. (2006, May). FAQ on Net Neutrality. Weblog entry. May 31.

Creighton, R. (1970). *Urban Transportation Planning*. Urbana: University of Illinois Press.

Croome, D., and A. Jackson (1993). *Rails Through the Clay*. Capital Transport. St. Leonards on Sea, UK

Cudahy, B. (1995). *Under the Sidewalks of New York: The Story of the Greatest Subway System in the World*. New York: Fordham University Press.

Cudahy, B. (2006). *Box Boats: How Container Ships Changed the World*. New York: Fordham University Press.

Cupper, D. (1990, May/June). The Road to the Future. *American Heritage*, 103–111.

Dabney, B. (1993). *The Silver Sextant: Four Men of the Enlightenment*. B. P. Dabney. Norfolk, Virginia.

Dahl, R., and C. Lindblom (1953). *Politics, Economics, and Welfare*. Transaction Pub. Piscataway, New Jersey.

Dahlgren, J. (1998). High Occupancy Vehicle Lanes: Not Always More Effective than General Purpose Lanes. *Transportation Research 32*(2), 99–114.

Daniels, R,. and M. Trebilcock (1996). Private Provision of Public Infrastructure: An Organizational Analysis of the Next Privatization Frontier. *University of Toronto Law Journal 46*(3), 375–426.

Danielson, B., and M. Hubbard (1998). A Literature Review for Assessing the Status of Current Methods of Reducing Deer-Vehicle Collisions. A Report Prepared for The Task Force on Animal Vehicle Collisions, The Iowa Department of Transportation, and the Iowa Department of Natural Resources. Ames Iowa.

Danielson, M. (1965). *Federal-Metropolitan Politics and the Commuter Crisis*. New York: Columbia University Press.

Darling, H. (1980). *The Politics of Freight Rates: The Railway Freight Rate Issue in Canada*. Toronto: McClelland and Stewart.

David, P. (1986). Narrow Windows, Blind Giants and Angry Orphans: The Dynamics of Systems Rivalries and Dilemmas of Technology Policy. CEPR Technological Innovation Project Working Paper (10). Stanford University, Palo Alto California.

De Tocqueville, Alexis and Mayer, Jacob Peter and Lefebvre, Georges and Jardin, Andre, *L'ancien regime et la revolution*, 55, 1967, Cambridge Univ Press.

De Luca, R. (2012). The League of American Wheelmen and Hartford's Albert Pope Champion the Good Roads Movement.
`http://connecticuthistory.org/the-league-of-american-wheelmen-and-the-good-roads-movement-how-popes-bicycles-led-to-good-roads/`

De Soto, H., and P. Instituto Libertad y Democracia (Lima (1989). *The Other Path: The Invisible Revolution in the Third World*. New York: Harper & Row.

De Tarde, G. (1890). *The Laws of Imitation*. H. Holt and Company (1903 republication). New york

De Tocqueville, A. (1831, March 2004). *Democracy in America*. Wiley Online Library.

Dellinger, M. (2010). *Interstate 69: The Unfinished History of the Last Great American Highway*. New York: Scribner's Book Company.

Deloche, J., and J. Walker (1993). *Transport and Communications in India Prior to Steam Locomotion*. Oxford: Oxford University Press.

Denneen and Company (2009, January). Practical Measures to Increase Transit Industry Advertising Revenues. Technical Report TCRP Report 133, Transportation Research Board.

Di Caro, M. (2012). ICC Meets Traffic Projections, But Opponents Still Critical. *WMAL*.

Diamond, J. (2006). *Collapse: How Societies Choose to Fail or Succeed*. Penguin Group USA. New York

Dickerson, O. (1951). *The Navigation Acts and the American Revolution*. New York: A. S. Barnes (reprinted 1974).

Dillard, A. (1990). *The Writing Life*. Harper Perennial. New York

Dobyns, K. (1994). *The Patent Office Pony: A History of the Early Patent Office*. Sergeant Kirkland's Museum and Historical Society. Fredericksburg, Virginia.

Doig, J. (2001). *Empire on the Hudson: Entrepreneurial Vision and Political Power at the Port of New York Authority*. New York: Columbia University Press.

Downey, G. (2002). *Telegraph Messenger Boys: Labor, Technology, and Geography, 1850–1950*. New York: Routledge.

Downs, A. (1992). *Stuck in Traffic: Coping with Peak-Hour Traffic Congestion*. Washington, DC, and Cambridge, MA: Brookings Institution.

Due, J. (1979). State Rail Plans and Programs. *Quarterly Review of Economics and Business 19*(2), 109–130.

Due, J., and S. Leever (1993). The Post-1984 Experience with New Small and Regional Railroads. *Transportation Journal 33*(1), 40–52.

Duncan, D., and K. Burns (2003). *Horatio's Drive: America's First Road Trip*. New York: Knopf.

Dunn, J., and G. Kemper (1919). *Indiana and Indianans: A History of Aboriginal and Territorial Indiana and the Century of Statehood*, Volume 1–3. Chicago and New York: The American Historical Society.

Dupuit, J. (1844). On the Measurement of the Utility of Public Works. *International Economic Papers 2*(1952), 63–110.

Durrenberger, J. (1931). *Turnpikes: A Study of the Toll Road Movement in the Middle Atlantic States and Maryland*. Columbia University, New York (Ph.D. Thesis).

Earle, H. S. (1929). *The Autobiography of "By Gum" Earle*. Lansing, MI: The State Review Publishing Company.

Edwards, D., and R. Pigram (1986). *London's Underground Suburbs*. Baton Transport. UK.

Efrati, A. (2012, October 12). Google's Driverless Car Draws Political Power. *Wall Street Journal*, B1.

Ekelund, R., and R. Hébert (1999). *Secret Origins of Modern Microeconomics: Dupuit and the Engineers*. Chicago: University of Chicago Press.

Ellis, C. (1996). Professional Conflict over Urban Form: The Case of Urban Freeways, 1930 to 1970. *Planning the Twentieth-Century American City*, JHU Press, Baltimore, 262–279.

Engel, E., R. Fischer, and A. Galetovic (2006). Privatizing Highways in the United States. *Review of Industrial Organization 29*(1), 27–53.

Eno, W. (1939). *The Story of Highway Traffic Control*. Saugatuck, CT: Eno Foundation for Highway Traffic Control.

Fallows, J. (2010, May 3). If the TSA Were Running New York. *The Atlantic*.

Fausset, R. (2009, November 19, 2009). Judge Says U.S. Liable in Katrina: Ruling Finds That 'Gross Negligence' by the Army Corps of Engineers led to Levee Breaks in New Orleans. *Los Angeles Times.*

Fawaz, Y., and W. Garrison (1994). Truck and Highway Combinations for Increasing Trucking Productivity in Market Niches. *Transportation Research Record* (1430), 10–18.

Federal Communications Commission (2006). FCC 06-189: Memorandum Opinion and Order In the Matter of AT&T and BellSouth Corporation Application for Transfer of Control. Technical report, Washington, DC: US Government Printing Office.

Federal Transit Adminsitration (2009, November). Second Avenue Subway Phase I, New York, New York: Full Funding Grant Agreement. Technical report.

Fellmeth, R. (1970). *The Interstate Commerce Omission: The Public Interest and the ICC.* New York: Grossman.

Ferejohn, J. (1974). *Pork Barrel Politics: Rivers and Harbors Legislation, 1947–1968.* Polo Alto, CA: Stanford University Press.

Ferguson, N. (2011). *Civilization: The West and the Rest.* The Penguin Press. New York.

Ferrara, P. (1990). *Free the Mail: Ending the Postal Monopoly.* Cato Institution. Washington, DC.

Fielding, G. J., and D. B. Klein (1993). How to Franchise Highways. *Journal of Transport Economics and Policy* (May), 113–130.

Fields, G., D. Hartgen, A. Moore, and R. Poole (2009). Relieving Congestion by Adding Road Capacity and Tolling. *International Journal of Sustainable Transportation 3*(5), 360–372.

Fisher, T. (2009). Fracture Critical. Design Observer http://places.design observer.com/feature/fracture-critical/11477/ Accessed April 19, 2012

Flinn, M. (1957). *The Law Book of the Crowley Ironworks,* Volume 167. Surtees Society, Durham England.

Flint, A. (2011). *Wrestling with Moses: How Jane Jacobs Took on New York's Master Builder and Transformed the American City.* New York: Random House Trade Paperbacks.

Flyvbjerg, B. (2007). Cost Overruns and Demand Shortfalls in Urban Rail and Other Infrastructure. *Transportation Planning and Technology 30*(1), 9–30.

Flyvbjerg, B., N. Bruzelius, and W. Rothengatter (2003). *Megaprojects and Risk: An Anatomy of Ambition.* Cambridge: Cambridge University Press.

Flyvbjerg, B., M. Holm, and S. Buhl (2002). Underestimating Costs in Public Works Projects: Error or Lie? *Journal of the American Planning Association 68*(3), 279–295.

Flyvbjerg, B., M. Holm, and S. Buhl (2005). How (In)accurate are Demand Forecasts in Public Works Projects?: The Case of Transportation. *Journal of the American Planning Association 71*(2), 131–146.

Flyvbjerg, B., M. K. S. holm, and S. Buhl (2003). How Common and How Large Are Cost Overruns in Transport Infrastructure Projects? *Transport Reviews 23*(1), 71–88.

Fogel, R. (1964). *Railroads and American Economic Growth.* Baltimore, MD.: Johns Hopkins University Press.

Fogelson, R. (2003). *Downtown: Its Rise and Fall, 1880–1950.* New Haven, CT: Yale University Press.

Foley, M. (1990). *Laws, Men, and Machines: Modern American Government and the Appeal of Newtonian Mechanics.* London and New York: Routledge.

Folwell, W. (1921). *History of Minnesota*, Volume 2. Minnesota Historical Society Press. St. Paul, Minnesota.

Center for Universal Design (2011, May 30). The Principles of Universal Design. Technical report, North Carolina State University.

Ford, G. (1911). The Relation of City Planning to the Municipal Budget. *American City 4*, 66–71.

Forkenbrock, D. (2008). Policy Options for Varying Mileage-Based Road User Charges. *Transportation Research Record: Journal of the Transportation Research Board 2079*(-1), 29–36.

Forrester, J., and G. Brown (1975). *Collected Papers of Jay W. Forrester*. Cambridge, MA: Wright-Allen Press.

Foster, B. (2012, October). Facebook User Growth Chart. `http://www.benphoster.com/facebook-user-growth-chart-2004-2010/`

Foster, M. (1981). *From Streetcar to Superhighway: American City Planners and Urban Transportation, 1900–1940*. Philadelphia: Temple University Press.

Foxell, C. (2005). *Rails to Metro-Land*. Chesham, Bucks, England: Clive Foxell.

Frankston, B. (2006). Sidewalks: Paying by the Stroll. Website.

Friedlaender, A. F. (1965). *The Interstate Highway System*, Volume 38. Amsterdam: North Holland.

Garbarino, E., and O. Lee (2003). Dynamic Pricing in Internet Retail: Effects on Consumer Trust. *Psychology and Marketing 20*(6), 495–513.

Garfield, S. (2002). *The Last Journey of William Huskisson*. Faber. London.

Garrison, W. (1974). Social, Economic, and Planning Impacts of a Rapid Excavation and Tunneling Technology. In *Proceedings of a Conference on Rapid Excavation and Tunneling*, pp. 313–324. New York: American Society of Civil Engineers.

Garrison, W. (1984). Transportation Technology. *Transportation Research*, Volume 18, 267–276.

Garrison, W. (1985). Technology and the Infrastructure Problem. In *Rebuilding America: Infrastructure Rehabilitation*, pp. 6–16. ASCE: Reston, VA: ASCE.

Garrison, W. (1989). Using Technology to Improve Transportation Services. *Transportation for the Future*, Batten DF: Springer Verlag, 87–119.

Garrison, W., B. Berry, D. Marble, J. Nystuen, and R. Morrill (1958). *Studies of Highway Development and Geographic Change*. Seattle: University of Washington Press.

Garrison, W., and E. Deakin (1992). Land Use. Public Transportation. Englewood Cliffs, NJ: Prentice-Hall, 527–550.

Garrison, W., and D. Marble (1965). A Prolegomenon to the Forecasting of Transportation Development. Technical report, United States Army Aviation Material Labs Technical Report. Washington, DC: Office of Technical Services, US Department of Commerce.

Garrison, W., and R. Souleyrette (1996). Transportation, Innovation, and Development: The Companion Innovation Hypothesis. *Logistics and Transportation Review 32*(1).

Garrison, W., and J. Ward (2000). *Tomorrow's Transportation: Changing Cities, Economies and Lives*. Boston: Artech House.

Geddes, N. (1932). *Horizons*. Dover. Mineola, New York.

Geddes, N. (1940). *Magic Motorways*. New York: Random House.

General Motors Corporation. (1974, April). The Truth about American Ground Transport. Submitted to the Subcommittee on Antitrust and Monopoly of the Committee on the Judiciary, US Senate.

Gerhold, D. (1993a). Packhorses and Wheeled Vehicles in England, 1550–1800. *Journal of Transport History 14*(1), 1–26.

Gerhold, D. (1993b). *Road Transport before the Railways: Russell's London Flying Waggons*. Cambridge: Cambridge University Press.

Gerondeau, C. (1997). *Transport in Europe*. Boston and London: Artech House.

Gifford, J. (1984). The Innovation of the Interstate Highway System. *Transportation Research 18*(4), 319–332.

Gifford, J., and W. Garrison (1993). Airports and the Air Transportation System: Functional Refinements and Functional Discovery. *Technological Forecasting and Social Change 43*(2), 103–123.

Gladwell, M., A. Lightman, and N. Wozny (2011, November 14). The Tweaker: The Real Genius of Steve Jobs. *New Yorker*.

Glaskowsky, N. (1986). *Effects of Deregulation on Motor Carriers*. Washington, DC: Eno Foundation for Transportation.

Gofman, S. (1997). My Own Private Idaho: Staking Claims to the Public Streets. *Journal of Transportation Law, Logistics and Policy 64*(2), 495–504.

Goodrich, C. (1961). *Canals and American Economic Development*. New York: Columbia University Press.

Goodwin, A. (2002, accessed June 25, 2002). Alameda Corridor: A Project of National Significance. http://gulliver,trb.org/publications/mb/2002Ports/06Goodwin.pdf

Goodwin, P., J. Dargay, and M. Hanly (2004). Elasticities of Road Traffic and Fuel Consumption with Respect to Price and Income: A Review. *Transport Reviews 24*(3), 275–292.

Gordon, J. (2002). *A Thread across the Ocean*. New York: Perennial.

Gould, S., and R. Lewontin (1979). The Spandrels of San Marco and the Panglossian Paradigm: A Critique of the Adaptationist Programme. *Proceedings of the Royal Society of London. Series B. 205*(1161), 581–598.

Graham, A., and P. Senge (1980). A Long Wave Hypothesis of Innovation. *Technological Forecasting and Social Change 17*(4), 283–311.

Graham, D., and S. Glaister (2002). The Demand for Automobile Fuel: A Survey of Elasticities. *Journal of Transport Economics and Policy (JTEP) 36*(1), 1–25.

Graham, D., and S. Glaister (2004). Road Traffic Demand Elasticity Estimates: A Review. *Transport Reviews 24*(3), 261–274.

Griffiths, I. (1999). Rebecca in Pontarddulais. Salt Lake City: Genealogical Society of Utah, 1999.

Grindon, L. (1892). *Lancashire: Brief Historical and Descriptive Notes*. Macmillan. London.

Grossman, J. (2012). *Charles Dickens's Networks: Public Transport and the Novel*. Oxford: Oxford University Press.

Grubler, A. (1990). *The Rise and Fall of Infrastructures: Dynamics of Evolution and Technological Change in Transport*. Heidelberg: Physica-Verlag.

Hafner, K., and M. Lyon (1996). *Where Wizards Stay Up Late: The Origins of the Internet*. New York: Simon and Schuster.

Haines, C. (1818). Considerations on the Great Western Canal. Brooklyn, Spooner & Worthington, printers.

Hall, P. (1998). *Cities in Civilization: Culture, Technology and Urban Order*. London: Weidenfeld and Nicolson.

Hall, P. (2009). Magic Carpets and Seamless Webs: Opportunities and Constraints for High-Speed Trains in Europe. *Built Environment 35*(1), 59–69.

Hamer, A. (1976). *The Selling of Rail Rapid Transit*. Lexington, MA: Lexington Books.

Hamer, M. (1987). *Wheels Within Wheels: A Study of the Road Lobby*. London: Friends of the Earth/Routledge & Kegan Paul.

Hands Off the Internet (2006). About Us. `http://www.internetofthefuture.com/` Accessed 2006

Hankla, J. (2001). The Alameda Corridor Project: Its Successes and Challenges. In *Hearing Before the Subcommittee on Government Efficiency, Financial Management, and Intergovernmental Relations of the Committee on Government Reform, 107th Congress, House Hearings*. Washington, DC: US Government Printing Office.

Harris, C., and E. Ullman (1945). The Nature of Cities. *Annals of the American Academy of Political and Social Science 242*, 7–17.

Hatoko, M., and D. Nakagawa (2007). Comparative Analysis of Swiss and Japanese Trunk Railway Network Structures. In *World Conference of Transportation Research WCTRS*. Berkeley, California.

Hauer, E. (1994). Can One Estimate the Value of Life or Is It Better to be Dead Than Stuck in Traffic? *Transportation Research Part A: Policy and Practice 28*(2), 109–118.

Hay, P. (1985). *Brunel, Engineering Giant*. London: BT Batsford.

Hayek, F. (1989). *The Fatal Conceit: The Errors of Socialism*. Chicago: University of Chicago Press.

Hegemann, W., and E. Peers (1922). *The American Vitruvius: An Architect's Handbook of Civic Art*. Architectural Book Publishing Co., New York.

Heller, M. (2008). *The Gridlock Economy: How Too Much Ownership Wrecks Markets, Stops Innovation, and Costs Lives*. Basic Books. New York

Hénard, E. (1904). *Études sur les Transformations de Paris*. Librairies-Imprimeries Réunis. Paris

Hensher, D. (2003). Congestion Charging: What Sydney Can Learn from London. *On Line Opinion: Australia's e-Journal of Social and Political Debate*. Posted Monday, September 22. `http://www.onlineopinion.com.au/view.asp?article=735` (accessed July 18, 2013)

Heppenheimer, T. (1995). *Turbulent Skies: The History of Commercial Aviation*. New York: J. Wiley & Sons.

Herlihy, D. (2004). *Bicycle: The History*. New Haven, CT: Yale University Press.

Heseltine, P., and D. Silcock (1990). The Effects of Bus Deregulation on Costs. *Journal of Transport Economics and Policy*, 239–254.

Hicks, J. (1932). *The Theory of Wages* (2nd edition). London: Macmillan

Hillman, J. (1968). *Competition and Railroad Price Discrimination*. Evanston, IL: Northwestern University Transportation Center.

Hillman, J. (1979). *Psychological Fantasies in Transportation Problems*. Dallas: The Center for Civic Leadership, the University of Dallas.

Hillsman, E. (1995). Transportation DSM: Building on Electric Utility Experience. *Utilities Policy 5*(3-4), 237–249.

Hilton, G. (1990). *American Narrow Gauge Railroads*. Stanford, CA: Stanford University Press.

Hilton, G. and J. Bier (1997). *The Cable Car in America: A New Treatise upon Cable or Rope Traction as Applied to the Working of Street and Other Railways*. Stanford, CA: Stanford University Press.

Hippel, E. v. (1988). *The Sources of Innovation*. Oxford: Oxford University Press.

Hirschman, A. (1970). *Exit, Voice, and Loyalty: Responses to Decline in Firms, Organizations, and States*. Cambridge, MA: Harvard University Press.

Hirschman, A. (1981). *Essays in Trespassing: Economics to Politics and Beyond*. Cambridge: Cambridge University Press.

Hirschman, A. (1985). *The Strategy of Economic Development*. Number 44. Cambridge, MA: Harvard University Press.

Hoffmann, P. (2002). *Tomorrow's Energy: Hydrogen, Fuel Cells, and the Prospects for a Cleaner Planet*. Cambridge, MA:MIT Press.

Hollander, J. (2003). *The Real Environmental Crisis: Why Poverty, not Affluence, Is the Environment's Number One Enemy*. Berkeley: University of California Press.

Holmes, E. H., and J. T. Lynch (1957). Highway Planning, Past, Present and Future. *Journal of the Highway Division, Proceedings of the ASCE 83*(HW3), 1298-1–1298-13.

Holt, W. (1923). *The Bureau of Public Roads: Its History, Activities and Organization*. Number 26. Baltimore, MD: The Johns Hopkins Press.

Hommels, A. (2005). *Unbuilding Cities: Obduracy in Urban Sociotechnical Change*. Cambridge, Massachusetts: MIT Press.

Hood, C. (2004). *722 Miles: The Building of the Subways and How They Transformed New York*. Baltimore, MD: Johns Hopkins University Press.

Horwood, E., and R. Boyce (1956). *Studies of the Central Business District and Urban Freeway Development*. Seattle: University of Washington Press.

Hourdakis, J., and P. Michalopoulos (2002). Evaluation of Ramp Control Effectiveness in Two Twin Cities Freeways. *Transportation Research Record 1811*(1), 21–29.

Howe, D. (2007). *What Hath God Wrought: The Transformation of America, 1815–1848*, Volume 5. New York: Oxford University Press.

Hoyle, S. (1982). The First Battle for London: A Case Study of the Royal Commission on Metropolitan Termini. *London Journal viii*, 140–155.

Hoyt, H. (1939). *The Structure of Growth of Residential Neighborhoods in American Cities*. Washington, DC: US Government Printing Office.

Hubert Jr, P. G. (1894, March). The Cable Street-Railway. *Scribner's Magazine 15 (3)*.

Hughes, T. (1989). *American Genesis: A Century of Invention and Technological Enthusiasm, 1870–1970*. New York: Viking.

Hughes, T. (2000). *Rescuing Prometheus*. New York: Vintage.

Humphreville, J. (2012, July 30). The Alameda Corridor Has Success Written All Over It . . . Except for that Boat Load of Debt. *Citywatch LA*.

Hunter-Zaworski, K. M. (2012). Impacts of Low-Speed Vehicles on Transportation Infrastructure and Safety. *Journal of Transport and Land Use 5(2)*, 68–76.

Indian Analyst, The (2007a). *South Indian Inscriptions: Inscriptions Collected during the Year 1906–07*. What Is India Publishers.
http://www.whatisindia.com/inscriptions/south_indian_inscriptions/volume_3/virarajendra_i.html (Accessed July 18, 2013)

Indian Analyst, The (2007b). *South Indian Inscriptions: Part II: Miscellaneous Inscriptions from the Tamil Country: VIII. Inscriptions of Virarajendra I*. What Is India Publishers. `http://www.whatisindia.com/inscriptions/south_indian_inscriptions/volume_3/virarajendra_i.html` (Accessed July 18, 2013)

Institute, R. M. (2003, 18 September). Transportation Fuel Savings.

Intergovernmental Panel on Climate Change (2007). Climate Change 2007: Working Group I: The Physical Science Basis: Glossary A-D. Technical report, Intergovernmental Panel on Climate Change.

Isard, W. (1942). A Neglected Cycle: The Transport-Building Cycle. *The Review of Economics and Statistics 24*(4), 149–158.

Issel, W. (1999). Land Values, Human Values, and the Preservation of the City's Treasured Appearance: Environmentalism, Politics, and the San Francisco Freeway Revolt. *Pacific Historical Review 68*(4), 611–646.

Iteris (2012, March 28). *National ITS Architecture* 7.0.

Jackson, A. A. (1986). *London's Metropolitan Railway*. David and Charles.

Jackson, A. A. (2006). *London's Metro-Land*. Newton Abbot: Capital History Publishing.

Jackson, G. (1983). *The History and Archaeology of Ports*. World's Work Tadworth, Surrey, UK.

Jacobs, J. (1961). *The Death and Life of Great American Cities*. New York: Random House.

Jacobs, J. (1984). *Cities and the Wealth of Nations: Principles of Economic Life*. New York: Random House.

Jacobs, J. (1992). *Systems of Survival: A Dialogue on the Moral Foundations of Commerce and Politics*. Cambridge: Cambridge University Press.

Jacobs, J. (2005). *Dark Age Ahead*. Vintage Canada.

Jacobs, J., et al. (1970). *The Economy of Cities*. New York: Random House.

Jakle, J. and K. Sculle (2002a). *Fast Food: Roadside Restaurants in the Automobile Age*. Baltimore, MD: Johns Hopkins University Press.

Jakle, J., and K. Sculle (2002b). *The Gas Station in America*. Baltimore, MD: Johns Hopkins University Press.

Jakle, J., and K. Sculle (2008). *Motoring: The Highway Experience in America*. Athens: University of Georgia Press.

Jakle, J., K. Sculle, and J. Rogers (2002). *The Motel in America*. Baltimore, MD: Johns Hopkins University Press.

James, E. W. (1916). Distribution of Traffic on a Rectangular System. *Engineering News Record 24*, 439–440.

Japanese Railway Engineering (JRE). (1994). No. 131, *Special Issue: 30 Years of Progress in the Shinkansen*. Tokyo: Japan Railway Engineer's Association.

Jeans, J. (1875). *Jubilee Memorial of the Railway System: A History of the Stockton and Darlington Railway and a Record of its Results*. London: Longmans.

Jenkins-Smith, H., and K. Herron (2004). *A Decade of Trends in Public Views on Security: US National Security Surveys 1993–2003*. College Station, Texas: The Bush School of Government and Public Service, Texas A&M University.

Jevons, W. S. (1866). *The Coal Question; an Enquiry Concerning the Progress of the Nation, and the Probable Exhaustion of our Coal-mines*. London: Macmillan.

John, R. (1998). *Spreading the News: The American Postal System from Franklin to Morse*. Cambridge, MA: Harvard University Press.

Johnson, K. (2009). Captain Blake versus the Highwaymen: Or, How San Francisco Won the Freeway Revolt. *Journal of Planning History 8*(1), 56–83.

Johnson, L., D. Rote, J. Hull, H. Coffey, J. Daley, and R. Giese (1989). MAGLEV Vehicles and Superconductor Technology: Integration of High-Speed Ground Transportation into the Air Transportation System. Technical report. Argonne, IL: Argonne National Lab.

Johnson, S. (2006). *The Ghost Map: The Story of London's Most Terrifying Epidemic–and How it Changed Science, Cities, and the Modern World.* Riverhead Books (Hardcover). New York.

Johnson, S. (2010). *Where Good Ideas Come From: The Natural History of Innovation*. ePenguin. New York: Penguin Books

Johnston, J. (1950–1951). Lord Gordon Gordon. *Transactions of the Manitoba Historical Society Series 3.*

Jolly, S., and B. Bayman (1986). *Docklands Light Railway Official Handbook* (1st Edition). London

Jomini, A. (1841). *Précis de l'art de la guerre*, Volume 2. J. B. Petit.

Jones, D. (1985). *Urban Transit Policy: An Economic and Political History*. Englewood Cliffs, NJ: Prentice-Hall.

Jones, D. (1990). Commercial Progress versus Local Rights: Turnpike Building in Northwest Rhode Island in the 1790s. *Rhode Island History 48*, 21–32.

Jones, P. (1978). *Under the City Streets*. Holt Rinehart and Winston. New York.

Journal, W. B. (1999, December 3). Greenway Finally Going Somewhere. *American City Business Journals, Inc.*

Kahn, M. (2010). *Climatopolis: How Our Cities Will Thrive in the Hotter Future*. Basic Books. New York.

Kain, J. (1990). Deception in Dallas: Strategic Misrepresentation in Rail Transit Promotion and Evaluation. *Journal of the American Planning Association 56*(2), 184–196.

Kain, J. (1997). Cost-Effective Alternatives to Atlanta's Rail Rapid Transit System. *Journal of Transport Economics and Policy*, 31 (1), 25–49.

Kasarda, J., and G. Lindsay (2011). *Aerotropolis*. New York: Farrar, Straus and Giroux.

Kaufman, A. (1997). Cultural and Ethical Underpinnings of the Navy's Attitude Toward Naval Mining. Technical report, DTIC Document.

Keay, J. (1991). *The Honourable Company: A History of the English East India Company*. London: HarperCollins

Kelly, K. (2010). *What Technology Wants*. New York: Viking Press.

Kennedy, D. (1995a). London Bus Tendering: A Welfare Balance. *Transport Policy 2*(4), 243–249.

Kennedy, D. (1995b). London Bus Tendering: An Overview. *Transport Reviews 15*(3), 253–264.

Kennedy, T., and P. McEnroe (2007, August 18, 4:36). Phone Call Put Brakes on Bridge Repair: Plans to Reinforce the Bridge Were Well Underway When the Project Came to a Screeching Halt in January Amid Concerns about Safety and Cost. *Star Tribune*.

Kihss, P. (1970). Benign Neglect on Race is Proposed by Moynihan. *New York Times*, A1.

Kilgore, E. (2012, July 23). The "Swap" Reagan Never Proposed. *The Washington Monthly*.

Kimball, J. (2011, September 23). Report Says 1,149 Minnesota Bridges are Deficient. *MinnPost*.

King, D. (2011). Developing Densely: Estimating the Effect of Subway Growth on New York City Land Uses. *Journal of Transport and Land Use 4*(2).

King, D. (2012, June). Tesla's Elon Musk: Most New Cars will be Electric by 2032, No Matter Who's in White House. *Autoblog Green*.

Klapper, C. (1978). *Golden Age of Buses*. London: Routledge and K. Paul.

Klein, D., A. Moore, and B. Reja (1997). *Curb Rights: A Foundation for Free Enterprise in Urban Transit*. Washington, DC: Brookings Institution Press.

Knight, R., and L. Trygg (1978). Urban Mass Transit and Land Use Impacts. *Transportation 5*, 12–24.

Knittel, C. (2009). *The Implied Cost of Carbon Dioxide under the Cash for Clunkers Program*. Institute of Transportation Studies, University of California, Davis.

Kolko, G. (1976). *Railroads and Regulation, 1877–1916*. Westport, CT: Greenwood Press.

Kondratieff, N. (1935). The Long Waves in Economic Life. *Review of Economic Statistics 17*(6).

Konvitz, J. (1978). *Cities and the Sea: Port Planning in Early Modern Europe*. Baltimore, MD: Johns Hopkins University Press.

Konvitz, J. (1985). *The Urban Millennium: The City Building Process from the Early Middle Ages to the Present*. Carbondale, IL: Southern Illinois University Press.

Koppel, T. (1999). *Powering the Future: The Ballard Fuel Cell and the Race to Change the World*. New York: Wiley.

Korotayev, A., and S. Tsirel (2010). A Spectral Analysis of World GDP Dynamics: Kondratieff Waves, Kuznets Swings, Juglar and Kitchin Cycles in Global Economic Development, and the 2008–2009 Economic Crisis. *Structure and Dynamics 4*(1).

Korr, J. (2002). *Washington's Main Street: Consensus and Conflict on the Capital Beltway, 1952–2001*. Ph. D. thesis, University of Maryland, College Park.

Koshland, D. (1989). A Tax on Sin: The Six-Cylinder Car. *Science 243*, 281.

Kotsialos, A., and M. Papageorgiou (2004, December). Efficiency and Equity Properties of Freeway Network-Wide Ramp Metering with AMOC. *Transportation Research Part C-Emerging Technologies 12*(6), 401–420.

Kozel, S. (1997, accessed January 15, 2004). Dulles Transportation Corridor.

Krumm, B. (1994). High Speed Ground Transportation Systems: A Future Component of America's Intermodal Network. *Transportation Law Journal 22*, 309–326.

Krylov, V. (2001). *Noise and Vibration from High-Speed Trains*. London: Thomas Telford.

Kuhn, T. (1957). *The Copernican Revolution*, Volume 16. Cambridge, MA: Harvard University Press.

Kwon, E., S. Nanduri, R. Lau, and J. Aswegan (2001). *Comparative Analysis of Operational Algorithms for Coordinated Ramp Metering*, Volume 1748. Washington, DC: Transportation Research Board.

Lamar, J. (1914). Dissent, *German Alliance Ins. Co. v. Lewis*, 233 U.S. 389. United States Supreme Court Case.

Langfitt, F. (2012, August 29). Chinese Blame Failing Bridges on Corruption. *National Public Radio*.

Latour, B. (1996). *ARAMIS or the Love of Technology*, Volume 102. Cambride, MA: Harvard University Press.

Latour, B. (2005). *Reassembling the Social: An Introduction to Actor-Network-Theory*. New York: Oxford University Press.

Lawson, D. (1993). "Passing the Test"–Human Behavior and California's Smog Check Program. *Air & Waste 43*(12), 1567–1575.

Lay, M. (1992). *Ways of the World: A History of the World's Roads and of the Vehicles That Used Them*. New Brunswick, NJ: Rutgers University Press.

Layton, E. (1971). *The Revolt of the Engineers*. Cleveland, OH: Case Western Reserve Press.

Lee Jr, D. (1973). Requiem for Large Scale Models. *Journal of the American Institute of Planners 39*(3), 163–178. (Based on Models and Techniques for Urban Planning, Cornell Aeronautical Lab, 1968, VY 2474-G-1).

LeHong, H., and J. Fenn (2012, September 18). Key Trends to Watch in Gartner 2012 Emerging Technologies Hype Cycle. *Forbes*.

Lessig, L., and R. W. McChesney (2006). No Tolls on the Internet. *Washington Post*, June 8, A23.

Levinson, D. (2001). *Financing Transportation Networks*. Northampton, MA: Edward Elgar.

Levinson, D. (2002a). *Financing Transportation Networks*. Northampton, MA: Edward Elgar.

Levinson, D. (2008). Density and Dispersion: The Co-Development of Land use and Rail in London. *Journal of Economic Geography*, 8, Oxford, Univ. Press, 2008.

Levinson, D. (2008). The Orderliness Hypothesis: The Correlation of Rail and Housing Development in London. *The Journal of Transport History 29*(1), 98–114.

Levinson, D. (2012). Accessibility Impacts of High-Speed Rail. *Journal of Transport Geography*. 22(5), 288–291.

Levinson, D., and A. El-Geneidy (2009). The Minimum Circuity Frontier and the Journey to Work. *Regional Science and Urban Economics 39*(6), 732–738.

Levinson, D., and R. Karamalaputi (2003a). Induced Supply: A Model of Highway Network Expansion at the Microscopic Level. *Journal of Transport Economics and Policy* 37(3), 297–318.

Levinson, D., and R. Karamalaputi (2003b). Predicting the Construction of New Highway Links. *Journal of Transportation and Statistics* 6(2/3), 81–89.

Levinson, D., and A. Kumar (1994). The Rational Locator: Why Travel Times Have Remained Stable. *Journal of the American Planning Association* 60(3), 319–332.

Levinson, D., and A. Kumar (1995). Activity, Travel, and the Allocation of Time. *Journal of the American Planning Association* 61(4), 458–470.

Levinson, D., J. Mathieu, D. Gillen, and A. Kanafani (1997). The Full Cost of High-Speed Rail: An Engineering Approach. *Annals of Regional Science 31*(2), 189–215.

Levinson, D., and A. Odlyzko (2008a). Too Expensive to Meter: The Influence of Transaction Costs in Transportation and Communication. *Philosophical Transactions of the Royal Society A: Mathematical, Physical and Engineering Sciences 366*(1872), 2033.

Levinson, D., and A. Odlyzko (2008b). Too Expensive to Meter: The Influence of Transaction Costs in Transportation and Communication. *Philosophical Transactions of the Royal Society A: Mathematical, Physical and Engineering Sciences*, 366, no. 1872, pp. 2033–2046. The Royal Society, 2008.

Levinson, D., and F. Xie (2011). Does First Last? The Existence and Extent of First Mover Advantages on Spatial Networks. *Journal of Transport and Land Use 4*(2).

Levinson, D., and B. Yerra (2002). Highway Costs and Efficient Mix of State and Local Funds. *Transportation Research Record 1812*, 27–36.

Levinson, D., and L. Zhang (2004). Evaluating Effectiveness of Ramp Meters. In D. Gillen and D. Levinson (Eds.), *Assessing the Benefits and Costs of Intelligent Transportation Systems*. Kluwer, Norwell, Massachusetts.

Levinson, D., and L. Zhang (2006). Ramp Meters on Trial: Evidence from the Twin Cities Metering Holiday. *Transportation Research Part A: Policy and Practice 40(10)*, 810–828.

Levinson, D. M. (1999, MAY). Space, Money, Life-Stage, and the Allocation of Time. *Transportation 26*(2), 141–171.

Levinson, D. M. (2002b). *Financing Transportation Networks*. Northampton, MA: Edward Elgar.

Levy, P. (2007, August 2). 4 Dead, 79 Injured, 20 Missing after Dozens of Vehicles Plummet into River. *Star Tribune*.

Lewis, N. (1916). *The Planning of the Modern City: A Review of the Principles Governing City Planning*. New York: John Wiley & Sons.

Li, S., J. Linn, and E. Spiller (2012). Evaluating "Cash-for-Clunkers": Program Effects on Auto Sales and the Environment. *Journal of Environmental Economics and Management*, 65(2), March 2013, 175–193.

Lienhard, J. (2006). *How Invention Begins: Echoes of Old Voices in the Rise of New Machines*. New York: Oxford University Press.

Lind, W. (1965). *Thomas H. MacDonald: A Study of the Career of an Engineer Administrator and His Influence on Public Roads in the United States, 1919–1953*. Ph. D. thesis, American University.

Lösch, A. (1967). *The Economics of Location*. New York: John Wiley.

Lowson, M. (2005). Personal Rapid Transit for Airport Applications. *Transportation Research Record: Journal of the Transportation Research Board 1930*(1), 99–106.

Lynn, G., and R. Reilly (2002). Growing the Top Line Through Innovation. *Chief Executive* (Fall).

Lyovic, J. (2000). Train Resistance, Load Modeling, AC Traction System Modeling. Technical report, University of Newcastle.

MacKaye, B. (1930). The Townless Highway. *The New Republic 62*, 93–95.

Mackie, P., J. Preston, and C. Nash (1995). Bus Deregulation: Ten Years On. *Transport Reviews 15*(3), 229–251.

Maguire, E., D. Gadian, I. Johnsrude, C. Good, J. Ashburner, R. Frackowiak, and C. Frith (2000). Navigation-Related Structural Change in the Hippocampi of Taxi Drivers. *Proceedings of the National Academy of Sciences 97*(8), 4398.

Maio, D. (1986). *The Feasibility of a Nationwide Network for Longer Combination Vehicles*. U.S. Federal Highway Administration. Washington, DC: U.S.

Marchetti, C. (1980). Society as a Learning System: Discovery, Invention, and Innovation Cycles Revisited. *Technological Forecasting and Social Change 18*(4), 267–282.

Marchetti, C. (1994). Anthropological Invariants in Travel Behavior. *Technological Forecasting and Social Change* 47(1), 75–88.

Markus, F. (2004). Toll Dispute Sparks Chinese Riot. *BBC News*. 16 November.

Marsh, G. (1885). *Man and Nature: The Earth as Modified by Human Action*. New York: C. Scribner's Sons.

Marshall, A. M. (2008). *Thailand's Moment of Truth, a Secret History of 21st Century Siam, Part One of Four*. Longform.org.

McClintock, M. (1926). *Report and Recommendations of the Metropolitan Street Traffic Survey*. Chicago Association for Commerce.

McCullough, D. (1992). *Truman*, Volume 1. New York: Simon & Schuster.

McCullough, R., and W. Leuba (1973). *The Pennsylvania Main Line Canal*. American Canal and Transportation Center, York, Pennsylvania.

McNeill, J. (2000). *Something New under the Sun: An Environmental History of the Twentieth-Century World*. W. W. Norton, New York.

Mech, L., D. Dawson, J. Peek, M. Korb, and L. Rogers (1980). Deer Distribution in Relation to Wolf Pack Territory Edges. *The Journal of Wildlife Management 44*(1), 253–258.

Meehan, S. (2002, June 20). Dulles Greenway Single-Day Traffic Exceeds 60,000.

Mensch, G. (1979). *Stalemate in Technology: Innovations Overcome the Depression*. Cambridge, MA: Ballinger and International Institute of Management.

Mertz, L. The Bragdon Committee. `http://www.fhwa.dot.gov/infrastructure/bragdon.cfm` Publisher US Federal Highway Administration, Washington DC

Metz, D. (2008). *The Limits to Travel: How Far Will You Go?* Earthscan/James & James. Abingdon, Oxford, UK

Meyer, E. (2008, December 16). A Shopping Nexus Outside Washington Plots a Future as an Urban Center. *New York Times*.

Mian, A., and A. Sufi (2012). The Effects of Fiscal Stimulus: Evidence from the 2009 Cash for Clunkers Program. *The Quarterly Journal of Economics 127*(3), 1107–1142.

Millard-Ball, A., and L. Schipper (2011a). Are We Reaching Peak Travel? Trends in Passenger Transport in Eight Industrialized Countries. *Transport Reviews 31*(3), 357–378.

Millard-Ball, A., and L. Schipper (2011b). Are We Reaching Peak Travel? Trends in Passenger Transport in Eight Industrialized Countries. *Transport Reviews 31*(3), 357–378.

Miller, G. (1971). *Railroads and the Granger Laws*. Madison: University of Wisconsin Press.

Mishra, L. (2000). *Child Labour in India*. New York: Oxford University Press.

Modelski, G., and W. Thompson (1996). *Leading Sectors and World Powers: The Coevolution of Global Politics and Economics*. Columbia: University of South Carolina Press.

Mohl, R. (2004). Stop the Road Freeway Revolts in American Cities. *Journal of Urban History 30*(5), 674–706.

Mohl, R. (2008). The Interstates and the Cities: The US Department of Transportation and the Freeway Revolt, 1966–1973. *Journal of Policy History 20*(02), 193–226.

Mohring, H. (1993, May). Land Rents and Transport Improvements: Some Urban Parables. *Transportation 20*(3), 267–283.

Mohring, H., and M. Harwitz (1962). *Highway Benefits: An Analytical Framework*.

Mokhtarian, P. L., and C. Chen (2004). TTB or Not TTB, That Is the Question: A Review and Analysis of the Empirical Literature on Travel Time (and Money) Budgets. *Transportation Research, Part A: Policy and Practice 38A*, 643–675.

Mokyr, J. (1993, May). Creative Forces. *Reason*.

Mokyr, J. (2004). *The Gifts of Athena: Historical Origins of the Knowledge Economy*. Princeton, NJ: Princeton University Press.

Moynihan, D. (1970). Benign Neglect for Issue of Race. *Wall Street Journal*, March 3, 1970, 357–366.

Mueller, J., and M. Stewart (2011). *Terror, Security, and Money: Balancing the Risks, Benefits, and Costs of Homeland Security*. Oxford University Press on Demand.

Munnell, A., and L. Cook (1990). How Does Public Infrastructure Affect Regional Economic Performance? In *Is There a Shortfall in Public Capital Investment? Proceedings of a Conference*. Harwich Port, Massachusetts, June.

Nadiri, M. I., and T. P. Mamuneas (1994). The Effects of Public Infrastructure and R&D Capital on the Cost Structure and Performance of U.S. Manufacturing Industries. *The Review of Economics and Statistics 76*, 22–37.

Nakicenovic, N. (1988). Dynamics and Replacement of US Transport Infrastructures. *Cities and Their Vital Systems*.

Nash, C. (1993). British Bus Deregulation. *The Economic Journal*, 1042–1049.

National Research Council. (1975). A Review of Short Haul Passenger Transportation. Washington, D.C.: Committee on Transportation Report.

National Transportation Safety Board. (2008). Collapse of I-35W Highway Bridge Minneapolis, Minnesota, August 1, 2007. Accident Report NTSB/HAR-08/03 PB2008-916203.

National Transportation Statistics, (2012). *U.S. Department of Transportation. 2012.* Washington, D.C.: Bureau of Transportation Statistics.

Nelson, J. (1959). *Railroad Transportation and Public Policy*. Brookings Institution.

Nelson, R., and S. Winter (1982). *An Evolutionary Theory of Economic Change*. Cambridge, Massachusetts: Belknap Press of Harvard University Press.

Nevin, R. (2000). How Lead Exposure Relates to Temporal Changes in IQ, Violent Crime, and Unwed Pregnancy. *Environmental Research 83*(1), 1–22.

Newbery, D., and G. Santos (1999). Road Taxes, Road User Charges and Earmarking. *Fiscal Studies 20*(2), 103–132.

Newman, P., and J. Kenworthy (1989). *Cities and Automobile Dependence: An International Sourcebook*. Gower Publishers, England.

News-Record, E. (1996, January 8). Light Traffic Chills Dulles Debut. *Engineering News-Record 236*(1), 13.

Newswire, P. (1999, April 29). Dulles Greenway Obtains Triple-A Bond Rating in $350 million Refinancing Project. Press release.

Noam, E. M. (1994, August). Beyond Liberalization II: The Impending Doom of Common Carriage. *Telecommunications Policy 18*, 435–452.

Norton, P. (2008). *Fighting Traffic: The Dawn of the Motor Age in the American City*. Cambridge, MA: MIT Press.

Nye, D. (1992). *Electrifying America: Social Meanings of a New Technology, 1880–1940*. Cambridge, MA: MIT press.

O'Connell, H., R. Dexter, and P. Bergson (2001). Fatigue Evaluation of the Deck Truss of Bridge 9340. Technical Report 2001-10, Minnesota Department of Transportation.

Odlyzko, A. (1997). The Slow Evolution of Electronic. *Electronic Publishing 97*, 4–18.

Odlyzko, A. (1999). Paris Metro Pricing: The Minimalist Differentiated Services Solution. *Quality of Service, 1999. IWQoS'99. 1999 Seventh International Workshop on*, 159–161.

Office of Technology Assessment (1983). US Passenger Rail Technologies. Technical Report OTA-STI-222, Office of Technology Assessment, Washington, D.C.

Olmsted Jr, F. (1910). Basic Principles of City Planning. *City Planning: A Series of Papers Presenting the Essential Elements of a City Plan*, 1–18.

Osnos, E. (2012, October 22). Boss Rail: The Disaster That Exposed the Underside of the Boom. *The New Yorker*.

Pacey, A. (1974). *The Maze of Ingenuity: Ideas and Idealism in the Development of Technology*. London: Allen Lane.

Parnell, H. (1837). *A Treatise on Roads: Where in the Principles on Which Roads Should Be Made Are Explained and Illustrated, by the Plans, Specifications, and Contracts*. Longman, London.

Parthasarathi, P., and D. Levinson (2010). Post-Construction Evaluation of Traffic Forecast Accuracy. *Transport Policy 17*(6), 428–443.

Patton, C., and D. Sawicki (1993). *Basic Methods of Policy Analysis and Planning*, Volume 7. Englewood Cliffs, NJ: Prentice Hall.

Perez, C. (2002). *Technological Revolutions and Financial Capital: The Dynamics of Bubbles and Golden Ages*. Northampton, MA: Edward Elgar.

Pew Research Center for the People and the Press (2010). Public Trust in Government: 1958–2010.

Phelps, M., L. Blanchard, J. Robertson, C. Buckner, R. Morse, and S. W. E. Dept (1978). *Public Works in Seattle: A Narrative History [of] the Engineering Department, 1875–1975*. Seattle Engineering Department.

Pickrell, D. (1992). A Desire Named Streetcar Fantasy and Fact in Rail Transit Planning. *Journal of the American Planning Association 58*(2), 158–176.

Pollakowski, H., and S. Wachter (1990). The Effects of Land-Use Constraints on Housing Prices. *Land Economics 66*(3), 315–324.

Poole, R. (1997). Privatization: A New Transportation Paradigm. *Annals of the American Academy of Political and Social Science 553*, 94–105.

Poole, R., C. Orski, and R. P. P. Institute (1999). *HOT Networks: A New Plan for Congestion Relief and Better Transit*. Reason Public Policy Institute.

Poole, R., and P. Samuel (2006). The Return of Private Toll Roads. *Public Roads 69*(5), 38.

Poole, R., and Y. Sugimoto (1995). Congestion Relief Toll Tunnels. *Transportation 22*(4), 327–351.

Poole Jr, R., and P. Fixler Jr (1987). Privatization of Public-Sector Services in Practice: Experience and Potential. *Journal of Policy Analysis and Management 6*(4), 612–625.

Pope, C., and B. Lomborg (2005). The State of Nature. *Foreign Policy 149*, 67–73.

Porter, M. (1990). *The Competitive Advantage of Nations: With a New Introduction*. Free Press. New York.

Potter, I. (1891). *The Gospel of Good Roads: A Letter to the American Farmer*. The League of American Wheelmen. Newport, Rhode Island.

Prendergast, L., and R. Williams (1981). Individual Travel Time Budgets. *Transportation Research Part A: General 15*(1), 39–46.

Preston, H. (1991). *Dirt Roads to Dixie: Accessibility and Modernization in the South, 1885–1935*. Knoxville: Univ of Tennessee Press.

Preston, J., and G. Wall (2008). The Ex-ante and Ex-post Economic and Social Impacts of the Introduction of High-Speed Trains in South East England. *Planning Practice and Research 23*(3), 403–422.

Public Port Facts. American Association of Port Authorities (AAPA). 2004. *Industrial Information*. http://www.aapa-ports.org/industryinfo/portfact.htm, accessed March 2004.

Pulver, J. (2006). The Jeff Pulber Blog, January 29, 2006.

Raitz, K., and G. Thompson (1996). *The National Road*. Baltimore, MD: Johns Hopkins University Press.

Rankin, W. (1997). *Bureau of Highway Traffic: A History, 1925–1982*. Bureau of Highway Traffic Alumni Assoc. Cheshire, Conn.

Reckdahl, K. (1999). The Cars of St. Bonifacius. *City Pages 20*(968).

Reid, H. (1840). *The Steam-Engine: Being a Popular Description of the Construction and Action of that Engine; with a Sketch of Its History, and of the Laws of Heat and Pneumatics...* William Tait. Edinburgh.

Reps, J. (1992). *The Making of Urban America: A History of City Planning in the United States*. Princeton, NJ: Princeton University Press.

Rice, R., and E. Rogers (1980). Re-Invention and Innovation: The Case of Dial-A-Ride. *Knowledge 1*(4), 499–514.

Richardson, P. (1980). Benjamin Franklin and Timothy Folger's First Printed Chart of the Gulf Stream. *Science 207*(4431), 643–645.

Richmond, J. (1998). The Mythical Conception of Rail Transit in Los Angeles. *Journal of Architectural and Planning Research 15*, 294–320.

Richmond, J. (2001). A Whole-System Approach to Evaluating Urban Transit Investments. *Transport Reviews 21*(2), 141–179.

Riley, R. Q. (1994). *Alternative Cars in the 21st Century: A New Personal Transportation Paradigm (R-227)*. Warrendale, Pa.: Society of Automotive Engineers.

Ringwalt, J. (1888). *Development of Transportation Systems in the United States*. Philadelphia: Railway World Office.

Ripley, W. Z. (1950). *Railroads, Rates and Regulations*. Beard Books. Frederick, Maryland.

Roads&Bridges (2008, November 17). NTSB Releases Report on I-35W Bridge Collapse.

Robinson, A. (1955). The 1837 Maps of Henry Drury Harness. *The Geographical Journal 121*(4), 440–450.

Robinson, C. C. (2011). *Pioneers of Transportation*. Institute of Transportation Engineers. Washington DC.

Rogers, E. (1995). *Diffusion of Innovations*. New York: Free Press.

Rogers, E., K. Magill, and R. Rice (1979). The Innovation Process for Dial-A-Ride. Technical report.

Rose, M. (1990). *Interstate: Express Highway Politics, 1939–1989*. Knoxville: University Tennessee Press.

Rosenstein, M. (2000). *The Rise of Maritime Containerization in the Port of Oakland: 1950 to 1970*. New York: New York University Press.

Saleth, R., and A. Dinar (1999). *Evaluating Water Institutions and Water Sector Performance*, Volume 23. World Bank Publications. Washington, DC.

Saleth, R., and A. Dinar (2004). *The Institutional Economics of Water: A Cross-Country Analysis of Institutions and Performance*. Northampton, MA: Edward Elgar.

Samuel, P., and R. Poole (2005). *Should States Sell Their Toll Roads?* Reason Public Policy Institute.

Samuel, P., R. Poole, and J. Holguin Veras. 2003. *Toll Truckways: A New Path Toward Safer and More Efficient Freight Transportation*. Policy Study 294. Los Angeles: Reason Public Policy Institute.

Sanger, M., and M. Levin (1992). Using Old Stuff in New Ways: Innovation as a Case of Evolutionary Tinkering. *Journal of Policy Analysis and Management 11*(1), 88–115.

Santini, D. (1988). A Model of Economic Fluctuations and Growth Arising from the Reshaping of Transportation Technology. *Logistics and Transportation Review 24*(2), 121–151.

Sapart, C., G. Monteil, M. Prokopiou, R. van de Wal, P. Sperlich, J. Kaplan, K. Krumhardt, C. van der Veen, S. Houweling, M. Krol, et al. (2012). Natural and Anthropogenic Variations in Methane Sources over the Last 2 Millennia. *Nature 490*, 85–88.

Sauerbier, C. (1956). *Marine Cargo Operations*, Volume 3. New York: Wiley.

Save the Internet (2006). Frequently Asked Questions. Webpage.

Schiffer, M., T. Butts, and K. Grimm (1994). *Taking Charge: The Electric Automobile in America*. Washington, DC: Smithsonian Institution Press.

Schneier, B. (2006, August 24). Refuse to Be Terrorized. *Wired*.

Schön, D. (1963). Invention and the Evolution of Ideas: The Displacement of Concepts. London: Tavistock Publications.

Schor, J. (1993). *The Overworked American: The Unexpected Decline of Leisure*. New York: Basic Books.

Schordock, J. (1985, May). Union Pacific Intermodal. *UP Magazine*.

Schrag, Z. (2004). The Freeway Fight in Washington, DC: The Three Sisters Bridge in Three Administrations. *Journal of Urban History 30*(5), 648–673.

Schrag, Z. (2006). *The Great Society Subway: A History of the Washington Metro*. Baltimore, MD: Johns Hopkins University Press.

Schultz, S. (1989). *Constructing Urban Culture: American Cities and City Planning, 1800–1920*. Philadelphia: Temple University Press.

Schumpeter, J. (1912). *Theorie der wirtschaftlichen Entwicklung*. Leipzig: Duncker & Humblot. English translation published in 1934 as *The Theory of Economic Development*.

Schumpeter, J. (1934). *The Theory of Economic Development: An Inquiry into Profits, Capital, Credit, Interest, and the Business Cycle*. Cambridge, Mass: Harvard University Press.

Schwartz, G. T. (1976). Urban Freeways and the Interstate System. *Southern California Law Review 49(3)*, 406–513.

Schwartz, J. (2012, May 7). Freight Train Late? Blame Chicago. *New York Times*, A15.

Scott, S. W. (1818). *The Heart of Midlothian*. Black, New York: D. Appleton and Company (1876 edition).

Seely, B. (1984). The Scientific Mystique in Engineering: Highway Research at the Bureau of Public Roads, 1918–1940. *Technology and Culture 25*(4), 798–831.

Seely, B. (1987). *Building the American Highway System*. Philadelphia: Temple University Press.

Sellnow, G. (2000, March 4). Coal Train Coming. *Rochester Post-Bulletin*.

Shallat, T. (1994). *Structures in the Stream: Water, Science, and the Rise of the US Army Corps of Engineers*. Austin: University of Texas Press.

Shea, A. (2010). *The Phone Book: The Curious History of the Book That Everyone Uses But No One Reads*. New York: Perigee Books.

Sheriff, C. (1997). *The Artificial River: The Erie Canal and the Paradox of Progress, 1817–1862*. New York: Hill and Wang.

Sherow, J. (1990). US Army Corps of Engineers: Albuquerque District, 1935–1985. *Environmental History Review 14*(1–2), 170–171.

Shoup, D. (2005). *The High Cost of Free Parking*. Planners Press, American Planning Association. Chicago.

Siddall, W. (1974). No Nook Secure: Transportation and Environmental Quality. *Comparative Studies in Society and History 16*(1), 2–23.

Sigafoos, R. (1983). *Absolutely Positively Overnight!: Wall Street's Darling Inside and Up Close*. Memphis: St. Luke's Press.

Simmons, J. (1978). *The Railway in England and Wales, 1830-1914*. Leicester: Leicester University Press.

Simmons, J. (1986). *The Railway in Town and Country*, Newton Abbot, U.K./North Pomfret , Vt.: David & Charles.

Simon, H. (1959). Theories of Decision-Making in Economics and Behavioral Science. *The American Economic Review 49*(3), 253–283.

Simon, H. (1972). Theories of Bounded Rationality. *Decision and Organization 1*, 161–176.

Sinclair, U. (1935). *I, Candidate for Governor: And How I Got Licked*. University of California Press on Demand.

Singh, V. (2008). *The Pearson Indian History Manual for the UPSC Civil Services Preliminary Examination. 2nd edition*. New York: Pearson Education.

Sivak, M. (2009). The Effect of the "Cash for Clunkers" Program on the Overall Fuel Economy of Purchased New Vehicles. University of Michigan, Ann. Arbor, Transporation Research Institute.

Slater, C. (2003). General Motors and the Demise of Streetcars. *Alfred P. Sloan: Critical Evaluations in Business and Management 2*(3), 414.

Slater, D. W. (1992). Transportation and Economic Development: A Survey of the Literature, Directions. In *The Final Report of the Royal Commission on National Passenger Transportation 3*. Ottawa, Ontario: The Royal Commission on National Passenger Transportation.

Smalkoski, B., and D. Levinson (2005). Value of Time for Commercial Vehicle Operators. *Journal of the Transportation Research Forum 44*(1), 89–102.

Small, K., C. Winston, and C. Evans (1989). *Road Work: A New Highway Pricing and Investment Policy*. Washington, DC: Brookings Institution Press.

Smiles, S. (1858). *The Life of George Stephenson*, Railway Engineer. Boston: Ticknor & Fields.

Smiles, S. (1884). *Men of Invention and Industry*. J. Murray. London.

Smith, C. (1990). The Longest Run: Public Engineers and Planning in France. *The American Historical Review 95*(3), 657–692.

Smyth, R., V. Mishra, and X. Qian (2008). The Environment and Well-Being in Urban China. *Ecological Economics 68*(1), 547–555.

Snell, B. C. (1973). *American Ground Transport: A Proposal for Restructuring the Automobile, Truck, Bus, and Rail Industries*. Presented to the Subcommittee on Antitrust and Monopoly of the Committee of the Judiciary, U.S. Senate, February 26, 1974. Washington, D.C.: U.S. Government Printing Office, 16–24.

Sobel, D. (1996). *Longitude: The True Story of a Lone Genius Who Solved the Greatest Scientific Problem of His Time*. Penguin. New York.

Solow, R. (1957). Technical Change and the Aggregate Production Function. *The Review of Economics and Statistics 39*(3), 312–320.

Sommer, M. (2002, March 11). DM&E - Resources and Links. *Minnesota Public Radio News*.

Spanhake, D. (2001). "Specialized Transit and Elderly, Disabled, and Families in Poverty Populations." Minneapolis: University of Minnesota, Center for Transportation Studies, March.

Spann, E. (1988). The Greatest Grid: The New York Plan of 1811. *Two Centuries of American planning*, 11–39.

Spengler, E. H. (1930). Land Values in New York in Relation to Transit Facilities. Technical report, New York: Columbia University Press.

Sperling, D., and D. Gordon (2009). *Two Billion Cars: Driving Toward Sustainability*. New York: Oxford University Press.

Spielberg, F., and P. Shapiro (2000). Mating Habits of Slugs: Dynamic Carpool Formation in the I-95/I-395 Corridor of Northern Virginia. *Transportation Research Record: Journal of the Transportation Research Board 1711*(1), 31–38.

St.Clair, D. (1986). *The Motorization of American Cities*. New York: Praeger Publishers, Incorporated.

Staley, S., and A. Moore (2009). *Mobility First: A New Vision for Transportation in a Globally Competitive Twenty-First Century*. Lanham, MD: Rowman & Littlefield.

Standage, T. (1998). *The Victorian Internet: The Remarkable Story of the Telegraph and the Nineteenth Century's Online Pioneers*. New York: Walker.

Starkie, D. (1988). The New Zealand Road Charging System. *Journal of Transport Economics and Policy 22*(2), 239–245.

Stigler, G. (1951). The Division of Labor Is Limited by the Extent of the Market. *The Journal of Political Economy 59*(3), 185–193.

Stigler, G., and C. Friedland (1962). What Can Regulators Regulate: The Case of Electricity. *Journal of Law and Econics 5*, 1.

Stine, J. K. (1981). *Nelson P. Lewis and the City Efficient: The Municipal Engineer in City Planning During the Progressive Era*. Essays in Public Works History No. 11. Chicago: Public Works Historical Society.

Stover, J. (1995). *History of the Baltimore and Ohio Railroad*. West Lafayette, IN: Purdue University Press.

Sullivan, D. (2000). "Dulles Greenway Begins Construction of Additional Lanes." Sterling, Va.: Dulles Greenway press release, June 2.

Summerton, J. (1994). *Changing Large Technical Systems*. Boulder, CO: Westview Press.

Sun-Herald, The (2004). Curse of the Stone Obelisk Dogs Doubting Author. *Sydney Morning Herald*.

Sunstein, C. (2003). Terrorism and Probability Neglect. *Journal of Risk and Uncertainty 26*(2), 121–136.

Supersonic Spies. Nova. 1998. *PBS Airdate*. `http://www.pbs.org/wgbh/nova/transcripts/2503supersonic.html`, accessed January 27, 1998. (1998).

Sutcliffe, A. (2005). *Steam: The Untold Story of America's First Great Invention*. New York: Palgrave Macmillan.

Swift, E. (2011). *The Big Roads: The Untold Story of the Engineers, Visionaries, and Trailblazers Who Created the American Superhighways*. Boston: Houghton Mifflin Harcourt (HMH).

Tabarrok, A. (2011). *Launching The Innovation Renaissance: A New Path to Bring Smart Ideas to Market Fast*. TED Books. New York.

Taggart, S. (1999, October). The 20-Ton Packet. *Wired Magazine 7*(10).

Tai, H., and C. Hwang (2005). Analysis of Hub Port Choice for Container Trunk Lines in East Asia. *Journal of the Eastern Asia Society for Transportation Studies 6*, 907–919.

Takeda, I., and F. Mizuoka (2003). The Privatization of the Japanese National Railways, The Myth of Neo-liberal Reform and Spatial Configurations of the Rail Net in Japan: A View from Critical Geography. In N. Low and B. Gleeson (Eds.), *Making Urban Transportation Sustainable*, pp. 149–164. New York: Palgrave Macmillan.

Tanner, J. (1981). Expenditure of Time and Money on Travel. *Transportation Research Part A: General 15*(1), 25–38.

Taylor, B., E. Kim, and J. Gahbauer (2009). The Thin Red Line: A Case Study of Political Influence on Transportation Planning Practice. *Journal of Planning Education and Research 29*(2), 173–193.

Taylor, G. (1951). *The Transportation Revolution, 1815–1860*, Volume 4. New York: Holt, Rinehart, and Winston.

Thorne, J. (1876) Handbook to the Environs of London: Alphabetically Arranged, Containing an Account of Every Town and Village, and of All Places of Interest, Within a Circle of Twenty Miles Round. London: John Murray.

Tugwell, F. (1988). *The Energy Crisis and the American Political Economy: Politics and Markets in the Management of Natural Resources*. Stanford, CA: Stanford University Press.

Turnbull, G. L. (1979). *Traffic and Transport: An Economic History of Pickfords*. London: George Allen & Unwin.

US Bureau of Public Roads (1939). Toll Roads and Free Roads. Technical report, 76th Cong., 1st sess., H. Doc. 272.

US Department of Transportation, Federal Highway Administration. (1977). *America's Highways, 1776–1976: A History of the Federal-aid Program*.

US Energy Information Administration. (2012, July 27). Monthly Energy Review. Technical report.

US Environmental Protection Agency (2012). National Ambient Air Quality Standards (NAAQS).

US Federal Railroad Administration (1976–1987). Northeast Corridor Improvement Project, Joint Annual Report. Technical report, Department of Transportation, Washington, DC.

US Federal Railroad Administration (1990). Assessment of the Potential for Magnetic Levitation Transportation in the United States. Technical report, US Federal Railroad Administration.

US Federal Railroad Administration and National Railroad Passenger Corporation (1981). Rail Passenger Corridors, Final Evaluation. Technical report, US Department of Transportation.

US General Accounting Office (1987). Highway Needs: An Evaluation of DOT's Assessment Process. Technical Report GAO/RCED-87-136.

Vaughan, A. (1991). Isambard Kingdom Brunel: Engineering Knight-Errant. London: John Murray, 2.

Verhoef, E. (1994). External Effects and Social Costs of Road Transport. *Transportation Research Part A: Policy and Practice 28*(4), 273–287.

Viner, J. (1926). Urban Aspects of Highway Finance. *Public Roads 6*, 233–235.

Vision of Britain, A. (2006). A Vision of Britain Through Time. `http://www.visionofbritain.org.uk/index.jsp` (Accesssed July 18, 2013)

Voith, R. (1991). Changing Capitalization of CBD-Oriented Transportation Systems: Evidence from Philadelphia, 1970–1988. Working Paper 91-19, Federal Reserve Bank of Philadelphia, Philadelphia, PA.

Von Hippel, E. (1986). Lead Users: A Source of Novel Product Concepts. *Management Science*, 32(7) informs, 791–805.

Wachs, M. (1982). Ethical Dilemmas in Forecasting for Public Policy. *Public Administration Review 42*(6), 562–567.

Wachs, M. (1990). Ethics and Advocacy in Forecasting for Public Policy. *Business & Professional Ethics Journal*, 9(1/2), 141–157.

Wachs, M. (2003). A Dozen Reasons for Gasoline Taxes. *Public Works Management & Policy 7*(4), 235.

Walmsley, D., and K. Perrett. (1992). The Effects of Rapid Transit on Public Transport and Urban Development. Wokingham, U.K.: Department of Transport, Transport Research Laboratory.

Ward, J. R. (1974). *The Finance of Canal Building in Eighteenth-century England*. Oxford: Oxford University Press.

Ward, R. (2003). *London's New River*. Phillimore & Company. Stroud, Gloucestershire, UK

Washington, Columbia Historical Society (1902). *Records of the Columbia Historical Society, Washington, D.C.*, Volume 5. The Society.

Watt, N. (2012). David Cameron Unveils Plan to Sell Off the Roads: Sovereign Wealth Funds to Be Allowed to Lease Motorways in England, Says Prime Minister. *The Guardian Newspaper*, Sunday 18 March 2012.

Webb, S., and B. Webb (1913). *The Story of the King's Highway*, Volume 5. London: Longmans, Green.

Webber, M. (1964). The Urban Place and the Nonplace Urban Realm. In *Explorations into Urban Structure'*. Philadelphia: University of Pennsylvania.

Webber, M. M. (1976, Fall). The BART Experience: What Have We Learned? *The Public Interest* (45).

Weiner, E. (2008). *Urban Transportation Planning in the United States: History, Policy, and Practice*. New York: Springer Verlag.

Weingroff, R. (1996). Creating the Interstate System. *Public Roads 60*(1), 10–17.

Weingroff, R. (1997). From Names to Numbers: The Origins of the US Numbered Highway System. *AASHTO Quarterly Magazine 76*(2).

Weingroff, R. (2002). The Man Who Loved Roads. *Public Roads 65*(6), 37–46.

Weisberg, S. (2009). *Barney Frank: The Story of America's Only Left-Handed, Gay, Jewish Congressman*. Amherst: University of Massachusetts Press.

Wellington, A. (1902). *The Economic Theory of the Location of Railways*. New York: John Wiley and Sons.

White, H. (1963). *A Regional History of the Railways of Great Britain*: Volume III, *Greater London*. London: Phoenix House.

White, J. (1966). Public Transport in Washington before the Great Consolidation of 1902. *Records of the Columbia Historical Society, Washington, DC 66*, 216–230.

White, P. (1990). Bus Deregulation: A Welfare Balance Sheet. *Journal of Transport Economics and Policy*, 311–332.

White, P. (1997). What Conclusions Can Be Drawn about Bus Deregulation in Britain? *Transport Reviews 17*(1), 1–16.

Whyte, W. (1993). *The Exploding Metropolis*, Volume 1. University of California Press on Demand.

Willeke, R. (1993). Zur Frage der externen Kosten und Nutzen des motorisierten Straßenverkehrs. *Zeitschrift für Verkehrswissenschaft 64*(4).

Wilson, J. (1980). The Politics of Regulation. New York. Basic Books

Wilson, W. (1989). *The City Beautiful Movement*. Baltimore, MD: Johns Hopkins University Press.

Winfrey, R. (1968). *Economics of the Maximum Limits of Motor Vehicle Dimensions and Weights*. Federal Highway Administration, Offices of Research and Development.

Wolf, C. (2012, March 22). America's Broken Bridges. *BusinessWeek*.

Wolmar, C. (2005). *On the Wrong Line: How Ideology and Incompetence Wrecked Britain's Railways*. Kemsing Publishing. London, Aurum Press.

Wolmar, C. (2010a). *Blood, Iron, & Gold: How the Railroads Transformed the World*. Public Affairs. London: Atlantic Books.

Wolmar, C. (2010b). The PPP Is the Scandal No One Noticed. `http://www.christianwolmar.co.uk/2010/05/the-ppp-is-the-scandal-no-one-noticed/` Accessed July 18, 2013

Wood, G. (2009). *Empire of Liberty: A History of the Early Republic, 1789–1815*, Volume 4. New York: Oxford University Press.

Wright, R. (2011). Container Shipping Alliances to Join Forces. *Financial Times*, Dec. 21, 2011.

Wu, T. (2003). Network Neutrality, Broadband Discrimination. *Journal of Telecommunications and High Technology Law 141*.

Wu, T. (n.d.). Network Neutrality. Webpage.

Xie, F., and D. Levinson (2011). Evaluating the Effects of the I-35W Bridge Collapse on Road-users in the Twin Cities Metropolitan Region. *Transportation Planning and Technology 34*(7), 691–703.

Xie, F., and D. M. Levinson (2012). *Evolving Transportation Networks*. Transportation Research, Economics, and Policy. New York: Springer.

Yago, G. (1984). *The Decline of Transit: Urban Transportation in German and US Cities, 1900–1970*. Cambridge: Cambridge University Press.

Yamamoto, L. (2003, Emeryville, California). Railroading BART. *The Monthly 33*(5).

Yamins, D., S. Rasmussen, and D. Fogel (2003). Growing Urban Roads. *Networks and Spatial Economics 3*(1), 69–85.

Yerra, B. and D. Levinson (2005). The Emergence of Hierarchy in Transportation Networks. *Annals of Regional Science 39*(3), 541–553.

Young, A. (1970). General View of the Agriculture of the County of Sussex. London: David & Charles Reprints (reprint of 1813 and 1817 editions).

Zahavi, Y. (1974). Travel Time Budgets and Mobility in Urban Areas. Technical Report PB-234145/1 DOTL NTIS, Federal Highway Administration.

Zahavi, Y. (1981). *The UMOT Project*. US Government Printing Office.

Zahavi, Y., and J. Ryan (1980). Stability of Travel Components over Time. *Transportation Research Record 750*, 19–26.

Zahavi, Y., and A. Talvitie (1980). Regularities in Travel Time and Money Expenditures. *Transportation Research Record 750*, 13–19.

Zhang, L. (2010). Do Freeway Traffic Management Strategies Exacerbate Urban Sprawl? *Transportation Research Record: Journal of the Transportation Research Board 2174*(1), 99–109.

Zhang, L., and D. Levinson. Travel Time Variability after a Shock: The Case of the Twin Cities Ramp Meter Shut Off. In Y. Iida and M. Bell (Eds.), *The Network Reliability of Transport*. Pergamon. Oxford, UK.

Zhang, L., and D. Levinson (2002). Estimation of the Demand Responses to Ramp Metering. In *Proceedings of the 3rd International Conference on Traffic and Transportation Studies*, Volume 1, pp. 674–681.

Zhang, L., and D. Levinson (2004). Optimal Freeway Ramp Control without Origin-Destination Information. *Transportation Research Part B: Methodological 38*(10), 869–887.

Zhang, L., and D. Levinson (2010, May). Ramp Metering and Freeway Bottleneck Capacity. *Transportation Research Part A 44*(4), 218–235.

Zhang, L., B. McMullen, D. Valluri, and K. Nakahara (2009). Vehicle Mileage Fee on Income and Spatial Equity. *Transportation Research Record: Journal of the Transportation Research Board 2115*(1), 110–118.

Zhao, Z., K. Das, and C. Becker (2010). *Funding Surface Transportation in Minnesota: Past, Present and Prospects*, Volume 10. University of Minnesota, Center for Transportation Studies.

Zingales, L. (2012). *A Capitalism for the People*. New York: Basic Books.

Zou, X., and D. Levinson (2003). Vehicle-Based Intersection Management With Intelligent Agents. In *ITS America Annual Meeting Proceedings*. Minneapolis.

Index

Printed in the USA/Agawam, MA
March 20, 2019